# International Symposium of the Society of Core Analysts (SCA 2021)

AA003529

E3S Web of Conferences Volume 366 (2023)

Online
13-16 September 2021

**Editor:**

**C.F. Berg**

ISBN: 978-1-7138-6642-8

**Printed from e-media with permission by:**

Curran Associates, Inc.
57 Morehouse Lane
Red Hook, NY 12571

**Some format issues inherent in the e-media version may also appear in this print version.**

This work is licensed under a Creative Commons Attribution 4.0 International License. License details: http://creativecommons.org/licenses/by/4.0/.

No changes have been made to the content of these proceedings. There may be changes to pagination and minor adjustments for aesthetics.

Printed with permission by Curran Associates, Inc. (2023)

For additional information, please contact EDP Sciences – Web of Conferences
at the address below.

EDP Sciences – Web of Conferences
17, Avenue du Hoggar
Parc d'Activité de Courtabœuf
BP 112
F-91944 Les Ulis Cedex A
France

Phone: +33 (0) 1 69 18 75 75
Fax: +33 (0) 1 69 28 84 91

contact-edps@webofconferences.org

**Additional copies of this publication are available from:**

Curran Associates, Inc.
57 Morehouse Lane
Red Hook, NY 12571 USA
Phone: 845-758-0400
Fax: 845-758-2633
Email: curran@proceedings.com
Web: www.proceedings.com

# TABLE OF CONTENTS

# SCA Symposium
# 2021
# *Virtual Conference*

SOCIETY OF
CORE ANALYSTS

## Introduction

The Society of Core Analysts (SCA), founded in 1986, is a chapter-at-large of the Society of Petrophysicists and Well Log Analysts (SPWLA). The SCA symposium is a four-day conference and exhibition, which takes place annually in different locations around the world with 2021 being its 35th year. The symposium features one online workshop during the first day and a plenary virtual live technical session showcasing 30 oral presentations and approximately 20 posters presentations.

The annual symposium has built a reputation over the last 30+ years for its high-quality technical presentations in the area of Core Analysis, as well as for giving the main players in the industry the opportunity to showcase their latest technical innovations and offerings in equipment and services. Technical delegates and service companies will be represented and will share their ideas, innovations, knowledge, best practices, products and services.

## PheedLoop Digital Conference Platform

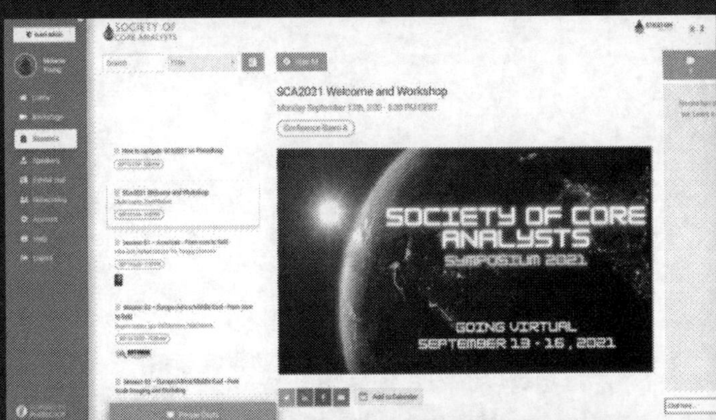


## Our Sponsors

CORE TECHNICAL SERVICES
ROTUNDA SCIENTIFIC TECHNOLOGIES www.RotundaSciTech.com
VINDUM ENGINEERING, INC.
GIT GREEN IMAGING TECHNOLOGIES, INC
DWL Surface Measurement While Drilling™
STRATUM RESERVOIR
CHANDLER ENGINEERING | AMETEK OIL & GAS
MATH 2 MARKET

SCA Symposium
2021
Virtual Conference
SCA
SOCIETY OF
CORE ANALYSTS
Speakers Line-Up
13-16 September 2021
Ahmed Zankoor
MSc.Research Assistant
University of Wyoming
UNIVERSITY of WYOMING
Evgeny Ivanov
PhD.Senior Research Scientist
Schlumberger
Schlumberger
Bergit Brattekaas
PhD. Researcher
University of Bergen
UNIVERSITAS BERGENSIS
Fabrice Pairoys
PhD. Special Core Analysis
TOTAL S.A.
TotalEnergies
Bo Gong
Petrophysicist
CHEVRON
Chevron
Fatimah Alzubaidi
Mcc. Researcher
University of New South Wales
University of South Wales
Bruno da Silva FalcÃo
MSc. Researcher
CSIRO Energy BU
CSIRO
Giorgio Volonte
Technical Leader, Rock mechanics
ENI SpA
Eni
David K. Potter
Professor
University of Alberta
UNIVERSITY OF ALBERTA
Holger Ott
Professor
MontanuniversitÃt Leoben
MONTAN UNIVERSITÄT
WWW.UNILEOBEN.AC.AT
Eva G. Vinegar
PhD. Researcher
The University of Texas at Austin
THE UNIVERSITY OF TEXAS AT AUSTIN
Igor Varfolomeev
Reseacher
Schlumberger
Schlumberger
Please visit us at : WWW. SCAWEB.ORG

# SCA Symposium 2021
*Virtual Conference*

SOCIETY OF CORE ANALYSTS

**Speakers Line-Up**

**13-16 September 2021**

**Ivan D. Perez Torrijos**
Researcher
**University of Stavanger**

**Mohamed Regaieq**
Digital Rock Physics Engineer
**TOTAL S.A.**

**Jorge Andres Tovar Munoz**
MSc.Production Ops.
**University of Campina**

**Mohamed Awad**
PhD. Student
**University of Houston**

**Malin Haugen**
PhD. Student
**University of Bergen**

**Mark Knackstedt**
Professor
**Australian National University**

**Matthieu Mascle**
Researcher
**IFPEN**

**Olivier Lopez**
Leading researcher
**Equinor**

**Mike Dick**
PhD. Principal Researcher
**Green Imaging Technology**

**Nele Wenck**
PhD. Student
**Imperial College London**

**Mohamed Khater**
PhD. Petroleum Engineer
**CSIRO**

**Nikolay Evseev**
Senior Research Scientist
**Schlumberger**

**Please visit us at : WWW. SCAWEB.ORG**

SCA Symposium
2021
Virtual Conference
SCA
SOCIETY OF
CORE ANALYSTS
Speakers Line-Up
13-16 September 2021
Quentin Danielczick
Engineer
TOTAL S.A.
TOTAL
Steffen Berg
PhD. Physics; Principal Science Expert
SHELL
Rafael Salazar-Tio
Industry Process Senior Specialist
Dassault Systemes
Tanguy Lhomme
Researcher
EPSLOG BV.
Raheleh Farokhpoor
Petrophysicist
Lundin-Energy Norway
Lundin
Energy
Zoya Heidari
Associate Professor
University of Texas
THE UNIVERSITY OF TEXAS
AT AUSTIN
Roland Lenormand
Researcher
CYDAREX
CYDAREX
Sandra Profice
Researcher
TOTAL S.A.
TotalEnergies
Subhash Ayirala
Champion of SmartWater flooding
Saudi Aramco
aramco
Please visit us at : WWW. SCAWEB.ORG

# SCA Symposium
## Virtual Workshop
### *13 September, 2021*

SOCIETY OF
CORE ANALYSTS

**Monday September 13th, 8:00 - 11:00 AM CST**

## From Core to Field:
## WHOLE CORE IMAGE DATA in PETROPHYSICAL and RESERVOIR ENGINEERING WORKFLOWS

**Mark Knackstedt**

Professor, Department of Applied Mathematics, Australian National University

**Multiscale imaging and the integration of small-scale heterogeneity into static & dynamic models**

**Olivier Lopez**

PhD. Fundamental and applied geochemistry; Leading researcher, Equinor ASA

**Core and borehole images analysis for automatic classification of lithòfacies and reservoir evaluation**

**Zoya Heidari**

Associate Professor, University of Texas

**Quantification of Rock Fabric and its Integration with Well Logs for Enhanced Formation Evaluation**

# From Core to Field

## Day 1 *(Session 1)*

## 13 September, 2021

SOCIETY OF CORE ANALYSTS

**Monday September 13th, 12:00 - 1:30PM CST**

**12:00 AM – 12:30 AM / CST**

### Rafael Salazar-Tio

### Dassault Systemes

**Multiscale Digital Rocks: Application of a multi-scale multi-phase workflow to a Carbonate reservoir rock**

**12:30 AM – 1:00 PM / CST**

### Tanguy Lhomme

### Epslog

**Combining high-resolution core data with unsupervised machine learning schemes for the identification of rock types and the prediction of reservoir quality**

**1:00 PM – 1:30 PM / CST**

### Mike J. Dick

### Green Imaging Technology

**NMR Core Analysis**

**On Whole Core Samples**

# From Core to Field
## Day 2 *(Session 2)*
## 14 September, 2021

SOCIETY OF CORE ANALYSTS

**Tuesday September 14th, 3:00 - 4:30AM CST**

**3:00 AM – 3:30 AM / CST**

Schlumberger

**Evgeny Ivanov**

**Schlumberger**

**Acid treatment optimization based on digital core analysis**

**3:30 AM – 4:00 AM / CST**

Imperial College London

**Nele Wenck**

**Imperial College London**

**Advanced Core Characterisation to Improve Multiphase Flow Prediction in Heterogeneous Sandstone and Carbonate Rocks**

**4:00 AM – 4:30 AM / CST**

Schlumberger

**Igor Varfolomeev**

**Schlumberger**

**3D multiclass digital core models via micro-CT, SEM-EDS and deep learning**

# Pore Scale Imaging and Modelling

## Day 2 *(Session 3)*

## 14 September, 2021

SOCIETY OF CORE ANALYSTS

**Tuesday September 14th, 8:00 - 9:30AM CST**

**8:00 AM – 8:30 AM / CST**

TotalEnergies

### Mohamed Regaieg

### Total S.E.

Large two phase DRP simulations for relative permeability uncertainty assessment

**8:30 AM – 9:00 AM / CST**

### Malin Haugen

### University of Bergen

Calcite-functionalized micromodels for pore-scale investigations of $CO_2$ storage security

**9:00 AM – 9:30 AM / CST**

Schlumberger

### Nikolay Evseev

### Schlumberger

The Problem of Stability of Gas-Condensate Mixture at Pore-Scale: The Study by Density Functional Hydrodynamics

# Pore Scale Imaging and Modelling

## Day 2 *(Session 4)*

## 14 September, 2021

SOCIETY OF CORE ANALYSTS

**Tuesday September 14th, 11:00 -12:30PM CST**

**11:00 AM – 11:30 AM / CST**

### Ahmed Zankoor

### University of Wyoming

In-situ characterization of capillary pressure during three-phase flow in porous media

**11:30 AM – 12:00 PM / CST**

### Jorge Andres Tovar Munoz

### University of Campina

Study of Relative Permeability Curves for a $CO_2$-Brine System at Reservoir Conditions in Carbonates

**12:00 PM – 12:30 PM / CST**

### David K. Potter

### University of Alberta

Comparing Predicted 3D Anisotropic Properties from Micro-CT Derived Digital Rock Physics with Experimentally Measured Anisotropy

# Advanced Imaging and EOR

## Day 3 *(Session 5)*

## 14 September, 2021

SOCIETY OF CORE ANALYSTS

**Tuesday September 14th, 6:00-7:30PM CST**

**6:00 PM – 6:30 PM / CST**

**Mohamed Khater**

**CSIRO**

Dissolution behavior in carbonate reservoirs during WAG injection: An experimental study

**6:30 PM – 7:00 PM / CST**

**Fatimah Alzubaidi**

**University of New South Wales, Sydney, Australia**

Improved fracture segmentation from unwrapped images of drill core using an innovative two-stage segmentation Mask R-CNN

**7:00 PM – 7:30 PM / CST**

**Bruno da Silva Falcão**

**CSIRO Energy BU**

Monitoring fluid migration using an in-situ nuclear magnetic resonance core flooding system integrated with fiber optic sensors: A proof of concept

# Improved SCAL Techniques and Interpretation

## Day 3 *(Session 6)*

## 15 September, 2021

SCA

SOCIETY OF CORE ANALYSTS

**Wednesday September 15th, 3:00-4:30AM CST**

**3:00 AM – 3:30 AM / CST**

Lundin Energy

**Raheleh Farokhpoor**

**Lundin-Energy Norway**

Core cleaning and wettability restoration,
selecting appropriate method

**3:30 AM – 4:00 AM / CST**

UiS University of Stavanger

**Ivan D. Pierez Torrijos**

**University of Stavanger**

Replicating Initial Wettability in Carbonates by
Advanced Core Analyses:
A Step Closer to Representative Reservoir Wettability

**4:00 AM – 4:30 AM / CST**

TotalEnergies

**Quentin Danielczick**

**TOTAL S.A.**

Wireless Acquisition For Resistivity Index in Centrifuge :
A New Method to Estimate Archies Law Parameters

# Improved SCAL Techniques and Interpretation

## Day 3 *(Session 7)*

## 15 September, 2021

SOCIETY OF CORE ANALYSTS

**Wednesday September 15th, 8:00 - 9:30 AM CST**

**8:00 AM – 8:30 AM / CST**

**Fabrice Pairoys**

**TOTAL SA**

Comparing Steady-State and Semi-Dynamic Methods for Relative Permeability Determination: New Insights

**8:30 AM – 9:00 AM / CST**

**Steffen Berg**

**Shell Global Solutions International B.V.**

Physical origin of pressure- and saturation fluctuations in steady-state core floods

**9:00 AM – 9:00 AM / CST**

**Matthieu Mascle**

**IFPEN**

In-situ investigation of salt precipitation dynamic induced by gas flow-through drying using simultaneous Neutron and X-ray dual-beam Radiography

# Poster Presentations
## Day 3 *(Session 8)*
## 15 September, 2021

SOCIETY OF CORE ANALYSTS

**Wednesday September 15th, 11:00-12:30PM CST**

**11:00 AM – 12:30 PM / CST**

## Poster Presentations

- Evaluation of in-Situ Gelation Behavior by a Slim Tube Technique
- Integration of core-scale logging, dual-energy computed tomographic imaging and geochemical and mineralogical analysis of a composite core
- Reactivation and branching by $CO_2$ injection in pre-faulted Berea and Boise sandstones. Laboratory micro-seismic imaging
- Novel analysis of molecular structures at oil-brine interface for Smartwater flooding.
- Digital extraction of core properties from whole core photos
- Developed emulsification in porous media flow
- The Effect of Microporosity and Oil Property on the Initial and Residual Oil Saturations of a Bimodal Carbonate
- Data rescue – maximising the value of archived core using non-destructive Multi-Sensor Core Logger (MSCL) and machine learning techniques
- Rate dependency of capillary trapping in gas-brine flow
- Multi Field Evaluation of $T_2$ Pore Size Distributions and $T_1$-$T_2$ 2D Maps
- Influence of Wettability on Molecular Adsorption and Kinetics Studied by In-Situ NMR
- Tortuosity and Cementation exponent as variables related to heterogeneity and the impact on Sw calculations in Tambaredjo field of Suriname.
- Elucidating Hydrocarbon and Brine Flow Paths in Source Rock Shales Using NMR Imaging
- Powder River Basin Integrated Hydrocarbon System Case Study
- Rate-Controlled Mercury Injection Experiments to Characterize Pore Space Geometry of Berea Sandstone
- Evaluation of the InterFoam Solver in the Prediction of Immiscible Two-Phase Flow in Imbibition and Drainage on the Pore-Doublet System.
- Dynamic pore-scale modeling of residual fluid configurations in disordered porous media.
- Evaluation the effectiveness of the different methods enhanced oil recovery (thermal and chemical) on plugs of the Cenomanian tier
- Oscillating Back Pressure Regulator (OBPR) for High-Pressure Oscillating Back Pressure Regulator (OBPR) for High-Pressure
- A Mechanistic Study of Wettability Alterations in Sandstone by Low Salinity Water Injection (LSWI) and $CO_2$ Low Salinity Water-Alternating-Gas (WAG) Injection

# Laboratory Core Analysis

## Day 4 *(Session 9)*

## 16 September, 2021

SOCIETY OF CORE ANALYSTS

**Thursday September 16th, 3:00 - 4:30AM CST**

**3:00 AM – 3:30 AM / CST**

### Holger Ott

### Montanuniversität Leoben

Material balance and mixing behavior during emulsification of crude oil by using micro-X-ray tomography

**3:30 AM – 4:00 AM / CST**

### Subhash Ayirala

### Saudi Aramco

SmartWater Synergy with Surfactant and Polymer: A Microscale Investigation at Crude Oil-Water Interface

**4:00 AM – 4:30 AM / CST**

### Roland Lenormand

### CYDAREX

DarcyShale: An improved GRI method for more reliable measurements on low permeability samples

# Core Analysis
## Day 4 *(Session 10)*
## 16 September, 2021

SOCIETY OF CORE ANALYSTS

**Thursday September 16th, 8:00-9:30AM CST**

**8:00 AM – 8:30 AM / CST**

### Sandra Profice
**TOTAL S.A.**

The KPC-Log: A new method for measuring the permeability of a non-cylindrical tight sample

**8:30 AM – 9:00 AM / CST**

### Bergit Brattekaas
**Dept. of Physics and Technology, University of Bergen**

Of rats and rocks: using pre-clinical PET imaging facilities in core analysis

**9:00 AM – 9:30 AM / CST**

### Giorgio Volonte
**Eni Spa – E&P Division**

Integrated workflow for rotary sidewall core orientation: best practices and examples from planning to execution

# Unconventional and $CO_2$

## Day 4 *(Session 11)*

## 16 September, 2021

SOCIETY OF CORE ANALYSTS

**Thursday September 16th, 11:00 - 12:30PM CEST**

**11:00 AM –11:30 AM / CST**

**Bo Gong**

**Chevron**

Rock Image Enhancement Using Super-Resolution Neural Networks

**11:30 AM – 12:00 PM / CST**

**Eva G. Vinegar**

**The University of Texas at Austin**

Evolution of NMR and Electrical Resistivity During Early Maturation of an Organic-Rich Chalk

**12:00 PM – 12:30 PM / CST**

**Mohamed M. Awad**

**University of Houston**

**Petroleum Engineering Department**

Study of Bakken and Three Forks Formations using NMR Core Analysis

# SCA Symposium 2021

# *Spotlight on our Speakers*

SOCIETY OF CORE ANALYSTS

**Olivier Lopez**
**PhD. Fundamental and applied geochemistry; Leading researcher**
**Equinor ASA**

Olivier Lopez is a leading researcher in geology and petrophysics in the research and technology department of Equinor ASA.
His work is mainly focusing on reservoir evaluation and characterization for siliciclastic, carbonate and shale systems and integration of new technologies.
Before joining Equinor ASA, Olivier was project manager in Numerical Rocks ASA.
Olivier holds a Ph.D. degree (2006) in fundamental and applied geochemistry

**Zoya Heidari**
**Associate Professor**
**University of Texas**

Zoya Heidari is an Associate Professor in the Hildebrand Department of Petroleum and Geosystems Engineering at The University of Texas at Austin. Before joining UT Austin, Zoya was an Assistant Professor at Texas A&M University from 2011 to 2015.
She has been the founder and director of the UT Austin Industrial Affiliate Research Program on "Multi-Scale Rock Physics" since 2016. She was also the founder and the director of the Texas A&M Joint Industry Research Program on "Multi-Scale Formation Evaluation of Unconventional and Carbonate Reservoirs" from 2012 to 2015.
Among her many awards, Zoya is the recipient of the 2021 SPWLA Distinguished Technical Achievement Award, the 2020 SPWLA Young Professional Technical Award, the 2019 EAGE Arie van Weelden Award, the 2019 AIME Rossiter W. Raymond Memorial Award, the 2019 SPE Distinguished Membership Award, and the 2017 SPE Cedric K. Ferguson Medal.
Her research interests include Petrophysics, Rock Physics, Multi-Scale Formation Evaluation, Borehole Geophysics, Integrated Reservoir Characterization, and Completion Petrophysics.

**Mark Knackstedt**
**Professor Department of Applied Mathematics**
**Australian National University**

Mark Knackstedt (BSc, Columbia; PhD, Rice) is a Professor at the Department of Applied Mathematics and researcher in the National CT Lab at the Australian National University. He was the 2019 Kimberley Clark Interpore Lecturer and has been an SPE Distinguished Lecturer (2015-16) and three-time SPWLA distinguished speaker. He was awarded the George C. Matson Memorial Award from the AAPG in 2009 and the ENI award for New Frontiers in Hydrocarbon Research in 2010.

# SCA Symposium 2021

# *Spotlight on our Speakers*

SOCIETY OF CORE ANALYSTS

**Mike Dick**
**PhD. Molecular physics; Principal Research Scientist**
**Green Imaging Technology**

Mike Dick has a Bachelor of Science degree from the University of New Brunswick. His post-graduate work focused primarily on molecular physics and included both a Master's degree from the University of New Brunswick and a Ph.D. from the University of Waterloo.
After graduation, Mike's career diversified working first in astrophysics while doing a post-doctoral fellowship at NASA's Jet Propulsion Lab. Following this position, Mike moved into the field of nuclear physics in Chalk River Ontario with Bubble Technology Industries (BTI) as a research scientist.
His work with BTI focused primarily on developing unique radiation detectors for defense and homeland security organizations, first responders, space agencies, regulatory/standards groups, and research institutions. This work has resulted in three patent filings.
Mike joined the Green Imaging Technologies in September 2015 as a Principal Research Scientist. Mike works closely with the research team innovating new NMR techniques and tools for our clients

**Rafael Salazar-Tio**
**Industry Process Expert Senior Specialist**
**Dassault Systemes**

Rafael Salazar-Tio holds a bachelor degree in Physics from the National University of San Marcos, Peru and a PhD degree in Physics from the University of the Balearic Islands, Spain.
He held Research Associate positions in the Eindhoven University of Technology, and in the Washington University in St. Louis, both in Physical chemistry.
He worked in Petrophysics R&D at Chevron for 12 years before joining Dassault Systemes in 2018. Rafael's research interests include multi-scale and multi-physics methods and applications in porous media.

**Tanguy Lhomme**
**PhD. Applied Earth Sciences and Director**
**Epslog**

Tanguy Lhomme is a director at Epslog. Before joining Epslog in 2013, Tanguy worked eight years as a Reservoir engineer with Shell International, including 4 years on reservoir surveillance and enhanced Oil Recovery projects for PDO in Oman.
He holds a PhD in Applied Earth Sciences from Delft University, with a research thesis on lab experiments of hydraulic fracturing and an MSc in Mining and Petroleum Engineering from the University of Minnesota, with a research thesis on fundamental aspects of rock strength testing.
His current interests are in innovation for core testing and analysis, unconventional reservoir characterization, multi-disciplinary data integration, Artificial Intelligence applied to E&P digital data processing and analysis.

# SCA Symposium 2021

## *Spotlight on our Speakers*

SOCIETY OF CORE ANALYSTS

**Evgeny Ivanov**
**Senior Research Scientist**
**Schlumberger**

Evgeny is a Senior research scientist in Schlumberger Moscow research center since 2012 and his areas of interest are thermodynamics, multiphase transport and numerical modelling.
In 2008 he finished a Masters degree in Physics from Moscow Institute of Physics and Technology, Russia.
He holds a Doctorate in Fluid Mechanics from Moscow Institute of Physics and Technology, Russia

**Igor Varfolomeev**
**Researcher**
**Schlumberger**

Graduated from Moscow Institute of Physics and Technology.
Came to Schlumberger Moscow Research (SMR) center in 2010.
Igor works both on 3D image processing algorithms (segmentation, spatial registration, etc.) and on image acquisition (microCT, SEM, etc.).

**Nele Wenck**
**PhD. Student**
**Imperial College London**

Nele is a second year PhD student at Imperial College supervised by Samuel Krevor, Ann Muggeridge and Samuel Jackson.
Her work focuses on investigating the impact of rock heterogeneity on multiphase flow properties, with particular attention to the observed fluid behavior at carbon capture and storage sites.

# SCA Symposium 2021

# *Spotlight on our Speakers*

SOCIETY OF CORE ANALYSTS

**Mohamed Regaieg**
**Digital Rock Physics engineer**
**TOTAL S.A.**

Mohamed Regaieg is a senior Digital Rock Physics engineer in TotalEnergies . In 2015, he was awarded a Ph.D. degree in petroleum engineering from Heriot Watt University for research in pore network modelling of fingering phenomena during unsteady-state waterflooding of heavy oils.
He then joined Total as a research reservoir engineer at the Geoscience Research Centre in Aberdeen, UK, where he has developed Total's in-house pore network simulator.
He focused on improving the physics and the computational efficiency of the simulator.
In 2019 he moved to TOTAL's E&P Scientific and Technical Center in Pau, France where he is involved in the industrialisation and the integration of Digital Rock Physics in reservoir engineering workflows.
He works on the full Digital Rock Physics simulation workflow from the image processing to multiphase flow simulations

**Nikolay Evseev**
**Senior Research Scientist**
**Schlumberger**

Nikolay Evseev is a senior research scientist at Schlumberger Moscow Research where he is a lead developer of the density functional hydrodynamics research code (DHD simulator) that is used for numerical simulation of multiphase flow in various applications such as SCAL and EOR on Digital Rock models.
Before joining Schlumberger back in 2008, he worked at the Schmidt Institute of Physics of the Earth of the Russian Academy of Sciences where he developed various scientific and engineering software used for geological modeling, hydrocarbon reservoir simulation, and 3D visualization.
He also participated in field development planning works for various Russian O&G companies and was responsible for geological modeling and reservoir simulation.
He is a co-author in more than 30 technical publications, including two books

**Maulin Haugen**
**PhD. Student**
**University Of Bergen**

Malin is a PhD student at the department of physics and technology at the University of Bergen.
Her research focuses on subsurface $CO_2$ storage and security on several scales; from pore scale to large-scale flow visualization (https://fluidflower.w.uib.no/).
Prior to shifting focus towards $CO_2$ storage she has 5 years experience as a reservoir engineer, working with optimizing oil and gas production, planning new wells to be drilled and generating short-term production budgets.
Malin completed her master's degree in reservoir physics in 2012, studying $CO_2$ injection in fractured chalk for enhanced oil recovery

# SCA Symposium
# 2021
# *Spotlight on our Speakers*

SOCIETY OF CORE ANALYSTS

**Ahmed Zankoor**
**MSc. Petroleum Engineering; Research Assistant**
**University of Wyoming**

Ahmed holds a MSc from the faculty of Petroleum and Mining Engineering, Suez University, Egypt.
He is a Research assistant at the Center of Innovation for Flow through Porous Media, University of Wyoming.

**David K. Potter**
**Professor of Geophysics and Petrophysics**
**University of Alberta**

David K. Potter is currently Professor of Geophysics and Petrophysics at the University of Alberta, Canada.
He was the Director of the Integrated Petroleum Geoscience (IPG) MSc program at the University of Alberta from 2009-2014.
Before that he was the Anadarko Professor of Formation Evaluation, and the Director of Research, at the Institute of Petroleum Engineering, Heriot-Watt University, UK.
David was the President of the Society of Core Analysts (SCA) in 2008-2009, and received the Darcy Award for technical achievement from the SCA in 2015.

**Jorge Andres Tovar Muñoz**
**MSc. Production Operations Assistant**
**University of Campinas**

Jorge Andrés is a Colombian Petroleum engineer (Universidad Surcolombiana, 2014) and has a M.Sc. in Petroleum Engineering (University of Campinas, 2020).
Currently working as Production Operations Assistant for an independent company in the East Llanos basin of Colombia

# SCA Symposium 2021

# *Spotlight on our Speakers*

SOCIETY OF CORE ANALYSTS

**Bruno da Silva Falcao**
**MSc. Petroleum Engineering Research Student**
**CSIRO**

Bruno graduated with a B.Sc. in Petroleum engineering from the State University of Northern of Rio de Janeiro (UENF) in 2019.
During his bachelor's degree he was granted a scholarship to study abroad for 16 months at the University of Alberta, Canada where he also worked as a summer intern at Alberta Innovates – Technology Futures. In addition, Bruno gained industrial experience in a 6-months internship program at Halliburton (Brazil) in the Production Enhancement Department.
Most recently, Bruno joined the Petroleum Engineering Research group at Edith Cowan University, Australia, as a Masters by Research student. His research, in collaboration with CSIRO (Kensington), focuses on the application of FBG sensors to map the deformation of rock reservoirs during core flooding tests

**Fatimah Alzubaidi**
**MSc. Petroleum Engineering**
**University of New South Wales**

Fatimah Al-Zubaidi is a Ph.D. candidate in Minerals and Energy Resources Engineering at the University of New South Wales, Australia.
She was awarded a M.Sc. in Petroleum Engineering from the same institution.
She holds an Australian Government Research Training Program Scholarship. Her major research interest is rock typing using machine learning

**Mohammed Kather**
**PhD. Petroleum engineering**
**CSIRO**

Mohamed holds a doctoral degree (PhD) in petroleum engineering from a Curtin university.
He is a petroleum engineer specialize in wells and reservoir performance analysis, Enhanced Oil Recovery, Special Core Analyses, Core flooding, Petroleum economics and Project management, Oil and gas-related operations.

# SCA Symposium 2021

# *Spotlight on our Speakers*

SOCIETY OF CORE ANALYSTS

**Quentin Danielczick**
**Engineer**
**TOTAL S.A.**

In 2019, Quentin Danielczick was awarded an engineer degree in industrial technologies from ENSGTI in Pau – France.
During his studies, he joined Total as an intern to develop and industrialize processes and in-house software's in the centrifuge and capillary pressure laboratory.
Since 2019 he is working for TOTAL's E&P Scientific and Technical Center in Pau, France, he is involved in the industrialization and the integration of innovations in the SCAL field

**Ivan Dario Pinerez Torrijos**
**PhD Student**
**University of Stavanger**

Iván D. Piñerez Torrijos holds a MS degree in Chemical engineering from the University of Barcelona, and a PhD from University of Stavanger.
He is now working as a postdoctoral researcher in Reservoir Chemistry at the University of Stavanger.
His main research topics are reservoir chemistry, SCAL studies with high focus on wettability, and Smart Water EOR processes in both carbonate and sandstone reservoirs.
Dr. Piñerez has participated in fundamental research to improve the chemical understanding of reservoir wettability and wettability alteration processes.
He has also contributed in industry projects sponsored by international oil companies with the objective to evaluate the EOR-potential for specific oil reservoirs

**Raheleh Farokhpoor**
**Petrophysicist**
**Lundin-Energy Norway**

Raheleh is a reservoir engineer with more than 10 years experiences in experimental and digital special core analysis.
She has a PhD in "geological $CO_2$ storage" with focus in $CO_2$-brine core analysis.
At the moment, she is working as senior reservoir engineer in Lundin-Energy where she is responsible for SCAL and PVT projects

# SCA Symposium
# 2021
# *Spotlight on our Speakers*

SOCIETY OF CORE ANALYSTS

**Fabrice Pairoys**
**PhD. Expert in Core and Special Core Analysis,**
**TOTAL S.A**

Fabrice Pairoys owns a Ph.D. in Multiphase flow through porous media.
He has a strong knowledge in Petrophysics, especially in core and special core analysis with over 15 years of experience.
He has been working as Principal research scientist and core analysis expert with SCHLUMBERGER before joining TOTALENERGIES as Petrophysics expert in 2019.
Fabrice has published over 20 papers and given numerous presentations/training in the special core analysis domain (Kr & Pc interpretations, wettability, core resistivity/NMR/dielectric...)

**Steffen Berg**
**PhD. Researcher,**
**Shell Global Solutions International BV**

Steffen Berg is a Principal Science Expert at Shell Global Solutions International B.V. in the Netherlands. His main research interests range from the fundamental aspects of multiphase flow in porous media to special core analysis, $CO_2$ sequestration and EOR using 3D imaging by X-ray computed tomography to gain insights into the transport mechanisms from pore to Darcy scale. He holds a master's degree in materials science from the University of the Saarland and a PhD in physics from the University of Mainz / Max Planck Institute for Polymer Research at Mainz, Germany.
After a postdoc at Princeton University he joined Shell as a research scientist. He is currently also a visiting reader in the Earth Science & Engineering and Chemical Engineering departments at Imperial College London

**Matthieu Mascle**
**Researcher**
**IFPEN**

Mathieu is working as a research scientist at IFP Energies Nouvelles, in the department of Earth Sciences and Environmental Technologies.
His focus of interest is the understanding of multi-phases flows at multi-scale.

# SCA Symposium 2021

# *Spotlight on our Speakers*

SOCIETY OF CORE ANALYSTS

**Subhash Ayirala**
**PhD. Petroleum engineering;**
**Champion of SmartWater flooding, Saudi Aramco**

Subhash Ayirala is the Champion of SmartWater flooding at Saudi Aramco's Exploration and Petroleum Engineering Center—Advanced Research Center, Dhahran, Saudi Arabia.
He has 15+ years reservoir engineering experience in upstream oil & gas industry.
He holds MSc and PhD degrees in petroleum engineering from Louisiana State University, Baton Rouge, USA.
He has written over 75 technical papers, 50+ journal publications, and holds 30 granted patents.
Subhash is SPE distinguished member and currently serves as co-Executive Editor for SPE Reservoir Evaluation & Engineering and Elsevier's Journal of Petroleum Science and Engineering.
He received SPE Outstanding Technical Editor recognition 9-times and is also a recipient of the 2017 SPE A Peer Apart Award

**Roland Lenormand**
**Cydarex**

Roland has a background in theoretical physics. He hold a thesis on multiphase flow in porous media and has worked on this subject during all his career in academic research (CNRS), Schlumberger and IFP.
He is now in charge of the company CYDAREX that commercializes the laboratory software CYDAR, SCAL teaching equipment and devices for permeability measurements on small samples.
Roland also provides teaching and consulting in special core analysis.

**Holger Ott**
**Professor of Reservoir Engineering**
**Montanuniversität Leoben**

Holger holds a PhD in physics from the Freie Universität Berlin in Germany, and is Professor of Reservoir Engineering at the Montanuniversität Leoben in Austria and head of the Department Petroleum Engineering.
Prior to his current position, Holger worked as senior scientist for Shell Global Solutions International in the Netherlands.
During his time at Shell, Holger was teaching as Honorary Reader at the Department of Chemical Engineering at Imperial College London in the United Kingdom. In 2015 he wrote his Habilitation thesis on geological carbon sequestration at RWTH Aachen University in Germany to obtain the Venia Legendi in Applied Geosciences.
His research interests are fluid displacement and reactive transport processes on various scales with application to geological hydrogen and $CO_2$ storage and enhance oil recovery.

# SCA Symposium
# 2021
# *Spotlight on our Speakers*

SOCIETY OF CORE ANALYSTS

**Bergit Brattekaas**
**PhD. Scientific Researcher;**
**Department of Physics and Technology, University of Bergen**

Bergit Brattekås has a PhD in Reservoir physics.
She is a Scientific Researcher for the Department of Physics and Technology, University of Bergen, Norway.
Her research interests are in-situ imaging of dynamic multi-phase flow, focusing on positron emission tomography (PET) and magnetic resonance (MR) imaging, image analysis, $CO_2$ utilization and storage (CCUS), hydrogen storage, conformance control including polymers, polymer gels and foams, flow mechanisms and flow properties in fractured and synthetic media.

**Giorgio Volonte**
**Technical leader,**
**Rock mechanics laboratory, ENI Spa**

Giorgio joined ENI in 2004 as reservoir engineer after a post-graduate specializing master in Petroleum Engineering and graduation in 2003 in Environmental Engineering at Politecnico di Milano.
From 2006 to 2017 he has worked on well and field scale finite element geomechanical modeling in the petroleum engineering department of Eni S.p.A, particularly for subsidence, cap rock integrity and sand production problems.
During last 4 years Giorgio has been working as Technical leader of the rock mechanics laboratory of Eni S.p.A on geomechanical experimental studies for both business activities and R&D projects on CCS and hydrogen storage.

**Sandra Profice**
**Reservoir engineering;**
**Total S.A.**

Sandra Profice is a reservoir engineer at Total. She has been working on studies and R&D projects focused on the interpretation of logs and core flood experiments in monophasic and diphasic conditions, over the last ten years. She holds a PhD on gas flows in very low permeable porous media.

# SCA Symposium 2021

# *Spotlight on our Speakers*

SOCIETY OF CORE ANALYSTS

**Bo Gong**
**Petrophysicist,**
**Chevron**

Bo Gong is currently a Research petrophysicist with the Chevron Technical Center.
She received her PhD degree in Electrical Engineering from the University of Houston in 2014.
Her research interests include borehole imaging technologies, image processing and interpretation techniques, and machine learning applications in petrophysics.
Bo was a Distinguished Speaker for SPWLA in 2019 - 2020.

**Mohamed M. Awad**
**Research Assistant**
**Egyptian Petroleum Research Institute**

Mohamed is a Fulbright fellow, he had a MSc in Petroleum Engineering from University of North Dakota and was a Visiting Student in the Hirasaki research group at Rice University.
He is working as a Research Assistant at the Egyptian Petroleum Research Institute since 2014.
His Master thesis research was focusing on NMR core-analysis and chemical enhanced oil recovery in unconventional reservoirs

# PREFACE

The 2021 International Symposium of the Society of Core Analysts (SCA), founded in 1986, is a chapter - at -large of the Society of Petrophysicists and Well Log Analysts (SPWLA) . Generally, SCA symposium is a four - day conference and exhibition, which takes place annually in different locations around the world with 2021 being its 35th year. In 2021, the symposium was held online, between September 13 and 16, featuring one online workshop during the first day and a plenary virtual live technical session showcasing 30 oral presentations and approximately 20 posters presentations.

The collection of articles published in the SHS Web of conferences consists of 24 articles prepared for oral presentations. Twenty-three (23) proceedings presented in this symposium, published in Petrophysics or other journals, and twenty (20) posters are not included in this collection. Here is the list of papers no included in this collection.

**Papers presented at the 2021 SCA Symposium published in other journals**

**Combining high-resolution core data with unsupervised machine learning schemes for the identification of rock types and the prediction of reservoir quality**
*C. Germay, T. Lhomme, and P. Bisset.*

**NMR Core Analysis On Whole Core Samples**
*M. J. Dick, D. Veselinovic, R. Krumm, R. Antle, and D. Green.*

**Advanced Core Characterisation to Improve Multiphase Flow Prediction in Heterogeneous Sandstone and Carbonate Rocks**
*N. Wenck, S. Jackson, A. Muggeridge and S. Trevor.*
**Large two phase DRP simulations for relative permeability uncertainty assessment**
*M. Regaieg, I. Bondino, C. Varloteaux, T. F. Faisal, J. Yang and R. Rivenq.*

**Study of Relative Permeability Curves for a CO2-Brine System at Reservoir Conditions in Carbonates**
*J. Tovar, T. Berna, T. Borges, J. Vidal, R. Valladares de Almeida, G. Soares Bassani, and E. Koroishi*

**Comparing Predicted 3D Anisotropic Properties from Micro-CT Derived Digital Rock Physics with Experimentally Measured Anisotropy**
*D. K. Potter, L. L. Belisle, A.L Shewchuk, B. C. Snow, A. Zhu and S. Zhang*

© The Authors, published by EDP Sciences. This is an open access article distributed under the terms of the Creative Commons Attribution License 4.0 (http://creativecommons.org/licenses/by/4.0/).

**Improved fracture segmentation from unwrapped images of drill core using an innovative two-stage segmentation Mask R-CNN**
F. Alzubaidi, H. Nalinakumar, S. R. Clark, J. E. Lie, P. MOstaghimi, and R. T. Armstrong

**Initial Wettability in Carbonates by Advanced Core Analyses: A Step Closer to Representative Reservoir Wettability**
*I. D. Pinerez Torrijos, T. Puntervold, S. Strand, P. Hopkins, and P. Aslanidis*

**Wireless acquisition for resistivity index in centrifuge: - WiRI: A new method to estimate Archies law parameters**
*Q. Danielczick, P. Faurissoux, and B. Nicot*

**Physical origin of pressure- and saturation fluctuations in steady-state core floods**
*S. Berg, H. Van der Linde, N. Brussee, M. Rucker, H. Dijk, E. Unsal, T. G. Sorup and M. Appel.*

**In-situ investigation of salt precipitation dynamic induced by gas flow-through drying using simultaneous Neutron and X-ray dual-beam Radiography**
*M. Mascle, O. Lopez, H. Deschamps, L. Rennan, N. Lenoir, A. Tengattini, S. Youssef.*

**Material balance and mixing behavior during emulsification of crude oil by using micro-X-ray tomography**
*M. Borji, A. Kharrat and H. Ott*

**SmartWater Synergy with Surfactant and Polymer: A Microscale Investigation at Crude Oil-Water Interface**
*S. Ayirala, A. AlSofi, Z. Li, R. Mariah, Z. Xu, and A. Yousef.*

**DarcyShale: An improved GRI method for more reliable measurements on low permeability samples**
*R. Lenormand and S. Profice*

**The KPC-Log: A new method for measuring the permeability of a non-cylindrical tight sample**
*S. Profice, R. De Loubens, D. Potenzoni, P. Chehade, B. Nicot and R. Lenormand*

**Rock Image Enhancement Using Super-Resolution Neural Networks**
*B. Gong and P. Duke*

**Study of Bakken and Three Forks Formations using NMR Core Analysis**
*M. M. Awad, P. Singer, Z. H. Xie, M. Puerto, H. Pu and G. Hirasaki*

**Core Restoration and its effect on Initial Wetting of NCS Sandstone Reservoir Cores**
*P. Aslanidis, I. D. Pinerez Torrijoz, S. Strand, and T. Puntervold*

**Novel 3D Characterisation of Rock Properties at Core Scale**
*H. Velcin, J. Dautriat, L. Esteban, M. D. Lindsay, A. D. George, T. Richards, and A. Ramage*

**Comparison of High Resolution X-Ray Fluorescence and Probe Magnetics on Core with Borehole Spectral Gamma Ray in 3 Oil Sands Wells**
*T. H. To and D. Potter*

**Developed emulsification in porous media flow**
*A. Kharrat, B. Brandstatter, M. Borji, G. Friz-Popovski, O. Paris and H. Ott*

**Data-Driven Methods to Predict Recovery Factor in Water-Alternating-Gas (WAG) Laboratory Coreflood Experiments**
*A. K, Idogun and L. A. James*

**Tortuosity and Cementation Exponent as Variables related to Heterogeneity and the Impact on Sw Calculations in Tambaredjo Field of Suriname**
*A. Mijland, E. R. Acosta, and M. Pinas*

## Posters presented at the 2021 SCA Symposium

**Evaluation of in-Situ Gelation Behavior by a Slim Tube Technique.**

**Integration of core scale logging, dual-energy computed tomographic imaging and geochemical and mineralogical analysis of a composite core**

**Reactivation and branching by CO2 injection in pre-faulted Berea and Boise sandstones. Laboratory micro-seismic imaging**

**Novel analysis of molecular structures at oil-brine interface for Smartwater flooding.**

**Digital extraction of core properties from whole core photos**

**Developed emulsification in porous media flow**

**The Effect of Microporosity and Oil Property on the Initial and Residual Oil Saturations of a Bimodal Carbonate**

**Data rescue – Maximising the value of archived core using non-destructive Multi-Sensor Core Logger (MSCL) and machine learning techniques**

https://doi.org/10.1051/e3sconf/202336600001

**Rate dependency of capillary trapping in gas-brine flow**

**Multi Field Evaluation of T2 Pore Size Distributions and T1-T2 2D Maps**

**Influence of Wettability on Molecular Adsorption and Kinetics Studied by In-Situ NMR**

**Tortuosity and Cementation exponent as variables related to heterogeneity and the impact on Sw calculations in Tambaredjo field of Suriname.**

**Elucidating Hydrocarbon and Brine Flow Paths in Source Rock Shales Using NMR Imaging**

**Powder River Basin Integrated Hydrocarbon System Case Study**

**Rate-Controlled Mercury Injection Experiments to Characterize Pore Space Geometry of Berea Sandstone**

**Evaluation of the InterFoam Solver in the Prediction of Immiscible Two-Phase Flow in Imbibition and Drainage on the Pore-Doublet System.**

**Dynamic pore-scale modeling of residual fluid configurations in disordered porous media.**

**Evaluation the effectiveness of the different methods enhanced oil recovery (thermal and chemical) on plugs of the Cenomanian tier**

**Oscillating Back Pressure Regulator (OBPR) for High-Pressure Oscillating Back Pressure Regulator (OBPR) for High-Pressure**

**A Mechanistic Study of Wettability Alterations in Sandstone by Low Salinity Water Injection (LSWI) and CO2 Low Salinity Water-Alternating-Gas (WAG) Injection**

# Multi-scale Digital Rock: Application of a multi-scale multi-phase workflow to a Carbonate reservoir rock

*Andrew* Fager[1], *Hiroshi* Otomo[1], *Rafael* Salazar-Tio[1], *Ganapathi* Balasubramanian[1], *Bernd* Crouse[1], *Raoyang* Zhang[1], *Hudong* Chen[1], *and Josephina* Schembre-McCabe[2]

[1]Dassault Systèmes, USA
[2]Chevron, USA

**Abstract.** In some of the challenging digital rock applications the trade-off between model resolution and representative elemental volume is not captured in a single resolution model satisfying the minimum requirements for both aspects. In the wide range of lithofacies found in carbonate reservoir rocks, some facies fall in this category, where large pores, ooids or vugs, are connected by small scale porous structures that could have orders of magnitude smaller pores. In these cases a multi-scale digital rock approach is needed. We recently developed an extension to a digital rock workflow that includes a way to handle sub-resolution pore structures in single phase and multi-phase flow scenarios in addition to regular resolvable pore structures. Here we present an application of this methodology to a multi-scale limestone carbonate rock. A microCT image captures the large pores for this sample, but does not resolve all the pores smaller than the pixel size. A three phase image segmentation that considers pore, solid and under-resolved pores or porous media (PM) is generated. A high resolution confocal image model is obtained for a representative region of the smaller pores or PM region. A set of constitutive relationships (namely permeability vs. porosity, capillary pressure vs saturation and relative permeability vs saturation) are obtained by simulation from the high resolution confocal model. The low resolution segmented image, a porosity distribution image, and the constitutive relationships for the PM are input in an extended LBM multi-scale multi-phase solver. First we present results for absolute permeability and show a parametric study on PM permeability. The model recovers the expected behaviour when the PM regions are considered pore or solid. A consistent value of permeability with experiments is obtained when we use the PM permeability from the high resolution model. To demonstrate the multi-phase behaviour, we present results for capillary pressure imbibition multi-scale simulations. Here a small model for a dual porosity system is created in order to compare single scale results with the multi-scale solver. Finally, capillary imbibition results for the whole domain are shown and different wettability scenario results are discussed. This application illustrates a novel multi-scale simulation approach that can address a long standing problem in digital rock.

## 1 Introduction

Petrophysical properties are critical to understand and forecast oil & gas production. In particular, properties related to single-phase and multi-phase fluid flow are required. These properties represent effective macroscopic constitutive relationships for flow; such as absolute permeability vs porosity, J-function capillary pressure and relative permeability vs. saturation. Laboratory measurements are typically done for this purpose, but another option is to use digital rock to simulate petrophysical properties directly using the particular rock 3D structure, obtained using x-ray micro-tomography (microCT) [1]. Properties directly depend on the internal pore structure, and different type of structures have different constitutive relationships; for example; random porous media, packed spheres, parallel planes, bundle of tubes, heterogeneous sandstones, shale rocks, carbonate rocks. No single property model is good for all porous media types, therefore the need of laboratory measurements or simulation on the actual 3D image of the particular rock. Samples that belong to one porous media type will share similar constitutive relationships. This is the reasoning behind core sampling per porous media type.

A typical microCT image resolution can be few micro-meter per pixel. For most sandstones this resolution may capture the main percolation pore space, while small crevices and some clay porosity are not captured. Since the main percolating path is primarily responsible for fluid flow, this image could still represent the flow properties of the rock. But in many other cases the pores (bodies and throats) that form the interconnected pore structure can vary in size in the same rock by orders of magnitude. This case can be found in some carbonate rock lithofacies, which are complex structures form by recrystallization of micro-organisms, where larger pores are many times connected only through pores smaller than the typical microCT image resolution, which is not able to capture the main percolating pore structure and is not suitable for flow simulation. An increase in imaging resolution can enable percolation, but with more magnification, the larger pores are not going to be captured properly by the image. This is a genuine example where a multi-scale approach is needed to enable simulation.

* Corresponding author: Andrew.FAGER@3ds.com

© The Authors, published by EDP Sciences. This is an open access article distributed under the terms of the Creative Commons Attribution License 4.0 (http://creativecommons.org/licenses/by/4.0/).

In the digital rock literature, there are several publications on multi-scale extensions of multi-phase pore-network models [26-27]. Regarding voxel based models, like LBM, there are several publications [2-6, 24-25] on single-phase multi-scale fluid flow simulations, where formulations for a viscous force with the permeability has been provided by the Brinkman equation and the permeability value was derived from simulations at finer resolutions, but extensions to the multi-phase flow simulation are not discussed. Two recent publications [7, 8] consider multi-phase fluid flow in multi-scale pore structures. In [7] relative permeability from under-resolved regions was computed by solving the transport equation for the total energy, Helmholtz free and kinetic energy, but without considering other important physical properties such as the capillary-pressure curves. On the other hand, in [8], the capillary saturations are calculated in the finer scale using a Young-Laplace relationship and used for estimating capillary pressure in under-resolved regions, but no fluid flow simulation was performed involving pore structure connectivity and relative permeability was also not considered.

We present in this paper an application of a recently developed [9-10] extension to the Shan-Chen multi-phase Lattice Boltzmann model [11-15] that introduces numerical models for treating multi-scale porous media, such as carbonate rocks, which includes under-resolved pores as well as resolved pores. The flow behaviour of under-resolved pores (porous media) is described by a complete set of multi-phase constitutive relationships, which includes single phase permeability-porosity correlation, relative-permeability-saturation, and capillary-pressure-saturation curves. This approach allows for local variability of porosity and saturation in under-resolved regions, while fluid forces are applied locally satisfying the constitutive relationships in the porous media. In the rest of this section we describe our general workflow for multi-scale simulations, as well as describe the carbonate sample selected for this application. In the next section we demonstrate the workflow steps, including image segmentation, porosity estimation, and porous media characterization, all required inputs for the multi-scale multi-phase simulation. Finally we discuss the results for single-phase and multi-phase and present our conclusions.

### 1.1 Multi-scale modelling

Current microCT technologies focus on imaging one single scale, and requires one to choose this scale by balancing the trade-off between image resolution and representative elemental volume (REV). Image resolution must be at least small enough to resolve a percolating pore structure, while REV must be at least large enough to statistically represent all sizes of pores found in the sample. Unfortunately, the number of pixels in microCT imaging systems is limited, and constrains the pixel size and REV to each other. In cases where pore sizes vary by orders of magnitude, like carbonate rocks, a compromise between pixel size and REV cannot be found, and porous media (PM) pixels have to be considered. PM pixels represent regions where pores are present but smaller than the pixel size. An example of a Carbonate rock [21] image is shown in figure 1 (top), where some regions showing as intermediate grey can be identified as PM, with pores sizes smaller than the pixel size. The intermediate grey value can be interpreted as the pore fraction or sub-resolution porosity, between 100% pore (darkest regions) and 0% pore (brightest regions). In general the sub-resolution porosity of PM pixels can be different, even if the pixels belong to the same PM-type. In a similar way that core samples can be classified by porous media types, PM pixels can also be classified by PM-types if they share similar constitutive relationships, or equivalently similar 3D pore structure, as illustrated in figure 1. High resolution imaging, like scanning electron microscopy (SEM) can be used to identify and select representative sample images for a PM-type, for instance in a carbonate rock like the one in figure 1, we could find that all the under-resolved pixels belong to a single PM-type, that we could call "carbonate PM-type 1" for instance.

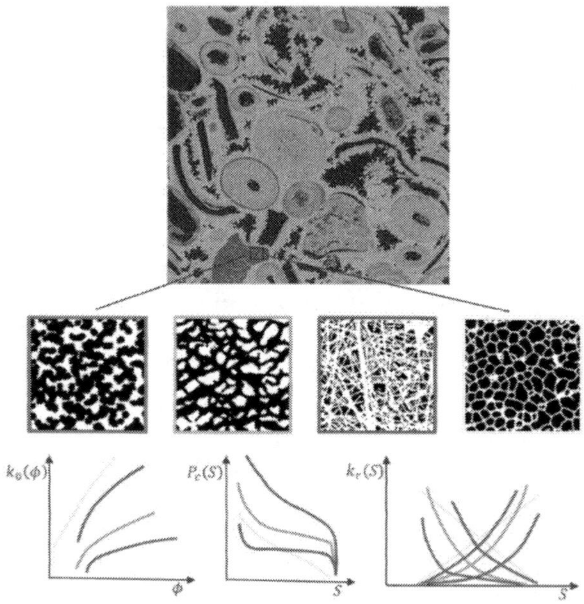


**Fig. 1.** Modelling of under-resolved pore structures using the concept of PM-types classification and characterization.

In general, it could be possible to have more than one PM-type present in an image, corresponding to different sub-resolution pore structures. A rock can have few clay minerals, with distinct microscopic pore structure, like kaolinite that exhibits typical booklet or book-like structures, grown diagenetically inside pores, Illite is hairy grain coating, and chlorite forms flakes. Microporous carbonates can be also present in rocks, as well as partially dissolved feldspars and other metamorphic microporous rocks. In shale rocks organic content, like kerogen also have a microporous structure. Each of these PM-types will behave differently under fluid flow and their identification is part of the image segmentation process. In many cases, like the carbonate rock example in figure 1, a single PM-type can be sufficient.

In order to sample 3D structures of a given PM-type, a combination of high resolution imaging techniques can be used, for example: SEM, focus ion-beam (FIB), laser confocal scanning microscopy (LCSM), X-ray nano-tomography are the most common. By performing simulations using 3D images of these PM samples, it is possible to obtain a suitable set of constitutive relationships that represents well the corresponding PM-type as illustrated in figure 1 (bottom). In a sense, the constitutive relationships for these PM-types can be seen in turn as the result of high resolution digital rock simulations of micron size models. These relationships, absolute permeability vs porosity, J-function capillary pressure and relative permeability vs. saturation, describe a continuum behaviour where porosity, saturation and other properties are described as continuum variables in contrast to a discrete representation where each pixel is pore or solid. It is feasible that some representative PM-types can be characterized and collected into PM libraries that could be used in the same way as core samples called "analogues" are used in core analysis, in this case as a convenient complement to high resolution imaging.

These PM pixels with their local porosity estimated from the images and corresponding constitutive relationships should be sufficient input for a continuous solver that simulates the interaction between such PM pixels. Moreover, the challenge is on extending the current digital rock solvers from only pore/solid pixels to include also PM pixels in the algorithms. Here, we use a solution based on the lattice Boltzmann method (LBM), although the methodology itself is not limited to LBM, for this issue by employing a new workflow [9-10] with effective computational models for flows in under-resolved porous regions. In a fluid-flow simulation at a certain resolution level, flow contributions from under-resolved porous regions are properly taken into account by applying numerical models at each location using local information of geometry and fluids. The models reproduce proper forces acting on the fluids such as viscous, pressure, and capillary forces, using local representative physical properties captured by the PM-type constitutive relationships.

There are various ways to define numerical models for fluid forces in the under-resolved region. We show here one possible example [9-10]. The viscous force $\vec{F}^{\alpha}_{PM\_vis}$ can be defined using the absolute permeability $K_0$ and relative permeability $K_r^{\alpha}$ as,

$$\vec{F}^{\alpha}_{PM\_vis} = -\frac{\nu^{\alpha}}{K_0 K_r^{\alpha}} \rho^{\alpha} \vec{u}^{\alpha} \qquad (1)$$

where $\alpha$ is an index for fluid phases and $\nu$ is the kinematic viscosity. Here, $\rho$ and $\vec{u}$ are density and velocity of fluid flow, respectively. Also, $K_0$ and $K_r^{\alpha}$ are functions of the porosity $\phi$ and density ratio of fluid phases under an assumption of homogeneous structure for under-resolved porous region. On the other hand, an example for definition of capillary force $\vec{F}^{\alpha}_{PM\_cap}$ can be written as,

$$\vec{F}^{\alpha}_{PM\_cap} = -\frac{2\sigma \cos\theta J}{\sqrt{K_0 K_r^{\alpha}/\phi}} \left(\widehat{\nabla \rho^{\alpha_1}}\right) \cdot H(At, |\partial_x(At)|) \qquad (2)$$

where the hat notation indicates the unit vector and $\theta$ is the fluid-solid contact angle in the under-resolved porous structure. Here, $J$ is the Leverett J-function, $P_c\sqrt{K_0 K_r^{\alpha}/\phi}/\sigma \cos\theta$, is the normalized capillary function of density ratio of fluid phases, and $\alpha_1$ is an index of a corresponding fluid phase. A function $H$ is a switch function depending on the local multi-phase interface information. Here, $At = (\rho^{\alpha_1} - \rho^{\alpha_2})/(\rho^{\alpha_1} + \rho^{\alpha_2})$. This switch function is necessary for the diffusive multi-phase model because its non-zero interface thickness may cause excessive artificial force. Moreover, this definition cannot cover a scenario where a fluid phase is confined in an under-resolved pixel. In order to mitigate this problem, an additional model can be implemented using the Leverett J-function and local pressure. Since the saturation is almost insensitive to the local pressure force in this scenario, the local expected saturation level can be computed accurately using the local pressure and the Leverett J-function. Accordingly, the local saturation is controlled without significant deviation from the expected value.
Accurate wettability can be realized with the Shan-Chen model in the LBM approach, which naturally constructs the inter-component force using the fluid potential from neighboring sites [9-13]. On the solid boundary, the proper wall potential is assigned so that the targeted contact angle is achieved. By simply extending this model, the wettability originated from solid parts in the porous media can be properly taken into account. Specifically, the potential in the porous media is constructed by both fluid density and solid wall potential weighted by local porosity.

### 1.2 Test sample

For this study, we use an Indiana limestone rock sample with porosity and permeability measurements in a 1.5 inch diameter core-plug sample, summarized in table 1. Mercury Intrusion Porosimetry (MIP) characterization was done in a 0.25 inch end-trim for this sample and results are shown in figure 2, including the saturation vs. pressure measured data and the derived pore throat size distribution, where two pore size families are clearly appreciated. To study the oil-water behaviour in this Indiana limestone core-plug sample, a spontaneous counter-current imbibition experiment in a decane-saturated and decane-aged sample was done, yielding a remaining decane saturation of 38%.

**Table 1.** Porosity, permeability laboratory measurements.

| | **un-stressed** | **800 psi** |
|---|---|---|
| Klinkenberg Permeability, mD | 12.5 | 6.68 |
| Permeability to Air, mD | 14.6 | 7.94 |
| Swanson Permeability, mD | 10.3 | - |
| He Porosity, fraction | 0.17 | 0.16 |

https://doi.org/10.1051/e3sconf/202336601001

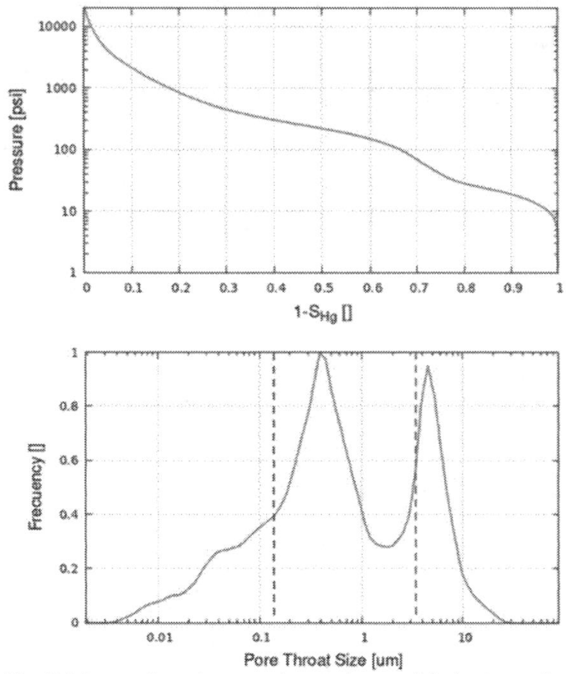


**Fig. 2.** Mercury intrusion porosimetry data, and derived pore throat size distribution of Indiana limestone sample

Imaging was acquired at two resolutions, 3.5 $\mu$m/pixel x-ray micro-tomography 3D imaging, shown in figure 3 for a selected $600^3$ volume, and laser confocal scanning microscopy at 0.128 $\mu$m/pixel, shown in figure 4. A statistical reconstruction algorithm [22] used a set of 2D confocal images focused at a typical micro-porous region as input to build a representative 3D model. The multi-point statistical (MPS) reconstruction method [22] was developed, validated and used extensively for geomodeling applications, and is well known by his capability to capture high order correlations beyond 2-point correlation functions or variograms. Although, we have not done and extensive validation for pore-scale modeling applications, the MPS reconstruction is not a required element in our workflow, which can use in general any other type of higher resolution 3D imaging, like nanoCT and FIB.

We show with vertical dashed lines in figure 2, these two imaging resolutions as a reference of what pore sizes can be captured and modelled using each of these imaging techniques.

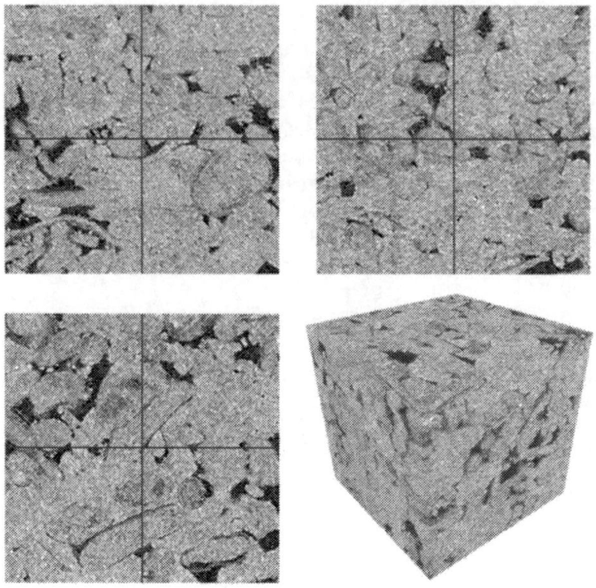

**Fig. 3.** Cross-sectional view and 3D view of the microCT image for our application sample, Indiana Limestone.

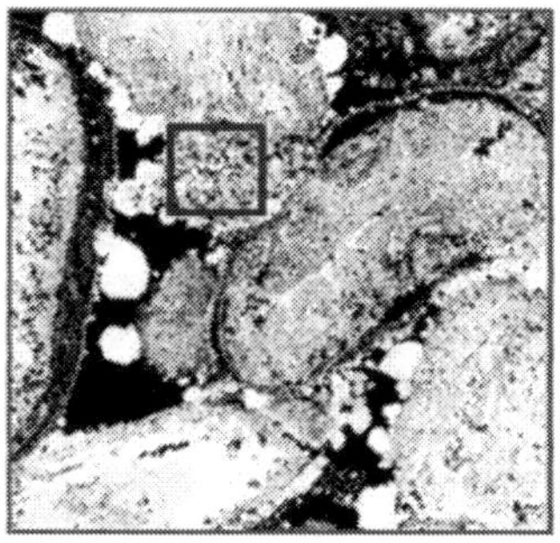

**Fig. 4** View of a slide of confocal image in the Indiana Limestone sample, at nominal resolution of 0.128 $\mu$m/pixel.

## 2 Multi-scale workflow

In this section we follow the multi-scale workflow steps for the carbonate sample described in the previous section. A summary of the workflow steps is shown in figure 5, which includes: large scale modelling: segmentation, porosity estimation, small scale modelling: PM samples selection, properties simulation using digital rock, and generation of constitutive relationships for the PM-type. We perform our multi-scale flow simulations using these inputs from both scales, and the results are presented and discussed in the following section.

 https://doi.org/10.1051/e3sconf/202336601001

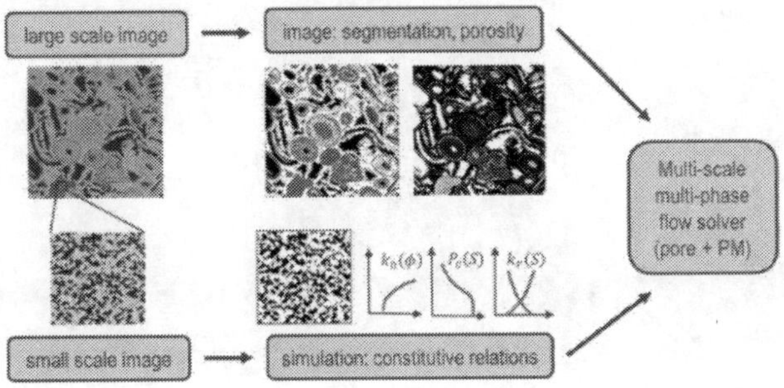


**Fig. 5.** Multi-scale workflow.

### 2.1 Image segmentation, porosity

MicroCT images show a grey scale that quantify the x-ray attenuation coefficient of the material contained in each pixel. Different pure materials (minerals and fluids) have different attenuations values and show as different grey scale values. If each pixel contains only one type of material, segmentation reduces to locate the best thresholds that separate grey scale values corresponding to different materials. However, when a pixel contains pores smaller than the pixel size, a.k.a PM pixel, there is a partial volume effect and the x-ray attenuation is an average of the solid and pore fractions within the PM pixel.

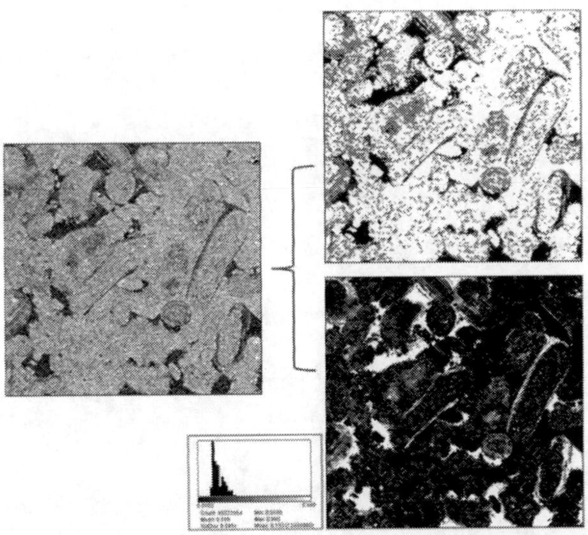

**Fig. 6.** Generation of image segmentation and image porosity.

Depending on the how different the PM pixel grey values are compared to other materials present on the image, it may be possible to use only thresholding to segment the PM pixels from the other pure materials. If grey scale is not enough, textural features can also be used to segment out PM pixels of different PM-types. In the Indiana Limestone sample presented in the previous section, we identify one PM-type, corresponding to a crystalline micro-structure sometimes referred as micrite. We show in figure 6 (top-right) a possible segmentation for our Indiana Limestone image, where pixels of pore/solid show as black/white, while pixels for the PM-type "micrite" are marked with a single shade of medium grey. The volume fraction of pore pixels in the image is 7%, while the volume fraction of PM pixels is 21.3%.

Next, we estimate the values for the local porosity in each of the pixels marked as PM. We use the segmented image to mask out each of the three sub-sets of pixels in the original grey scale image in figure 6 (left). We compute the average grey scale value for the two sub-sets corresponding to pore and solid pixel labels. These two grey scale values should correspond to the values of porosity 1.0 and 0.0 respectively. A linear transform between grey scale values and porosity is calibrated using these two reference points. Notice that this transformation is only possible in cases with only one type of mineral. In general, a linear transform can also be applied using an estimation for the mean value and standard deviation of the porosity in the PM region. In addition to the calibration of grey scale to porosity, a masked Gaussian smoothing filter can be applied to reduce the noise level for the PM porosity. The result for the local porosity image is shown in figure 6 (bottom-right), with a PM porosity mean and standard deviation of 0.186 and 0.05 respectively and a histogram distribution shown next to the porosity image. Notice that the porosity values in the pixels labelled as pore and solid are 1.0 and 0.0 respectively, this is enforced by the masking criteria. A quick calculation of the total porosity for this sample image, including the two contributions from pore and PM is: 7% + 0.186*21.3% = 11%. A more accurate way to map the under-resolved porosity is by direct measurement, using an x-ray attenuating contrast agent like Iodine in a brine solution that is vacuum saturated in the sample. A reference dry-microCT (unsaturated) is performed first, and used later to subtract from the wet-microCT (saturated). This differential imaging technique has demonstrated quantitative accuracy for estimating the under-resolved porosity [30-32] and robustness regarding different mineralogy, therefore it is not limited by lithology. However, the availability of this imaging technique seem to be mainly in research laboratories, being a less common capability in commercial microCT facilities. If only dry microCT is available, the method described earlier can still be used.

These two images: (1) PM-type label and (2) sub-resolution porosity, can be sufficient input for a complete description of flow behaviour of the PM pixels; assuming that a PM library of constitutive relationship curves is accessible for the LBM solver to fetch the corresponding flow response of each PM pixel given its pair values of: PM-type and sub-resolution porosity. In the next sub-section we present the required PM library of constitutive relationship.

### 2.2 Porous media models

Based on [28-29], "micrite" particles are classified in three classes; fine, medium and coarse micrites, with a diameter (or maximum side length of the crystal) of 0.1-3μm, 3-6μm and 6-10 μm, respectively. They exhibit a relation between micropore classes and micrite microtextures: Very fine micropores (0.1 – 2 μm) are found in compact anhedral micrites (like shown in figure 7), coarse micropores (6 – 10 μm) are found in microrhombic and polyhedral micrites. Care should be taken to choose the correct PM-type representative sample if direct imaging is not available. In this application, for modelling the identified micrite PM-type, a model was obtained by LCSM microscopy imaging in a selected representative location of the identified micrite PM-type in the Indiana Limestone sample. A 3D model was constructed from those confocal images with a resolution of 200 nanometres. An image of the confocal model is shown in figure 7.

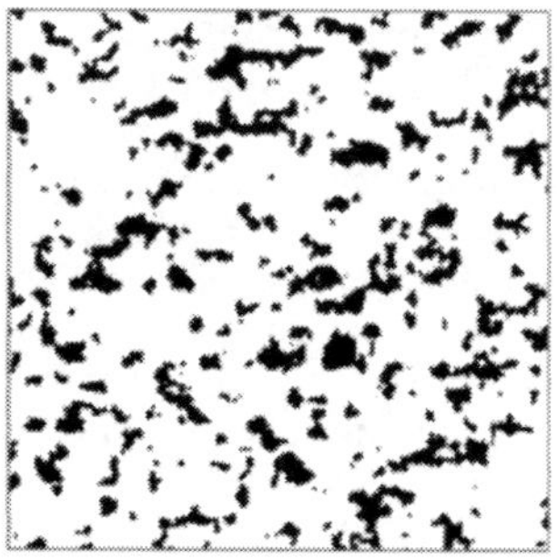

**Fig. 7.** PM model for the micrite PM-type, based on confocal imaging

The porosity, permeability values for the confocal model are 18.2%, 1.7mD. We use the confocal model for micrite as it comes from a high resolution image of the actual sample. Fluid flow simulations using our single-scale LBM solver produce a set of constitutive relationships, as illustrated in figure 8, absolute permeability vs porosity, capillary pressure vs. saturation and relative permeability vs. saturation, for the micrite PM-type represented by the confocal PM model, shown in figure 7.

In order to capture the single phase absolute permeability correlation with porosity we consider the simulation results from eight sub-volumes and fit a parametric relationship as show in figure 8 (top). The fitting quality can be affected by the functional form used, and the limited data range, however the application range is also limited, 12%-24% from the porosity distribution in figure 6, with the maximum of the distribution where the actual data is found. For two-phase flow simulations, we consider in this case an imbibition setup (inlet water, outlet oil) and fit similar parametric relationships in order to capture continuous functions of saturation conveniently, figure 8 (Pc middle, Kr bottom). The wettability condition was set to water wet, with a water-contacted angle of 10°, oil-contacted angle 30°, a water-oil interfacial tension of 30 dynes/cm, and an initial water saturation of approximately 35%. The set of curves in figure 8 will be use as the constitutive relationships that represent the flow behaviour of the PM pixels classified as belonging to the PM-type identified as micrite for our Indiana Limestone sample in the multi-scale simulations presented in the next section.

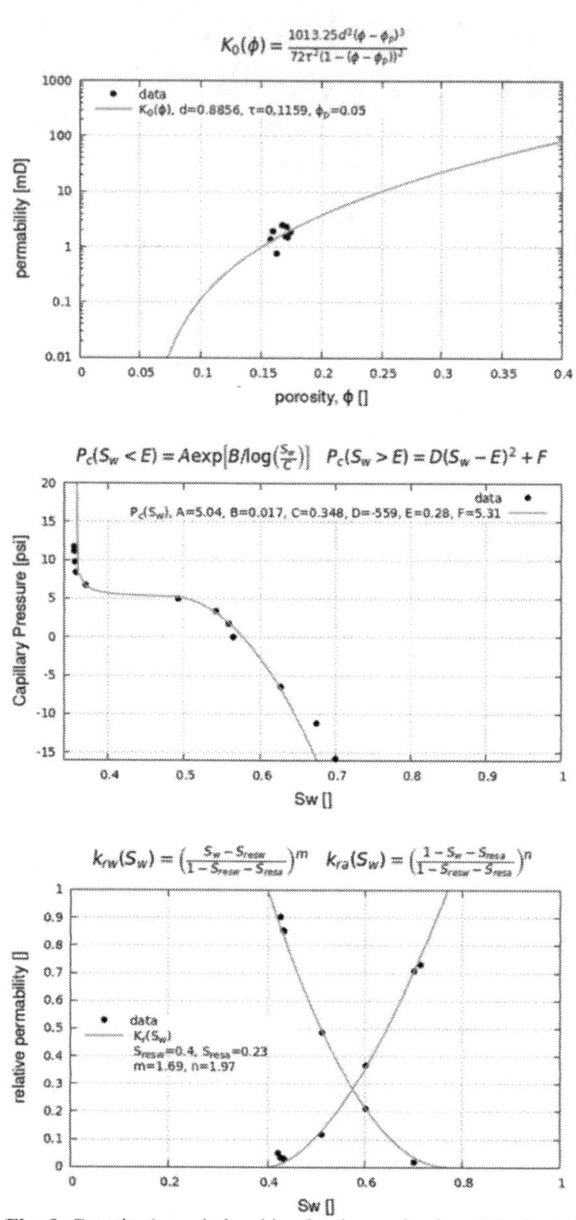


**Fig. 8.** Constitutive relationships for the confocal model of micrite PM-type.

## 3 Results and discussion

As a baseline for comparison we would like to consider two references: (1) A single resolution model, where in figure 6 (top-right) all the pixels marked as PM are re-labelled as solid, and (2) a registered model, which is a full resolution dual porosity model, constructed by scaling the microCT model by a factor of 10×, and replace the PM regions by copies of the 3D PM model, shown in figure 7, using periodic mirror boundaries that enforce full connectivity of the PM pore structure.

In figure 9 we show an illustration of this registered model constructed using the confocal model pore structure. The top image shows a slice from the resulting $6000^3$ pixels model, with a resolution of 0.35 $\mu$m/pixel, which is 10 times smaller than the original microCT image resolution. The full model shown in figure 9 is too large to be used in a simulation, therefore we will use sub-volumes or coarsened versions of this model for comparison.

In this section we discuss the single-phase and multi-phase multi-scale simulation results for our Indiana Limestone sample and compare them with the available experimental measurements and our baseline simulation results.

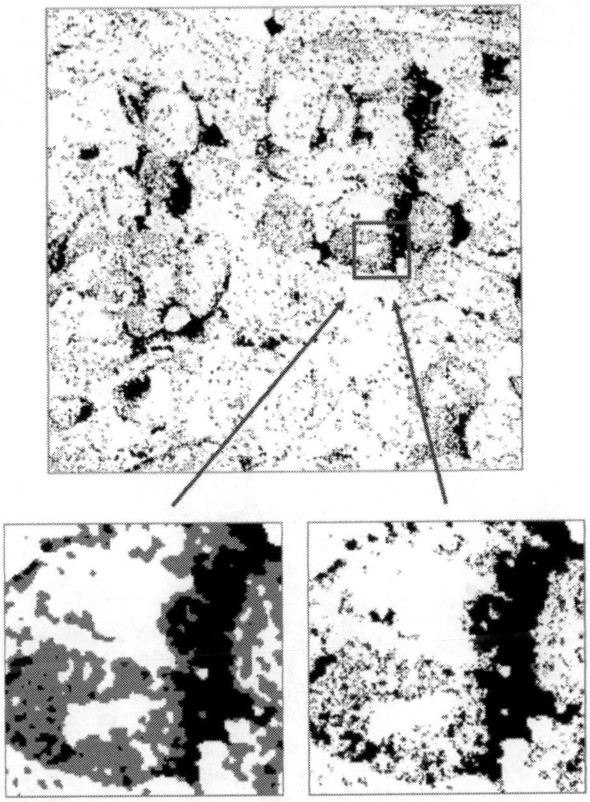

**Fig. 9.** Registered confocal model, $6000^3$.

### 3.1 Single phase

As shown in the pore size distribution from figure 2 (bottom), the microCT resolution of 3.5 $\mu$m can only fully resolve the larger pores of this Indiana Limestone sample, not the micro-pores. Moreover, when we analyse the connectivity of the fully resolved percolating pore structure, we find that the critical pore throat is just 1 pixel. This means that the single-scale pore structure (PM considered solid), is barely connected. In order to obtain a numerically stable estimation of the baseline permeability for the single resolution model, we increase the resolution in the model by a factor of 2×. The single-scale single-phase LBM solver simulation result for absolute permeability using the 2× model is 7.7 mD (the 1× model result is 6.0 mD). We select a representative direction ($k_x$) in order to facilitate the comparison between different results. The velocity flow paths for this single resolution model are show in figure 10 (left). Next, we consider the registered large size model described in the introduction of this section. We utilize a surface element based LBM modelling capability [16-20] to capture features below the nominal pixel resolution by using as input geometry a grey scale image of the pore-solid boundary. This grey scale boundary interface is actively used to locate a pore-solid surface out of the original pixel restricted grid, allowing us effectively to resolve pore structures smaller than the nominal pixel size, and avoid "stair case" related artefacts in the flow. We nominally coarsen our $6000^3$ registered by a 5× factor, however the grey scale resulting model could be considered equivalent to a $2400^3$ model. The registered confocal model simulation result for absolute permeability is 8.2 mD. Velocity flow paths for this registered model are shown in figure 10 (right). This small increment in the registered model result relative to the single resolution model permeability indicates that the role of the under-resolved porosity for single-phase flow is secondary in the case of this sample.

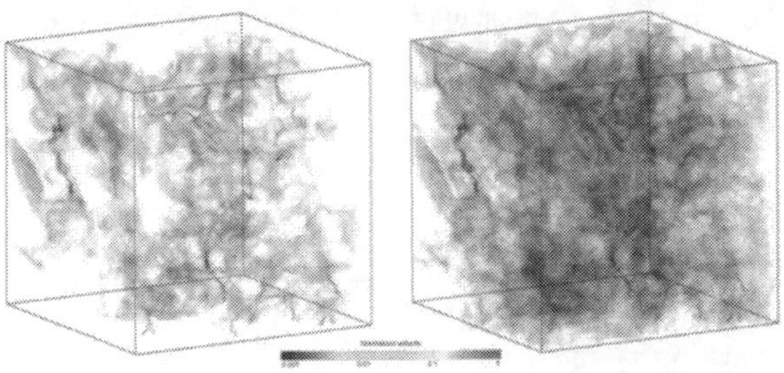

**Fig. 10.** Velocity flow lines for single resolution model (left) and registered confocal model (right).

For the multi-scale single-phase simulation in this Indiana Limestone sample, following the workflow in figure 5, we use as geometry input the pair of images for the segmentation and porosity, figure 6 (right), and the porosity vs permeability relationship for the confocal model of micrite PM-type shown in figure 8 (top). The multi-scale simulation result for absolute permeability is 7.8 mD, which is a value consistent with the results from the single resolution model and registered model simulations.

Finally we want to discuss a parametric study for the overall single-phase flow in this sample as a function of a varying value of the PM permeability. The goal of this exercise is to verify the consistency between our single-scale solver and our multi-scale solver: for a small enough value of $k_{PM}$ used in our multi-scale solver we should recover the same results as in our single-scale solver when PM pixels are re-labelled as solid. Similarly, for a large enough value of $k_{PM}$ used in our multi-scale solver, we should recover the same results as in our single-scale solver when PM pixels are re-labelled as pore. The results of this parametric study are shown in figure 11. As expected we recover with our multi-scale solver the respective end member results of the single-scale solver for the overall permeability. Moreover, we show in figure 11 that the range of the permeability expected from laboratory measurements, table 1, is consistent with the result obtained with the confocal model we use for micrite PM-type.

This kind of response function in figure 11, can also be useful to understand the ranges of permeability for the under-resolved porosity where a significant impact in the overall flow can be expected. This response function is not trivial and depends on the complexity of the PM plus pore connected structure.

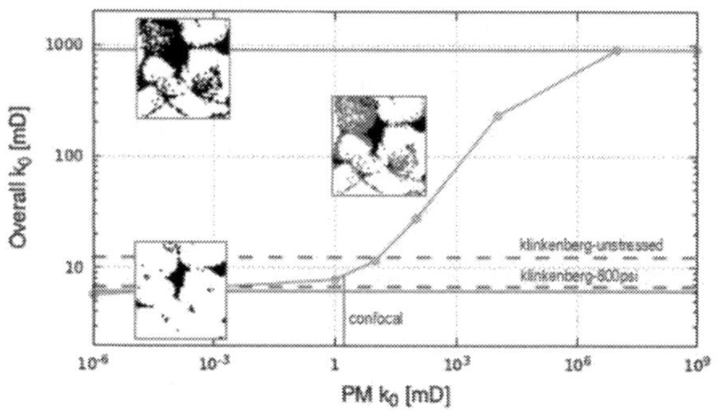


**Fig. 11.** Single-phase multi-scale flow permeability prediction as function of PM permeability

### 3.2 Multi-phase drainage

Although it is entirely possible to use it for drainage, we will delay the use of our multi-scale multi-phase LBM extension until the next sub-section for imbibition, which represent a more interesting scenario for oil-water behaviour, and has some experimental data to compare with, while in the other hand for drainage we have only MIP experimental data to compare with, therefore we take this as an opportunity to describe our MIP multi-scale extension for this type of simulation. We use the well-known percolation invasion method for MIP simulation [23], which uses the maximal inscribed sphere map on the pore space and the Young-Laplace inverse relationship between pore radius and pressure for capillary equilibrium. In the maximal inscribed sphere map, each pore pixel is assigned a radius value equal to the largest sphere radius that can be inscribed in that region of the pore space and includes this pixel inside the sphere. This map can be used to represent the local capillary equilibrium inter-phase between ideally wetting and non-wetting fluids, given that the contact angle of the inscribed sphere is 180°. This is a good approximation for mercury intrusion porosimetry (MIP), between the inter-phase of liquid mercury and vapour mercury. In order to minimize the surface-to-volume ratio effect given the small size of digital rock models as compared to real core samples used in the laboratory, we do not surround the whole model with mercury, but we restrict the mercury to be in contact with only one face of the rectangular prism model. For an initial low applied pressure (large applied radius), all pixels marked with radius larger or equal to the "applied radius" and connected with the inlet are marked as invaded by mercury. The applied pressure is increased and this percolation invasion process is repeated until all the pore structure is invaded. A saturation can be computed as the pore pixels invaded fraction, and recorded for each pressure. In figure 12 we show the results of this method applied to the single resolution case in our Indiana Limestone model. For the registered cases, due to the size limitations of the method, we show a 2× coarsened version of the registered model using the confocal micrite PM-type model, figure 12. For comparison, we include the experimental MIP data. We also use an experimental total porosity value of 16.5% to normalize the simulated saturations.

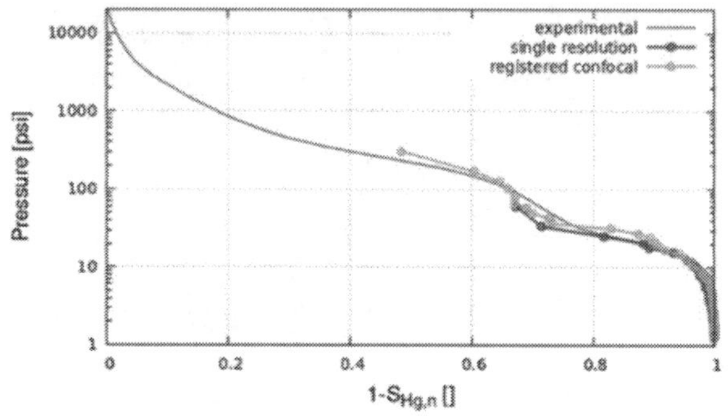


**Fig. 12.** Drainage results, pressure vs. normalized mercury saturation for single resolution model and registered model compared to experimental MIP.

Notice in figure 12 that the simulation results have an intrinsic maximum pressure limit associated with the minimum pixel size of the model, due to the Young-Laplace relationship. In order to extend this method to include the PM pixels, we consider the extension of the Young-Laplace relationship to radius smaller than the pixel size for a saturation $S$ and pressure $P$

$$P(\phi_i, S) = \frac{2\sigma \cos \quad (S)}{r_e(\phi_i)} \tag{3}$$

where the effective radius of a PM pixel of porosity $\phi_i$, is $r_e(\phi_i) = \sqrt{k(\phi_i)/\phi_i}$, $\sigma$ is the interfacial tension, $\theta$ is the contact angle, and $J(S)$ is the Leverett J-function that can be computed from the single-scale MIP simulation results on a 3D PM sample, like the micrite PM-type sample shown in figure 7. The results of this extended method that considers capillary behavior in PM pixel is shown in figure 13, where we show the MIP-PM simulation results using the confocal PM model.

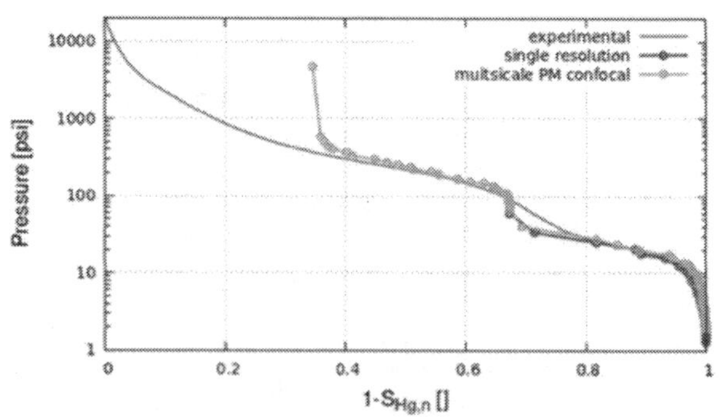


**Fig. 13.** Drainage results, pressure vs. normalized mercury saturation for single resolution model and multi-scale PM confocal MIP model compared to experimental MIP.

Similarly to what was done for the registered model, the saturation calculation is done using a total experimental porosity value of 16.5%, instead the total image porosity, which is 11% when one uses the micrite PM-type confocal model. The difference in total image porosity comes from a rescaling of the porosity input image, figure 6 (bottom-right), to have a mean value of 0.186 in order to be consistent with the porosity (and resolution) of the micrite confocal PM model used. In the case shown in figure 13, we were able to recover the location of the second entry pressure corresponding to the under-resolved micrite porosity, down to the limit imposed by the resolution of the PM model used, which is around 0.2 $\mu$m.

### 3.3 Multi-phase imbibition

This sub-section covers the application of our multi-scale multi-phase LBM method to the Indiana Limestone model. A full size single-scale multi-phase simulation was not possible due to the limited connectivity of the full size microCT model (critical radius of 1 pixel). A full size registered model is also impractical. Therefore, we decide to test first a smaller $100^3$ sub-volume for methodology comparison, as shown in figure 14, microCT image, multi-scale PM model segmentation and porosity, single resolution model, and registered confocal model.

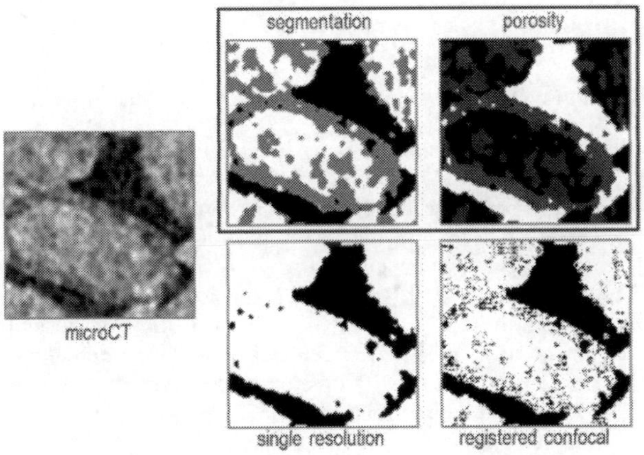


**Fig. 14.** Model images for the sub-volume $100^3$

We considered an imbibition setup (inlet water, outlet oil) and the wettability condition was set to water wet, with a water-contacted angle 10°, oil-contacted angle 30°, and an interfacial tension water-oil 30dynes/cm. It was difficult to have a comparable initial water saturation value between models, since each one has a different effective porosity, but we tried to set it around 10%, relative to a common total porosity value used to normalize saturations to be comparable with each other. For the single resolution model and registered confocal model the 10% water saturation was set from a maximal inscribed sphere based drainage (section 3.2) and we use our single-scale LBM solver. For the multi-scale PM model the initial conditions were set with no water in the resolvable pore and a partial constant initial water saturation around 40% in the PM and our multi-scale LBM solver extension was used, with inputs: segmented and porosity images, figure 14 (top-middle and top-right), constitutive relationships for imbibition in the confocal PM model for micrite PM-type, figure 8.

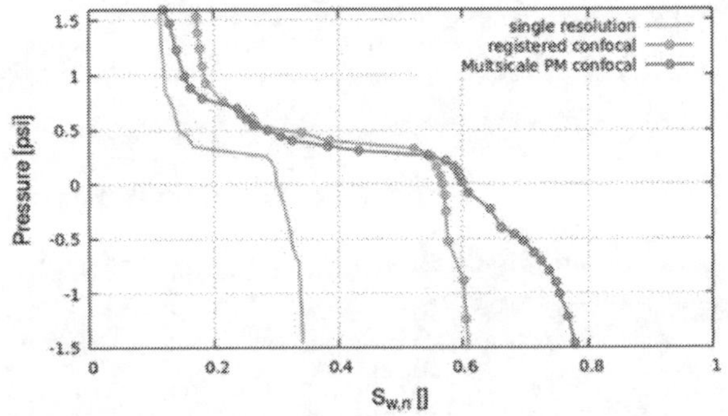


**Fig. 15.** Comparison of imbibition results for the sub-volume $100^3$

A comparative plot with the three results is shown in figure 15, where there is a clear distinction between the single resolution model result that ignores under-resolved porosity contributions and the other two results that consider the under-resolved porosity contribution. The main difference relates to the capacity to mobilize more of the remaining oil in the pore space when the additional connectivity provided by the under-resolved porosity is considered. Notice that even though there is some variability with the registered result, there is an important consistency trend with the results obtained using the multi-scale PM-pore-solid model. The results in figure 15 provide some validation that our multi-scale multi-phase LBM extension is reasonably capturing the multi-phase behaviour as compared with some high resolution dual porosity model results, registered-confocal.

Moving up in model size, we consider next a $250^3$ sub-volume scaled by 2× to a $500^3$ size that shows more heterogeneity, but still keeping a minimal resolved pore connectivity, with a critical radius of 2 pixels. While not yet of REV size to compare with laboratory data, this model still gives us some additional insight on the fluids distribution at different stages of the imbibition process. This time we run only single resolution model and multi-scale model simulations. Registered models become impractical at this point. Same wettability conditions were set as in the previous case, and initial conditions were set with no water in the resolvable pore and a constant initial water saturation around 40% in the PM. In this case we use the laboratory total porosity, 16.5% to normalize all saturations.

The simulation results are shown in figure 16. First, we compare the single resolution model result (dotted yellow line) to the multi-scale result (green line), similarly to what we observed in the previous $100^3$ case, the addition of the PM connectivity enables a better oil mobilization in the resolvable pores. Next, we calculate separately the saturation contributions to the multi-scale total saturation results from the resolvable pore (purple) and from PM (blue). The pore contribution starts at zero saturation, but for better visualization we move the curve to start at the same location as the PM curve. We can see that at the beginning of the simulation the PM saturation increases more than the pore saturation, while at later stages of the simulation the order is reversed. In the bottom of figure 16 we show four stages of the fluid distribution corresponding to the green circles that also show visually the same interplay between the pore and PM water saturation evolution along the simulation.

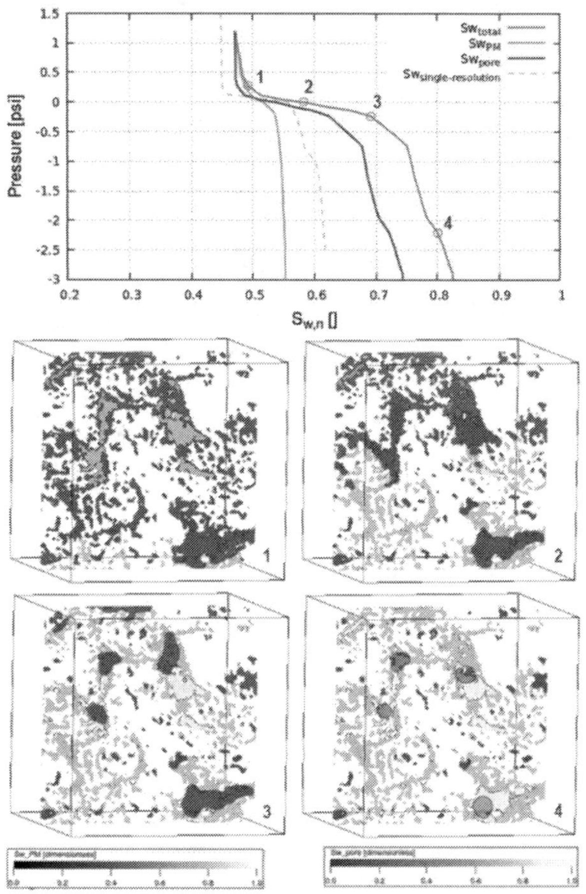


**Fig. 16.** Comparison of imbibition results for the $500^3$ case.

Finally we run our multi-scale multi-phase simulation in the whole model $600^3$ at the original resolution 3.5 $\mu$m/pixel, where the structure of only resolvable pore is barely connected by 1 pixel. We use the same initial conditions and porosity normalization as in the previous sub-volume $500^3$ case. In addition to the same water-wet setup used in the two previous cases ($100^3$ and $500^3$), we also test a mix-wet condition in which the resolvable pore is considered oil-wet, with a water-contacted angle 10°, oil-contacted angle 150°, while the PM wettability is maintained as water-wet, this is done by using the same constitutive relationships as in figure 8. In general if a different wettability condition for the PM is to be used, a new set of constitutive relationships would need to be created where the desired wettability conditions for PM are used in the simulation of the corresponding PM model.

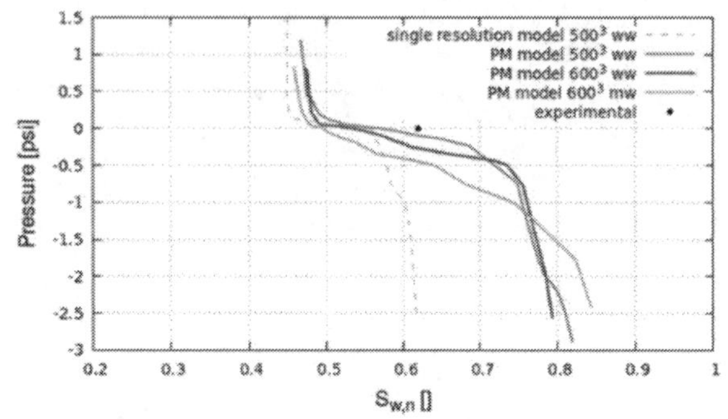


**Fig. 17.** Comparison of imbibition results for the $600^3$ case.

Four results are shown in figure 17, a $600^3$ water-wet case result (purple), two results for the sub-volume high resolution $500^3$ water-wet case: single resolution model (dashed yellow) and multi-scale (green), and the $600^3$ mix-wet case just described (blue). As for the water-wet cases, they are deemed to be in good agreement for the whole volume $600^3$ and the $500^3$ sub-volume as well as consistent with the water saturation of 62% from laboratory spontaneous imbibition measurement. Next, the mix-wet scenario behaves in the expected way relative to the water-wet case, most of the blue curve is below the purple curve, except at the end of the curve where the order is inverted. Finally, our initial observation holds: the multi-scale results show a larger mobilization of remaining oil as compared with the single-scale results.

## 4 Conclusions

In this paper we show a new method for multi-scale flow simulations, single-phase and multi-phase, and an application to a multi-scale carbonate rock, where under-resolved pores are believed to play an important role for oil production. We show consistency between our single-scale LBM solver and the new multi-scale LBM solver extension. In addition to an extension to the LBM solver, we also show a possible extension to include under-resolved pores to the percolation invasion method for MIP simulation that uses the maximal inscribed sphere map. It is worth noticing too that we discuss a complete workflow where a systematic identification and characterization of PM-types play a central role. Additionally, this framework allows for internal variation of porosity within the same PM-type. We believe this complete workflow represent a significant advancement in the application of digital rock methods to a difficult class of multi-scale rocks, such as carbonates. By including the simulation of flow in under resolved pores, the LBM extension solver is not limited to a minimal pixel size anymore. Given the relaxation of the resolved percolating pore space, which extend now to the join pore + PM space, other possible applications of this workflow could include plug-size sample simulation and whole core medical CT based simulation when heterogeneity is not captured properly by plug-size samples. A critical requirement could be, being able to identify PM-type classes that can be represented well by few samples at the right scale. A future workflow for core-plug and whole-core would definitely require several scales for which this workflow can be used progressively to connect pairs of adjacent scales. Multiple scales of images and other experimental data would be necessary for testing and validation of this workflow at core-plug scale and above. The current multi-scale workflow is more suitable for isotropic uniform cases, when the under-resolved pores are much smaller than the pixel-size. Cases where the under resolved pores are close to the pixel-size would need local directionality. This limitation can be important to capture properly for instance under-resolved pore throats. We are working on an improvement that uses the image itself to compute local directionality and rotate the PM-type permeability tensor accordingly. This extension to the workflow will be the subject of a follow up publication.

**References**

1. G.R. Jerauld, J.T. Fredrich, N.M. Lane, Q. Sheng, B. Crouse, D.M. Freed, A. Fager, R. Xu, Abu Dhabi Int. Pet. Exhib. Conf., Society of Petroleum Engineers SPE-188688-MS (2017)
2. D.M. Freed, International Journal of Modern Physics C **09**, 1491 (1998)
3. N.S. Martys, J.G. Hagedorn, Materials and Structures **35**, 650 (2002)
4. I. Ginzburg, G. Silva, L. Talon, Phys Rev E **91**, 023307 (2015)
5. J. Li, D. Brown, Geofluids, 1740693 (2017)
6. J. Zhu, J. Ma, Scientific Reports **8** 5826 (2018)
7. A. Demianov, O. Dinariev, N. Evseev, The Canadian journal of chemical engineering **89**, 206 (2011)
8. M. Suhrer, X. Nie, J. Toelke, Proceedings of international petroleum technology conference, 20035 (2020)
9. H. Otomo, H. Chen, R. Zhang (to be published)
10. H. Otomo, R. Salazar-Tio, H. Chen, R. Zhang, A. Fager, G. Balasubramanian, B. Crouse, H. Fan, J. Yang, US patent application filing number 17/136,259 (2020)
11. H. Otomo, H. Fan, R. Hazlett, Y. Li, I. Staroselsky, R. Zhang, H. Chen, C. R. MECANIQUE **343** 559 (2015)
12. H. Otomo, H. Fan, Y. Li, M. Dressler, I. Staroselsky, R. Zhang, H. Chen, J. Comput. Sci. **17** 334 (2016)
13. H. Otomo, B. Crouse, M. Dressler, D.M. Freed, I. Staroselsky, R. Zhang, H. Chen, Comput. & Fluids **172** 674 (2018)
14. H. Chen, R. Zhang and P. Gopalakrishnan, US Patent 9576087 (2013)
15. H. Chen, R. Zhang and P. Gopalakrishnan, Phys. Scr. **95** 034003 (2020)
16. H. Chen, C. Teixeria, K. Molving, Int J Mod Phys C **9** 1281 (1998)
17. Y. Li, R. Zhang, R. Shock, H. Chen, Eur. Phys. J. Special Topics **171** 91 (2009)
18. Y. Li, R. Shock, R. Zhang, H. Chen, J.Fluid Mech **519** 273 (2004)
19. H. Fan, R. Zhang, H. Chen, Phys. Rev. E **73** 066708 (2006)
20. B. Crouse, D. M. Freed, N. Koliha, G. Balasubramanian, R. Satti, D. Bale, S. Zuklic, The Society of Core Analysis conference SCA2016-058
21. T. Bultreys. Savonnières carbonate. Digital Rocks Portal http://www.digitalrocksportal.org/projects/72 (2016)
22. S. Strebelle, Mathematical Geology **34** 1 (2002)
23. D. Silin, T. Patzek, Physica A **371** 336 (2006)
24. A. Fager, R. Salazar-Tio, G. Balasubramanian, and B. Crouse, RMAG/DWLS FALL SYMPOSIUM. Denver (2019)
25. M. Zhang, R. Salazar-Tio, A. Fager, B. Crouse, Unconventional Resources Technology Conference URTeC (2020)
26. S. K. Masalmeh, X. Jing, S. Roth, C. Wang, H. Dong, M. Blunt, Society of Petroleum engineers SPE-177572-MS (2015)
27. L. C. Ruspini, G. Lindkvist, S. Bakke, L. Alberts, A. M. Carnerup, and P. E. Øren, Society of Petroleum Engineers SPE-180268-MS (2016)
28. R. L. Folk, American Association of Petroleum Geologist Memoir **1** 62 (1962)
29. M. H. Rahman, B. J. Pierson, W. I. Yusoff, International Petroleum Technology Conference, IPTC 14583 (2011)
30. A. Ghous, T. J. Senden, R. M. Sok, A. P. Sheppard, V. W. Pinczewski, M. A. Knackstedt, Society of Petrophysicists and Well Log Analysts Symposium, Abu Dhabi, SPWLA-MERS-2007-E (2007)
31. M. A. Boone, T. De Kock, T. Bultreys, G. De Schutter, P. Vontobel, L. Van Hoorebeke, V. Cnudde, Materials Characterization, **97**, 150 (2014)
32. Q. Lin, B. Bijeljic, H. Rieke, M. J. Blunt, Water Resources Research **53**, 7457 (2017)

https://doi.org/10.1051/e3sconf/202336601002

# Digital Core Analysis as an Efficient Tool for Acid Treatment Optimization

Evgeny Ivanov[1], Dimitry Korobkov[1], Igor Varfolomeev[1], Alexander Demianov[1], Alexander Sidorenkov[1], Anna Beletskaya[2], and Mikhail Stukan[1]

[1]Schlumberger, 125171 Leningradskoe shosse 16A, build. 3, Moscow, Russian Federation
[2]Schlumberger, 555 Industrial Boulevard, Sugar Land, TX 77478, USA

**Abstract.** Digital rock workflow for acid treatment optimization is developed. It includes construction of a digital rock model based on X-ray microtomography, running a set of multiscale digital experiments and a number of acid flooding experiments on rock samples for the model tuning. Here, we present results of application of the developed workflow for Austin chalk rock treatment by hydrochloric acid on core samples of 8 mm in diameter. Pore-scale simulations were performed using high-resolution 3D microCT models to determine the dependencies of rock permeability, active surface area and effective reaction rate on actual porosity. These properties were used to populate the laboratory-size core models, and Darcy-scale numerical approach was applied to simulate the dissolution at different injection rates. The core-scale simulations demonstrated inhomogeneous character of dissolution process leading to wormhole development in a certain range of injection rates. In parallel to numerical simulations, laboratory experiments on the same rock samples were performed at flow rates close to the optimal regimes followed by microCT imaging of wormhole structures. This information was then used to tune the model parameters. After that, the dependencies of the number of pore volumes injected until the breakthrough (PVBT) on the injection rate were obtained numerically for all the cores considered.

## 1 Introduction

Acidizing is one of the approaches widely used to increase the rate of hydrocarbons production from carbonate reservoirs. The effect is achieved due to partial dissolution of matrix rock by acid solution injected under pressures below the formation fracture pressures. Depending on parameters, such as reactive fluid composition, rate of fluid injection, and rock mineral composition, different dissolution patterns can be created. Face dissolution is observed when the acid is spent before penetrating deep into the rock. While uniform dissolution takes place, when the acid is evenly distributed along the treatment zone leading to uniform increase of rock porosity. The optimal inflow of hydrocarbons toward the wellbore or fracture is achieved, when the acid creates a set of thin channels, or so-called wormholes.

The treatment efficiency is crucially dependent on the acid composition and injection scenario. Since the optimal parameters are determined by geology and lithology of the reservoir, to achieve the best effect, each treatment should be preceded by experiments on representative rock samples.

One of the key characteristics determined in the framework of acidizing job design is the pore volume to breakthrough (PVBT). Which is defined as the amount of treatment fluid (measured in core pore volumes) required to be injected before the appearance of macroscopic channel on the opposite side of the core. PVBT dependence on the treatment fluid injection rate goes through the minimum, which reflects the optimum injection rate [1, 2].

Standard design of acidizing job heavily relies not only on laboratory experiments, but also on the simulation of treatment fluid impact on transport properties of the rock. Among the parameters to be optimized are the treatment fluid composition, sequence of injected fluids, amount of the fluid to be injected, and the injection rate. These simulations base on geology model of the near wellbore zone of the collector and account for fluid-rock interactions via such effective parameters as PVBT, which have to be determined in core experiments. Unfortunately, the amount of the core material available is usually not sufficient for comprehensive laboratory analysis of various fluids and treatment scenarios. Moreover, the destructive nature of acidizing experiments imposes the fundamental limitation: experiments are performed on different core samples, which makes impossible any direct comparison of different acidizing scenarios. Conventional laboratory acidizing experiments performed using insufficient number of samples are less conclusive, since it is hard to distinguish the effects caused by the changes in the treatment scenario from those related to the variation in sample properties without averaging over a sufficiently large collection of samples. Therefore, a potential added value of simulation-based screening of acidizing scenarios for performance of acidizing technology can hardly be overestimated. Development of numerical models applicable for acidizing optimization have been of high interest during the past decades [3, 4]. According to the recent review [5], most available acidizing models do not provide acceptable accuracy in prediction of the optimum

© The Authors, published by EDP Sciences. This is an open access article distributed under the terms of the Creative Commons Attribution License 4.0 (http://creativecommons.org/licenses/by/4.0/).

injection rate. One of the possible reasons is that due to the limitations in computational resources, reactive transport models cannot simultaneously account for spatial heterogeneity at different scales: from macro scale, down to the pore scale. Thus, to make predictions on scales interesting from a practical point of view, the well-mixed condition is often assumed at scales smaller than the model discretization size. Similarly, many semi-empirical models assume the existence of dominant wormhole and ignore fine-scale acidizing patterns [3]. Therefore, using those models, one cannot properly reproduce the actual wormhole patterns and hence identify the optimum acid composition and fluid injection rates. We believe that a proper digital acidizing model should be able to reproduce the laboratory results scales, under a wide range of experimental conditions — only then such a model can be considered as a verified one and used at practically interesting scale.

There are definite indications that rock pore structure is a critical factor in carbonate acidizing job design [6, 7, 8]. Dubetz et al. [6] have shown that pore size distribution functions can correlate with optimal acidizing parameters better than porosity or permeability.

In acidizing processes, the transport phenomenon interferes with the chemical reaction. This determines the size and the resolution of a model required — it must capture the pore space geometry before and after the treatment and describe the whole physical sample.

Here, we propose a combined two-scale approach based on detailed digital rock models obtained using X-ray microCT. The pore scale approach used in this work is based on a combination of principles of thermodynamics and density functional theory applied for hydrodynamics (DFH) [9]. This approach allows to estimate the effective reactive and transport properties of the rock directly from the pore geometry and the treatment fluid properties [10] and is not restricted to any specific treatment fluid or mineral. The inhouse simulation software (Direct Hydrodynamics (DHD)) we use for pore-scale simulations, proved to be applicable in many problems of HC recovery from the porous media [11-13].

In our previous work [12], we described implementation of the heterogeneous chemical reactions at pore scale accounting for the evolution of pore geometry during reactive flow. However, accounting for all the fundamentals of treatment process makes the developed algorithm computationally expensive. Therefore, it was applied at small models of porous media sufficient to capture just few connected pore channels.

In the current work the pore-scale simulations were used in assumption of fixed pore geometry just to calculate the transport and reactive properties in each individual sub model, obtained from detailed microCT scan of carbonate rock. Results of numerous calculations were used to correlate the relative active surface and permeability to porosity in order to use these correlations in core-scale modeling.

The continuum model was utilized to simulate the reactive fluid flow using the Darcy equation on core scale. To capture the mineral dissolution process, the closing relations for the effective reaction rate and evolution of rock transport properties were defined from the direct hydrodynamics modeling at pore scale.

The developed approach allows to effectively account for the core-scale phenomena and the presence of fine pores in low-permeable matrix in the models, which contain high permeable structures like fractures and wormholes. It also allows to upscale the treatment fluid–mineral rock chemical reaction from pore scale to the continuous Darcy-scale modeling.

In this work, we demonstrate an application of the suggested approach on example of Austin chalk core samples treatment with aqueous solution of hydrochloric acid.

This paper is organized as follows. First, the summary of laboratory core tests is presented. Then the microCT scan equipment and digital model construction technique is described, followed by numerical model description. Finally, the results of laboratory tests together with numerical simulations are discussed and followed by conclusions.

## 2 Experimental setup

While analyzing heterogeneous reservoir rocks, it is important to perform laboratory measurements and microCT imaging on the same core plug. Taking into account the constrains of both techniques the Ø8 mm plug with 10-20mm length looks like an optimal tradeoff between the image quality and the suitability of the sample for the laboratory experiments. We use existing routine core analysis (RCAL) and special core analysis (SCAL) methodology and equipment adapted for the cores of such size [14].

A special configuration of 3-phase core flooding setup was assembled for the test (Fig. 1). For flooding experiments on mini-plugs, we built a Hassler-type core holder with electrodes for electrical properties measurements by two-wire method. In the mini plug flooding experiment, the pressure drop can be measured between inlet and outlet only because of small plug length.

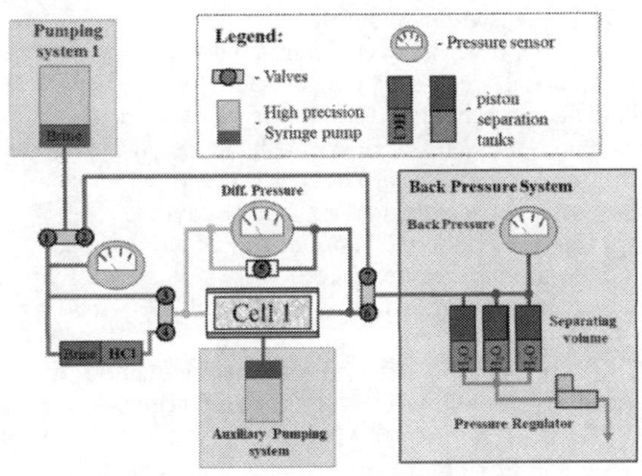


**Fig. 1.** Hydraulic scheme of coreflooding setup.

The outer part of the core holder is connected to auxiliary pumping system to maintain confining pressure.

Core flooding setup consists of the following main parts (see Fig. 1):

• Main pumping system with two precision syringe pumps for constant pulseless fluid injection with the flow rate range 0.001 to 30 ml/min at pressure up to 40 MPa.

• Separation tank connected through valve 4 to core inlet. Prior to the experiment, it is filled with acid. During the experiment, acid is displaced into the line and comes to the core.

• Inlet fresh line connected through valves 1 and 3 to core inlet. It is used for the initial core flooding, stabilization of the flow and for flash after acid pumping.

• Auxiliary line between valves 2 and 7. It is used to fill up back pressure system before the test.

• Back pressure system to maintain pore pressure at core outlet at constant level in range 1 to 40 MPa.

• Differential pressure measurement system (includes a couple of gauges and safety bypass with valve 5) placed between inlet and outlet of the core holder for precision pressure drop measurements (±0.0003 MPa) along the studied core.

Multiphase core flooding system was simplified for the acidizing test. It was found that a single pumping system that provides the constant flow rate directly (for brine injection) or through a separation tank (for acid) is much better than two independent pumping systems for injection of brine and acid. At the single pump configuration, one can benefit from pulseless swap of flow from brine to acid and back.

## 3 Acid flooding experiments on Austin Chalk core samples

For acidizing experiments, we chose well known carbonate outcrop called Austin chalk. This rock is rather homogeneous in terms of mineralogy and void space geometry at least on millimetre to centimetre scale. Gas permeability values of the studied carbonate miniplugs were in range 11.3 to 18.0 mD at porosity 23.2 to 28.6 %.

As a model brine we used 2 %wt solution of NaCl, while solution containing 15 %wt of HCl and 2 %wt of NaCl was used for acidizing.

Before acidizing test, a mini-plug was fully saturated with model brine using vacuum saturator and then loaded into the core holder. Temperature, as well as confining and pore pressures were gradually increased to the target values in a course of one hour. The described series of the experiments was done at ambient temperature (~27° C), 70 bar pore pressure (i.e. back pressure in Fig. 1) and 110 bar confining pressure. Listed pore pressure was chosen to reduce the effect of the evolved $CO_2$ gas while acidizing.

Experiment was started with measurement of single-phase water permeability (100 % of brine in flow) at different flow rates. After flow stabilization at the required flow rate (i.e. pressure drop along the core became constant), the inlet valves 3 and 4 (Fig. 1) were switched to allow acid flow instead of brine flow.

After injecting the required amount of acid solution, the valves were switched again to injection of brine to flash out and stop the acidizing process. Upon completion of the experiment and drying the sample, amount of dissolved/removed mineral was calculated based on the sample weight before and after the experiment.

Electrical resistivity of the core was measured by two-point method during all flooding experiments and the measured value was used to estimate the time when the acid front came to the miniplug inlet.

It is extremely important for miniplug acidizing to reduce dead volume of tubes at core inlet, i.e. between outlet of valves 3, 4 and miniplug inlet. Otherwise, mixing of the acid solution and brine leads to the decrease of the acid concentration at the front. This can significantly influence the results of the rock dissolution. In our test we minimized the dead volume to 0.3 ml to guarantee close to a step-function rapid increase of acid concentration at the core inlet.

Mutual analysis of pressure and electrical resistivity data during the experiment allowed to determine the basic parameters of carbonate acidizing process, including the beginning of the wormhole generation and its breakthrough at the opposite end of the sample. When the acid front approaches the core inlet, the electrical resistance begins to decay. In general, the acid injection is stopped as soon as the differential pressure drops to zero, indicating that the core is fully penetrated by high permeable dissolved wormhole. Depending of the flooding test purpose, the acid flow could be stopped early. Early stop of the acid injection gave us opportunity to study using microCT the wormhole structure at different stages of its evolution.

Two carbonate miniplugs (Exp2, Exp3 – target samples) were chosen for laboratory test and for digital model constructions and simulation, while one additional miniplug (Exp1 – trial sample) was prepared from a similar rock sample for preliminary measurements of PVBT and possible adjustment of experiment scenario for the target cores. Parameters of the core samples and experiment scenarios are listed in Table 1.

**Table 1**. Description of core samples and laboratory tests

| | **Exp1*** | **Exp2** | **Exp3** |
|---|---|---|---|
| Core size L / Ø mm | 17.7 / 7.6 | 16.7 / 7.7 | 16.4 / 7.7 |
| Porosity | 23.2 % | 24.3 % | 26.6 % |
| Permeability | 9.1 mD | 12.2 mD | 15.8 mD |
| Pore volume | 0.187 ml | 0.186 ml | 0.204 ml |
| Pre-test measurements | | High-res. microCT | |
| Acid flooding test scenario | 1) saturation with brine | | |
| | 2) flow stabilization | | |
| | 3) injection of 15 % HCl @ 0.4 ml/min | | |
| | until BT reached | 0.2ml injected | 0.1ml injected |
| | 4) cleanup | | |
| Post-test measurements | Porosity measurement and microCT to determine wormholes structure | | |

*Trial sample

Standard PVBT measurement implies acid injection until the rapid pressure drop at inlet gauge, which conforms to corresponding permeability increase. Unfortunately, the acid located within the core and line at that moment dissolves additional mineral during the cleanup stage, resulting in overdeveloped wormholes structure. To minimize this effect, we measured an approximate value of PVBT on the trial sample and adjusted the volumes of acid injected on the target cores.

Electrical resistivity measurement allows to detect acid front approaching the face of core sample, where electrode is placed, and thus eliminates the effect from dead zone of holder before the core. Fig. 2 demonstrates a resistivity decrement after approximately 0.33 to 0.35 ml of acid is injected.

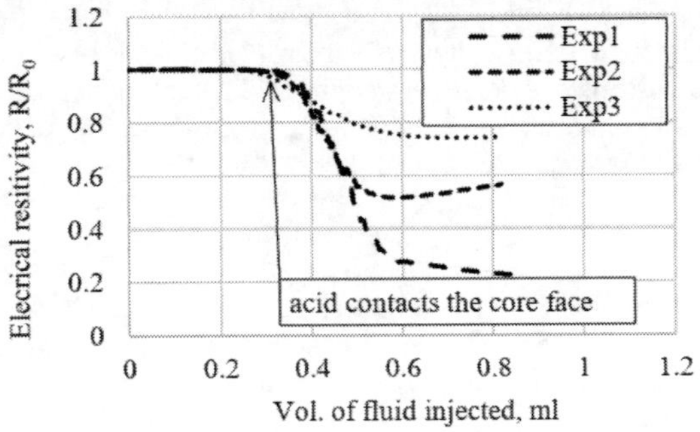


**Fig. 2.** Electrical resistivity of the cores during fluid injection. The start of resitivity decrease corresponds to the contact of acid front with the core face (shown by callout).

Pressure profiles during acidizing process (difference between inlet and outlet valves) are shown in Fig. 3. The moments the acid front contacted the cores faces are marked to demonstrate the difference between the amount of acid required to breakthrough the core and the total volume of acid injected to the moment. For the trial sample, BT was achieved at 0.58 ml of acid injected, but only ~0.24 ml actually came into the core at that time. Experiments on the target cores were adjusted and only 0.2 and 0.1 ml of acid were injected; sharp pressure leaps at the corresponding moments (see Fig. 3) indicate valves switching. The actual acidizing of target cores took place during the cleanup phase, when brine was injected into core holder. Brine injection lasted for about tens of pore volumes until the core properties became stable.

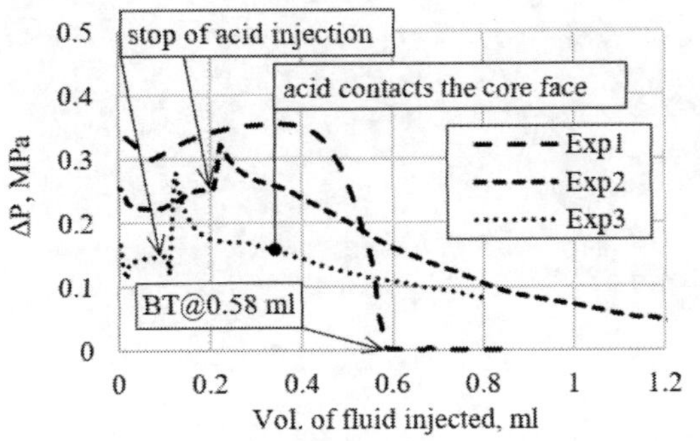


**Fig. 3.** Differential pressure (between inlet and outlet valves) as measured during acidizing process in all experiments. Acid injection was started at zero and stopped at specific for each core volume (shown by callouts).

The post-acidizing core parameters were measured after full cleanup on dried cores (see Table 2).

**Table 2.** Core measurements after acid flooding experiments

| | Exp1* | Exp2 | Exp3 |
|---|---|---|---|
| Core size L / Ø mm | 17.7 / 7.6 | 16.7 / 7.7 | 16.4 / 7.7 |
| PVBT | 2.4 ml (1.3 PV) | | |
| Porosity (increment) | 31.7 % (+8.5 %) | 26 % (+1.7 %) | 27.6 % (+1 %) |
| Permeability (relative incr.) | >10 D ×1000 | 1.4 D ×115 | 980 mD ×62 |

* For the trial sample the values of porosity and permeability don't correspond to the wormhole breakthrough, since the acid injection wasn't stopped at BT, full amount of acid injected was ~3.5PV

## 4 Core imaging and pore-scale calculations

For all samples 3D images both before and after the acid treatment were obtained using Bruker™ SkyScan™ 1172 microCT scanner with a maximum 10 W tube power and 100 kV tube voltage, at "step and shoot" sample rotation type and 0.1° rotation of sample per step. For the Ø = 7.7 mm, L = 16.4 mm miniplug, we achieved ~2.2 μm voxel size. It means that the whole sample comprises of ~4000×4000×8000 voxels in total. It should be noted that the charge-coupled device (CCD) detector installed in this microCT got 4000×2096 pixels, and the conventional circular scanning trajectory [15] was used – each sample was scanned as a sequence of overlapping vertical segments that were stitched together after the Filtered back projection (FBP) reconstruction. We compensated for gentle slow changes in X-Ray tube brightness and beam hardening artefacts using spatially smoothed histogram matching for large cylindrical regions.

The core flooding process was performed outside of the microCT; thus, the acquired images were not spatially aligned. We used in-house developed software [16] to spatially register the "before acid" and "after acid" images. Spatial registration allows direct voxel-to-voxel image comparison (Fig. 4), as well as comparison of actual and simulated wormhole geometries.

We built binary pore/void models (Fig. 5) for both "before acid" and "after acid" images using a modified image segmentation method, based on indicator kriging [17]. In this work, we assumed that the regions of the sample that were far enough from the wormhole were not affected by the acid. This allowed to automatically optimize the "after acid" image segmentation parameters by minimizing the pixelwise mismatch between both binary images within such a region. The binary models of the wormholes were built based on the difference between

https://doi.org/10.1051/e3sconf/202336601002

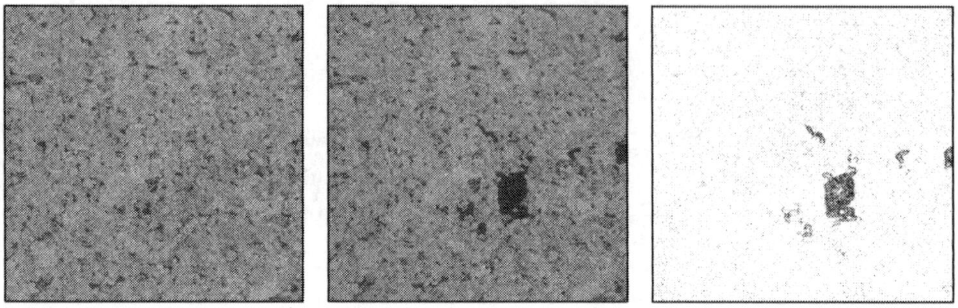

**Fig. 4.** Fragments of the spatially registered microCT images (from left to right): before acid treatment, after the treatment, and their absolute difference.

spatially registered models of core before and after acidizing experiment (Fig. 4).

Each full-sample model was cropped into a regular grid of 425 non-overlapping $396^3$ sub-blocks, which were used to model numerically the rock properties using the in-house developed software for high-resolution digital model analysis based on the principles of DFH [9]. An example of $396^3$ block with noticeable changes on its top face is shown in Fig. 5.

The permeability calculation implies obtaining the steady-state solution for velocity distribution within resolved pores with proper (no slip) boundary conditions on the faces of rock cells together with holding fixed overall flow rate through the model. Such calculation requires iterations and may be computationally expensive for big models of large core sample. But, for the needs of the current work, we calculated velocity distribution on cubic models of moderate size (see Fig. 6) to investigate statistics of properties distribution, so permeability calculations were relatively cheap in terms of computational resources and could be processed in parallel. In addition, we calculated porosity and active surface of mineral over the same sub-models, to construct proper correlations in order to utilize them for construction of the core-scale model.

Automatic parameters finetuning improves overall results consistency. Even though for some individual sub-blocks the data obtained looked counterintuitive, e.g., the open porosity decreased after the acid treatment (Fig. 7). Although this is physically possible, e.g., due to non-dissolvable particles migration, in the current study it is likely that such events should be attributed to image imperfections (e.g., noise). Provided that such effects are inevitable, fairly large statistics are required to mitigate their influence on the final result.

It should be noted that all "before acid" blocks had porosity between 7.5 to 15% (Fig. 8). That is significantly below the lab measurement results (Table 1), which can be associated with large amount of below-resolution porosity. However, the calculated permeability remained within 5 to 50 mD range (Fig. 8), which aligns with the laboratory data fairly well. Thus, we assumed that the larger-pores network, responsible for the most part of the flow rate was resolved well enough for adequate transport modeling. This allowed us to limit ourselves to simple binary models on this stage, avoiding additional higher-resolution scanning, multiscale model construction, and calibration steps [13].

An example of $396^3$ block with significantly altered porous structure is shown in Fig. 5 and Fig. 6. The "before acid" permeability of this block is only 8 mD, while the "after acid" permeability is 75 mD. Meanwhile, the resolved porosity changed from 10 % to 13 %, and the open porosity changed from 6 % to 11 %. Note the high-velocity region next to the (local) tip of the wormhole — this was the major flow channel even before the acid treatment, and the local flow rate significantly increased after the treatment.

**Fig. 5.** 3D renders of the binarized microCT images of Exp3 sample (from left to right): full ~4000×4000×8000 voxel "before acid" state, spatially registered "after acid" state (with wormhole shown in green), example of inner $396^3$ voxel cube in "before acid" state, the same cube in "after acid" state.

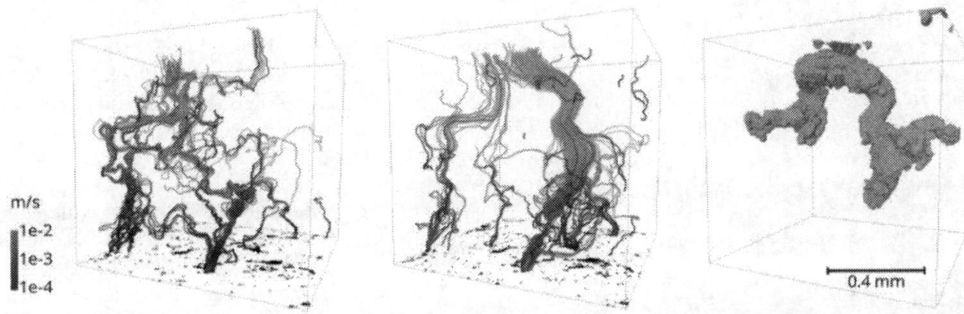


**Fig. 6.** Streamlines from single phase digital flow simulations on two spatially registered $396^3$ voxel binary models, shown in Fig. 5: "before acid" (left) and "after acid" (middle). Semi-transparent part of the flowlines next to the top face of the cube correspond to the flow source. Fluid velocity is color-coded in logarithmic-scale units. The dissolved volume is visualized on the right.

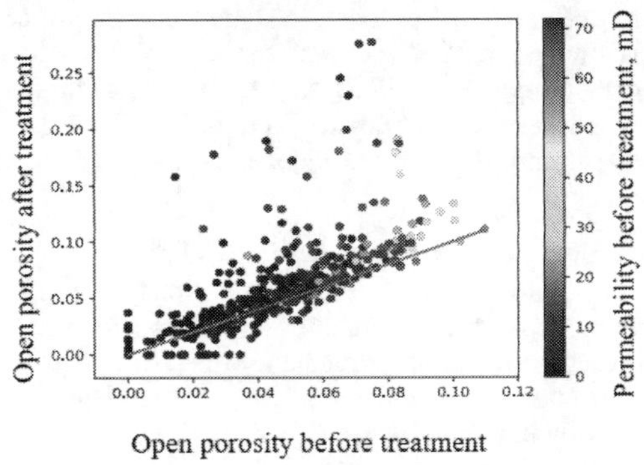


**Fig. 7.** Crossplot open porosity before and after the acid treatment, with color-coded permeability for individual $396^3$ blocks. Note that mostly-dissolved blocks featured higher porosity and permeability even before the acid treatment. Red line is $y = x$.

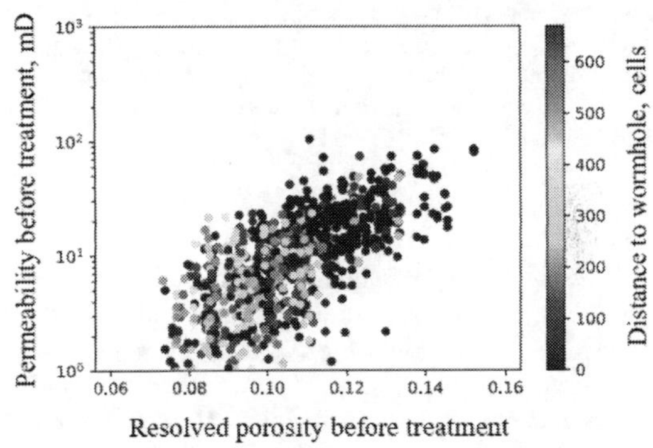


**Fig. 8.** Permeability vs porosity crossplot, with color-coded distance to wormhole for individual $396^3$ blocks. Note that wormhole is generally passes through blocks with high porosity and high permeability.

## 5 Pore-scale model

Reactive flow modeling is computational resource intensive. Thus, we downscaled full sample ~4000×4000×8000 voxel binary models to just ~200×200×400 voxel models, where each voxel encoded local porosity (Fig. 9) (that is, the average of the corresponding 20×20×20 region of the original image). Instead of binary pore/mineral representation of the core, each voxel in such a model is associated with an initial permeability value in accordance with its intrinsic porosity, based on the porosity-permeability correlation plots obtained on pore-scale models.

As it was mentioned above, parameters of the pore-scale model were adjusted to represent spatial inhomogeneity of core properties and its evolution during dissolution process.

Resolved porosity ($\varphi_{res}$), as it follows from the results of numerical analysis of core models, constitute only ~1/3 of total porosity ($\varphi_{tot}$) measured. Deficient unresolved porosity ($\varphi_{unres}$) was distributed uniformly over the solid mineral; the new total porosity, given by the following equation, should match the value measured in laboratory experiment (see Table. 1):

$$\varphi = \varphi_{res} + (1 - \varphi_{res}) \cdot \varphi_{unres}$$

In actual work, we adopted the correlation proposed by Maheshwari et al. [2]. The properties of porous media were assumed to be linked to the open porosity according to the two-parametric dependence:

$$\frac{K}{K_0} = \left(\frac{\varphi}{\varphi_0}\right)^{\gamma} \left(\frac{\varphi(1-\varphi_0)}{\varphi_0(1-\varphi)}\right)^{2\beta} \quad (1)$$

$$\frac{r_p}{r_{p0}} = \sqrt{\frac{K\varphi_0}{K_0\varphi}} \quad (2)$$

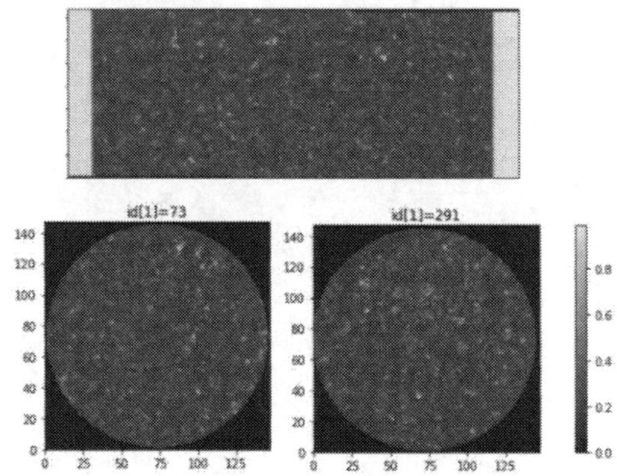

**Fig. 9.** Downscaled model of Ø8 mm core (Exp3) before acizing. The porosity (after correction to account for the unresolved pores) distribution is shown by color within longitudinal and cross sections, dark zones are impermeable walls of coreholder, bright yellow are spacers between inlet/outled sides of coreholder and core faces. Voxel size is 50 µm.

$$\frac{a_v}{a_{v0}} = \frac{\varphi r_{p0}}{\varphi_0 r_p} \quad (3)$$

Here, $\varphi$ is porosity including unresolved one, $K$ is absolute permeability, $r_p$ is average pore radius, and $a_v$ is active surface area available for heterogeneous reaction per unit volume of porous media. Variables with index 0 denote values with respect to initial (reference) state of the porous media. Exponents $\beta$ and $\gamma$ are pore-broadening and pore-connectivity parameters; they are estimated using the properties of the rock. It should be noted that permeability increases infinitely while mineral dissolving, while porosity converges to 1.0 as defined by eq. (1). In order to avoid infinite values in equations, the actual value of permeability should be limited at some reasonable level. In the current work we applied upper limit for permeability at level of the 75-90% of dissolved porosity.

**Table 3.** Parameters of the pore-scale model described by eqs. (1-3)

| | Exp2 | Exp3 |
|---|---|---|
| $\varphi_{unresolved}$ | 18.3% | 20.2% |
| $\varphi_0^*$ | 27% | |
| $K_0$, mD | 19 | |
| $a_0$, m$^{-1}$ | $1.8 \cdot 10^4$ | |
| $\beta$ | 2 | |
| $\gamma$ | 1 | |

## 6 Core-scale reactive flow model

For core scale modeling we utilized continuous media approach instead of direct hydrodynamics within complex pore space geometry, used for pore scale simulations.

In the current work, reactive flow modeling is formulated in accordance with well-known continuous model of transport in porous media, assuming the local mass velocities to be defined by Darcy law. Our formulation accounts for chemical reactions, which are responsible for the porosity increase due to dissolution of mineral, and changes in the solution composition, due to spent of active component of acid. The reactions are governed by kinetic law defined by reaction rate in dependence on local concentration of acid, relative area of active surface of the mineral in contact with acid, and in general it may vary with flow rate and pore channels structure.

Before providing governing equations, let us specify basic variables used in continuum mechanics to describe instantaneous state of mixture: component molar densities $n_i$ (index $i$ denotes the component number ($i$ = *Mineral, Water, Acid, reaction Product*)), pressure $P$, spatial Cartesian coordinates $a = x, y$ or $z$, and time coordinate $t$. We considered isothermal processes, so the temperature was assumed to be fixed, and dependence of certain variables on temperature was omitted. We will use the shortened notation $\partial_t B \stackrel{\text{def}}{=} \partial B/\partial t$ for partial derivatives with respect to coordinates and time, as well as the summation over repeated indexes is implied hereinafter (e.g., $\partial_a I_{ia} \stackrel{\text{def}}{=} \frac{\partial}{\partial x} I_{ix} + \frac{\partial}{\partial y} I_{iy} + \frac{\partial}{\partial z} I_{iz}$).

Upscaling from distinguished individual pores to Darcy description implies reformulation of transport properties, reaction kinetics, and mineral dissolution to continuous representation of porous media [18, 19]. In the framework of this approach, the common conservation equations are the following:

$$\partial_t(\varphi n_i) + \partial_a I_{ia} = q_i = \eta_i r_{eff}, \quad (4)$$

$$I_{ia} = n_i U_a + Q_{ia},$$

$$U_a = -\frac{K_{ab}}{\mu} \partial_b P,\ Q_{ia} = -\varphi D_{i,ab} \partial_b n_i,$$

$$\partial_t \varphi = -\eta_M v_M r_{eff}, \quad (5)$$

where $r_{eff} = r_{eff}(n_i, S_a, u)$ is the rate of the effective chemical reaction of mineral dissolution, which accounts for mixture composition, mineral-acid contact area $S_a$, and the flow rate $u$. Here, $U_a$ is the seepage velocity component along the axis $a$; $D_{i,ab}$ is the effective diffusion tensor, including dispersion of velocity within porous media; $\eta_M$ and $v_M$ are stochiometric coefficient and mineral molar volume, respectively.

In the current work, we used a simplified model of fluid components (pseudo components) and reaction kinetics. First, we assumed the reaction rate to be dependent on a single component (acid) only. Therefore, calculation of the transport equation for the other components could be avoided, since the distribution of reaction products didn't affect the result. Also, the fluid was considered as incompressible; hence, any change of the fluid volume is either due to porosity increase or due to generation of new volume of fluid during reaction. This can be expressed by a source term in continuous equation:

$$\partial_t \varphi + \partial_a U_a = \eta_i v_i r_{eff},\ i \in \{W, A, P\}$$

That is solved in form of general elliptic problem on pressure distribution:

$$\frac{1}{\mu}\partial_a(K_{ab}\partial_b P) = -\eta_i v_i r_{eff}, i \in \{M, W, A, P\} \quad (6)$$

The sum $V_r = \eta_i v_i$ could vary depending on the experiment conditions and, moreover, during the dissolution process (if non-ideal solution model is applied), since molar volumes of components are functions of mixture composition.

The certain form of permeability $K_{ab}$ and dispersion $D_{i,ab}$ tensors, as well as the effective reaction rate as a function of rock properties and flow rate serve as the closing relation for eqs. (4)-(6) and should be obtained from experiments and pore-scale modeling.

## 7 Numerical approach to core-scale reactive flow modeling

Our implementation of the reactive flow in porous media modeling can be divided into two computationally expensive parts: pressure distribution calculation and solution of mass transfer equations with chemical reactions. The problem of pressure determination inside each voxel of the simulation domain can be reduced to the solution of a system of linear equations (6). In the current implementation, we used generalized minimal residual method (GMRES) solver with multigrid preconditioner from NVIDIA AmgX library, which gives an opportunity to utilize GPU accelerators.

We utilized an explicit algorithm for mass transfer calculations. The developed algorithm had first-order approximate precision for space and time steps and was implemented both for parallelized calculations on CPUs or GPUs. CPU realization used Numba library for python and was tested on 48 cores. GPU version utilized a single graphical card for better performance – enabling parallelization on several GPUs allowed to enlarge the grid size, but on the price of calculation speed. The CPU version showed good performance while using up to 12 cores. With larger core count, it became less effective. GPU utilization significantly improved mass transfer calculation performance: modern GPUs showed approximately 9 times speed up compared to 12 CPU cores (see Table 4 and Fig. 10).

**Table 4.** GPU calculation performance.

| GPU | Execution time (ms per call) | GPU memory usage (GB) |
|---|---|---|
| **Pressure calculation step** | | |
| Tesla V100 | 180 | 1.065 |
| Tesla P100 | 260 | 1.052 |
| Tesla K40 | 670 | 0.666 |
| **Transport+reaction calculation step** | | |
| Tesla V100 | 0.97 | 0.125 |
| Tesla P100 | 1.10 | 0.125 |
| Tesla K40 | 3.98 | 0.125 |

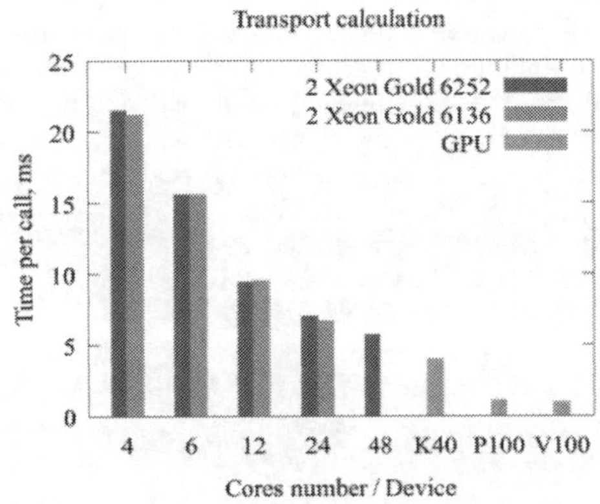


**Fig. 10.** Time required for transport equations calculation procedure at different CPUs and GPUs.

We performed several tests to determine performance of our code on different hardware. For tests, we used model with 184×75×75 grid cells. Since pressure calculation is the most computationally expensive part of the algorithm (see Table 4), it was performed once per certain number of simulation steps according to the changes in permeability due to chemical reactions. In the test case, 10000 transport steps were made, and it required 54 pressure updates. This proportion (54:10000) heavily depends on the simulated domain structure and can vary starting from 1:10000 to 1:10 reflecting the wormhole propagation.

According to the tests performed, a GPU version is preferable for better calculation speed, however, the grid size is limited by GPU's memory capacity. For modern P100 and V100 accelerators, it is about 13.8M (~440*176*176) voxels in a simulation domain. Preferable solution to overcome this limitation is optimization of algorithm for pressure solver to minimize possible performance lost on parallelized GPUs.

## 8. Acidizing modeling and results analysis

Core-scale digital models of each of the sample cores (see Table 1) were constructed and tuned as described in the previous sections. Results of laboratory tests were used to adjust such parameters as effective reaction rate and diffusion, which are hard to be measured experimentally. Although it is possible to estimate the dependence of the effective reaction rate on active surface area (and indirectly, porosity), acid concentration and fluid injection rate from the digital models, the exact value of the effective reaction rate cannot be defined without additional experiments. Similarly, the construction of acid diffusion/dispersion model, which accounts for heterogeneity of fluid velocity and mixing of fresh acid and reaction product with brine, also requires labor-intensive experiments. Though such highly elaborated approach, which requires a series of laboratory and numerical experiments for parameters tuning is known [20], in the current work, we decided to utilize simplified reaction/diffusion models, taking into account only the mobility of the components and effective interaction between them at given experimental condition, without considering their chemical nature. This enabled building a practically applicable model that reproduced the experimental result without numerous laboratory experiments; however, its application is limited to certain combination of acid, mineral and experiment conditions. Therefore, the respective models should be adjusted for each study.

Equations for reaction and diffusion applied in the current work are as follows:

$$r_{eff} = k_{eff} S_a n_A \quad (7)$$

$$D_{i,ab} = D_{acid}\, I_{ab} \quad (8)$$

Where $k_{eff}$ is the heterogeneous reaction rate coefficient, $S_a$ is the active surface area per unit volume of porous media, while $n_A$ is calculated as concentration of acid per unit volume of fluid, $D_{acid}$ is the diffusion coefficient of the effective component of the acid solution, and $I_{ab}$ is the identity matrix.

The simulation parameters (see Table 5) were determined based on the experiment conditions or tuned iteratively for the best match between the simulation results and respective experimental data for treated cores.

**Table 5.** List of simulation parameters.

| Parameter | Unit | Value |
|---|---|---|
| brine/acid solution viscosity, $\mu$ | mPa*s | 0.84 |
| acid component diffusion, $D_{acid}$ | $m^2/s$ | $1.5\cdot10^{-9}$ |
| stochiometric coefficients<br>mineral $\eta_M$<br>acid $\eta_A$<br>water $\eta_W$<br>reaction products $\eta_P$ | | <br>-1<br>2<br>-4<br>1 |
| molar volume<br>mineral $v_M$<br>acid $v_A$<br>water $v_W$<br>reaction products $v_P$ | $m^3$/mole | <br>0.06357<br>0.03064<br>0.018<br>0.1501 |
| heterogeneous reaction rate, $k_{eff}$ | m/s | $4\cdot10^{-3}$ |

In the simulations the moment wormhole break through the core outlet was determined geometrically, unlike to the experimental approach, where PVBT was measured by pressure difference. The connected region(s) of core, where porosity is increased up to the limit (determined in the separate sensitivity study) of 95-99%, was defined as wormhole(s), and the length of such region(s) from the inlet face to its farthest tip was the parameter indicating the wormholed part of the core.

Simulation of the acidizing process on each of digital core models was performed in accordance with experimental workflow. Trial sample was acidized at fixed acid flow rate (0.4 ml/min) and PVBT value was measured and compared with the experimental value. Variations of the reaction rate and acid component diffusion coefficients allow to amend the wormholes structure and resulting PVBT value in a wide range. However, in the current study the values close to the reported in literature [19,21] worked rather well.

After the acidizing model parameters were determined at the cores-scale digital model of the first trial core, experiments on target cores were performed with limitation of injected acid volume in agreement with the laboratory workflow.

As the result for each core model, the PVBT value, final porosity distributions, total amount of dissolved rock, permeability, and wormhole geometry were obtained and compared with the experimental data.

Table 6 summarizes the result of experiments and respective simulations in terms of PVBTs and core after-treatment porosities and permeabilities. The observed systematic deviations between experimental and simulated permeability may be caused by neglecting transition from Darcy to Navier-Stokes flow in big caverns and dissolved wormholes. The correct accounting for viscous effects in open channels would require solution of vector non-linear Navier-Stokes equation in such regions which would result in overconsumption of memory and computational time. The necessary optimizations and improvements are the subject of further research.

**Table 6.** Simulation results and respective experimental data

| | Exp1 | Exp2 | Exp3 |
|---|---|---|---|
| Volume of acid injected | PVBT measure | 0.2 ml<br>1.1 PV | 0.1 ml<br>0.5 PV |
| Laboratory measurements on treated core | | | |
| PVBT<br>Porosity<br>Permeability | 1.3PV | <br>26%<br>1.4 D | <br>27.6%<br>980 mD |
| Simulation | | | |
| Porosity<br>Permeability | | 25.6%<br>1.1 D | 27.5%<br>760 mD |
| PVBT at 0.4ml/min | | 1.1PV | 1.34PV |

The final wormhole structures obtained for two target cores in simulations and in experiments (build based on microCT interpretation) are shown in Figs. 11 and 12.

Pairwise comparison of wormholes images corresponding to simulations and experiments shows qualitatively similar patterns, however plenty of disturbances in details are also clear. In general, it may be explained by insufficient quality (resolution and noise) of models obtained from microCT and applied for numerical simulation and due to the fact that in numerical simulations we did not account for Navier-Stokes flow in channels. However, it is possible to assume that the major heterogeneous features of rock were properly captured in the digital models, since the trends are reproduced fairly well, and integral characteristics after the acid treatment are close to the ones measured on physical core samples.

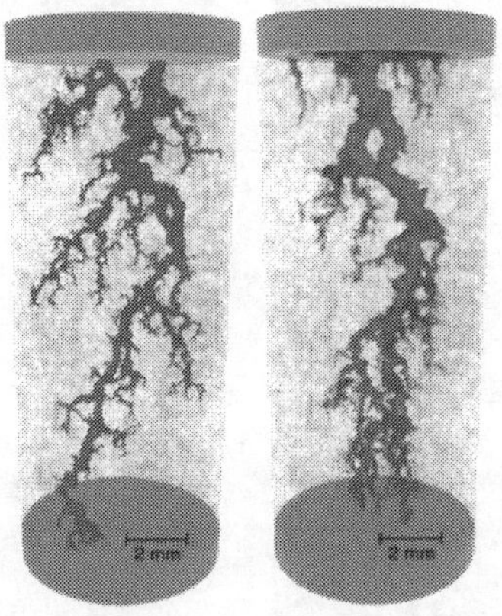


**Fig. 11.** Wormhole patterns obtained after Exp2 laboratory acidizing experiment (left) and the corresponding digital simulation (right). Yellow zones represent inlet/outlet space.

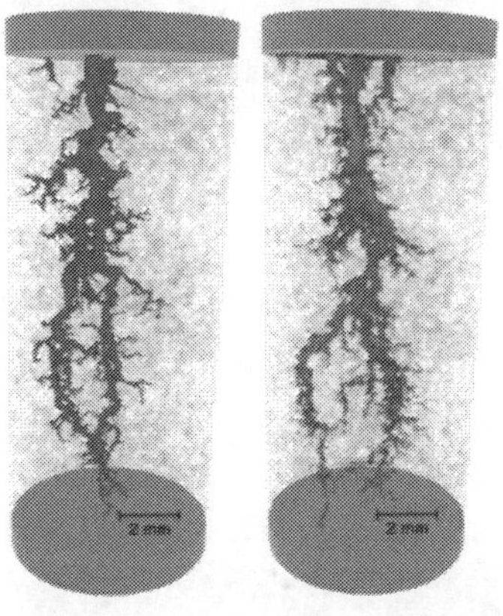


**Fig. 12.** Wormhole patterns obtained after Exp3 laboratory acidizing experiment (left) and the corresponding digital simulation (right). Yellow zones represent inlet/outlet space.

The simulated streamlines (Fig. 13) within core/wormhole show the flow distribution during the acidizing experiment. As the wormhole propagation approached the end, the maximum velocity speed increased for several orders of magnitude. Because of that, the calculation performance also drastically decreased closer to the end of simulation, due to limitation on the time step of explicit scheme.

Digital experiment allows to repeat simulations at varying fluid injection rates in a wide range from 0.04ml/min to 1ml/min. As the result, the PVBT curves for two cases were obtained (see Fig. 14).

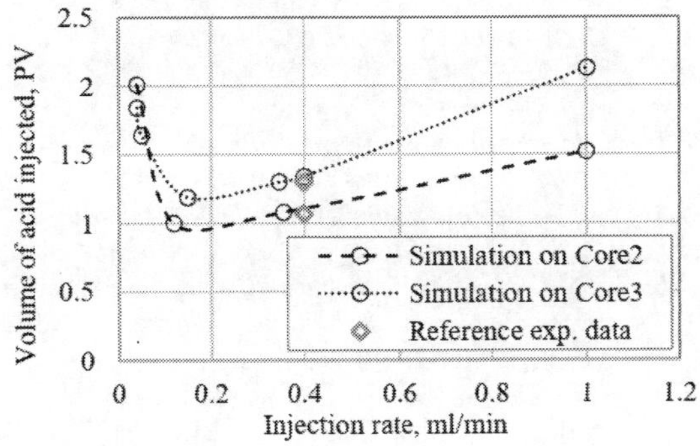


**Fig. 14.** PVBT curves obtained on digital cores together with reference laboratory data.

## 9. Conclusions

In this study, we analyzed carbonate acidization process using the two-scale continuum model. We simulated reactive flow of acid in carbonate rock at Darcy scale, while the accounting for the underlying physics at the pore level was performed through direct pore-scale simulation of transport properties on micromodels of real rocks. To tune rock models and demonstrate predictive capability of the present approach, the PVBT and dissolution patterns were obtained from numerical simulations and compared with experiments on Ø8mm cores of Austin Chalk. Also, PVBT curves in a wide range of injection rates were calculated on the same core models, which would be impossible in the laboratory due to the destructive nature of the experiments. The suggested approach opens the way for improvements in acidizing job design by increasing the consistency between the models used the for reactive flow modeling and actual pore-scale structure of the real rocks.

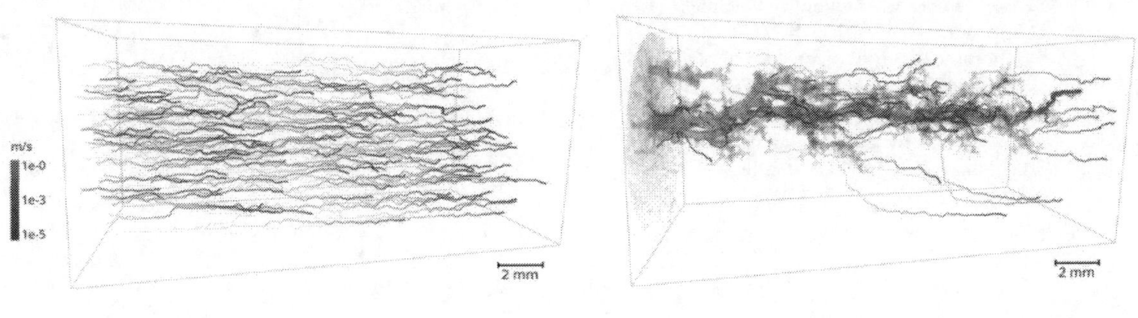


**Fig. 13.** Simulated streamlines from single-phase core-scale flow modelling (Exp3) with color-coded velocity and semi-transparent simulated wormhole (in green), before the acid treatment (left) and after (right).

## References

1 C.N. Fredd, H.S. Fogler, *SPE Formation Damage Control Symposium*, SPE-31074-MS (1996)
2 P. Maheshwari, R.R. Ratnakar, N. Kalia, V. Balakotaiah, Chem. Eng. Science 90 258–274 (2013)
3 G. Glasbergen, N. Kalia, M.S. Talbot, *8th European Formation Damage Conference, Scheveningen*, SPE-121464-MS (2009)
4 C.I. Steefel, et al, Computat Geosci. 19 (3): 445–478 (2015)
5 O.O. Akanni, H.A. Nasr-El-Din, *SPE Middle East Oil & Gas Show and Conference*, SPE-172575-MS (2015)
6 D. Dubetz, et al, SPE Annual Technical Conference and Exhibition, SPE-181725-MS (2016)
7 J. Etten, D. Zhu, A.D. Hill, *EUROPEC* SPE-174314-MS (2015)
8 M.E. Ziauddin, E. Bize, SPE Middle East Oil and Gas Show and Conference, SPE-104627-MS (2007)
9 A. Demianov, O. Dinariev, N. Evseev, Introduction to The Density Functional Method in Hydrodynamics (Moscow: Fizmatlit, 2014)
10 E. Ivanov, et al., *SPE Russian Petroleum Technology Conference*, SPE-202016-MS (2020)
11 A. Demianov, O. Dinariev, N. Evseev, Can. J. Chem. Eng. 89 (2): 206–226, (2011)
12 A. Beletskaya, et al., *SPE Russian Petroleum Technology Conference*, SPE-187805-RU (2017)
13 O. Dinariev, N. Evseev, D. Klemin., *E3S Web Conf.* 146 01001 (2020)
14 D. Korobkov, A. Goncharov, *Society of Core Analysts*, SCA2014-081, (2014)
15 Andrew M. Kingston, et al., Proceedings Volume 9967, Developments in X-Ray Tomography X; 996712 (2016)
16 A. Nadeev, "Method For Evaluating Material Sample Structural Changes Resulting From Acting Upon Sample", WO2016013955
17 W. Oh, B. Lindquist, IEEE Transactions on Pattern Analysis and Machine Intelligence, Volume: 21, Issue: 7, (1999)
18 M.K.R. Panga, M. Ziauddin, V. Balakotaiah, AIChE Journal 51 3231–3248 (2005)
19 M. Ghommem, et al., Journal of Petroleum Science and Engineering 131: 18-33 (2015)
20 M. Ghommem, W. Zhao, S. Dyer, Journal of Modeling and Simulation Vol.5: 63-71 (2014)
21 M. Gautelier, E.H. Oelkers, J. Schott, Chemical Geology 157 (1-2): 13–26 (1999)

# 3D Multiclass Digital Core Models via microCT, SEM-EDS and Deep Learning

*Igor* Varfolomeev[1], *Vladimir* Svinin[1,2], and *Ivan* Yakimchuk[1]

[1]Schlumberger Moscow Research, Reservoir Potential Department, 125171 Moscow, Russia
[2]NRNU MEPHI, ICIS, 115409 Moscow, Russia

**Abstract.** We describe an integrated methodology for constructing a 3D multiclass model of a rock sample, based on X-ray microtomography (microCT) and quantitative evaluation of minerals (QEMSCAN) by automated SEM-EDS (Scanning Electron Microscopy, Energy Dispersive Spectroscopy). We focus on building an automated operator-independent workflow, allowing to distinguish between voxels featuring substantially different physical properties, such as void, quartz, denser and less dense clay aggregates. The workflow is demonstrated using a set of five ⌀8 mm Berea sandstone miniplugs. For each miniplug, a ~$4000^3$ voxel microCT image is acquired. Next, each miniplug is cut into smaller pieces, and the 45 resulting polished surfaces are subjected to the QEMSCAN analysis, producing ~$4000^2$ pixel mineral maps. Each mineral map is automatically spatially registered with the corresponding microCT image using an in-house surface-based algorithm. Further, the ground truth images for the supervised multiclass segmentation are constructed from the mineral maps. We compare 3D and 2D convolutional neural network (CNN) architectures with the baseline Naïve Bayes classifier, which is roughly equivalent to the approaches commonly used in practice today. We find that supervised CNN-based segmentation is fairly stable, despite microCT image quality non-uniformness and achieves higher quality scores compared to feature based and baseline approaches.

## 1 Introduction

Nowadays, the Digital Rock approach is a well-known technique for core analysis. Generally, it consists of numerical simulation of physical phenomena on a digital representation of a core sample.

Centimetre-scale Digital Rock models are usually created via X-ray microtomography (microCT) imaging and subsequent image segmentation. Binary solid/void segmentation using methods like Indicator Kriging [1] and Active Contours [2] is still most commonly used in practice. Such models proved to be suitable for estimating single-phase and multi-phase permeabilities of some rock samples, like Berea sandstone [3], where spatial resolution of the modern microCT scanners easily allows to resolve the pore throats. One major drawback, associated with these segmentation methods is the significant operator involvement; the operator's choices might greatly influence the ultimate result, and it might be not clear which segmentation parameters are more appropriate. Due to the nature of the hydrodynamics, the required parameter selection precision increases dramatically as the pore throat sizes get closer to the microCT resolution limit. Another drawback is that the substances featuring sub-resolution porosity could not be appropriately represented in such models. This might limit the ability to model, for example, clayey sandstones or carbonates. Both difficulties could be handled through the use of multiclass operator independent segmentation process and a suitable physical properties simulator.

Modern sophisticated simulation techniques allow to take into account a sizeable set of physicochemical properties (e.g., mineralogy, sub-resolution porosity, wettability) with respect to their volumetric spatial distributions [4,5]. In this paper, we focus on the segmentation part.

## 2 Related works

One of the first works, discussing the possibility of combining SEM and microCT images is [6,7]. The authors acquired microCT and QEMSCAN images of a sandstone sample, and spatially registered them. However, the segmentation technique, described by the authors is limited to a direct correlation between the microCT grayscale values and mineral types. Although such approach allows to distinguish minerals with significantly different X-ray effective linear attenuation coefficients, like quartz and pyrite, it is usually unreliable in most practical cases, like when distinguishing quartz and dense kaolinite. The contrast between different practically important minerals tends to be low on regular microCT systems, equipped with ~100 kV X-Ray tubes, although it could be significantly higher on monochromatic lower-kV synchrotron-based systems.

© The Authors, published by EDP Sciences. This is an open access article distributed under the terms of the Creative Commons Attribution License 4.0 (http://creativecommons.org/licenses/by/4.0/).

### 2.1 Extracting features

Some works highlight the possibility to use additional features instead of just the grayscale value for CT image segmentation, for example, image gradients in [8]. The operator selects an area, attributed to a specific class on a 2D intensity-gradient histogram, which improves segmentation quality, compared to grayscale-only methods.

It is natural to assume that the segmentation quality could be further improved by increasing the number of the features (e.g. local average, median, variance, ...) taken into account. However, as the number of features grows, it becomes problematic to manually select ranges, attributed to specific classes. Thus, instead, operator directly specifies a set of voxels, attributed to each specific class, and the machine learning (ML) supervised segmentation task arises. The [9] was one of the first works to utilize the Fiji Trainable Weka Segmentation tool [10] for the microCT image segmentation task. The authors demonstrate that this approach allows better particle separation than the traditional, grayscale-based binary segmentation.

Nowadays, multiple software packages, similar to the Fiji Trainable Weka Segmentation are available, namely ilastik [11] and Zeiss Zen Intellesis™[12]. However, they share the same basic idea — using manual brush strokes as a training set for a ML classifier. Each tool provides a different set of feature extraction filters. Their common drawback is increased memory requirements due to in-memory architecture. Two more tools to mention are Thermo Scientific™ Amira-Avizo™ (including the XImage PAQ plugin) and ORS™ Dragonfly™, which also provide the ability to train pre-configured CNN models.

### 2.2 Training data

Some authors use semi-automatic segmentation results to train their ML models. In [13] authors train 2D SegNet CNN [14] using a set of 20 images, 256×256 each. The classes they consider are mostly easily distinguishable by their grayscale values, except a single phase, that got its own distinct microstructure. In [15] authors also use a "mainly grayscale-based" method to create a training set for their CNN, even though they do have two QEMSCAN images of the same sample. The motivation for such a decision is that those two slices do not provide a large enough training set. Some image registration difficulties are also mentioned. The CNN is able to almost perfectly reproduce that ground truth labeling, achieving ~99% accuracy. However, it is not entirely correct to directly compare this score with the score achieved with an independent ground-truth data (such as QEMSCAN-based).

Considerable efforts were spent on inventing a better ground truth segmentation, suitable for training ML models. Some works, discussing microCT image segmentation use an "image degradation" approach to create a ground truth segmentation and the "image to be segmented" from a single source. In [12] an already segmented volume is forward-projected into the projection domain, then shadow projections are blurred, and Gaussian noise is added to them; and finally a new volumetric image is reconstructed from these shadow projections. According to the authors, this emulates the real noise generating processes, taking place in a microCT device. However, it should be mentioned that this simple model does not take many other possible imperfectnesses of the real microCT device into account. In [16] the ground truth segmentation is produced from real full-quality X-ray projections, while the "image to be segmented" is produced from the decimated or downscaled subset of the same projections — this emulates faster microCT acquisition.

Although such approaches are well-suited for development purposes, we believe that the usage of a naturally higher-resolution SEM data represents a more promising approach.

In [17] authors use SEM-EDS to create a mineral map of the edge of a rock sample cylinder. This might not be the best option, because microCT images typically feature significant artefacts next to the outer edges of the sample being scanned, especially flat edges. In [18] both sample cylinder edge and a flat cut surface are scanned before the microCT, which might produce similar issues. The authors of the [19,20] also perform SEM-EDS on the top part of the cylindrical sample after the microCT measurement. They mention the imperfectness of the result, even after the histogram matching [19]. The authors utilize 2D SURF-based image registration [20] and also mention the imperfectness of the result, even though their sample contains large distinctive features. They conclude [19] that the feature based ML segmentation was unable to improve the quality, as compared to the grayscale-based segmentation. This result is also partially associated with the classes in question — all 3 classes being considered are easily distinguishable by their grayscale level.

In [21] authors use spatially registered QEMSCAN images of internal miniplug cross-sections to construct the ground truth labelling. They mainly focus on a feature-based segmentation of a coarse-grained sandstone via per-grain classification. Indeed, for larger, easily separable (using a watershed transform) and fairly uniform grains, this provides an ability to accumulate a massive amount of statistics that characterizes each grain. The non-zero threshold rotation-invariant Local Binary Patterns (LBP) [22] feature extraction technique was recognized as one of the most robust, allowing to distinguish even such a similar minerals, like quartz and albite. The demonstrated proof-of-concept per-pixel feature-based segmentation result considers 5 classes, and the whole experiment uses a single QEMSCAN image.

## 3 Imaging

### 3.1. X-ray microCT imaging

We use Bruker™ SkyScan™ 1172 cone-beam microCT scanner to obtain images of the five similar ⌀8 mm Berea sandstone miniplugs with ~2.2 μm voxel size, using 100 kV tube voltage and the maximum available 10 W tube power. Each sample is also ~8 mm high, which results in ~$4000^3$ voxel microCT image. It must be noted that the CCD detector being used got 4000×2096 pixels, and the conventional circular trajectory is used. Thus, the sample is scanned as two independent vertical segments, that are later stitched together by the scanner's image reconstruction software. We would later discuss how this affects the segmentation quality.

### 3.2 Sample slicing

After the acquisition of the microCT images, we use a precision diamond saw to cut each sample into smaller blocks and embed them into epoxy resin. Each block undergoes standard SEM-EDS sample preparation procedures, namely polishing and carbon deposition. The resulting geometry is illustrated in Fig. 1. Each cutting plain is roughly orthogonal to the plug cylinder axis, but up to 5° errors are possible, due to overall mechanical tolerances. We avoid using original outer miniplug edges for the SEM-EDS mapping, not only because these areas are associated with significant artifacts, but also because we would like to use 3D CNN receptive field later.

**Fig. 1.** Positions of the surfaces scanned with the SEM and the miniplug, imaged with microCT (purely schematic).

Furthermore, once the first set of the SEM images is acquired, we re-polish all blocks, to remove few extra microns, and repeat the imaging process. This allows us to obtain numerous slices from a limited physical volume. In this paper, about 9 slices were imaged from each of 5 miniplugs. A total of 45 slices allows to investigate the segmentation quality as a function of the size of the training set, and the effects, associated with non-uniformity of the microCT image quality.

In a daily practice, the "densely packed cross-sections" geometry allows to efficiently utilize the volume of the scanned sample, to save the most of the miniplug intact, e.g., for the subsequent laboratory experiments.

### 3.3 SEM imaging

We use Thermo Scientific™ QEMSCAN™ 650F SEM to acquire ~4000×4000 mineral maps of each cross-section. The bundled iMeasure® software acquires EDS spectra with about 1000 counts for each pixel of the mineral map and independently classifies each pixel in accordance with the pre-built mineral library [23]. We would refer to this image as QMS-mineralogy. Simultaneously, the BackScattered Electrons (BSE) image with the same resolution is acquired, we would refer to it as QMS-BSE. The software automatically moves the SEM table to sequentially cover the whole ⌀8 mm area of the sample with ~0.3×0.3 mm frames. The example mineral map is shown in Fig. 4a.

We also utilize Thermo Scientific™ Maps™ software, to obtain a higher-resolution BSE (Backscattered Electrons) image of the same surface, we would refer to it as Maps-BSE.

## 4 Image registration

To utilize classic supervised ML segmentation approaches, we must first spatially register the microCT and the SEM images. This task is complicated by the large size of the data (a single $4000^3$ 8-bit image takes 64 GB), and the lack of reliable "special" points — the whole image could be viewed as a semi-stochastic texture. Thus, most modern image registration techniques either incur a large computation cost [24], associated with the direct area-based registration, or require a rather fine manually selected initial starting point.

In this work, each mineral map is automatically spatially registered with the corresponding microCT image using the following in-house registration approach, which consists of the three steps.

### 4.1 Surface-based registration

The aim of the initial step is to roughly locate the global minima, with the precision, comparable with the size of the grain (or pore), as required for the subsequent area-based optimization process to start in the vicinity of the global minima.

The general rigid-body image registration problem could be formulated as an optimization of six translation-rotation parameters, and a scale parameter:

$$T_m = [x_m, y_m, z_m, \varphi_m, \theta_m, \psi_m, s_m].$$

Here, $x_m, y_m, z_m$ are related to image shift along corresponding axes; $\varphi_m, \theta_m, \psi_m$ are x-y-z Euler angles; and $s_m$ is image scaling.

In this notation, we can significantly reduce the volume of the parameter space to look through. For the binarized versions of the images, namely $I_{3D}$ and $I_{2D}$, we extract the "center of mass" points. That is $(x_{2Da}, y_{2Da})$ for the 2D image. For 3D image, we use a number of z-slices and the least-squares approach to estimate the cylinder axis position $\left(x_{3Da}(z_{3D}), y_{3Da}(z_{3D})\right)$ for each $z_{3D}$ within the 3D image.

The key point is the extraction of the outer contours of the images, as illustrated in Fig. 2. For 3D image, they could be formalized as

$$R_{2D}(\varphi_{2D}) = max(r):(I_{2D}(x_{2D}, y_{2D}) = 1)$$
$$R_{3D}(\varphi_{3D}, z_{3D}) = max(r):(I_{3D}(x_{3D}, y_{3D}, z_{3D}) = 1)$$
$$x_{2D} = x_{2Da} + rsin(\varphi_{2D})$$
$$y_{2D} = y_{2Da} + rcos(\varphi_{2D})$$
$$x_{3D} = x_{3Da}(z_{3D}) + rsin(\varphi_{3D})$$
$$y_{3D} = y_{3Da}(z_{3D}) + rcos(\varphi_{3D})$$

Having the surface profiles, we can estimate $s_m$ using the average radius values. The $x_m$ and $y_m$ could be calculated based on the axis position, once the rest of the parameters would be known.

Thus, we would only need to optimize 4 parameters: $[z_m, \varphi_m, \theta_m, \psi_m]$, out of which for $\varphi_m$ and $\theta_m$ we may consider only a quite limited discrete set of their values, because, in accordance with the assumption about the cutting geometry, they should not exceed 5°.

By the simple geometrical means, the $[z_m, \varphi_m, \theta_m, \psi_m]$ parameterize the sine-like $R_{2D}$ curve position on the $R_{3D}$ surface (Fig. 3), and the aim is to find the matching set of parameters, that would define the correct position of the $R_{2D}$ curve (some additional details could be found in [21]).

Due to relatively small size of the $R_{2D}$ and $R_{3D}$ arrays, which is unlikely to exceed 10 Mbytes, and a limited number of options to consider, even the exhaustive grid search only takes a few seconds on a modern CPU.

According to our practice, the surface profiles are good unique descriptors, and the rock sample surface is never smooth, which allows to automate the whole image registration process.

Moreover, the method works even if the outer surface of the sample is partially damaged. One could notice a perfectly smooth and round part of the surface in the top-right area of the sample shown in Fig. 4b. This artefact resulted from a sample being slightly outside the cylinder-shaped microCT reconstruction volume. Nevertheless, the surface-based registration works in this case as well.

### 4.2 Area-based registration

The aim of the second image registration step is to improve the result of the first step, still sticking to the rigid-body assumption. This is implemented as a direct iterative "black-box" optimization of all $T_m$ parameters,

**Fig. 2.** Surface profiles of a miniplug sample (coarse-grained sandstone is used for readability).

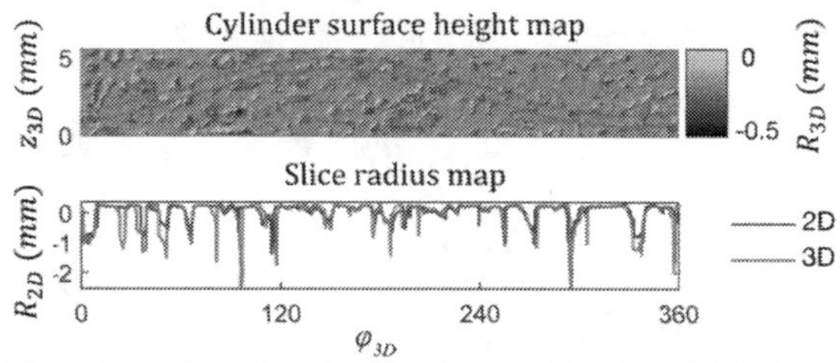


**Fig. 3.** Matching 2D and 3D surface profiles of a miniplug

using area-based registration norm. To compare discrete-index mineral map with continuous grayscale microCT image, we use mosaic-image Put'ev matching score [25] (or similarly, the explained variance score, calculated after assigning each mineral the grayscale value equal to the average of all the microCT image pixels, matching to that mineral, according to current $T_m$).

### 4.3 Smooth non-rigid distortions compensation

Although the miniplug usually represents a fairly good example of a rigid body, the rigid body approximation is insufficient for pixel-perfect SEM-microCT image registration. The issue arises mainly due to acquisition hardware imperfectnesses — even tiny sub-pixel SEM lens distortions tend to accumulate to several pixels–large smooth distortions, when multiple SEM images are stitched together.

One possible workaround is to calibrate the SEM distortions using a special regular sample, and later apply the corresponding de-warping to all SEM images [7]. One of the drawbacks of this method is that it assumes that the distortions would not change after the calibration. Another drawback is that this method is unable to compensate the distortions, that arise from a limited SEM stage movement precision and the SEM image stitching algorithm glitches. Instead, we directly apply an Optical Flow estimation method [26] to the pair of images in question. We use smooth large-scale components of the resulting vector field to de-warp the SEM image. This ensures that small artefacts, like mechanically damaged grains would not result in major image distortions. The transformation is so smooth, that for each small sub-area, containing a few grains, it could be viewed as a plain translation.

It must be noted that this de-warping step only considers the 2D distortions in the plane of the SEM image. Thus, it is unable to fully replace

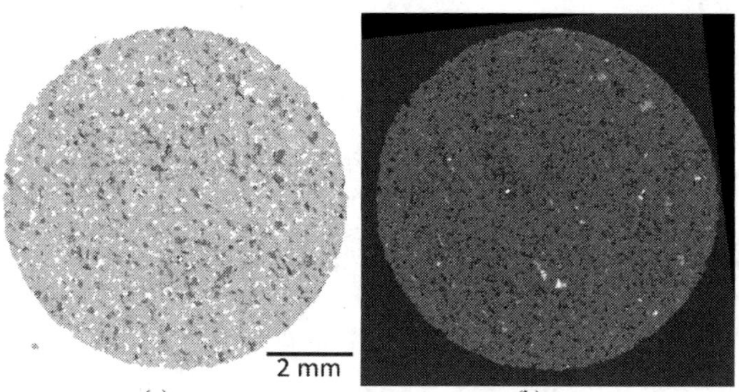


(a) (b)

**Fig. 4.** Example 3D-to-2D image registration procedure result: (a) de-warped QMS-mineralogy image; (b) spatially corresponding microCT cross-section

the area-based registration step, which is fully 3D, even if the surface-based registration step provides a very good initial approximation.

The ultimate result of the image registration procedure is shown in Fig. 4. Finally, to allow calculating features on the original microCT data, we scale the SEM image to match the original microCT voxel size. Also, for ML needs, we not only export a single matching microCT cross-section (Fig. 4b) but a 401-slices thick 3D block (e.g. ± 200 slices around the matching cross-section).

## 5. Relabelling

In [27] it was demonstrated that the QEMSCAN classification result could be misleading sometimes and not agree with the optical petrography results — the issues generally arise for minerals that are very close in their elemental concentration proportions, but considered very different from a petrography point of view. One simple example is the quartzite and quartz pair. These two minerals are usually easily visually distinguishable on a BSE image, but might be indistinguishable from the QEMSCAN's "independent pixels" point of view. This particular pair could be separated via a joint QMS-mineralogy and QMS-BSE images post-processing, honouring some minimal neighbourhood of the pixel in question. In this work, we do not question the accuracy of the QEMSCAN image. However, the original QEMSCAN image is still poorly suitable for direct use as a ground truth labelling, mainly due to the noise-like patterns, shown in Fig. 5.

Firstly, we note a number or single-pixel grains, that are definitely too small to be recognizable on a microCT image (Fig. 5a) — even though QEMSCAN pixel size (2 µm) is very close to microCT voxel size (~2.3 µm), the physical resolution of the SEM is significantly higher. Moreover, such objects would hardly influence any physical properties of the Digital Rock model anyway. However, the underlying grain still might be real, and might influence the microCT image somehow. Thus, the general principle we use is not to "smooth-out" such objects with a median filter or equivalent, but to re-assign them to the "Unknowns" class, which is later ignored by all the ML classifiers we use. The mask, highlighting such objects, is shown in the rightmost column in Fig. 5a.

Areas featuring high concentrations of such single-pixel grains are trickier. We found that removing relatively large areas around them stabilizes many classifiers, most probably because such objects are usually outliers in the ML sense. One might note that the

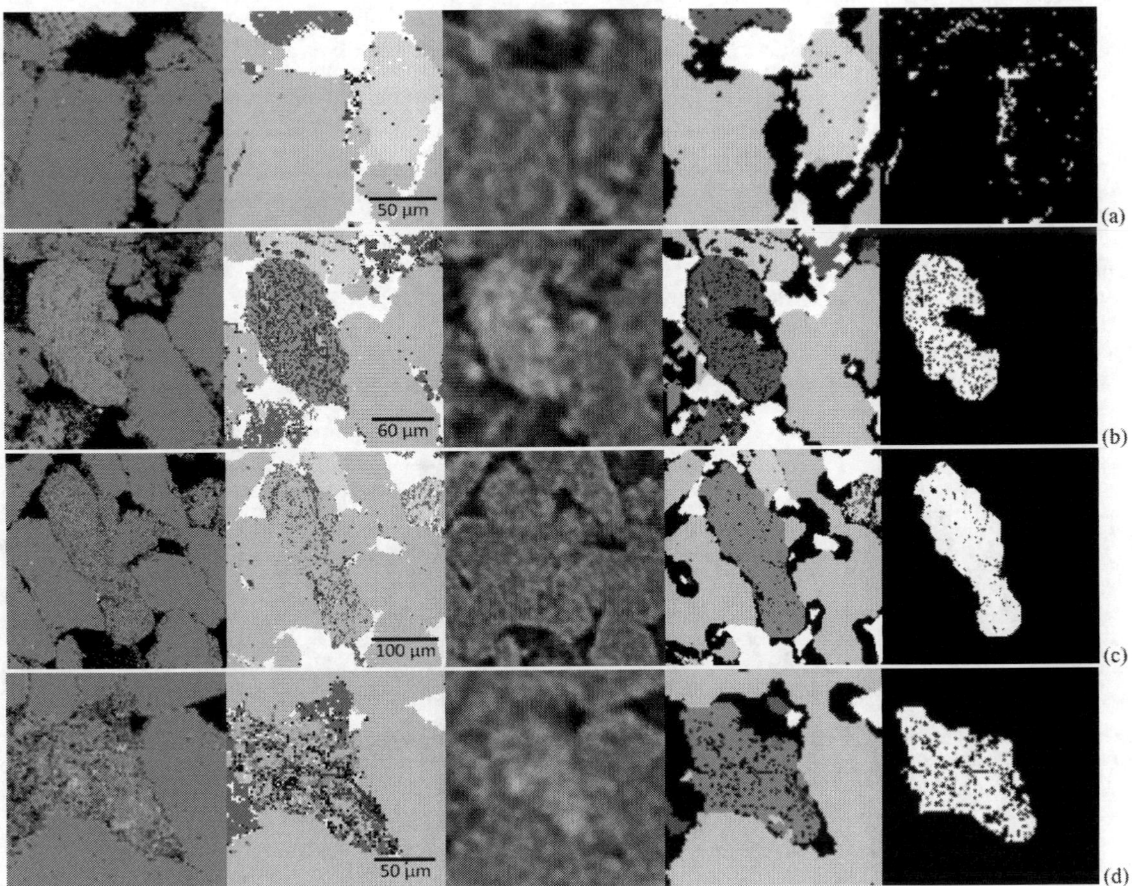


**Fig. 5.** QEMSCAN mineral map relabeling examples: 1st column - QMS-BSE image; 2nd column - QMS-mineralogy image; 3rd column – spatially registered microCT image; 4th column – relabeling result; 5th column – the relabeled pixels in question (see text); (a) small objects; (b) Clay_mixture class; (c) Quartz_with_clay class; (d) Quartz_clay_cement class.

https://doi.org/10.1051/e3sconf/202336601003

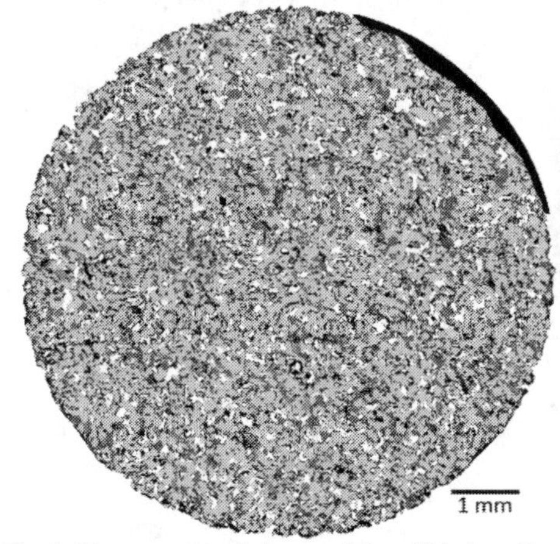


| Color | Class | Area % |
|---|---|---|
| | Quartz_dense | 64.99 |
| | Background / pore | 20.56 |
| | K-Feldspar | 04.15 |
| | Quartz_with_clay | 03.20 |
| | Albite | 01.99 |
| | Kaolinite_degraded | 01.24 |
| | Quartz_clay_cement | 01.23 |
| | Muscovit | 00.84 |
| | Dolomit | 00.39 |
| | Chlorit | 00.33 |
| | Kaolinite_dense | 00.32 |
| | Clay_mixtu | 00.28 |
| | Siderite | 00.23 |
| | Rutile | 00.22 |
| | Zircon | 00.03 |
| | Pyrite | 00.01 |
| | Unknown | |

**Fig. 6.** The ground truth image of the validation slice — the result of the relabeling procedure for the image, shown on Fig. 4a. The legend result and the area percentage for all the classes on the relabeled image.

segmentation task — now pixelwise classification losses begin to make sense.

All the said relabelling was implemented using basic morphological image processing operations, and carefully hand-tuned for the specific dataset in question (but not for individual slices). Theoretically, such a "clustering" operation could be automated to avoid operator involvement, but that would require considerable additional research.

Finally, slices are inspected manually, and the broad Unknown class strokes (black) are placed where major artefacts are noticed. An example of such a manual intervention is clearly visible in Fig. 6 on the top-right part of the image. The reason for this is the same artefact (resulting from a sample being slightly outside the cylinder-shaped microCT reconstruction volume) that was already discussed above. We also automatically detect the sample circle radius and fill the out-of-sample-circle area with Background/pore class colour (white). We ignore this area in all further area-percentage calculations, including segmentation accuracy calculations.

relabeled image in Fig. 5a contains more Unknowns than just the "small objects" mask, discussed in the previous paragraph.

More importantly, there is a multitude of clay-related classes. The Fig. 5b-d demonstrate that some grains look fairly uniform on the scale of tens of pixels (especially on the BSE image) but produce single-pixel noise-like patterns on both mineral map and BSE.

We relabel such objects, aiming to also distinguish the most important types of conglomerates. Most of them could be easily categorized into broad classes, each containing at least tens of such objects per a single QEMSCAN image. Namely, the grain shown in Fig. 5b is assigned to the Clay_mixture class. The grain, shown in Fig. 5c is assigned to the Quartz_with_clay class. The grain, shown in Fig. 5d is assigned to the Quartz_clay_cement class. Such relabelling not only makes sense from the further Digital Rock physical properties modelling point of view (and could be viewed as an upscaling step), but also greatly simplifies the

## 6. Feature-based segmentation

For the feature-based segmentation results, demonstrated below, we build on the results, demonstrated in [21,28]. We only use 2 QEMSCAN slices for training and a single QEMSCAN slice (shown in Fig. 6) for validation — feature-based approaches require less data before the result quality stops to increase, compared to the CNN-

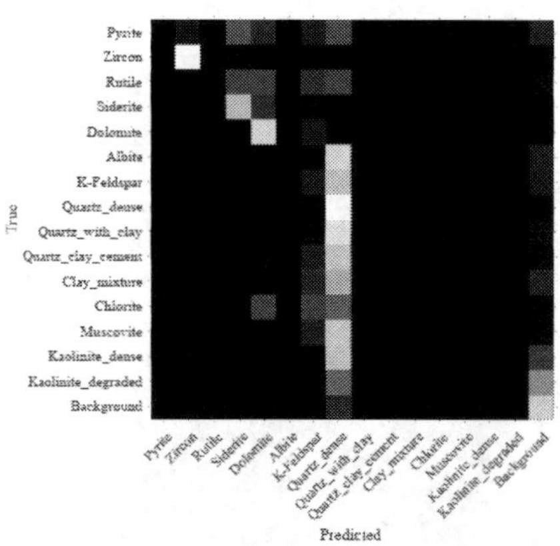


| Color | Area(%) | Prec. | Recall |
|---|---|---|---|
| | 64.99 | 0.818 | 0.978 |
| | 20.56 | 0.890 | 0.847 |
| | 4.15 | 0.227 | 0.093 |
| | 3.20 | 0.000 | 0.000 |
| | 1.99 | 0.000 | 0.000 |
| | 1.24 | 0.000 | 0.000 |
| | 1.23 | 0.000 | 0.000 |
| | 0.84 | 0.000 | 0.000 |
| | 0.39 | 0.486 | 0.897 |
| | 0.33 | 0.000 | 0.000 |
| | 0.32 | 0.000 | 0.000 |
| | 0.28 | 0.000 | 0.000 |
| | 0.23 | 0.668 | 0.793 |
| | 0.22 | 0.000 | 0.000 |
| | 0.03 | 0.755 | 0.973 |
| | 0.01 | 0.000 | 0.000 |
| | f1 | 0.256 | |
| | accuracy | 0.819 | |

**Fig. 7.** The result of the "Naïve Bayes" segmentation approach: validation slice segmentation (see legend in Fig. 6), confusion matrix (row-normalized), Precision and Recall (green-yellow-red palette highlights good-average-poor values).

 https://doi.org/10.1051/e3sconf/202336601003

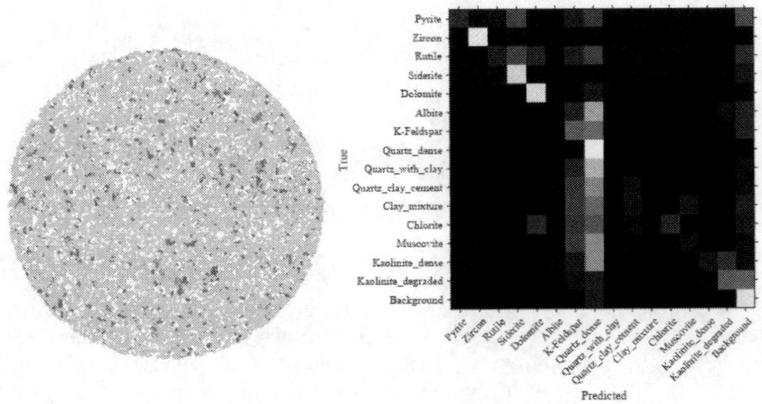


| Color | Area(%) | Prec. | Recall |
|---|---|---|---|
| | 64.99 | 0.871 | 0.976 |
| | 20.56 | 0.878 | 0.907 |
| | 4.15 | 0.487 | 0.406 |
| | 3.20 | 0.210 | 0.002 |
| | 1.99 | 0.389 | 0.007 |
| | 1.24 | 0.510 | 0.403 |
| | 1.23 | 0.355 | 0.054 |
| | 0.84 | 0.365 | 0.051 |
| | 0.39 | 0.743 | 0.917 |
| | 0.33 | 0.530 | 0.142 |
| | 0.32 | 0.220 | 0.063 |
| | 0.28 | 0.078 | 0.006 |
| | 0.23 | 0.616 | 0.895 |
| | 0.22 | 0.407 | 0.085 |
| | 0.03 | 0.728 | 0.944 |
| | 0.01 | 0.460 | 0.124 |
| | f1 | 0.372 | |
| | accuracy | 0.851 | |

**Fig. 8.** The result of the "LightGBM" segmentation approach. Designations are equal to Fig. 7.

based. Depending on the circumstances, this could be either a strength or a weakness.

### 6.1 Naïve Bayes

Our baseline "Naïve Bayes" approach actually utilizes only a single feature channel – the original microCT image, filtered with the manually tuned bilateral filter to suppress noise. We use parametric Naïve Bayes classifier assuming normal distribution. This is roughly equivalent to manually choosing a set of a threshold values after image pre-filtering. The result, shown in Fig. 7 highlights a set of substances that are easily distinguishable by their grayscale values with green background in Precision and Recall columns (definitions of the "Precision", "Recall", "accuracy" and "$F_1$" terms could be found in [29]). A visual comparison with Fig. 4b confirms this assumption.

Still, Naïve Bayes looks significantly better than the binary Otsu segmentation approach (Fig. 15, Fig. 16) which only achieves 0.813 accuracy and 0.109 $F_1$, (we use micro-averaging) assuming that it distinguishes Quartz_dense and Background classes.

### 6.2 LightGBM

LightGBM [30] is a modern gradient boosting framework, providing both high performance and state-of-the-art classification accuracy in many benchmarks. We combine it with two 3D 8-bit cube-shaped rotation-invariant LBP patterns and a local window statistics feature vector (e.g. average, median, variance, skew, kurtosis, …). We fine-tune LBP meta-parameters using cross validation (namely histogram windows size, pattern scale and threshold value). Also, we post-process the soft-classification result, provided by the LGBM using the Graph Cut approach [31] to make it a bit smoother. The result demonstrated in Fig. 8 achieves higher scores compared to Fig. 7.

## 7. CNN-based segmentation

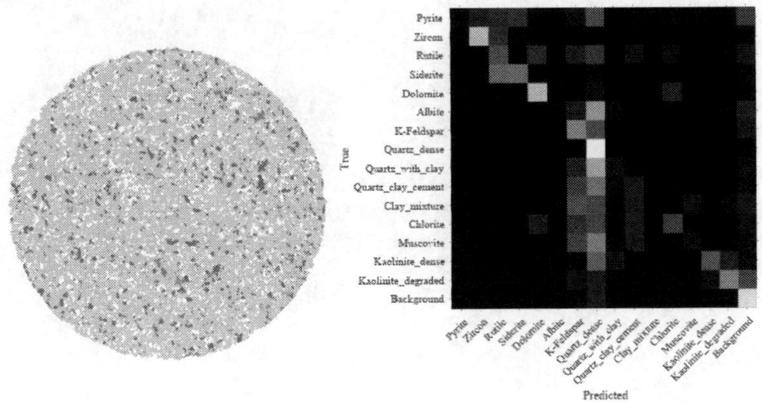


| Color | Area(%) | Prec. | Recall |
|---|---|---|---|
| | 64.99 | 0.887 | 0.979 |
| | 20.56 | 0.911 | 0.897 |
| | 4.15 | 0.554 | 0.570 |
| | 3.20 | 0.305 | 0.054 |
| | 1.99 | 0.368 | 0.023 |
| | 1.24 | 0.579 | 0.544 |
| | 1.23 | 0.304 | 0.092 |
| | 0.84 | 0.354 | 0.047 |
| | 0.39 | 0.742 | 0.817 |
| | 0.33 | 0.436 | 0.249 |
| | 0.32 | 0.300 | 0.356 |
| | 0.28 | 0.000 | 0.000 |
| | 0.23 | 0.803 | 0.473 |
| | 0.22 | 0.294 | 0.323 |
| | 0.03 | 0.860 | 0.800 |
| | 0.01 | 0.097 | 0.024 |
| | f1 | 0.407 | |
| | accuracy | 0.862 | |

**Fig. 9.** The result of the "3-slice U-net" segmentation approach. Designations are equal to Fig. 7.

 https://doi.org/10.1051/e3sconf/202336601003

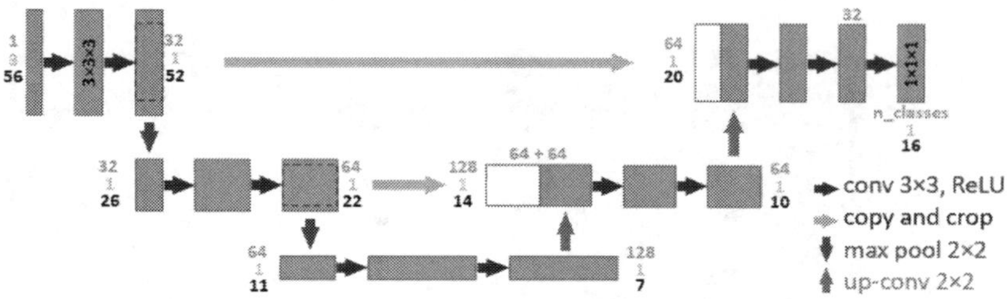


**Fig. 10.** The 3-slice U-net architecture. The number of feature channels is indicated in green; **z**-size is indicated in gray; **x** and **y** sizes are equal and are indicated in black.

The CNN-based segmentation results, demonstrated below, utilize the U-net architecture [32]. For the experiments in this section, we use the same single QEMSCAN slice for validation, and 44 slices for training. We use 3D receptive field, valid-mode convolutions and dense 2D output labelling (except that some pixels belong to the Unknown class). We argue that such a geometry is more efficient than sparse 3D labelling [33]. The training epoch corresponds to a number of randomly selected patches with a total area equal to the total area of the training set. We do not perform training data rebalancing or hard negative mining, despite severe imbalance. We use cross-entropy loss and Adam optimizer with learning rate 0.0002, β1=0.5, β2=0.99. Our experience suggests that further fine-tuning only provides specific minor improvements. Although they might be valuable in specific cases, they are out of the scope of the present article, aiming at a more general comparison.

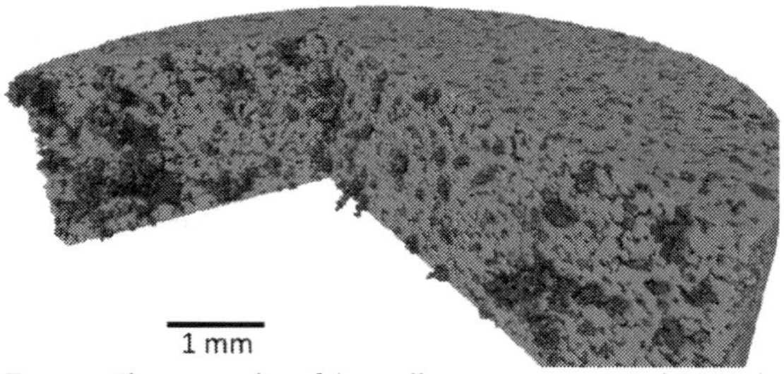


**Fig. 11.** The 3D render of the 3-slice U-net segmentation result

### 7.1 3-slice U-net

Our first, most lightweight U-net model uses 3-slice [34] architecture (Fig. 10), that is — only the first single convolution is 3×3×3, while the rest are 3×3, thus being almost 2D. Some additional details are available in [28]. The training performance is about 100 epochs per day, on a single GK210 chip of the Nvidia® K80 GPU. The Fig. 9 demonstrates the 1000-epoch training result, but scores only marginally improve after 200 epochs. On the large scale, this result is visually similar to the LightGBM result (Fig. 8), but the scores are significantly higher. Some differences could be seen in the Fig. 15 – the CNN generally produces less small infeasible details. The 3D rendering (Fig. 11) demonstrates adequate 3D geometry of the grains despite the "almost 2D" nature.

### 7.2 3D U-net

Our second U-net model (Fig. 14) is fully 3D and all 3 spatial dimensions are (almost) equal. It uses 204×204×92 input patches and 112×112 output patches. Specific patch sizes are used to ensure full symmetry in max pooling and up-convolution layers. Additional 3×3×1 convolution layer is added for the same purpose. Each convolution, except the last one, is followed by PReLU activation with 0.25 activation parameter. The last convolution uses softmax activation. Up-convolution layers use transposed convolutions and halve the number of the feature channels. The training performance is, again, about 100 epochs per day, but now a GPU sever with 7×RTX2080Ti is used. The training batch size is 7 (a single patch per a GPU) due to GPU memory limitations. After the first

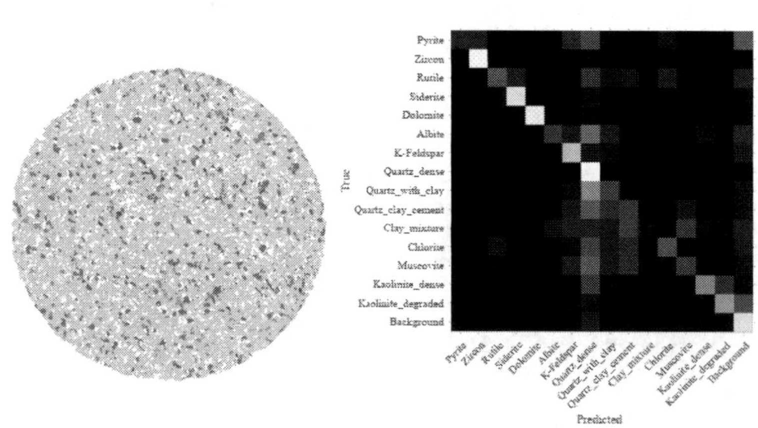


| Color | Area(%) | Prec. | Recall |
|---|---|---|---|
| | 64.99 | 0.921 | 0.973 |
| | 20.56 | 0.906 | 0.901 |
| | 4.15 | 0.779 | 0.796 |
| | 3.20 | 0.408 | 0.246 |
| | 1.99 | 0.547 | 0.200 |
| | 1.24 | 0.601 | 0.632 |
| | 1.23 | 0.371 | 0.222 |
| | 0.84 | 0.428 | 0.211 |
| | 0.39 | 0.882 | 0.939 |
| | 0.33 | 0.494 | 0.329 |
| | 0.32 | 0.415 | 0.542 |
| | 0.28 | 0.445 | 0.039 |
| | 0.23 | 0.741 | 0.907 |
| | 0.22 | 0.628 | 0.341 |
| | 0.03 | 0.818 | 0.955 |
| | 0.01 | 0.587 | 0.100 |
| | f1 | 0.504 | |
| | accuracy | 0.885 | |

**Fig. 12.** The result of the "3D U-net" segmentation approach. Designations are equal to Fig. 7.

100 epochs, accuracy score, calculated on our only validation slice, oscillates in [0.088, 0.0885] range.

For the Fig. 12, we select a result with the maximum accuracy observed. This result demonstrates a significant step forward, as compared to our previous results — the confusion matrix now looks much more like a clean diagonal. One exception is the most-common Quartz_dense class that is thus considered "the safest choice" when the U-net is not certain about a specific pixel. The Precision score is above 0.37 for all classes, but the Recall is still below 0.25 for the most difficult-to-distinguish classes, including clays. However, the confusion matrix suggests that in many cases clays are mostly mixed with other clays, which might be acceptable for many practical scenarios. Classes like Quartz_with_clay and Quartz_clay_cement are similar in their nature — there is no clear boundary between them.

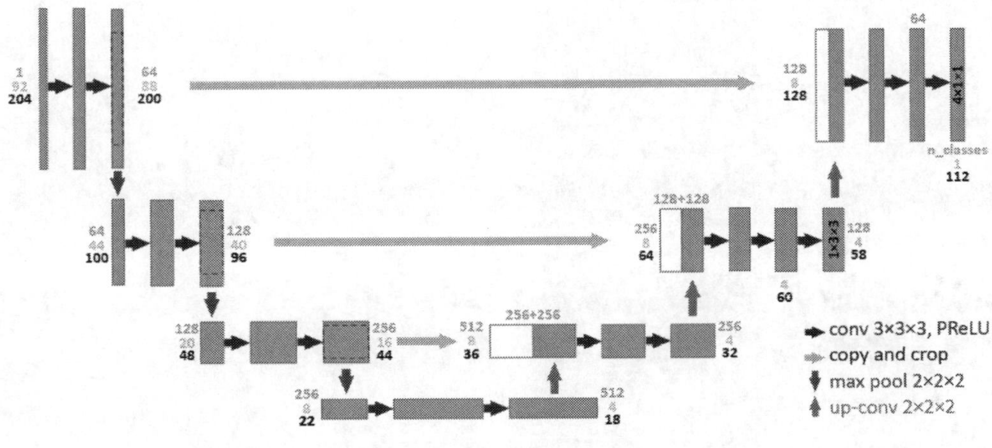


**Fig. 13.** The 3D U-net architecture. Designations are equal to Fig. 10.

The fact that Pyrite is commonly mixed with Quartz and Background could be attributed to the small size of the Pyrite grains and mechanical damage of the sample surface during polishing.

### 7.3 3D U-net+GAN

The Generative Adversarial Networks (GAN) are algorithmic architectures that use two neural networks, competing against each other, e.g., while the first one attempts to generate realistic synthetic images, the other one attempts to distinguish between the synthetic and real images. This concept, originally proposed in [35], proved to be very successful for continuous data generation tasks, like grayscale or RGB image generation, including image-to-image translation [36]. But only a limited success should be expected, when directly applying GANs to discrete data generation tasks, like the image segmentation task in question. The issue arises from a limited amount of information, available to the generator network through the argmax layer.

Last but not least, almost all ground truth patches contain at least one Unknown (black) pixel (see Fig. 16b), while our segmentation methods are designed to produce none (see Fig. 16e-k). If we blindly pass such "real" and "fake" patches to the discriminator, it would be able to distinct between them by just that. Thus, it would not provide any meaningful information about the "fake" patch quality to the generator. To fix this, we make generated patches look like real in terms of the Unknown class distribution, by copying the Unknown pixels from the ground truth image.

Our third U-net model utilizes effectively the same architecture as the previous one, but features an additional 14-layer discriminator network, attached to its output.

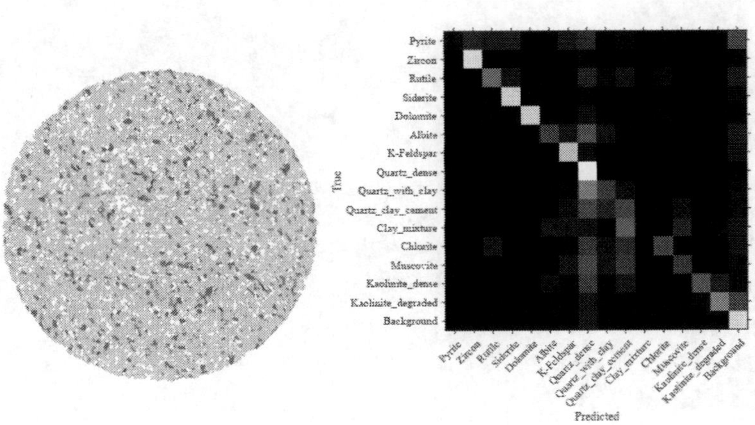


| Color | Area(%) | Prec. | Recall |
|---|---|---|---|
| | 64.99 | 0.924 | 0.973 |
| | 20.56 | 0.905 | 0.897 |
| | 4.15 | 0.797 | 0.826 |
| | 3.20 | 0.470 | 0.265 |
| | 1.99 | 0.622 | 0.301 |
| | 1.24 | 0.650 | 0.553 |
| | 1.23 | 0.305 | 0.313 |
| | 0.84 | 0.454 | 0.217 |
| | 0.39 | 0.929 | 0.886 |
| | 0.33 | 0.507 | 0.308 |
| | 0.32 | 0.493 | 0.380 |
| | 0.28 | 0.260 | 0.030 |
| | 0.23 | 0.765 | 0.888 |
| | 0.22 | 0.531 | 0.462 |
| | 0.03 | 0.862 | 0.920 |
| | 0.01 | 0.727 | 0.043 |
| | f1 | 0.542 | |
| | accuracy | 0.887 | |

**Fig. 14.** The result of the "3D U-net+GAN" segmentation approach. Designations are equal to Fig. 7.

https://doi.org/10.1051/e3sconf/202336601003

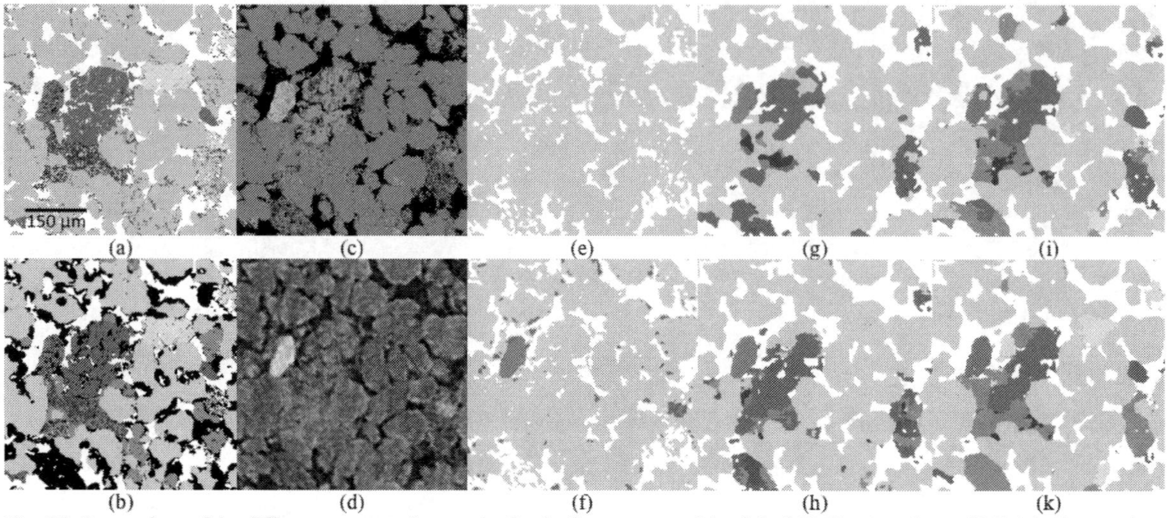


**Fig. 15.** Comparison of the different segmentation results for the fragment one: (a) original QMS-mineralogy; (b) Relabeling result (segmentation ground truth); (c) original QMS-BSE; (d) microCT; (e) Otsu; (f) Naïve Bayes; (g) LightGBM; (h) 3-slice U-net; (i) 3D Unet; (k) 3D Unet+GAN.

Only 2D 3×3 convolutions are used, because we lack 3D ground truth data. For the 4th, 7th and 10th convolutions, stride equals 2, for the rest stride equals 1. The number of feature channels first doubles after each filter, starting from 32 and ending up at 512, and then halves back down to 32. Each convolution is followed by a LeakyReLU activation with 0.2 parameter and instance normalization. The last convolution is rather followed with a sigmoid activation. During the training phase, generator and discriminator are updated with 1:1 ratio, using minimax GAN loss. Dropout and hard-negative mining are not used. The training performance drops to about 60 epochs per day.

Once again, in the Fig. 13 we just present the best result achieved. However, we must admit that in this case, the training performance is less stable, and results vary way more across different epochs, which is a typical for the architecture used. One promising point, however, is that this result is significantly better than those on Fig. 12 in terms of the $F_1$ score. Another promising point is that careful inspection of the smaller crops (Fig. 15) indicate some progress towards smoother, more realistic grains, with less unfeasible mineral mixtures. Even though this "smoothing" hardly ever provides a truly realistic result and sometimes results in an even more incorrect classification, like for a K-Feldspar grain in Fig. 16.

## 8. Image quality uniformness

To assess the microCT image quality uniformness, and the possible associated effects, we train 3-slice U-net on various sets of slices, and simultaneously evaluate it on all available slices (Fig. 17). One simple effect to mention is that training sets below 10 slices seem to be insufficient in terms of the $F_1$ score.

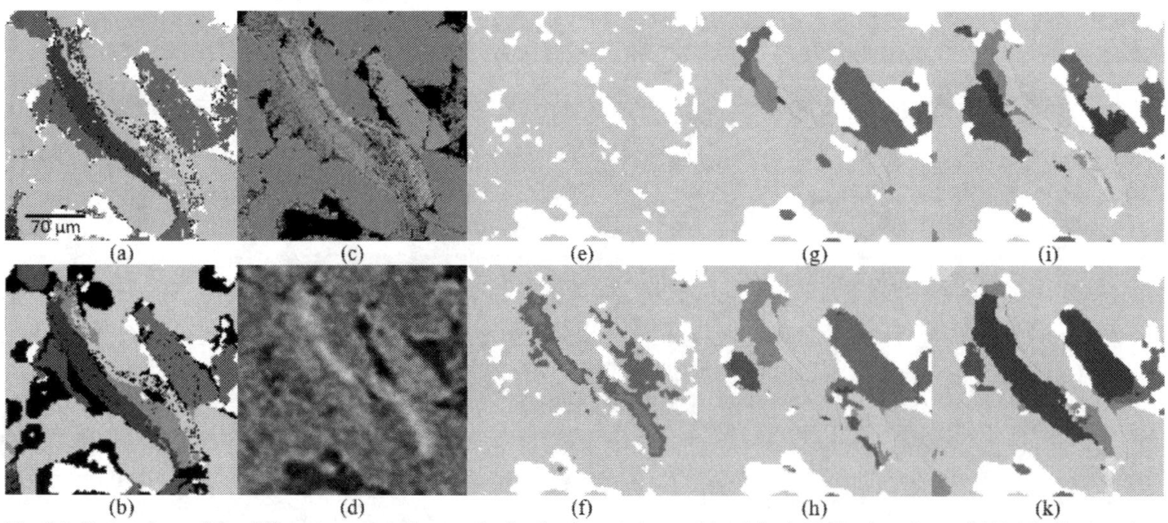


**Fig. 16.** Comparison of the different segmentation results for the fragment two: (a) original QMS-mineralogy; (b) Relabeling result (segmentation ground truth); (c) original QMS-BSE; (d) microCT; (e) Otsu; (f) Naïve Bayes; (g) LightGBM; (h) 3-slice U-net; (i) 3D Unet; (k) 3D Unet+GAN.

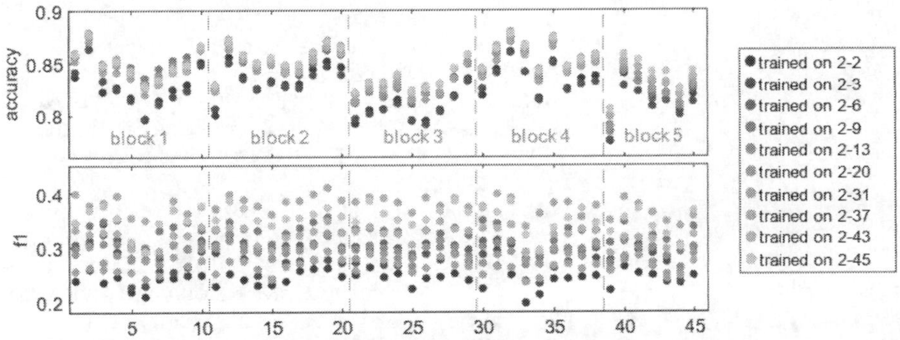


**Fig. 17.** The effect of the training set selection. Each block corresponds to a physical miniplug.

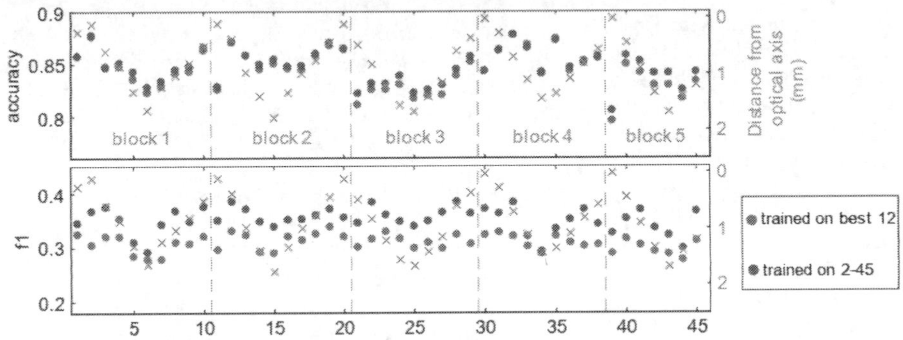


**Fig. 18.** The "distance from the optical axis" effect.

The second, more interesting effect we notice in Fig. 17 is that the segmentation quality is not growing monotonously when adding training slices. In particular, it significantly drops (in terms of the $F_1$ score) for most slices, when we add slices 38-43 to the training set. This might suggest that there are some "bad" slices, negatively affecting the segmentation quality. To assess this hypothesis, we train the same classifier using 12 best-accuracy slices, however, the segmentation quality decreases on all slices (Fig. 18). Thus, one might conclude that there are no "bad" slices, just an intrinsic non-uniformness, which is partially associated with microCT image quality non-uniformness.

We found a strong correlation between the distance from a slice to the microCT optical axis (central slice) and the segmentation quality for that slice (the fact that the first slice from each block demonstrates lower segmentation quality than the second one should be probably attributed to the sample preparation issues). Gradual minor quality drop in the areas further from the optical axis is the expected effect for conventional cone-beam circular-trajectory microCT scanners, utilizing Filtered Back Projection (FBP) reconstruction algorithm. The multiclass segmentation relies on barely visible tiny features and thus it is highly sensitive to such effects. However, this is certainly not the only source of imperfectnesses. We assume that the only feasible mitigation measure is to train models on a larger datasets, covering more-or-less all possible imperfectnesses. Later, such model could be fine-tuned for a specific smaller practical job.

## 9. Learned super-resolution

The ability of the CNN to resolve meaningful details in the microCT image could be also studied by using effectively the same CNN architecture and input data for a learned super-resolution task. For this purpose, we use a single full-slice Maps-BSE image for training, and another Maps-BSE image, from a different slice, for the visual comparison (Fig. 19). The result demonstrates that although our model is unable to reconstruct the Maps-BSE image exactly, and locations of specific tiny features may not match real in a pixel-perfect manner, on a larger-scale, the overall morphology and substance types are mostly reproduced correctly (which is actually more important than pixel-perfect positions for the subsequent physical properties modeling). Moreover, infeasible patches are uncommon. This is partially associated with the fact that continuous data generation task is more suitable for the modern GANs. This looks promising in a sense that there is still potential for segmentation quality improvement.

It should be noted that such a super-resolution tool is also valuable by itself. In [37] authors discuss microCT-to-microCT super-resolution and provide the visual comparison of the result with an SEM image. This super-resolution task is also close to the conditional image generation [38], which could be viewed as effectively the same task with a weaker preconditioning.

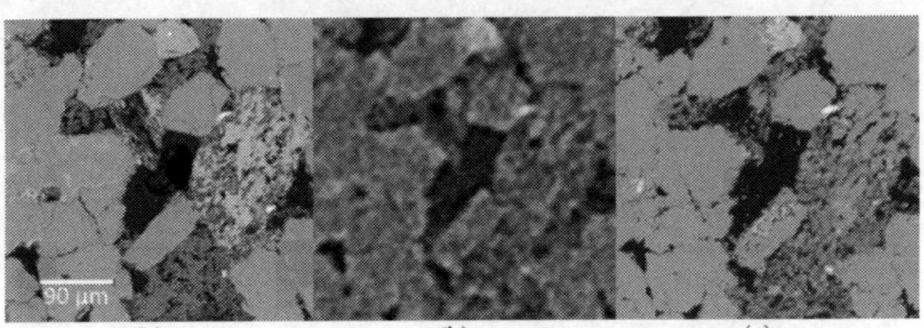


(a) (b) (c)

**Fig. 19.** Super-resolution experiment image fragments comparison: (a) original Maps-BSE image; (b) microCT image; (c) super-resolution result

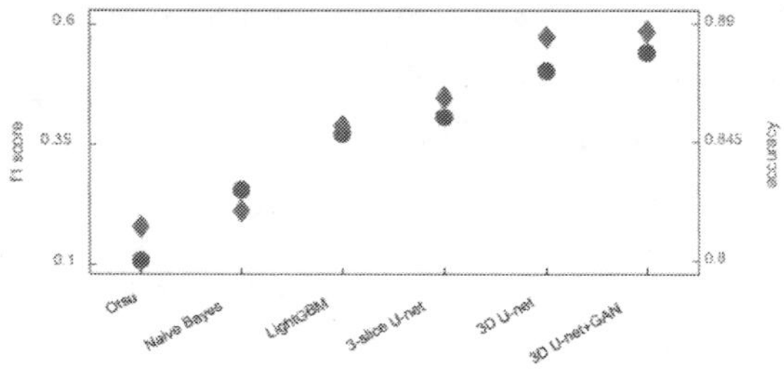


**Fig. 20.** The comparison of the segmentation quality scores.

## 10. Conclusions

Convolutional neural networks allow multiclass microCT image segmentation with a quality far beyond more traditional methods (Fig. 20). Larger 3D receptive fields are beneficial for the segmentation quality, but the associated computation cost is higher. Basic supervised training could be considered fairly stable and thus production ready. However, a tightly spatially matching sizable operator-independent ground truth labelling is required. The 3D GAN-based approaches have not demonstrated their full potential yet and are currently less stable in general. However, they already provide slight improvements and represent a promising subject for further research.

## References

1. W. Oh et al., IEEE Transactions on Pattern Analysis and Machine Intelligence **21**, 590–602 (1999)
2. A. P. Sheppard et al., Physica A: Statistical Mechanics and Its Applications **339**, 145–151 (2004)
3. D. A. Koroteev et al., in *Proc. of International Symposium of the Society of Core Analysts* (California, USA, 2013), 12
4. O. Dinariev et al., in *Proc. of International Symposium of the Society of Core Analysts* (Pau, France, 2019)
5. J. Goral et al., Sci. Rep. **10**, (2020)
6. M. Knackstedt et al., in *Soc. Pet. Eng. - SPE Asia Pac. Oil Gas Conf. Exhib., APOGCE* (2010), 1589–1597
7. T. K. Varslot et al., US20110181701 A1 (2011)
8. A. C. Jones et al., Biomaterials **28**, 2491–2504 (2007)
9. Y. Wang et al., Minerals Engineering **83**, 185–191 (2015)
10. I. Arganda-Carreras et al., Bioinformatics **33**, 2424–2426 (2017)
11. C. Sommer et al., in *2011 IEEE International Symposium on Biomedical Imaging: From Nano to Macro* (2011), 230–233
12. M. Andrew, Computational Geosciences **22**, 1503–1512 (2018)
13. S. Karimpouli et al., Computers & Geosciences **126**, 142–150 (2019)
14. V. Badrinarayanan et al., IEEE Transactions on Pattern Analysis and Machine Intelligence **39**, 2481–2495 (2017)
15. Y. Da Wang et al., ArXiv:2002.05322 (2020)
16. I. A. Varfolomeev et al., Computers **8**, 21 (2019)
17. M. Andrew et al., Microscopy and Microanalysis **23**, 156–157 (2017)
18. M. R. Ball et al., Microscopy and Microanalysis **25**, 410–411 (2019)
19. G. Tiu, Classification of Drill Core Textures for Process Simulation in Geometallurgy: Aitik Mine, Sweden, Master's thesis in Natural Resources Engineering, Luleå University of Technology, (2017)
20. P. I. Guntoro et al., Minerals Eng **142**, 19 (2019)
21. I. A. Varfolomeev et al., in *Proceedings of the 3rd IAPR Asian Conference on Pattern Recognition – ACPR* (Kuala Lumpur, Malaysia, 2015), 346–350
22. M. Pietikäinen et al., in *Computer Vision Using Local Binary Patterns* (Springer, London, 2011), 13–47
23. A. R. Butcher et al., in *Seventh Mill Operators' Conference* (The Australasian Inst. of Mining & Metallurgy, 2000), 267–271
24. S. Latham et al., in *Proc. of International Symposium of the Society of Core Analysts* (Society of Core Analysts, Abu Dhabi, UAE, 2008)
25. Yu. P. Pyt'ev et al., *Methods of Morphological Analysis of Images [In Russian]* (Fizmatlit, Moscow, 2010)
26. C. Liu, Beyond Pixels: Exploring New Representations and Applications for Motion Analysis, PhD thesis in Electrical Engineering and Computer Science, Massachusetts Institute of Technology, (2009)
27. I. A. Varfolomeev et al., in *Proceedings of the SPE Russian Petroleum Technology Conference and Exhibition* (Society of Petroleum Engineers, 2016)
28. I. A. Varfolomeev et al., VKIT [In Russian] (7), 3–9 (2019)
29. D. M. W. Powers, ArXiv:2010.16061 (2010)
30. G. Ke et al., in *Advances in Neural Information Processing Systems* (2017), 3149–3157
31. Y. Boykov et al., Pattern Analysis and Machine Intelligence, IEEE Transactions On **23**, 1222–1239 (2001)
32. O. Ronneberger et al., in *Medical Image Computing and Computer-Assisted Intervention – MICCAI 2015* (Springer, Cham, 2015), 234–241
33. Ö. Çiçek et al., ArXiv:1606.06650 (2016)
34. M. Lai, ArXiv:1505.02000 (2015)
35. I. J. Goodfellow et al., ArXiv:1406.2661 (2014)
36. P. Isola et al., in *IEEE Conference on Computer Vision and Pattern Recognition – CVPR* (2017), 5967–5976
37. Y. D. Wang et al., Water Resources Research **56**, e2019WR026052 (2020)
38. L. Mosser et al., in *Proc. of 80th EAGE Conference and Exhibition* (EAGE, 2018), 1–5

# Calcite-functionalized micromodels for pore-scale investigations of $CO_2$ storage security

*Malin* Haugen[1,*], *Benyamine* Benali[1], *Tore* Føyen[1,2], *Wen* Song[3], *Martin* A. Fernø[1], and *Bergit* Brattekås[1]

[1]Department of Physics and Technology, University of Bergen, Norway
[2]SINTEF Industry, Norway
[3]Hildebrand Department of Petroleum and Geosystems Engineering, University of Texas at Austin, USA

**Abstract.** Carbon capture and subsequent storage (CCS) is identified as a necessity to achieve climate commitments. Permanent storage of carbon dioxide ($CO_2$) in subsurface saline aquifers or depleted oil and gas reservoirs is feasible, but large-scale implementation of such storage has so far been slow. Although sandstone formations are currently most viable for $CO_2$ sequestration, carbonates play an important role in widespread implementation of CCS; both due to the world-wide abundancy of saline aquifers in carbonate formations, and as candidates for $CO_2$-EOR with combined storage. Acidification of formation brine during $CO_2$ injection cause carbonate dissolution and development of reactive flow patterns. Using calcite-functionalization of micromodels we experimentally investigate fundamental pore-scale reactive transport dynamics relevant for carbonate $CO_2$ storage security. Calcite-functionalized, two-dimensional and silicon-based, pore scale micromodels were used. Calcite precipitation was microbially induced from the bacteria *Sporosarcina pasteurii* and calcite grains were formed *in-situ*. This paper details an improved procedure for achieving controlled calcite precipitation in the pore space and characterizes the precipitation/mineralization process. The experimental setup featured a temperature-controlled micromodel holder attached to an automatic scanning stage. A high-resolution microscope enabled full-model (22x27 mm) image capture at resolution of 1.1 μm/pixel within 82 seconds. An in-house developed image-analysis python script was used to quantify porosity alterations due to calcite precipitation. The calcite-functionalized micromodels were found to replicate natural carbonate pore geometry and chemistry, and thus may be used to quantify calcite dissolution and reactive flow at the pore-scale.

## 1 Introduction

Commitments to reach climate goals require us to reduce greenhouse gas concentrations in the atmosphere. Carbon capture, utilization, and storage (CCUS) is identified to play a critical role in achieving these goals. In the Clean Technology Scenario presented by IEA, CCUS is the third-largest contributor (13%) to the cumulative emission reduction until 2060 [1]. Among other, $CO_2$ can be stored in deep saline aquifers within sandstone or carbonate formations. However, two of the possible risk aspects related to $CO_2$ storage are the potential for leakage to the surface and, for carbonate formations, dissolution of the reactive calcite rock minerals. The storage potential in carbonate reservoirs is large, but the reactive transport and mechanisms must be accounted for.

Biomineralization is the formation of minerals by living organisms. Microbial induced calcite precipitation (MICP) has been researched in conjunction with different engineering applications, for instance within construction engineering to seal fracture or strengthen concrete [2], for enhanced oil recovery [3, 4], to constrain contaminated groundwater [5], for cementation of unconsolidated porous media [6], and for $CO_2$ sequestration by reduction of permeability in fractured rock to potential seal leakage paths in the near well area and/or in the cap-rock [7-9].

Precipitation of calcium carbonate ($CaCO_3$) is controlled by four key factors: (1) concentration of calcium ($Ca^{2+}$) from the cementation solution, (2) concentration of dissolved inorganic carbon as produced by ureolysis, see equation 1-5 in section 2.2, (3) pH value, and (4) availability of nucleation sites [10]. $CaCO_3$ precipitation is dependent on presence of sufficient calcium and carbonate ions to ensure that the ion activity product exceeds the solubility product, making the system oversaturated and precipitation of calcium carbonate is likely. Microbial metabolic activity influence key factor no. 2 and 3, and also no. 4 by function as nucleation site [2]. Calcium carbonate is a common element found in many rocks, such as chalk and marble, it is also the main component in shells, pearls, and corals. $CaCO_3$ can crystallize in different polymorphs, seen as minerals in the nature. This includes calcite, vaterite and aragonite, with calcite as the most stable, and two hydrated polymorphs in addition to several amorphous calcium carbonate (ACC) phases [11]. Vaterite is the least thermodynamically stable polymorph and transforms rapidly to calcite or aragonite in aqueous solutions. According to Ostwald's rule of stages, metastable polymorphs are first formed and then transform into thermodynamically more stable polymorphs [12]. For high supersaturation values it is reported that calcite is initially precipitated as the hydrous form of ACC (tens to hundreds of nanometers) which are subsequently transformed into spherical vaterite and later calcite, while lower supersaturation values lead to direct calcite precipitation [13, 14].

A bacteria often used for studying MICP in the laboratory is the soil bacteria *Sporosarcina pasteurii* [3, 15-18]. In this research these bacteria have also been utilized, with the goal to build upon work by Song et al. 2018 [18] and present an improved procedure to functionalize the pore space in a micromodel with $CaCO_3$ and enable recording of continuous full-model images. A micromodel is a device with a porous pattern from a

* Corresponding author: malin.haugen@uib.no

© The Authors, published by EDP Sciences. This is an open access article distributed under the terms of the Creative Commons Attribution License 4.0 (http://creativecommons.org/licenses/by/4.0/).

https://doi.org/10.1051/e3sconf/202336601004

representative rock matrix etched into a silicon wafer before an optical transparent glass plate is bonded on top. It has four fluid flow ports, one in each corner, with an open fracture connecting two pairs of flow ports (Fig. **1**). This device enable direct study of microscale interactions between fluids and rocks. By functionalizing micromodels with an existing pore-network made of silica grains, it is possible to both precipitate and dissolve $CaCO_3$ minerals and still re-use the micromodel afterwards. This starting point is therefore of great interest during study of fundamental pore-level reactive transport dynamics that determine $CO_2$ storage security in carbonate formations. MICP in micromodels have also been studied by Wang [19, 20], where, among other, the effect of bacterial cells was studied as well as evolution of shape and size of the $CaCO_3$ precipitates during the MICP process. In this paper we are adding quantification of change in porosity due to $CaCO_3$ precipitation and pressure development throughout experiments.

The experimental set-up was designed to facilitate flow experiments at pressures between 0-120 bars and temperatures between 25-40 °C. Because of the dependency on the biological process, it is important to regulate the temperature to adjust the growth rate of the bacteria. Further, the tubing and valves included in the set-up provide a low, constant, dead volume. The possibility of independent injection and production from each of the four ports in the micromodel, combined with injection of bacterial solution through a loop with constant volume provide excellent experimental conditions. The result is a permeable porous model with carbonate grains distributed throughout the whole pore volume.

By utilizing an in-house built program, we can calculate change in porosity, and among other, identify area of representative elementary volume. The utilized microscope has an automatic scanning stage and a fluorescence module, which enable us to capture the complete porous network (22x27 mm) with resolution of 1.1 µm/pixel. This paper starts with an overview of the micromodel and bacteria utilized to functionalize the micromodel with calcium carbonate. Then, the experimental set-up and equipment is described, followed by the experimental procedures. Results in the form of images, porosity development due to in-situ $CaCO_3$ precipitation and relevance to $CO_2$ sequestration are presented, before concluding remarks.

## 2 Materials and methods

### 2.1 Micromodel

Direct study of microscale interactions between fluids and rocks at elevated reservoir temperature and pressure were enabled by microfluidic devices, further referred to as micromodels. The micromodels consist of two parts, an optical transparent borosilicate glass plate bonded on top of a silicon wafer with etched pattern of a representative porous rock matrix. The state-of-the-art etching method provides a two-dimensional porous system with representative pore-scale geometry with vertical grain walls, sharp edges, and a hydrophilic surface.

The porous network is based on thin sections from natural sandstone, a rock type widely used in laboratory flow experiments for enhanced oil recovery and/or $CO_2$ storage. Dead-end and isolated pores in the thin section were modified in the micromodel to enable flow across the whole pattern, resulting in higher micromodel porosity and permeability compared to the natural sandstone sample. A unique porous network (6.74 x 2.50 mm) was extracted from the thin section and etched in the silicon wafer with a constant depth of 30 µm. The unique pore pattern was repeated 36 times (4 x 9), resulting in a total porous network which is 27.0 x 22.5 mm large (Fig. **1**). The grains are discontinuous and irregularly shaped, providing flow tortuosity. In each corner, a port through the silicon wafer can connect the pore network to external fluid flow tubes. Two open channels between ports improve flow communication and sweep of the pore space. Further details about design and production procedures can be found in [21] and [22].

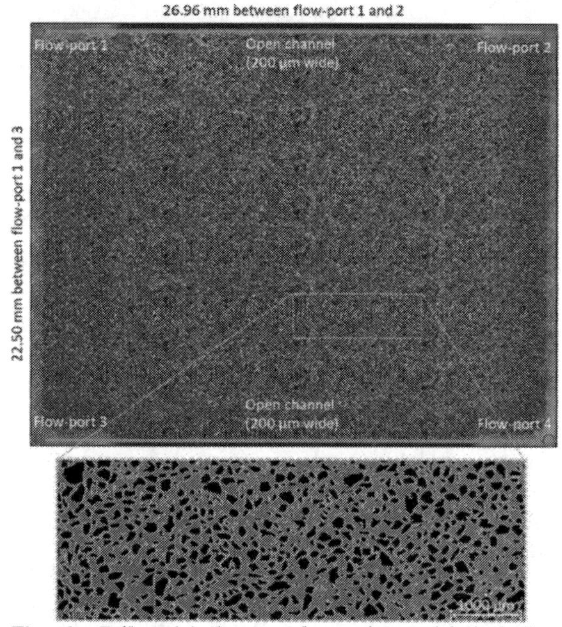


**Fig. 1.** Full-model image of a micromodel including dimensions. The four flow-ports and the two open-channel fractures between two pairs of ports are highlighted. A pattern generated from thin sections of natural sandstone is repeated 36 times to create the micromodel porous network. A zoom-in of the repeated pattern is highlighted in blue. The full-model image is generated from 121 individual images automatically captured by the microscope software with 10% overlap (shown in red dotted lines), stitched together to one image.

Micromodel properties are listed in Table **1**, as previously reported by Benali et.al (2021) [23]. Stitched full-model images were analyzed to determine grain size distribution, pore throat length distribution (defined as the shortest pore space distance between two neighboring grains) and porosity. Porosity was quantified to 0.61 based on amount of pore space to the total area (grains + pore space) and the total micromodel pore volume (PV) was 11.1 µL. Fig. **2** shows the distribution of grains and pore throat length in the micromodel. Average grain size is calculated to 6 464 µm$^2$ and average pore throat length is 41 µm, with the shortest pore throat being 0.7 µm with square or rectangular cross-section.

**Table 1.** Micromodel properties

| **Parameter** | **Value** |
|---|---|
| Width | 27.0 mm |
| Length | 21.4 mm |
| Depth | 30 µm |
| Porosity | 0.61 |
| Pore volume | 11.1 µL |
| Permeability | 2.97 D |
| Repetition of pattern | 36 |
| Grain size | 0.5 - 78 366 µm$^2$ |
| Pore throat length | 0.7 - 194 µm |

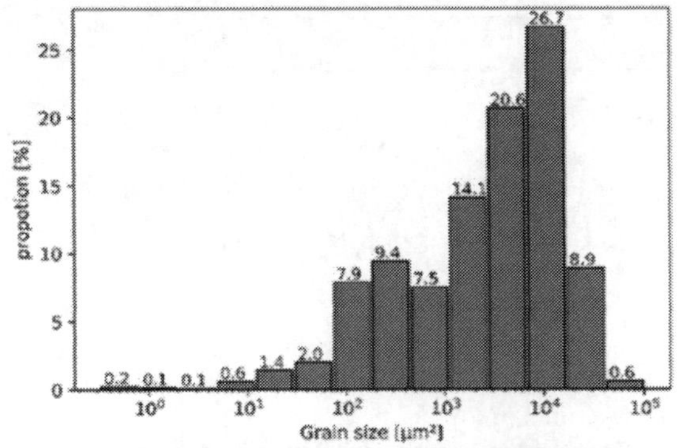


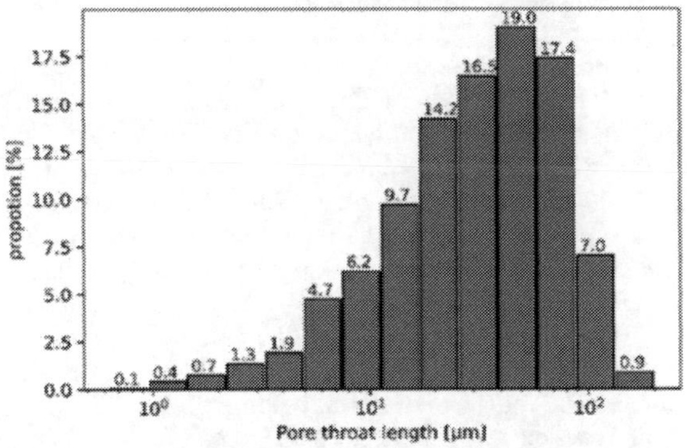


**Fig. 2.** Distribution of grain size (top) and pore throat length (bottom) in one of the 36 pore network repetitions which make up the micromodel.

The work presented here describes the experimental set-up and method to functionalize micromodels with calcium carbonate grains utilizing bacteria. This enables experimental investigations of fundamental pore-scale reactive transport dynamics in carbonate rocks.

## 2.2 Sporosarcina pasteurii bacteria

To functionalize micromodels with calcium carbonate for pore-scale investigations of $CO_2$ storage security, microbially induced calcite precipitation (MICP) by urea hydrolysis have been utilized. The bacteria *Sporosarcina pasteurii*, formerly *Bacillus pasteurii*, (Miquel 1889) DSM33 [24] are urease positive [25] and have a known ability to enable calcite precipitation [15] and growth on solid surfaces [2, 3]. The bacterial cells are gram positive, motile and rod shaped with size in the range of 0.5-1.2 x 1.3-4.0 µm [25]. This is an aerobic bacteria with optimum growth temperature and pH of 30 °C and 9, respectively [25].

The MICP process has been studied by several researchers and the following equations (Eq.1-5) have typically been used to describe this process [2, 7, 17, 26]. The bacteria enable hydrolysis of urea ($CO(NH_2)_2$) into ammonia ($NH_3$) and carbonic acid ($H_2CO_3$), see Eq.1. In solution this leads to production of ($OH^-$) and accumulation of ammonium ($NH_4^+$). As a result of increasing ($OH^-$) concentration, the pH in the microenvironment is increasing (Eq. 2 and 3), and subsequently causes carbonic acid to be converted to bicarbonate ($HCO_3^-$) and further to carbonate ions ($CO_3^{2-}$) (Eq.4). When calcium ions ($Ca^{2+}$) are added to the solution, the increasing pH also leads to precipitation of calcium carbonate ($CaCO_3$) [3, 7] when the solution is supersaturated with calcium- and carbonate- ions (Eq.5). Precipitation of calcium carbonate can occur in the bulk phase, or on the bacteria cell wall because it is negatively charged and will therefore attract the positive charged calcium ions [2, 16]. If the bacterium is covered in calcium carbonate the cell will die as nutrient transfer is prevented [2, 26].

$$CO(NH_2)_2 + 2H_2O \xrightarrow{urease} 2NH_3 + H_2CO_3 \quad (1)$$

$$2NH_3 + 2H_2O \leftrightarrow 2NH_4^+ + 2OH^- \quad (2)$$

$$H_2CO_3 + 2OH^- \leftrightarrow HCO_3^- + H_2O + OH^- \quad (3)$$

$$HCO_3^- + H_2O + OH^- \leftrightarrow CO_3^{2-} + 2H_2O \quad (4)$$

$$Ca^{2+} + CO_3^{2-} \rightarrow CaCO_3(s) \quad (5)$$

Details about preparing bacteria and fluids, and procedure for functionalization of micromodels are described in section 2.4.

## 2.3 Experimental set-up

A schematic of the experimental set-up for *in situ* visualization of calcite-functionalized micromodels is presented in Fig. **3**. The system also contains equipment to inject $CO_2$ or $CO_2$-saturated fluids for pore-scale investigations of $CO_2$ storage security at elevated reservoir pressure and temperature (see section 3.3). To facilitate injection of these fluids, the set-up is designed with all tubes and equipment (except autoclave-valves) in materials which are chemical compatible with these fluids. Polyether ether ketone (PEEK) has been found to resist all the chemicals and fluids used in this work and is therefore the main material used in this design.

On the inlet side of the micromodel, a high precision plunger pump (Quizix 5000-2.5K) is operated in independent mode and used for injection of water and $CO_2$. There is also an option to inject $CO_2$ into an in-house PEEK accumulator (30 ml) for $CO_2$-saturated fluids. As mentioned in section 2.1, the smallest pore throat in the micromodel is calculated to be 0.7 µm. To avoid injection of unwanted particles, like dust, that clogs up the micromodel, all fluids were filtered prior to use and flow

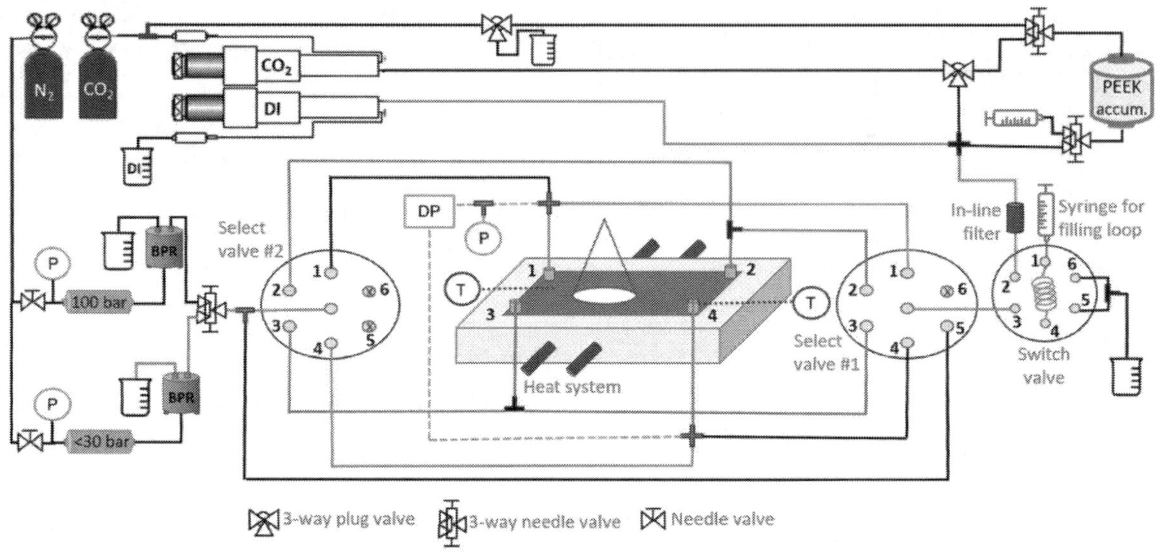


**Fig. 3.** Schematic of the experimental set-up for calcite-functionalization of micromodels. During injection of bacterial- or cementation solution, tubing marked in green were open/connected. Fluids were injected into a 1.0 ml loop in the switch valve and chased by DI from a pump. The micromodel was placed directly under a microscope, enabling in-situ visualization of $CaCO_3$ precipitation in the pore space. All tubing and tee/cross (purple) are in PEEK from Idex.

through a 0.5 µm inline filter assembly (Idex, A-431) prior to entering the micromodel.

The low pore volume (11.1 µL) makes it important to have a constant system dead-volume to ensure that fluids enter the micromodel when anticipated. This is achieved by utilizing two 6-ports select valves (Idex, MXP7970-000), one at inlet and one at the outlet. When valves #1 and #2 are connected to the fluid-flow-ports (Fig. **3**) it is possible to designate injection and production ports with a single mouse-click. When changing injection fluid, both select valves were set to position 5 for the fluids to by-pass the micromodel until pressure stabilized. During calcite generation experiments small amounts of different fluids will be injected into the model at various times. To facilitate variable fluid injection without air entering the tubes, a two-position switch valve (Idex, 9725) is included in the design. The switch valve has two positions: load and inject. At "load" position, fluids from the pump flow into port 2, via port 3 and to Select Valve #1. A coiled tube (referred to as "loop", see Fig. **3**) with a constant volume could then be filled with e.g. bacteria-solution using a luer-lock syringe. At "inject" position, flow from the pump is rooted through the loop and into Select Valve #1. Other valves used in the set-up are Autoclave valves as described in Fig. **3**.

The system is pressurized by two back pressure regulators (Equilibar), where one is kept at 100 bars for high pressure experiments, and one at lower pressure depending on experimental design. The pressure differential between inlet and outlet is logged with differential pressure transmitter (APLISENS PRE-28 SMART) with a measuring range of 0-2.5 bar and a static pressure limit of 250 bar. In addition, absolute pressure transmitter (ESI), with a range from 0-250 bar, is connected to flow port 1.

The micromodel-holder in the middle of Fig. **3** is designed and machined in-house from PEEK material. It consists of four drilled holes, made with a 10-32 coned drill, which is aligned with the fluid-flow-ports underneath the micromodel, two Omega temperature probes close to port number 1 and 4, and tracks for two 1/8-inch copper tubing for heating of the micromodel by circulation of warm water. Fluid flow tubes are connected to the micromodel-holder by finger tight fittings 10-32

**Fig. 4.** Image sequence showing steps for mounting micromodel in holder. Left to right: The first image shows only the holder with machined groove for the micromodel, O-rings around the four fluid-flow-ports, two temperature gauges and copper tubing for the heating system. In the second image, the O-rings are in place and thermal paste are covering the temperature gauges and copper tubing to ensure optimal heat transfer. In the third image, the micromodel is in place and in the fourth image the black-painted aluminum plate is placed on-top of the micromodel and tightened with 0.20 Nm force. When everything is in place, the system can be pressurized to 150 bar.

coned (Idex, F-120X) screwed directly into the four drilled holes in the PEEK holder. Mounting of micromodel in the holder is shown in detail in Fig. **4**.

As indicated by the cone in Fig. **3**, the micromodel-holder is placed directly underneath a microscope to enable *in situ* visualization of calcite generation and subsequent investigation of $CO_2$ storage in carbonate formations. The microscope is a Zeiss, Axio Zoom.V16 Microscope with two light sources: a cold-light (CL 9000 LED) source with diffuser S, and a fluorescence (FL) module. It has an automatic scanning-stage which enables high-resolution imaging of the entire micromodel within 82 seconds (with the CL 9000 LED source) and automatic stitching of the images to a full-model image. When generating images with both the CL 9000 LED source and one FL light channel for each tile, a full-model image is acquired within 293 seconds.

## 2.4 Experimental procedure

### *2.4.1 Preparation of fluids*

Bacterial growth medium and cementation solution (reactant) were prepared as described by [18]:

Bacterial growth medium: 47g of brain heart infusion broth (BHI Broth, 53286 Sigma-Aldrich) was mixed with 900 ml of deionized (DI) water, obtained from Milli-DI synthesis system including 0.2 µm filter (172-5109, Merck). Then the solution was sterilized in 15 minutes at 121 °C in an autoclave before cooled to ~30 °C. Next, a concentrated urea solution was added to make a 2 wt. % urea broth. This was achieved by mixing 20g of urea into 100 mL of DI water. Before injecting this urea solution into the growth medium, it was filtered through a 0.2 µm syringe filter (514-0061, VWR). The finished growth medium (BHI+U) was also filtered through the syringe filter prior to adding bacteria.

Reactant was prepared by mixing urea (urea, U5378, Sigma-Aldrich) and calcium chloride (calcium chloride dehydrate, 31306, Sigma-Aldrich) in DI water. As for the growth medium, the reactant was filtered through a 0.2 µm syringe filter prior to use to minimize the risk of small particles entering the micro-model.

### *2.4.2 Cultivation of bacteria*

Bacterium *Sporosarcina pasteurii* DSM 33 [24] were received freeze-dried from DSMZ GmbH supplier and cultivated as follows: First the pellet was removed from the sealed glass and added to 0.5 ml growth medium where it was allowed to rehydrate for 30 minutes. Second, 0.2 ml of the bacterial solution was moved to a centrifuge tube containing 10 ml growth medium and placed in a heating cabinet at 30 °C for 24 hours. Growth was detected after 24 hours, but the procedure was repeated one more time by adding 0.2 ml of the bacterial solution into a new centrifuge tube with 10 ml growth medium and stored in 30 °C for another 24 hours. After 24 hours, sufficient bacterial growth was observed, and the bacterial solution was ready to be used. Every 7 days, 0.2 ml of bacterial solution is moved to a new centrifuge tube containing 10 ml growth medium to prevent death of bacteria. In case bacteria die, an inventory of glycerol stock was prepared by adding 0.2 ml of bacterial solution and 0.2 ml of 30 % glycerol into 6 pendlorf microtubes for long term storage at -80 °C.

### *2.4.3 Calcite-functionalization of micromodels*

The experimental set-up (Fig. **3**) is designed to withstand pressure up to 150 bar, but the MICP process was performed at atmospheric pressure or slightly elevated pressure conditions (25 bars). The latter based on experience of when all $CO_2$ generated from the bacterial metabolism during settling time is completely dissolved in aqueous phase, to mitigate gas-filled pores that influence fluid flow. As the MICP process is developing to be utilized in more extreme environment [27], studying this process at elevated pressures is relevant. Several factors impact the precipitation of calcium carbonate during the MICP process and variables can be adjusted depending on the goal. Details about the experimental procedure used to generate calcium carbonate in a micromodel are explained below.

1. Initial conditions.

The micromodel is initially filled with deionized water (DI). DI is miscibly displaced by bacterial growth medium before 20 µl (2 PV) bacterial solution is injected at 6 µl/min (flow port 3 as inlet and 2 as outlet). 2 PV bacteria solution is consistent with results by Wang et.al (2018) [19], who have reported that bacterial suspension was homogeneously distributed in a microfluidic chip after 1.25 PV injected.

2. Bacterial growth in pore space.

Tubing used for bacterial solution are flushed with sterile bacterial growth medium before the micromodel is shut in. This enables time for bacteria to grow, settle and attach to the pore walls.

3. Prepare for injection of cementation solution.

Cementation solution is loaded in the loop of the switch valve and injected through the bypass next to flow port 1 (with both select-valves at position 1). When the bypass is filled with cementation solution, flow is routed at 1 µl/min to flow-port 1 in the micromodel, and flow port 2 is used as outlet.

4. Cementation solution into micromodel.

After the dead-volume from bypass to flow-port 1 has been injected, the outlet flow-port is changed from 2 to 4. *In situ* growth of calcium carbonate is visualized using a high-resolution microscope with automatic scanning stage, enabling full-model images of 1.1 µm/pixel within 82 seconds or 293 seconds if one FL channel is included.

## 2.5 Quantification of porosity

Changes in porosity from $CaCO_3$ precipitation were quantified from image analysis (using an in-house python script) on sub-sections during time-lapse imaging, and in the entire pattern for the last time step, see section 3.

Two different light sources (CL 9000 LED and FL) were available for in-situ imaging of the micromodel, each with its own method for calculating the area of precipitated $CaCO_3$ minerals. With the CL 9000 LED light source, calcium carbonate minerals appear white and can be quantified with simple segmentation. The white reflection edge around each silica grain were omitted from

 https://doi.org/10.1051/e3sconf/202336601004

the areal calculations by subtracting the initial image (experiment A). When both light sources were utilized, two approaches were developed, depending on when the FL tracer was injected: 1) displace the aqueous phase with a FL tracer after the MICP process. In the FL light image, the $CaCO_3$ minerals then appear as shades from green to black, where black minerals were interpreted as occupying the whole height in the micromodel; 2) include a FL tracer in the cementation solution during the MICP process. Calcite minerals then appear orange with the CL9000 LED light source (Fig.**6**, B.1). If the FL tracer is replaced with aqueous phase after the MICP process, only $CaCO_3$ minerals are visible in the FL light image. By combining threshold CL 9000 LED and FL images, the result contains the area of all precipitated $CaCO_3$ (Experiment B).

Porosity changes over time can be studied by dividing the full-model images into sub-sections (experiment A). This is achieved by adding the total silica grain area to the calcium carbonate area for each timestep, then normalizing against initial white areas related to reflections around silica grains and bacterial aggregation or potential biofilm. With this approach we mitigate areas related to white reflections which otherwise would impact the porosity as the software/script identify all white colors as porosity. The final fraction of calcite minerals in the pore space was calculated as total calcite grains area divided by the total area of each image, both in the sub-sections and the full-model image. Some calcite grains (part of a whole grain) appear transparent and were not detected by the software. Hence, the calculated calcite fraction is expected to be lower than the real fraction.

# 3 Results and discussion

## 3.1 Experimental conditions

Two experiments, with the same bacterial culture and procedure, constitute the experimental results in this study. The pore space bacterial settling times were different between the two experiments (Table **2**), but the initially high pH and bacterial movement observed indicated that they were in the active phase of the bacterial growth curve, performing ureolysis, see section 2.2. The procedure to generate calcite in the micromodel is presented in section 2.4. All experimental conditions for the two experiments are listed in Table **2**.

**Table 2.** Properties of bacterial- and cementation solution of experiments included in this paper.

| Experiment | A | B |
|---|---|---|
| **Properties of bacterial solution** | | |
| OD600 | ~0.6 * | ~0.6 * |
| Initial pH | ~ 9.2 * | ~ 9.2 * |
| PV injected | 2 | 2 |
| Injection rate [µl/min] | 6 | 6 |
| Growth time in micromodel [h] | 7 | 21 |
| Temperature [°C] | 31 | 30 |
| Pressure [bar] | 25 | atm |
| **Properties of cementation solution** | | |
| Urea [M] | 1 | 0.5 |
| $CaCl_2$ [M] | 1 | 0.5 |
| Injected [µl] | 244.2 | 221.8 |
| PV injected | 22.0 | 20.0 |
| Injection rate [µl/min] | 1 | 1 |
| Temperature [°C] | 31 | 30 |
| Initial pressure [bar] | 25 | atm |
| Pressure when pump stopped [bar] | 50 | 10 |
| FL included | No | Yes |

** Based on measurements on three bacterial solutions made with the same procedure/volumes*

During cementation solution injection (experiment A) the injection and differential pressure across the micromodel increased rapidly after 4.2 PV injected (Fig. **5**), possibly related to large amount of calcium carbonate precipitated. The PV injected increased linearly until the injection pump reached 50 bars (after 22.0 PV injected) due to blockage of the outlet tubing. Continued calcium carbonate growth was observed after injection stopped, reducing the porosity of the pore space (detailed later). For experiment B, the observed pressure behavior was similar.

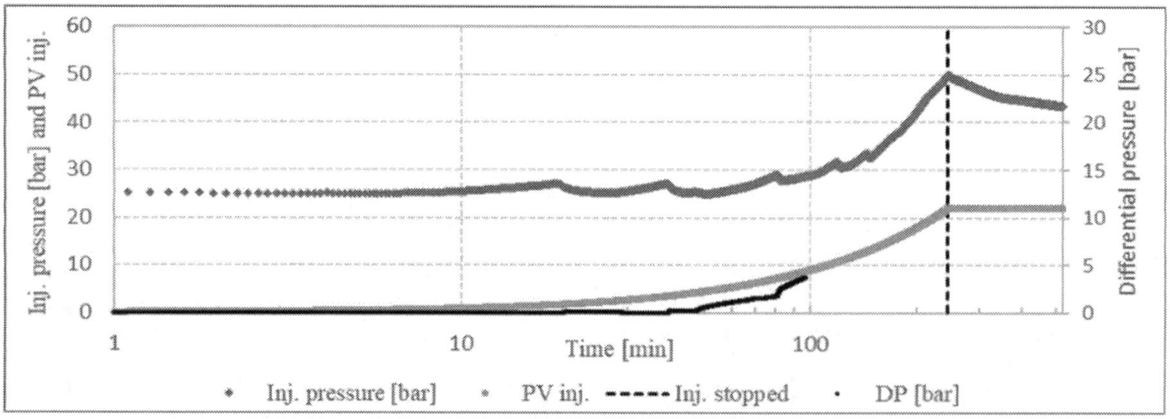


**Fig. 5.** Injection pressure (blue, left y-axis), differential pressure (black, right y-axis) and pore volumes injected (orange, left y-axis) during injection of cementation solution in experiment A. Injection stopped when the injection pressure reached 50 bars, indicated with vertical dashed line; measurements of differential pressure stopped at 3.7 bars because of limitation in measuring range. Calcite precipitation occurred both during injection and in a period afterwards.

### 3.2 Calcite precipitation

#### *3.2.1 Single pore*

The calcium carbonate grains have a variety of shapes, ranging from square/cubic, spherical to irregular (Fig. **6**). Calcite grains precipitated both close to pore walls and in the pore center, and in some cases calcite grains encapsulated silica grains. The *in-situ* development of calcium carbonate through the MICP process was further studied utilizing a fluorescence module that revealed color differences between minerals. Black calcite grains were interpreted as filling the entire height in the pores space (30 μm) and block flow. The internal structure of other minerals was visible in the FL image, in contrast to images generated with CL 9000 LED. Hence, combining FL and CL 9000 LED images have a large potential in characterizing the development of calcite minerals.

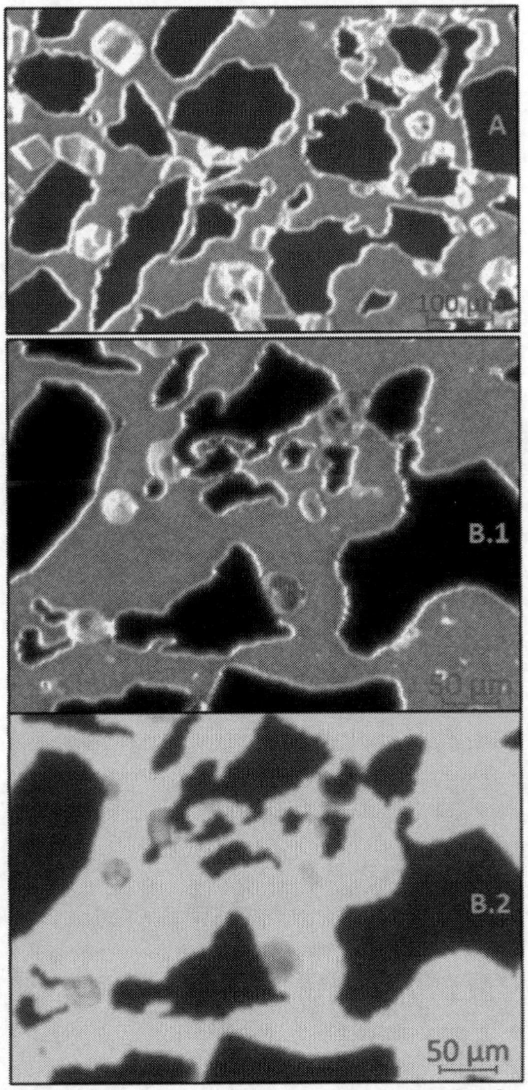


**Fig. 6**. Top image (A) show example of grains from experiment A, while bottom two (B) show image from experiment B generated with CL 9000 LED light source in B.1 and with FL light source in B.2.

#### *3.2.2 Sweep/areal*

The *Sporosarcina pasteurii* bacteria perform urease when settling in the micromodel, prior to injection of cementation solution. During this process, the pH is increasing and subsequently leads to carbonate ions being present in the pore space fluids. When calcium ions are added to the micromodel, calcium carbonate precipitate when the solution is supersaturated with these ions. To study the development of calcium carbonate in more detail, the micromodel is divided in three subsets (Fig. **7**). The change in porosity due to calcium carbonate precipitation (Fig. **8**) was calculated using segmented images (described in section 2.5). The fraction of calcium carbonate (area of $CaCO_3$ / area of image) is summarized in Table **3**. The procedure was compared with manual measurement (using a functionality in the Zeiss software) of the total calcium carbonate grain area divided by total area in the final image (at 520 minutes) in subset 1.

**Fig. 7.** Full-model image at the end of experiment A (520 min), including subset 1, 2 and 3.

**Table 3.** Summary of $CaCO_3$ fractions for experiment A.

| Method | Area | $CaCO_3$ fraction |
|---|---|---|
| Polygons in Zeiss | Subset 1 | 0.15 |
| Python script | Subset 1 | 0.13 |
| Python script | Subset 2 | 0.04 |
| Python script | Subset 3 | 0.02 |
| Python script | Full-model | 0.04 |

The development of calcium carbonate in subset 1 is detailed in Fig. **9**. Coinciding with cementation solution entering the inlet-flow-port, a small white irregularly-shaped $CaCO_3$ precipitate (< 30 μm$^2$) formed close to inlet and reached subset 1 after 1.4 min (0.1 PV injected). The initial porosity reduction (Fig. **8**) for subset 1 relates to these irregularly-shaped $CaCO_3$ precipitates, also seen in the image at 5 minutes in Fig. **9**. The amount of $CaCO_3$ precipitates were relatively constant until 19.3 min of injection (1.7 PV injected), when a small number of additional larger (~60-100 μm$^2$) irregularly-shaped $CaCO_3$ precipitates formed. Some of the irregularly-

https://doi.org/10.1051/e3sconf/202336601004

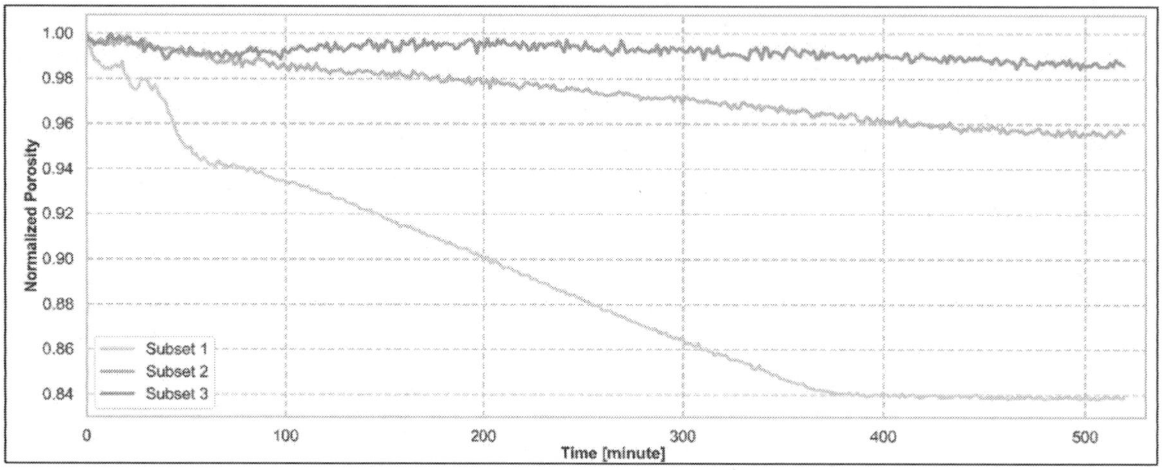


**Fig. 8.** Development of normalized porosity ($\Phi/\Phi_{initial}$) in each of the three subsets in experiment A. Porosity ($\Phi$) in each timestep was calculated as: $\Phi = 1 - (areal_{silica\ grains} + areal_{CaCO_3})/total\ area$.

shaped $CaCO_3$ precipitates continued to grow at apparently random places in the pore space, located mainly in the middle/right side in the field of view where the pores were largest. These precipitates dissolved as

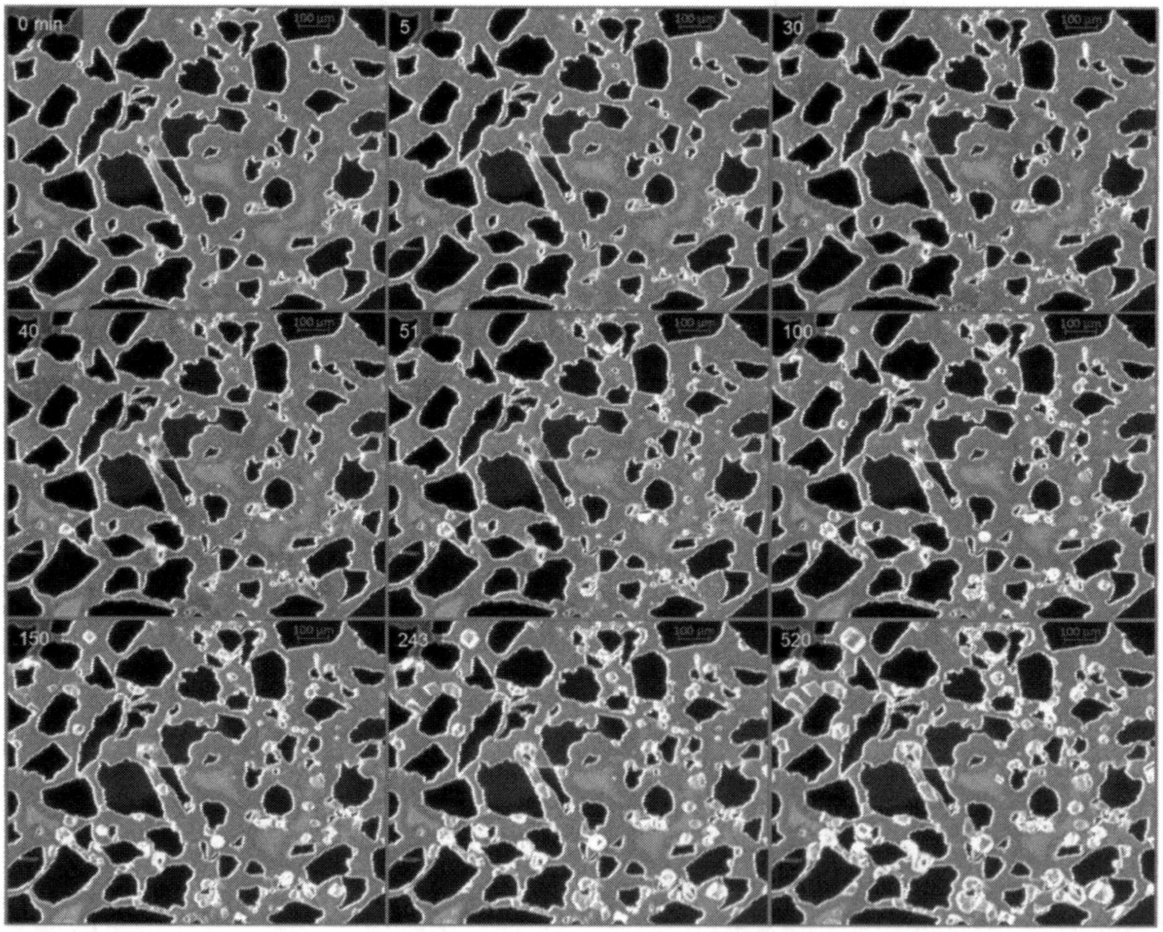


**Fig. 9.** Calcium carbonate precipitation in subset 1 of experiment A (Fig. 7) at different timesteps. Yellow number represent minutes from cementation solution entering the micromodel from the top left corner. This equal to PV injected until the pump stopped after 244 minutes.

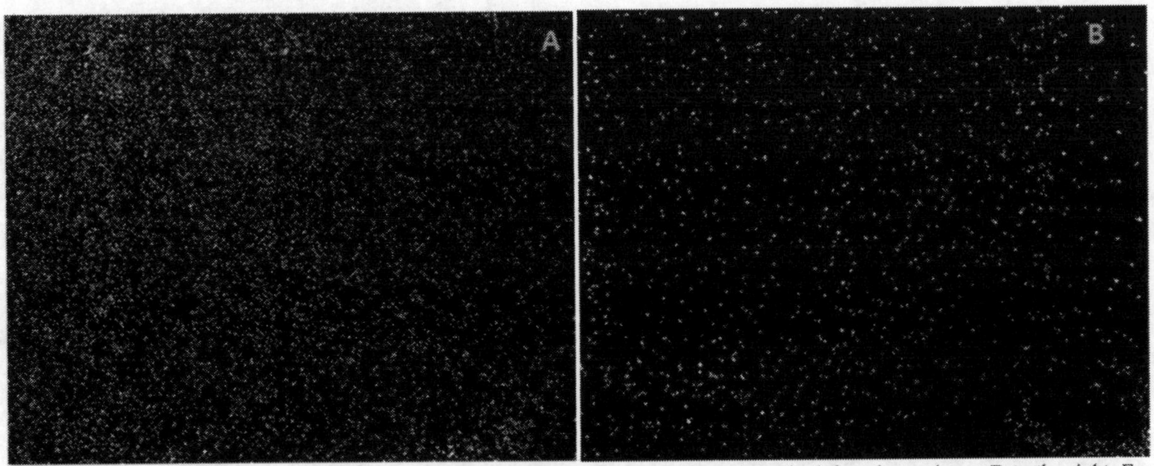


**Fig. 10.** Segmented images showing grains of calcium carbonate after experiment A to the left and experiment B to the right. For experiment A, the injection is at top left and outlet is bottom right, for experiment B it is opposite, with inlet at bottom right corner and outlet at top left. Width and height of both images are 24.9 mm and 19.9 mm respectively.

larger $CaCO_3$ crystals (spherical or rhombohedral) formed. These observations are consisted with findings from Wang et. al (2019) [20], reporting irregularly-shaped $CaCO_3$ dissolving in a circular trend with increasing radius outwards from the growing crystals. They further assumed that the rhombohedral crystals were calcite, spherical crystals were vaterite, and irregularly-shaped precipitates were amorphous $CaCO_3$ precipitates [20].

For the first 81 minutes, precipitating grains were predominantly spherical and/or irregularly shaped (not square or cubical). These continue to grow, but additional grains forming from this time were mainly square or cubical, with a larger area compared with the spherical grains. This change in grain precipitation coincided with a steeper slope in the differential pressure (Fig. **5**). After 244 minutes the cementation solution injection stopped (22.0 PV injected), but the calcite minerals continued to grow until 378 min. Final $CaCO_3$ fraction in subset 1 was 0.13 (Python script) or 0.15 (manual polygons drawn around all calcite minerals in the Zeiss software), confirming that the Python script could satisfactorily capture porosity reduction.

Many parameters influence the precipitation of calcium carbonate, and concentration of calcium chloride in the cementation solution is one of them. Compared to experiment A, the concentrations of calcium chloride and urea were reduced by 50% in experiment B. From the literature it is expected that lower concentration enables the cementation solution to spread out better in the pore space because the reaction is slowed down. Even though the final $CaCO_3$ fraction in subset 1 (experiment A) was 0.13, the full-model fraction was only 0.04. This indicates an uneven distribution of the carbonate minerals throughout the micromodel, and it is also visible in the left section of Fig. **10**. In experiment B the full-model $CaCO_3$ fraction was calculated to 0.006, it appears to be more evenly distributed, as seen in Fig. **10**.

#### 3.2.3 *Multiphase effects*

A product of bacterial respiration and urea hydrolysis is $CO_2$. The dissolved $CO_2$ form carbonic acid ($H_2O + CO_2$) and precipitates as a part of calcite (see eq 1-5), but also function as a buffer where it slows down the increasing pH as ammonia is produced by the urease [15]. At atmospheric pressure conditions some of the $CO_2$ develop into bubbles occupying relatively large areas in the pore volume. Because of the low injection rate, these bubbles impacted saturation distribution in the micromodel (Fig. **11**). As the permeability was reduced from precipitation of $CaCO_3$, the pore-pressure increased, and the bubble area reduced. Once a bubble was removed by dissolution in the aqueous phase, $CaCO_3$ precipitated in the area previously occupied by a bubble. It is expected that the change in pore-pressure, and hence $CO_2$ solubility, has an impact on the local pH value. This should be further studied by adding a pH sensitive FL tracer to the system.

### 3.3 Relevance to $CO_2$ sequestration

$CO_2$ sequestration in carbonate formations causes acidification of the brine and subsequent carbonate dissolution and development of reactive flow patterns. $CO_2$ is also a reaction product from the dissolution of calcium carbonate. Depending on the solubility and saturation of $CO_2$ in the pore fluid, it will go into solution or become bubbles that can eventually form a free gas phase. Song et al. 2018 presented the phenomena with a gas phase shielding carbonate minerals from dissolution as "grain engulfment" [18]. The functionalized micromodel in Experiment A was pressurized to 100 bars before $CO_2$ saturated HCl was injected. Fig. **12** shows $CO_2$ bubbles generated around $CaCO_3$ minerals during pore-level dissolution. Experiments with full model images of this dynamic reactive flow are ongoing.

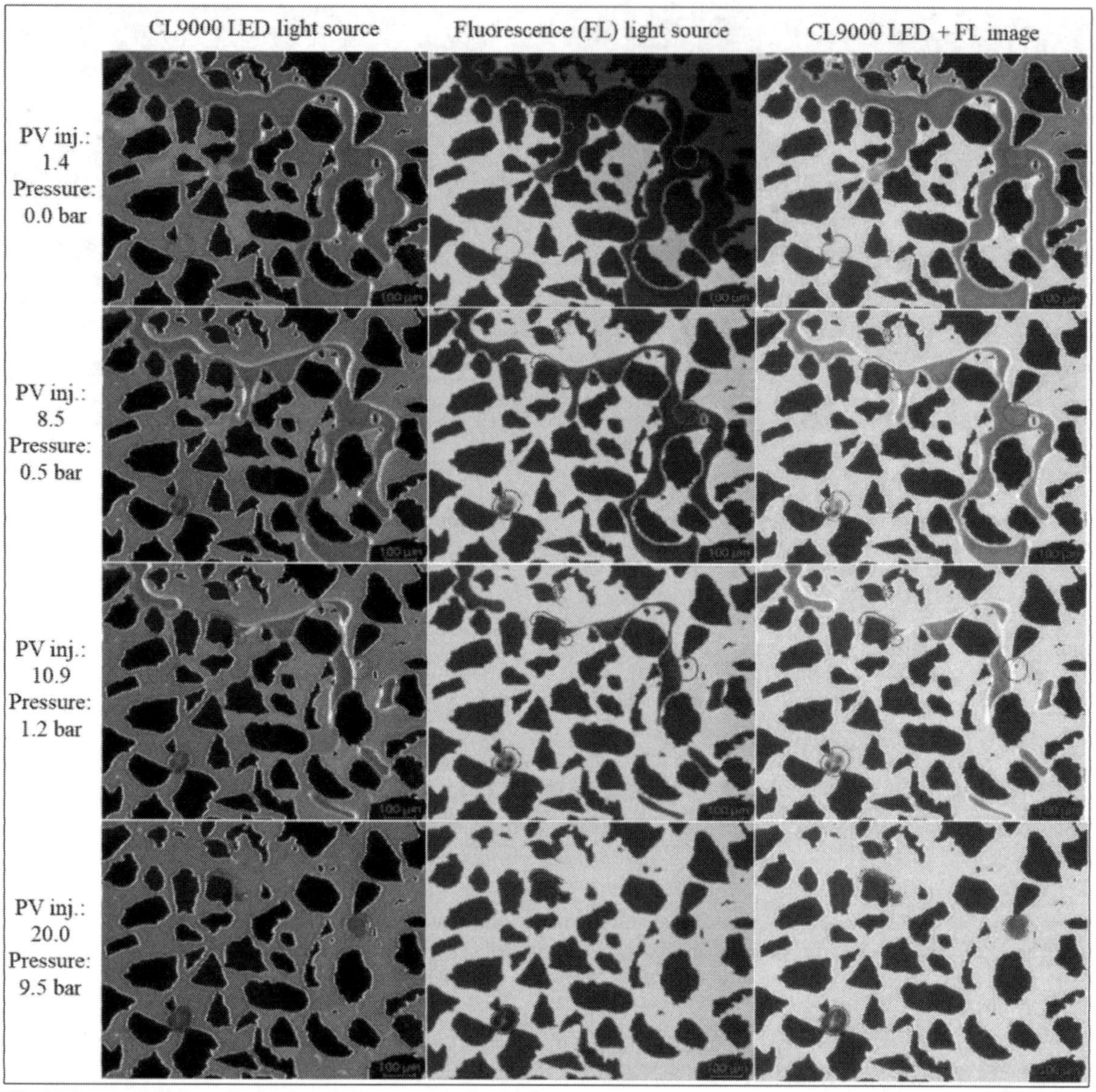


**Fig. 11.** Image sequence of calcium carbonate precipitation in a subset of experiment B. Pore volume injected is from cementation solution entering the micromodel in the bottom right corner (outside this field of view). Time step from top to bottom row of images are 15 min, 95 min, 122 min and 242 min, respectively. Pump stopped at 222 min (20 PV inj.). Red polygons are drawn in the Zeiss program based on image at 242 min.

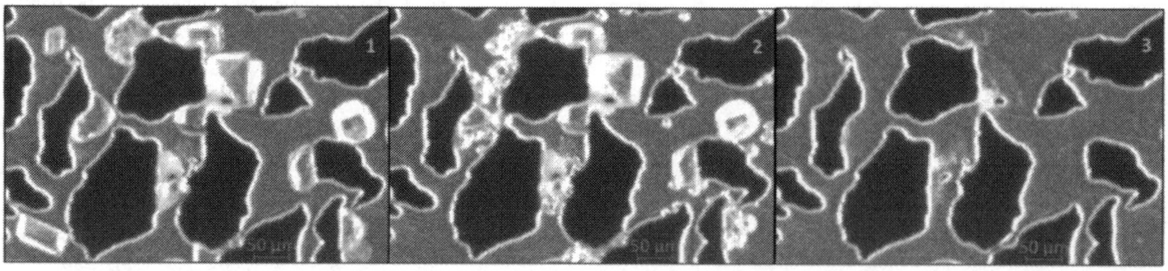


**Fig. 12.** Image sequence of $CaCO_3$ dissolution due to injection of $CO_2$ saturated HCl into a micromodel at 100 bars. Image 1) is from the functionalized micromodel in Experiment A, 2) show $CO_2$ bubbles as the minerals dissolve and 3) is after all the minerals have been dissolved.

## 4 Conclusions

Calcite-functionalization of micromodels was investigated to enable experimental investigation of fundamental pore-scale reactive transport dynamics; eventually determining carbonate $CO_2$ storage security. The system is very sensitive to external factors, and small procedural changes may result in large variations in results. Throughout this study we have demonstrated that bacteria are still active at 25 bar pore-pressure, and precipitation of calcium carbonate minerals occur at elevated pressures. To obtain the best conditions for segmenting of images, it is optimal to have the system under pressure and thereby limit gas-bubbles caused by the metabolism of bacteria.

This experimental set-up and methodology provide a unique method for studying the MICP process, where quantification of full-model calcite porosity development over time is demonstrated. This work also demonstrates that a micromodel can be functionalized with calcium carbonate minerals within a few days. Full-model images and elevated pressures provide the opportunity to support several research opportunities and investigations of fundamental pore-scale reactive transport dynamics:

- Carbon sequestration and dissolution of calcium carbonate
- Water weakening of chalk
- Injection rate dependency
- pH dependency using a pH sensitive FL tracer

MICP is a complex, biochemical process, and future work is planned to include a pH sensitive fluorescence tracer to improve understanding in how pH is changed throughout the different steps and how it influences the precipitation process. This focus aligns with future opportunities presented by Tang et. al (2020) [27] and would build upon work done by Zehner et.al (2020) [28] by introducing the FL tracer in a micromodel with representative pore scale geometry.

During injection of cementation solution, a significant proportion of the bacteria are flushed out, as have also been reported previously [19]. This could lead to preliminary clogging of outlet tube and thereby limit amount of $Ca^{2+}$ injected into the pore space. Future work is therefore planned with increased diameter of the outlet tube, expecting to increase calcite fraction in experiment equal to B.

**Nomenclature and abbreviations**

| | |
|---|---|
| CCS | Carbon, capture and storage |
| $CO_2$ | Carbon dioxide |
| MICP | Microbially induced calcite precipitation |
| BHI | Brain Heart Infusion |
| BHI+U | Brain Heart Infusion including urea |
| DI | Deionized water |
| OD600 | Optical density measured at a wavelength of 600 nm |
| PEEK | Polyether ether ketone |
| $CaCO_3$ | Calcium carbonate |
| $CO(NH_2)_2$ | Urea |
| $CaCl_2$ | Calcium chloride |
| $NH_3$ | Ammonia |
| $NH_4^+$ | Ammonium |
| $H_2CO_3$ | Carbonic acid |
| $HCO_3^-$ | Bicarbonate |
| $CO_3^{2-}$ | Carbonate ions |
| $Ca^{2+}$ | Calcium ions |
| FL | Fluorescent |
| CL 9000 LED | Cold-light source with diffuser S |
| $\Phi$ | Porosity |
| HCl | Hydrochloric acid |
| $H_2O$ | Water |

This work was supported by Norwegian Research Council under project 280341. The authors thank Dhanasekaran Sivakumaran for assistance in cultivating bacteria, and Tove Skogmo Folkestad and Sonja Ljostveit for autoclaving the bacterial growth medium at Haukeland University Hospital.

**References**

1. IEA, *The Role of CO2 Storage.* (2019).
2. W. De Muynck, N. De Belie, and W. Verstraete, *Microbial carbonate precipitation in construction materials: A review,* Ecological Engineering, **36**(2): p. 118-136 (2010)
3. F.G. Ferris, et al., *Bacteriogenic Mineral Plugging,* Journal of Canadian Petroleum Technology, **36**(09): p. 5 (1996)
4. J. Wu, et al., *Microbially induced calcium carbonate precipitation driven by ureolysis to enhance oil recovery,* RSC Advances, **7**(59): p. 37382-37391 (2017)
5. Y. Fujita, et al., *Evaluating the Potential of Native Ureolytic Microbes To Remediate a 90Sr Contaminated Environment,* Environmental Science & Technology, **44**(19): p. 7652-7658. (2010)
6. F.D. Meyer, et al., *Microbiologically-Induced Soil Stabilization: Application of Sporosarcina pasteurii for Fugitive Dust Control,* Geo-Frontiers, Vol. 2011, p.4002-4011 (2011)
7. A.C. Mitchell, et al., *Microbially Enhanced Carbon Capture and Storage by Mineral-Trapping and Solubility-Trapping.* Environmental Science & Technology, **44**(13): p. 5270-5276 (2010)
8. A. Phillips, et al., *Potential CO2 Leakage Reduction through Biofilm-Induced Calcium Carbonate Precipitation,* Environmental Science & Technology, **47**: p. 142-149 (2013)

9. A.B. Cunningham, et al., *Microbially enhanced geologic containment of sequestered supercritical CO2,* Energy Procedia, **1**(1): p. 3245-3252 (2009)
10. F. Hammes, and W. Verstraete*, *Key roles of pH and calcium metabolism in microbial carbonate precipitation.* Reviews in Environmental Science and Biotechnology, **1**(1): p. 3-7 (2002)
11. B. Krajewska, *Urease-aided calcium carbonate mineralization for engineering applications: A review.* Journal of Advanced Research, **13**: p. 59-67 (2018)
12. J.W. Mullin, *Crystallization*, 4th ed., Oxford: Butterworth-Heinemann (2001)
13. W. Zhang, et al., *In Situ Real-Time Study on Dynamics of Microbially Induced Calcium Carbonate Precipitation at a Single-Cell Level,* Environmental Science & Technology, **52**(16): p. 9266-9276 (2018)
14. L.A. Van Paassen, *Biogrout, ground improvement by microbial induced carbonate precipitation*, in *TU Delft, Applied Sciences*, TU Delft: The Netherlands (2009)
15. S. Stocks-Fischer, J.K. Galinat, and S.S. Bang, *Microbiological precipitation of CaCO3.* Soil Biology and Biochemistry, **31**(11): p. 1563-1571 (1999)
16. G.E. Mountassir, et al., *Hydrodynamic coupling in microbially mediated fracture mineralization: Formation of self-organized groundwater flow channels,* Water Resources Research, **50**(1): p. 1-16 (2014)
17. C. Wu, et al., *Microbially induced calcite precipitation along a circular flow channel under a constant flow condition,* Acta Geotechnica, **14**(3): p. 673-683 (2019)
18. W. Song, et al., *Mechanisms of Multiphase Reactive Flow using Biogenically Calcite-Functionalized Micromodels,* Lab on a Chip, **18**, (2018)
19. Y. Wang, et al., *A microfluidic chip and its use in characterising the particle-scale behaviour of Microbial-Induced Carbonate Precipitation (MICP),* Géotechnique, **69** (2018)
20. Y. Wang, et al., *Microscale Visualization of Microbial-Induced Calcium Carbonate Precipitation Processes,* Journal of Geotechnical and Geoenvironmental Engineering, **145**: p. 04019045 (2019)
21. M. Buchgraber, et al., *Creation of a dual-porosity micromodel for pore-level visualization of multiphase flow,* Journal of Petroleum Science and Engineering, **86-87**: p. 27-38 (2012)
22. M. Buchgraber, A.R. Kovscek, and L.M. Castanier, *A Study of Microscale Gas Trapping Using Etched Silicon Micromodels,* Transport in Porous Media, **95**(3): p. 647-668 (2012)
23. B. Benali, et al. *[Preprint] Pore-scale Bubble Population Dynamics of CO2-Foam at Reservoir Pressure.* https://essoar.org (2021)
24. J.H. Yoon, et al., *Sporosarcina aquimarina sp. nov., a bacterium isolated from seawater in Korea, and transfer of Bacillus globisporus (Larkin and Stokes 1967), Bacillus psychrophilus (Nakamura 1984) and Bacillus pasteurii (Chester 1898) to the genus Sporosarcina as Sporosarcina globispora comb. nov., Sporosarcina psychrophila comb. nov. and Sporosarcina pasteurii comb. nov., and emended description of th.* International Journal of Systematic and Evolutionary Microbiology, **51**(3): p. 1079-1086 (2001)
25. P. De Vos, G.M. Garrity, and D.H. Bergey, *Bergey's manual of systematic bacteriology: Vol. 3 : The Firmicutes*. 2nd ed. Vol. 3., New York: Springer (2009)
26. M.S. Reddy, *Biomineralization of calcium carbonates and their engineered applications: a review,* Frontiers in Microbiology, **4**(314) (2013)
27. C.-S. Tang, et al., *Factors affecting the performance of microbial-induced carbonate precipitation (MICP) treated soil: a review.* Environmental Earth Sciences, **79**(5): p. 94 (2020)
28. J. Zehner, et al., *Microbial-induced calcium carbonate precipitation: an experimental toolbox for in situ and real time investigation of micro-scale pH evolution,* RSC Advances, **10**(35): p. 20485-20493 (2020)

# The Problem of Stability of Gas-Condensate Mixture at Pore-Scale: The Study by Density Functional Hydrodynamics

*Oleg* Dinariev[1], *Nikolay* Evseev[1,*], and *Denis* Klemin[2]

[1] Schlumberger, Schlumberger Moscow Research, 16A/3, Leningradskoe shosse, Moscow 125171, Russia
[2] Schlumberger, Schlumberger Reservoir Labs, 6350 West Sam Houston Pkwy N, Houston, TX 77041, Houston, USA

**Abstract.** The method of the density functional hydrodynamics (DFH) is used to model compositional gas-condensate systems in natural cores at pore-scale. In previous publications, it has been demonstrated by the authors that DFH covers many diverse multiphase pore-scale phenomena, including fluid transport in RCA and SCAL measurements and complex EOR processes. The pore-scale modeling of multiphase flow scenarios is performed by means of the direct hydrodynamic (DHD) simulator, which is a numerical implementation of the DFH. In the present work, we consider the problem of pore-scale numerical modeling of three-phase system: residual water, hydrocarbon gas and hydrocarbon liquid with phase transitions between the two latter phases. Such situations happen in case of gas-condensate or volatile oil deposits, in oil deposits with gas caps or in EOR methods with gas injection. The corresponding field development modeling by the conventional reservoir simulators rely on phase permeabilities and capillary pressures, which are provided by laboratory core analysis experiments. But the problem with gas-liquid hydrocarbon mixtures is that in laboratory procedures it may be difficult or even impossible to achieve full thermodynamic equilibrium between phases as it must be under the reservoir conditions of the initial reservoir state. However, reaching the said equilibrium is quite possible in numerical simulation. In this work, the gas-liquid mixture, after being injected into core sample, would slowly undergo the rearrangement of the phases and chemical components in pores converging to the minimum of the Helmholtz energy functional. This process is adequately described by DFH with consequent impact on phase permeabilities and capillary pressure. We give pore-scale numerical examples of the described phenomena in a micro-CT porous rock model for a realistic gas-condensate mixture with quantitative characterization of phase transition kinetic effects.

## 1 Introduction

Core analysis methods based on using digital rock models (DRMs) gradually become more and more accepted as a complimentary tool to traditional lab core analysis. Although not free from its own difficulties, digital rock approaches can help to obtain data and reduce uncertainties in cases where traditional lab-based core analysis becomes especially challenging to perform, e.g., poorly consolidated cores, low permeable cores, complex fluids like live oils and gas-condensates. Currently there are known many alternative methods used for numerical simulation of dynamic processes on DRMs and we refer the reader to reviews in [1-5]. The modeling method developed by the authors is called density functional hydrodynamics (DFH).

In previous publications it has been demonstrated that various multiphase compositional problems can be modeled in the frame of the density functional hydrodynamics (DFH) [7-17], including practical digital rock applications [18-26]. In the present work we discuss the pore-scale modeling of gas-condensate mixtures in digital rock models obtained by X-ray microCT. The following are specific features of this class of problems: a) the thermodynamic properties of hydrocarbon mixtures are related to the multicomponent nature of such mixtures, so it is important to retain in the modeling the influence of different components; b) gas and liquid phases undergo chemical component exchange, which can be slow in comparison to the transport processes observed at the laboratory scale. The latter problem can affect the applicability of laboratory results to the gas-condensate reservoir modeling. Namely in the reservoir far from the production and injection wells the transport processes in pores are sufficiently slow, so the gas-condensate mixture at any moment can be close to the minimum of the Helmholtz energy functional, while under laboratory protocols the state of the mixture can be in some metastable state far from the true equilibrium state. In order to demonstrate this problem, we perform the simulation of transport of gas-condensate mixtures in two different scenarios: a) in accordance with laboratory protocols; b) using equilibrium gas-condensate states as initial input for the computations of phase permeabilities and capillary pressures. The results highlight the possible difference between two approaches, which in our

* Corresponding author: nevseev@slb.com

© The Authors, published by EDP Sciences. This is an open access article distributed under the terms of the Creative Commons Attribution License 4.0 (http://creativecommons.org/licenses/by/4.0/).

opinion proves the unique value of digital rock analysis in this class of problems.

The summation over repeated indices is implied everywhere. The indices $a,b,c=1,2,3$ are related to Cartesian coordinates $x^a$, the indices $i,j,k=1,...,M$ are related to chemical components in the fluid mixture. We consider isothermal processes, so the temperature is assumed to be fixed, and dependence of certain variables on temperature is omitted. We use short symbols for partial derivatives: $\partial_a=\partial/\partial x^a$ and $\partial_t=\partial/\partial t$, where $t$ is time.

## 2 Theoretical concepts and equations

### 2.1 Reminder of density functional hydrodynamics

Here we provide the reminder of some of the basic concepts of DFH, which are necessary for the modeling of gas-condensate mixtures. A detailed exposition of the DFH can be found in [7-10].

We consider continuum mechanics description of a mixture of $M$ chemical components present inside a spatial region $D$ with overall volume $V_D$. The region contains $N_{iD}$ molecules of each type. To avoid large numbers, the quantities $N_{iD}$ are measured in moles. In case of homogeneous and static mixture we define chemical component molar densities by $n_i=N_{iD}/V_D$. If the mixture is inhomogeneous and possibly evolving, one can define $n_i$ as a dynamic variable at time $t$ and spatial point $x^a$ by establishing a small volume limit, such as $n_i=n_i(t,x^a)=\lim_{V_D\to 0}(N_{iD}/V_D)$. Like elsewhere in continuum mechanics, the small volume limit is understood as the convergent procedure with $V_D$ being small, but still large in comparison to the molecular volume. The molar densities $n_i$ can be used to calculate overall molar density $n=\sum_j^M n_j$ and molar concentrations $c_i=n_i/n$.

By counting the flow rate of molecules through a small area inside the mixture, one can define the component flux $I_{ia}=I_{ia}(t,x^b)$. The component fluxes are used to calculate the mass flux $I_a=m_iI_{ia}$, where $m_i$ is the molar mass of the $i^{th}$ component. By introducing mass density $\rho=m_in_i$ it is possible to define mass velocity $v_a=\rho^{-1}I_a$. Component flux $I_{ia}$ can be represented as a combination of transport term $n_iv_a$ and diffusion flux $Q_{ia}$: $I_{ia}=n_iv_a+Q_{ia}$.

By definition diffusion flux does not influence net mass transfer, i.e., $m_iQ_{ia}=0$. We assume the existence of the Helmholtz energy functional

$$F_D=F_D[n_i]=\int_D \omega dV+\int_{\partial D} f_* dA, \tag{1}$$

$$\omega=f(n_i)+2^{-1}\nu_{ij}(n_k)\partial_a n_i\partial_a n_j, \tag{2}$$

where $\partial D$ is the boundary surface of the region $D$ (when the region is finite), $f=f(n_i)$ is the bulk Helmholtz energy density of homogeneous mixture, $\nu_{ij}$ is the positive-definite symmetric matrix, and $f_*=f_*(n_i)$ is the surface Helmholtz energy density, which is not equal to zero if $\partial D$ is a contact surface with some immobile solid. It is convenient to recollect certain thermodynamic equations involving Helmholtz energy (here and below $\kappa_i$ is the chemical potential of the $i^{th}$ component, $p$ is the hydrostatic pressure):

$$f=\kappa_in_i-p \quad, df=\kappa_idn_i \ , dp=n_id\kappa_i \ . \tag{3}$$

The model in Eqs. (1) and (2) is adequate for description of many important phenomena involving multiphase multicomponent mixtures. Up to now, it was successfully used to simulate multiphase multicomponent phenomena with or without phase transitions, surfactants, and mixtures with solid phases such as gas hydrates or solid particles [3-11].

The multiphase compositional transport in case of isothermal flow of fluids with Newtonian rheology is governed by the following equations written in a form of conservation laws for chemical components and momentum

$$\partial_t n_i+\partial_a(n_iv_a+Q_{ia})=0, \tag{4}$$

$$\partial_t(\rho v_a)+\partial_b(\rho v_av_b-P_{ab})=0, \tag{5}$$

together with the relations

$$P_{ab}=\sigma_{ab}+\tau_{ab}, \tag{6}$$

$$\sigma_{ab}=(\omega-\Phi_in_i)\delta_{ab}-\nu_{ij}\partial_a n_i\partial_b n_j, \tag{7}$$

$$\tau_{ab}=\eta_v\delta_{ab}\partial_cv_c+\eta_s(\partial_av_b+\partial_bv_a-\frac{2}{3}\delta_{ab}\partial_cv_c), \tag{8}$$

$$Q_{ia}=-D_{ij}\partial_a\Phi_j, \tag{9}$$

$$\Phi_i=\kappa_i+2^{-1}\frac{\partial\nu_{jk}}{\partial n_i}\partial_an_j\partial_an_k-\partial_a(\nu_{ij}\partial_an_j), \tag{10}$$

and subject to the boundary conditions

$$v_a=0 \ , \tag{11}$$

$$l^a Q_{ia} = 0, \quad (12)$$

$$\frac{\partial f_*}{\partial n_i} - l^a \nu_{ij} \partial_a n_j = 0, \quad (13)$$

where $p_{ab}$ is the stress tensor, $\sigma_{ab}$ is the static stress tensor, $\tau_{ab}$ is the viscous stress tensor, $\Phi_i$ is the generalized chemical potential of the $i^{th}$ mixture component, $\eta_v$ and $\eta_s$ are nonnegative bulk and shear viscosity coefficients, respectively, $D_{ij}$ is the nonnegative definite symmetric matrix subject to $D_{ij} m_j = 0$, $l^a$ is the internal normal unit vector at the boundary surface $\partial D$, and $\delta_{ab}$ is the Kronecker symbol.

In static equilibrium state (whether it is one-phase or multiphase) the generalized chemical potentials are constant all over the region $D$

$$\Phi_i = \lambda_i = const, \quad (14)$$

Since there exists the identity $\partial_b \sigma_{ab} = -n_i \partial_a \Phi_i$, the thermodynamic equilibrium conditions (14) provide the mechanical equilibrium conditions $\partial_b \sigma_{ab} = 0$. In case of two-phase one-dimensional static equilibrium distribution of components $n_i = n_i(x^1)$ (phases A and B, like gas and condensate or condensate and water), when $\lim_{x^1 \to -\infty} n_i = n_{iA}$ and $\lim_{x^1 \to +\infty} n_i = n_{iB}$, one can use Eq. (7) to derive the expression for interfacial tension (IFT), which helps to tune the matrix $\nu_{ij}$ to experimental IFT data

$$\gamma_{AB} = \int_{-\infty}^{+\infty} (\sigma_{22} - \sigma_{11}) dx^1 = \int_{-\infty}^{+\infty} \nu_{ij} \partial_1 n_i \partial_1 n_j dx^1 . \quad (15)$$

If both phases A and B are in contact with solid surface and $\left| f_*(n_{iA}) - f_*(n_{iB}) \right| < \gamma_{AB}$ then the wetting angle $\theta$ is determined by Young equation $\cos\theta = (f_*(n_{iA}) - f_*(n_{iB}))/\gamma_{AB}$. If $\left| f_*(n_{iA}) - f_*(n_{iB}) \right| \geq \gamma_{AB}$, the phase with lower surface Helmholtz energy density will form a layer in contact with solid surface.

## 2.2 Modeling of gas-condensate mixtures by density functional hydrodynamics

Our aim is to apply DFH for the numerical modeling of gas-condensate mixtures inside of the pore system of natural rock sample using 3D microCT X-ray models of the rock microstructure. Previously we have demonstrated how the DFH method can be used to model condensate banking [15] though in less rigorous mixture composition formulation than the one proposed in this work.

Here we use a Cartesian grid with cubic cells in one-to-one correspondence with 3D voxels of segmented microCT images. The numerical realization of DFH equations is described in previous publications [10, 13]. Here we discuss specific numerical features associated with gas-condensate modeling problem.

First, it is necessary to specify the bulk Helmholtz energy density $f = f(n_i)$, which represents bulk mixture thermodynamics including the phase behavior. Currently it is widely accepted, that the thermodynamic behavior of hydrocarbon mixtures is adequately described by Peng-Robinson equation of state (PR EOS) [27], so it is natural to rely on the corresponding analytic expression of Helmholtz energy. The main computational problem with gas-condensate mixtures is that the large number $M$ of chemical components leads to prohibitively large computational time. The way to remedy this problem is to perform computations in terms of smaller number of pseudocomponents, which are artificially constructed from the initial set of chemical components. In our case the transition from the true components to pseudocomponents (lumping) must be compatible with the DFH equations (4)-(13). Now we shall describe the mathematical procedure, which helps to reduce the number of effective components.

Let us consider linear transformation of molar densities with some invertible matrix $\Lambda^i_\alpha$ (here and below index $\alpha = 1,...,M$, and the summation is assumed over the repeated indices) as follows

$$n'_\alpha = \Lambda^i_\alpha n_i . \quad (16)$$

We define an auxiliary vector $g_\alpha$ with all components equal to 1. Since the matrix $\Lambda^i_\alpha$ is invertible there exists the inverse matrix $\Omega^\alpha_i$ such that $n_i = \Omega^\alpha_i n'_\alpha$. We assume that the matrix $\Lambda^i_\alpha$ satisfies the following additional condition

$$g_\alpha \Lambda^i_\alpha = g_i . \quad (17)$$

This condition means that the new overall molar density $n' = g_\alpha n'_\alpha$ coincides with the old one $n$. Therefore, there exists the linear relation between the new and the old concentrations, as follows

$$c'_\alpha = n'_\alpha / n' = \Lambda^i_\alpha c_i , \quad (18)$$

with the normalizing condition for concentrations $\sum_{\alpha=1}^{M} c'_\alpha = 1$. The transformation (16) can be applied to the DFH equations (4)-(13). The mass velocity $v_a$ and mass density $\rho = m_i n_i = m'_\alpha n'_\alpha$ are not transformed given we transform molar masses $m'_\alpha = \Omega^\alpha_i m_i$. The representation of the DFH problem (4)-(13) in new variables is evidently equivalent to the old representation, and any solution in terms of $n'_\alpha, \nu_\alpha$ can be used to compute solution in terms of $n_\alpha, \nu_\alpha$. So far, the transformation (16) is purely formal and has no physical sense, but it can help to reduce the number of effective components. Indeed, let us consider the phase transitions in the narrow pressure range, when the mixture with overall composition $c_{i0}$ is decomposed into gas and liquid (condensate) with respective compositions $c_{ig}$, $c_{il}$. In terms of new concentrations the mixture $c'_{\alpha 0}$ is decomposed into phases with compositions $c'_{\alpha g}$, $c'_{\alpha l}$. Imagine, that three specific concentrations $c'_{M0}$, $c'_{Mg}$, $c'_{Ml}$ coincide or are very close to each other

$$c'_{M0} \approx c'_{Mg} \approx c'_{Ml}. \tag{19}$$

In this case the hydrodynamic processes for the considered mixture, i.e., phase transitions, convective transport, and mixing, negligibly change the quantities (19). Then one can assume that there is a fixed quantity $c'_M$, and the set of unknown variables is reduced to $n'_\alpha, \nu_\alpha (\alpha=1,...,M-1)$ with corresponding reduction in system of Eqs. (4). If there are similar situations with other concentrations $c'_\alpha$, the number of effective component densities $n'_\alpha$ can be reduced even further. The problem is to find the matrix $\Lambda^i_\alpha$, which provides as many approximations like (19) as possible. Here is the mathematical procedure for the construction of this matrix.

In digital rock transport problems, the variations of pressure are small because of the small size of microCT rock models, so the matrix $\Lambda^i_\alpha$ is constructed for certain fixed pressure. At the same time variations of local composition can be significant, so we consider variations in the range

$$\left| c_i - c_{i0} \right| \le h c_{i0} \tag{20}$$

with fixed positive parameter $h$. Note that the concentration vector $c_i$ belongs to the hyperplane $\sum_{i=1}^{M} c_i = 1$ in *M*-dimensional space with restrictions $c_i \ge 0$. We shall use the notation $d\mu(c_i)$ for the respective Lebesgue measure at this hyperplane and $W_h$ for the corresponding volume of subset $C_h$ of concentrations (20). Using PR EOS for any vector $c_i$ we can calculate corresponding gas and condensate concentrations $c_{jg}(c_i)$, $c_{jl}(c_i)$. If by chance we have one-phase system, we put both compositions equal to $c_i$. Let us define by induction the sequence of vectors $e^i_\alpha (i=1,...,M-1)$ in accordance to the following conditions for the subsequent vector $e^i_m$: a) $g_i e^i_m = 0$; b) $e^i_m e^i_n = 0 \; (n=1,..,m-1)$; c) $e^i_m e^i_m = 1$; d) $e^i_m$ provides the minimum of the following error function

$$J_{M-m} = \left( W_h^{-1} \int_{C_h} d\mu(c_i) \sum_{n=1}^{m} \left( e^j_n (c_{jg}(c_i) - c_{jl}(c_i)) \right)^2 \right)^{1/2} \tag{21}$$

Minimization problem for the error function (21) is solved numerically. Using this solution, which is dependent on parameter $h$, it is possible to define the matrix $\Lambda^i_\alpha$ as follows

$$\Lambda^i_\alpha = e^i_{M-\alpha+1} \; (\alpha=2,...,M), \;\; \Lambda^i_1 = g_i - \sum_{m=1}^{M-1} e^i_m. \tag{22}$$

Now if we assume $c'_\alpha = c'_{\alpha 0} = const (\alpha = m+1,...,M)$, the DFH problem is reduced to the variables $n'_\alpha, \nu_\alpha (\alpha=1,...,m)$, while the parameters $n'_\alpha$ can be interpreted as new effective pseudocomponent densities. The cumulative error of this approximation is characterized by the function $J_m$. The analysis of this function can help to determine the best minimum number $m$, when the error is still sufficiently small.

Thus, we described a new lumping technique that is consistent with the DFH. The technique is purely mathematical and can be classified as a space reduction approach.

In this work, we chose the definition of $\Lambda^i_\alpha$ with $h=1$. After defining the new pseudocomponents, it is necessary to calculate the PR EOS Helmholtz energy function in terms of these new variables. To achieve further reduction in computational time, we approximated this function with Padé-like rational approximation, which is used in numerical solution of the DFH equations. Besides being a thermodynamic description of bulk fluid, this final version of the Helmholtz energy function is used to fit the matrix $\nu_{ij}$ to experimental IFT data.

It is necessary to express the surface Helmholtz energy density $f_*$ and transport coefficients $D_{ij}$ through the new variables. We use a linear approximation for $f_*$, which is compatible with Young's equation as it has been discussed in Section 2.1 (see also [16]). It is necessary to note that in digital rock problems the surface Helmholtz energy function $f_*$ must also compensate the difference between actual specific surface and smaller specific surface in microCT rock models. The transport matrix $D_{ij}$ is initially specified using known data on diffusion [28] and then it is transformed in accordance with Eq. (16) using the relation $D'_{\alpha\beta} = \Lambda^i_\alpha \Lambda^j_\beta D_{ij}$.

## 3 Numerical results

In interpretation of the numerical results for final static states it is necessary to remember that in accordance with equilibrium conditions (14) the chemical potentials for gas and liquid are equal and constant everywhere except for interfacial regions and regions in vicinity of the rock surface:

$$\kappa_{ig} = \kappa_{il} = const \tag{23}$$

Among other things this means that the composition of gas phase is the same everywhere where the conditions (23) take place, and the same statement is true for liquid phase. But variations of composition in grid cells, which cover the interfacial regions or are adjacent to the rock surface are permissible.

In this work, we studied a rich gas-condensate system corresponding to a gas-condensate field in southern Russia. The reservoir pressure is 50.6 MPa and temperature is 82 C. The dew point is at 48.7 MPa. A somewhat simplified composition of the reservoir gas, i.e., above dew point, is listed in Table 1.

**Table 1.** Initial reservoir gas composition.

| Mixture component | Molar fraction |
|---|---|
| CO2 | 0.0061 |
| N2 | 0.0015 |
| C1 | 0.702 |
| C2 | 0.083 |
| C3 | 0.033 |
| iC4 | 0.0085 |
| nC4 | 0.0079 |
| iC5 | 0.019 |
| nC5 | 0.011 |
| C6 | 0.015 |
| C7+ | 0.113 |

We begin by applying the space reduction technique described in Sec. 2.2 to the 11-component reservoir mixture listed in Table. 1. The calculated numerical values of the criterion $J_m$ of Eq. (21) are presented at Fig. 1. One can see that if maximum relative error at $h = 1$ (that is at 100% deviation from the original mixture composition) is set to be 5%, then the reasonable choice is $m=4$, because for $m<4$ the error grows rather drastically and goes well above 5%.

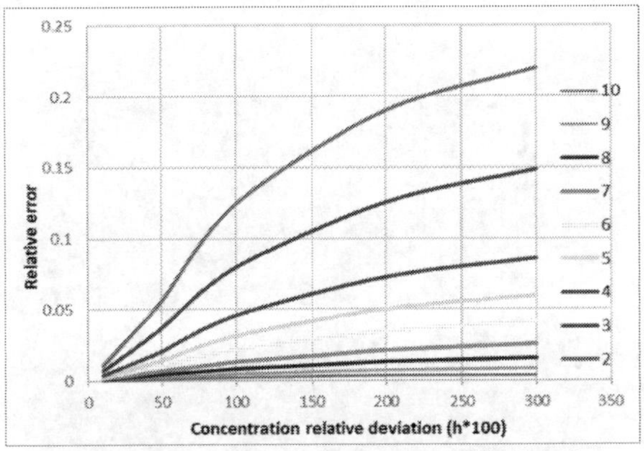


**Fig. 1.** Relative approximation error measured by the $J_m$ criterion versus maximum deviation from the original mixture composition measured in $h$. The different lines correspond to specified number of pseudocomponents.

Thus, we settle for a 4-pseudocomponent space that will be used in numerical simulation of gas-condensate flow on a digital rock model.

A core sample used in this work is a low-permeable, but relatively well sorted carbonate rock. The sample was X-ray scanned at 0.82 um/voxel spatial resolution. Image segmentation was based on an extended version of the Indicator Kriging approach [29]. An example of a 2D cross-section of the sample as well as a 3D view of the DRM used for numerical simulation are shown in Fig. 2. The porosity of the DRM is 0.175 and permeability is 1.8 mD. The DRM resolved porosity is somewhat lower than the 8 mm diameter core plug measured porosity, which is 0.225. The core plug measured permeability is 2.5 mD. This is a rather typical situation, especially in low-permeable rocks, when a considerable portion of porosity appears below the X-ray microCT resolution. For such cases, we have developed a multiscale workflow based on resolving a core plug at a range of different resolutions and using high-resolution models for determining effective properties that are then used in a combined multiscale low-resolution model [17]. That workflow is fully compatible with the compositional gas-condensate modeling in the frame of the space reduction approach, but we do not use it in the present work to avoid unnecessary complication and blurring the focus of this paper.

As the first step in numerical simulation on the DRM, we have modeled initial saturation of the reservoir gas and connate water (Fig. 3) by solving a full thermodynamic equilibrium problem (global minimum of Helmholtz free energy) following the methodology described in [25]. Spatial distribution of wettability was established by surface Helmholtz energy density field ($f_*$) with correlated stochastic noise representing expected natural variability in rock surface properties. The connate water saturation is $S_w$=0.18. According to the mixed-wet nature of the core connate water is primarily accumulated within small pores.

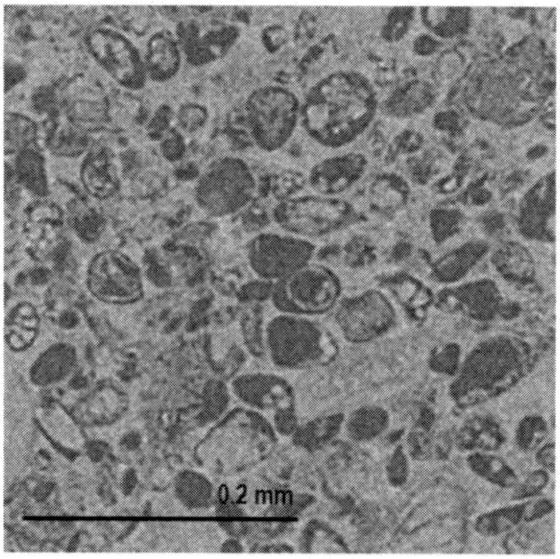

(a)

(b)

**Fig. 2.** (a) Micro-CT grayscale cross-section of the carbonate core sample at 0.82 υm resolution, (b) 3D view of the segmented pore space of a cubic DRM with dimensions $500^3$ at 0.82 υm resolution.

**Fig. 3.** 3D view of the distribution of the reservoir gas shown in green and connate water shown in semitransparent blue within the pore space of the DRM.

As the next step of numerical simulation, we have evaluated gas-condensate relative permeabilities at connate water. In order to obtain different condensate saturations, the simulations were conducted at different pressures below dew point. Also, two cases distinct in the simulation workflow have been modeled.

In **Case 1**, gas-condensate mixture corresponding to selected pressure below dew point was flowing through the DRM until stabilization of phases saturations within the model and then gas-condensate phase permeabilities were evaluated for each saturation point similarly to how we usually do this in case of immiscible phases [23]. In **Case 2**, after stabilization of phases saturation, we have waited until thermodynamic equilibrium takes place and pseudocomponents redistribute to minimize global Helmholtz free energy. In each case, the injection was arranged at a constant rate corresponding to a low capillary number flow. The mixture was injected at the bottom side of the model and arriving fluids were removed at the top side to sustain constant overall pressure. The simulations were carried out using the following parameters (inferred for the considered deposit at 42 MPa): $\eta_{sc} = 0.92\,\mathrm{mPa{\times}s}$, $\eta_{sg} = 0.037\,\mathrm{mPa{\times}s}$, $\gamma_{cg} = 0.0012\,\mathrm{N/m}$, $\gamma_{cw} = 0.02\,\mathrm{N/m}$, $\gamma_{gw} = 0.05\,\mathrm{N/m}$, where $\eta_{sc}, \eta_{sg}$ are the condensate and gas viscosities, and $\gamma_{cg}, \gamma_{cw}, \gamma_{gw}$ are the condensate-gas, condensate-water, and gas-water interfacial tensions, respectively.

Some examples of phases saturation and pseudocomponent distributions obtained during simulations are presented in Figs. 4 through 7. Fig. 4 compares gas, condensate and connate water saturations in Cases 1 and 2 obtained after injection of gas-condensate mixture at 42 MPa pressure. Although the distributions are generally similar, there are visible differences in details. Fig. 6 compares distributions of pseudocomponents that correspond to the saturations from Fig. 4. Similarly, Fig. 5 and 7 show distribution of phases and pseudocomponents, respectively, for lower value of pressure equal to 33 MPa.

As can be seen from the pictures, the heavy pseudocomponents (shown in red and yellow colors in Figs. 6 and 7) tend to group closer to connate water and occupy smaller pores, while bigger pores are mostly occupied by the lighter pseudocomponents (shown in green and aqua colors in Figs. 6 and 7). This is the consequence of the fact that condensate, which is dominated by heavy pseudocomponents, has lower interfacial tension with water than gas that is mostly formed by light pseudocomponents. The said details are seen clearer from the 2D cross-sections in Figs. 6 and 7.

 https://doi.org/10.1051/e3sconf/202336601005

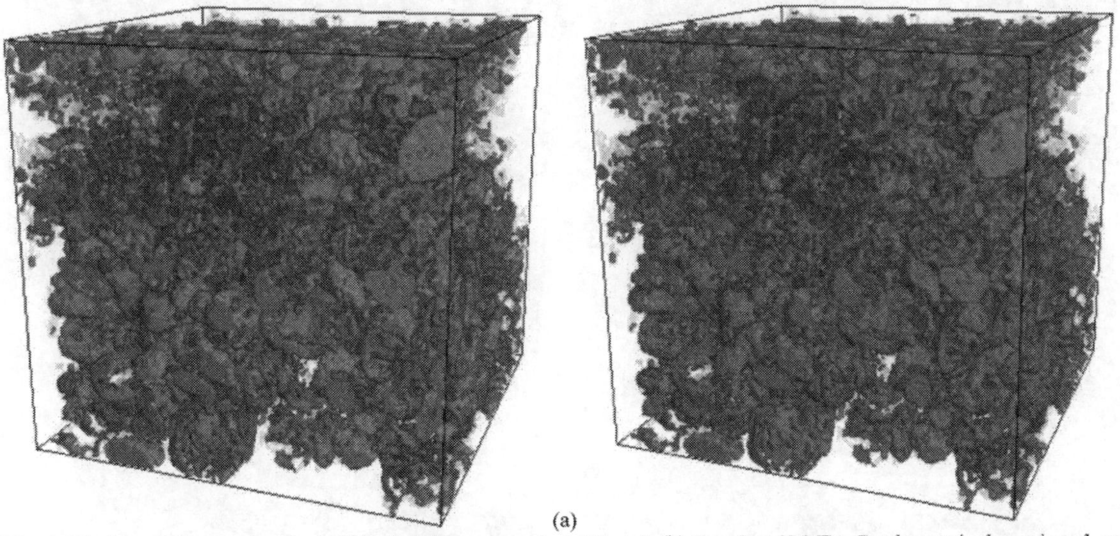

(a) (b)

**Fig. 4.** 3D view of phases saturation within DRM pore space in (a) Case 1, (b) Case 2 at 42 MPa. Condensate is shown in red, gas is shown in green, and connate water is shown in semitransparent blue.

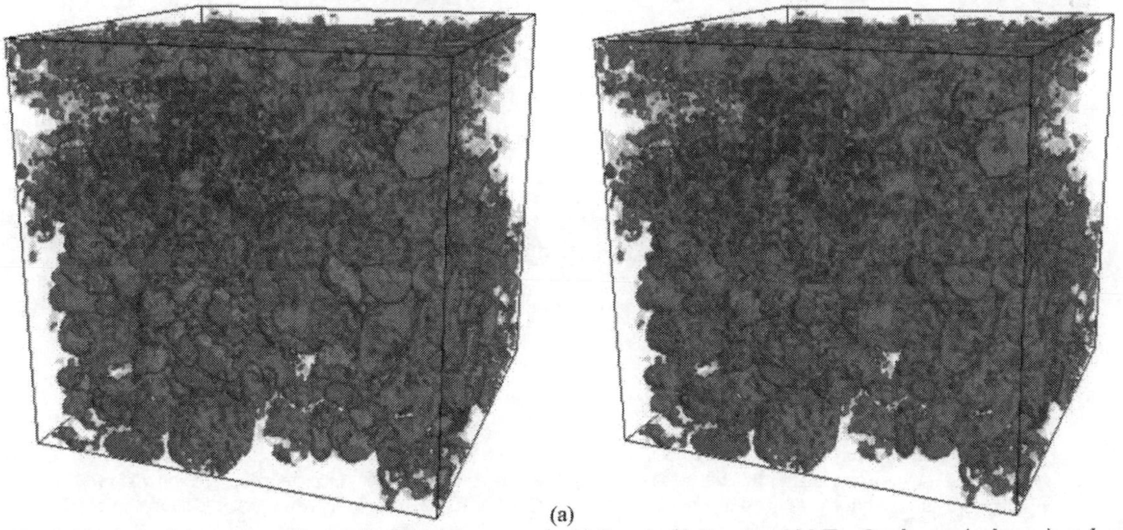

(a) (b)

**Fig. 5.** 3D view of phases saturation within DRM pore space in (a) Case 1, (b) Case 2 at 33 MPa. Condensate is shown in red, gas is shown in green, and connate water is shown in semitransparent blue.

https://doi.org/10.1051/e3sconf/202336601005

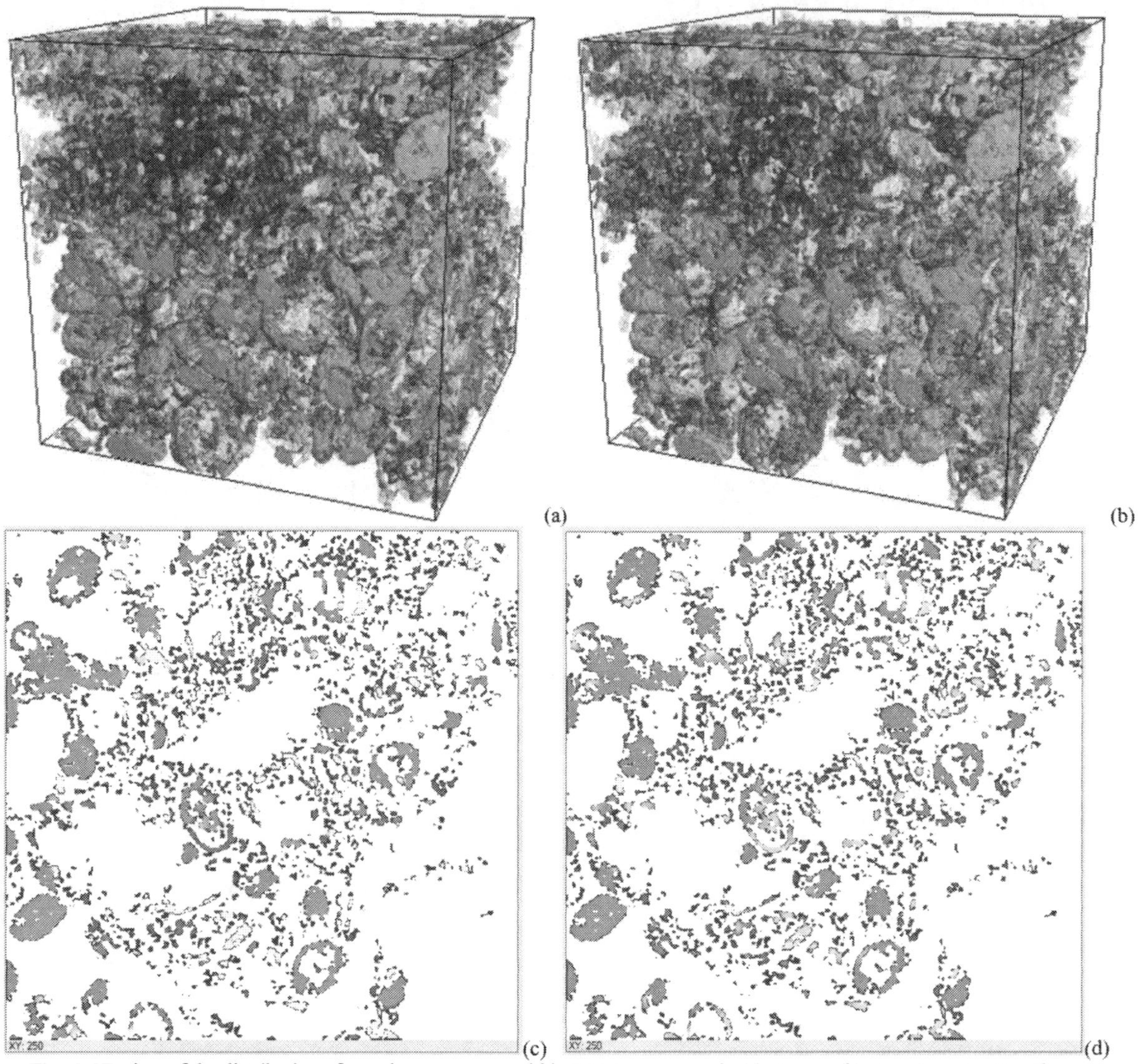


**Fig. 6.** 3D view of the distribution of pseudocomponents within DRM pore space in (a) Case 1, (b) Case 2 at 42 MPa. The same distributions within 2D cross-sections in (c) Case 1, (d) Case (2). The 4 hydrocarbon pseudocomponents are shown in red (1st ), yellow (2nd ), green (3d ), aqua (4th ), and intermediate colors indicate mixture of these; connate water is shown in blue.

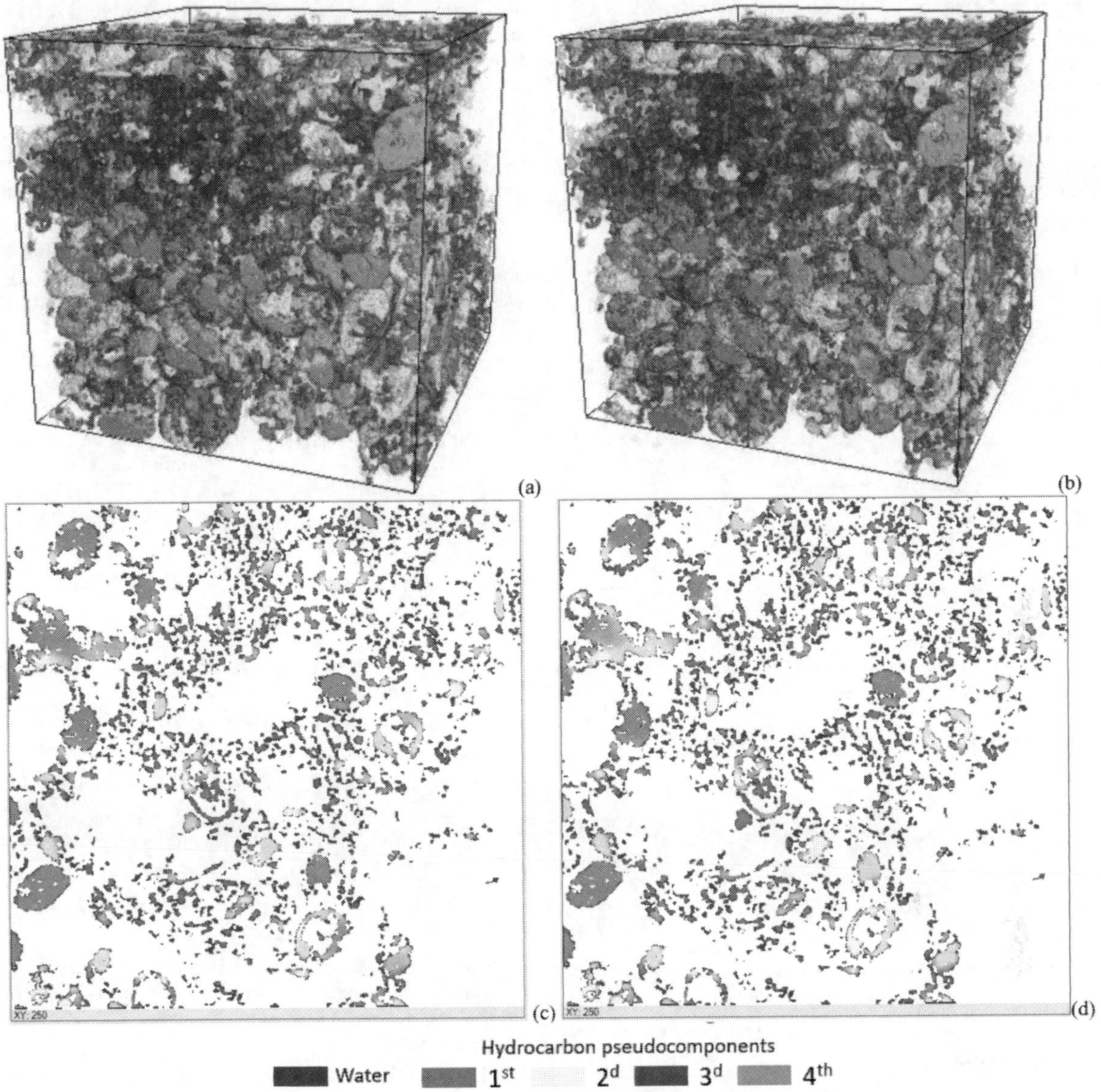


**Fig. 7.** 3D view of the distribution of pseudocomponents within DRM pore space in (a) Case 1, (b) Case 2 at 33 MPa. The same distributions within 2D cross-sections in (c) Case 1, (d) Case (2). The 4 hydrocarbon pseudocomponents are shown in red ($1^{st}$ ), yellow ($2^{nd}$ ), green ($3^{d}$), aqua ($4^{th}$ ), and intermediate colors indicate mixture of these; connate water is shown in blue.

During the flow gas-condensate mixture is not necessarily in equilibrium state everywhere (here it is worth stressing that the DFH equations are compositional and allow for non-equilibrium flows; thermodynamic equilibrium is produced as a result of global minimization of the Helmholtz free energy and the minimization process is coupled with the transport phenomena). Therefore, it is necessary to establish definition of phases for arbitrary mixture composition [10, 12]. This is done by the bulk Helmholtz energy analysis, which has minima corresponding to the equilibrium phases. If mixture composition appears within convex part of the Helmholtz energy around a phase then this composition corresponds to this phase. If, however, the composition appears outside any convex regions then such composition is unstable and does not belong to any of the phases.

A comparison between gas-condensate relative permeabilities is presented in Fig. 8. The gas and condensate saturations have been normalized relative to the total volume occupied by the hydrocarbon phases (connate water was excluded). It is noticeable that in Case 2 the two-phase mobility region is somewhat narrower than in Case 1. For example, condensate mobility threshold, $S_{c0}$, is equal to 0.28 in Case 2, while in Case 1 it is 0.23. This is explained by the fact that after fluid injection to saturation stabilization, the non-wetting phase tends to form a cluster connecting the injection side of the model with the opposite side. But after further stabilization towards thermodynamic equilibrium that cluster can deform, become less connecting, or can even break entirely. It is only because of low interfacial tension between gas and condensate that the difference in relative permeabilities between Cases 1 and 2 appeared to be rather modest; it might have been far greater have the phases had bigger IFT.

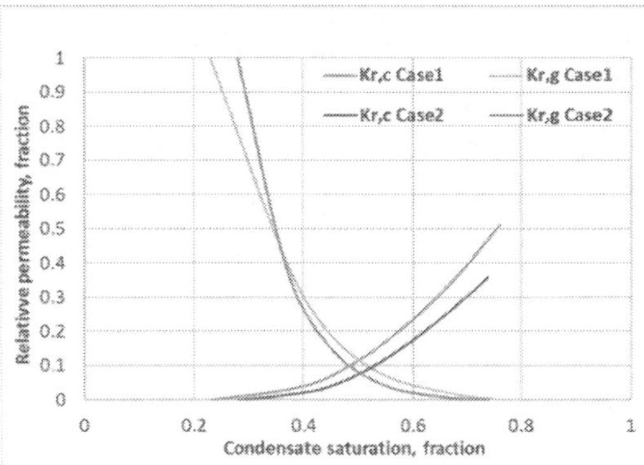


(a)

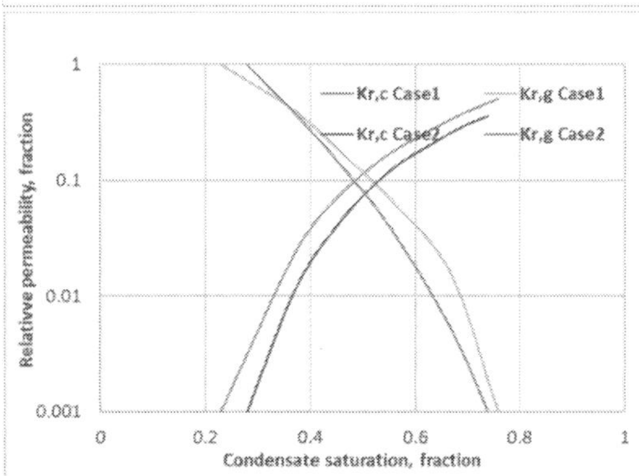


(b)

**Fig. 8.** Comparison between steady-state gas-condensate relative permeabilities calculated in Cases 1 and 2; (a) linear scale, (b) logarithmic scale.

## 4 Conclusion

In this work, we developed a compositional mixture lumping technique that is fully compatible with the DFH. This new lumping technique is a purely mathematical procedure that can be classified as a space reduction approach. As a result of its application, a pseudocomponent molar density space is built that has fewer dimensions than the space of the original molar densities. However, the inverse transformation exists that allow to reconstruct the original components from the new pseudocomponents. The error of such transformation can be estimated beforehand and can be used to select the optimal number of pseudocomponents.

Starting from an 11-component gas-condensate system we have arrived to a 4-pseudocomponent reduced system yielding maximum error of 5% within a subspace specified by a 100% deviation from the original mixture composition. Using this reduced system, we have conducted digital rock simulations by the DFH and obtained gas-condensate relative permeabilities. The simulations have been carried out in two different statements – a classical one similar to a standard laboratory protocol, and a new one attempting to mimic thermodynamic equilibrium conditions that can form over prolonged periods of time. It was observed that the second statement provides somewhat poorer overall mobility of phases as compared to the classical formulation disregarding thermodynamic equilibrium.

We thank Schlumberger for permission to publish this work. We also recognize helpful comments from the reviewers that enabled us to improve exposition of the material.

## References

1. Blunt, M.J., *Multiphase Flow in Permeable Media: A PoreScale Perspective*, (Cambridge University Press, Cambridge, 2017). ISBN 978-1-107-09346-1
2. C.F. Berg, O. Lopez, H. Berland, J. Petrol. Sci. Eng. **157**, 131 (2017)
3. T. Bultreys, W. De Boever, V. Cnudde, Earth Sci. Rev., **155**, 93 (2016)
4. M.J. Blunt, B. Bijeljic, H. Dong, O. Gharbi, S. Iglauer, P. Mostaghimi, A. Paluszny, C. Pentland, Adv. Water Resour. **51**, 197 (2013)
5. V. Joekar-Niasar, M.I.J. van Dijke, S.M. Hassanizadeh, Transp. Porous Med. **94**, 461 (2012)
6. P. Meakin, A.M. Tartakovsky, Rev. Geophys., **47**, RG3002 (2009)
7. O. Dinariev, J. Appl. Math. Mech. **59**, 745 (1995)
8. O. Dinariev, J. Appl. Math. Mech. **62**, 397 (1998)
9. A. Demianov, O. Dinariev, N. Evseev, Can. J. Chem. Eng., **89**, 206 (2011)
10. A.Yu. Demianov, O.Yu. Dinariev, N.V. Evseev, *Introduction to the density functional method in hydrodynamics* (Fizmatlit, Moscow, 2014). ISBN 978-5-9221-1539-1
11. D. Koroteev, O. Dinariev, N. Evseev, D. Klemin, A. Nadeev, S. Safonov, O. Gurpinar, S. Berg, C. van Kruijsdijk, R. Armstrong, M. T. Myers, L. Hathon, H. de Jong, Petrophysics **55**, 294 (2014).
12. R.T. Armstrong, N. Evseev, D. Koroteev, S. Berg, Adv. Water Resour., **77**, 57 (2015)
13. R.T. Armstrong, S. Berg, O. Dinariev, N. Evseev, D. Klemin, D. Koroteev, S. Safonov, Transp. Porous Med., **112**, 577 (2016)
14. N. Evseev, O. Dinariev, M. Hürlimann, S. Safonov, Petrophysics, **56**, 32 (2015)
15. O. Dinariev, N. Evseev, Computat. Geosci., **20**, 835 (2016)
16. O. Dinariev, N. Evseev, Phys. Rev., **E95**, 063307 (2017)
17. O. Dinariev O., N. Evseev N. and D. Klemin D. E3S Web of Conferences, 146, 01001 (2020)

18. A. Shandrygin, V. Shelepov, R. Ramazanov, N. Andrianov, D. Klemin, A. Nadeev, Yakimchuk, SPE-176629-MS (2015)
19. A. Shandrygin, V. Shelepov, R. Ramazanov, N. Andrianov, D. Klemin, A. Nadeev, S. Safonov, I. Yakimchuk, SPE-182037-MS (2016)
20. S.I. Geetan, R.M. MacDonald, D. Klemin, 2461115-MS URTeC Conference Paper (2016)
21. S.I. Geetan, R.M. MacDonald, D. Klemin, 2671283-MS URTeC Conference Paper (2017)
22. R.M. MacDonald, S.I. Geetan, D. Klemin, B. Hill, 2902911-MS URTeC Conference Paper (2018)
23. I. Yakimchuk, N. Evseev, D. Korobkov, I. Varfolomeev, O. Dinariev, V. Khan, D. Koroteev, D. Orlov, E. Muravleva, B. Belozerov, V. Krutko, SPE-196928-MS (2019)
24. D. Klemin, A. Serebryanskaya, O. Savelev, S. Melnikov, SPE-198610-MS (2019)
25. I. Yakimchuk, N. Evseev, D. Korobkov, O. Dinariev, O. Ridzel, V. Khan, V. Semkov, A. Zhonin, D. Kravets, SPE-202015 (2020)
26. I. Yakimchuk, N. Evseev, D. Korobkov, O. Ridzel, V. Pletneva, M. Yaryshev, I. Ilyasov, N. Glushchenko, A. Orlov, SPE-202013 (2020)
27. D.Y. Peng, D.B. Robinson, Ind. Eng. Chem. Fundam. **15** 59–64 (1976)
28. R.C. Reid, J.M. Prausnitz, B.E. Poling, *The Properties of Gases and Liquids* (Mc-Graw Hill, New York, 1987).
29. I. Varfolomeev, I. Yakimchuk, I. Safonov, Computers, **8**(4), 72 (2019).

# In-situ characterization of capillary pressure during three-phase flow in porous media

*Ahmed* Zankoor[1, *], *Mahdi* Khishvand[1], *Bradley* McCaskill[1], and *Mohammad* Piri[1]

[1] Center of Innovation for Flow through Porous Media, Department of Petroleum Engineering, University of Wyoming, Laramie, WY, USA.

**Abstract.** We characterize the pore-scale fluid distributions, phase connectivity, and local capillary pressures during three-phase flow in a water-wet Berea sandstone sample. In this investigation, we use a set of x-ray micro-tomography images acquired during three-phase core-flooding experiments conducted on a miniature core sample. We use several image analysis techniques to analyze the pore-scale fluid occupancy maps and use this information to develop several insights related to pore occupancy, oil and gas cluster distribution, and interfacial curvature during the gas injection process. The results of our investigation show that the large-, intermediate-, and small-sized pores are mostly occupied with gas, oil, and brine, respectively, which is consistent with the wetting order of the fluids (i.e., gas, oil, and brine are the nonwetting, intermediate wetting, and wetting phases, respectively). In addition, the connectivity analysis reveals that a significant amount of the gas phase was in the form of disconnected ganglia separated from the connected invading cluster. The presence of these trapped nonwetting phase clusters during the drainage process is presumably attributed to Roof snap-off and Haines jump events, as well as the anti-ripening phenomenon. Moreover, the average local oil-water capillary pressures are found to be greater than the gas-oil counterparts. This observation is then related to the relative location of the interfaces in the pore space and the threshold capillary pressures at which the various displacement events take place.

## 1. Introduction

Simultaneous flow of gas, oil, and aqueous phases in natural porous media are encountered in various areas of science and engineering, such as enhanced oil recovery processes and $CO_2$ sequestration in depleted oil reservoirs. The flow behavior in these systems is more complicated than two-phase flow conditions [1, 2] and is controlled by numerous factors including, but not limited to, wettability, interfacial forces between each fluid pairing, and the saturation history.

Previously, many studies have been dedicated to investigating the macro- and micro-scale behavior of three-phase flow systems. While the macro-scale studies in this field have mostly focused on three-phase relative permeability and oil recovery as functions of the key parameters (e.g., spreading coefficient, wettability, and saturation history) [3, 4], the initial micro-scale research attempts, mostly conducted using micromodels [5, 6, 7, 8], have been directed toward presenting detailed descriptions of the pore-scale displacement mechanisms and fluid distribution in the pore space. The information generated during these experimental works has then been used in developing pore-network models to predict fluid occupancy, relative permeability, and capillary pressure in natural rock samples and under various three-phase flow scenarios [9, 10].

The insights provided by the micro-scale studies have been frequently used to enrich pore-scale physics and improve the predictive capabilities of pore-scale models. In recent years, the advent of x-ray microtomography imaging and integration of these systems with core-flooding apparatuses has opened a new window of opportunity to directly observe the multi-phase flow phenomena in natural rock samples and further advance the understanding of such flow behavior on a pore-by-pore basis. Additionally, the advancement of image-based analysis has allowed for better characterization of such phenomena on a quantitative basis [11, 12, 13]. Several researchers have used these tools to conduct pore-scale studies for three-phase systems. For instance, Iglauer et al. [14, 15] have examined the impact of secondary and tertiary gas injection processes on the residual oil and gas trapping in water- and mixed-wet sandstones. The micro-CT imaging technique has also been used to investigate the interrelationship of pore-scale fluid configuration, in-situ wettability, and displacement mechanisms in water-wet sandstone [16], oil-wet carbonates [17], and fractured porous media [18]. The research studies in this field have recently been extended to other subjects, including the gas-oil miscibility [19], the dynamics of three-phase flow [20], and $CO_2$ storage in oil fields [21].

The efforts mentioned above have provided valuable insights into the micro-scale behavior of three-phase systems. Nonetheless, there have always existed some uncertainties in describing the pore-scale displacement mechanisms for three-phase flow due to the limited tools available for image analysis and the complexities associated while performing the flow experiments in natural core samples. In this study, we use improved methodologies to more accurately quantify the multi-phase flow parameters and shed more light on the pore-scale subtleties of a secondary gas injection experiment conducted in a miniature core sample. Specifically, we investigate the gas trapping behavior during a steady-state gas injection (drainage) process and highlight the possible pore-scale displacement mechanisms behind this. Moreover, we characterize the interfacial curvature and the respective in-

* Corresponding author: azankoor@uwyo.edu

© The Authors, published by EDP Sciences. This is an open access article distributed under the terms of the Creative Commons Attribution License 4.0 (http://creativecommons.org/licenses/by/4.0/).

 https://doi.org/10.1051/e3sconf/202336601006

situ capillary pressure to examine the pore-scale displacement mechanisms and the conditions for the formation of oil layers. The results of this study are of high importance for future pore-scale experimental and modeling research.

## 2. Experiments and data acquisition

In this study, we analyzed the micro-CT images obtained during a three-phase flow experiment in a water-wet Berea sandstone sample [16]. The diameter, length, porosity, and permeability of the sample were 81 mm, 4.47 mm, 0.186, and 584.46 mD, respectively. A synthetic brine solution, decalin, and nitrogen gas were used as, respectively, the wetting, intermediate-wetting, and nonwetting phases in this rock-fluid system. The three phases were brought into equilibrium at the experimental conditions (i.e., 5.52 MPa and 20° C) before injection into the rock sample to minimize mass exchange between the phases during the flow processes. The equilibrium spreading coefficient of the oil phase was measured to be -5.04 mN/m, which implies that the oil phase had non-spreading behavior. The experiment was conducted using a core-flooding apparatus integrated with a high-resolution micro-CT scanner. As shown in Figure (1), the core sample was first fully saturated with brine and then subjected to a drainage process using the oil phase to establish a target initial water saturation. The experiment was then followed by a secondary gas injection process through simultaneous injection of all three fluid phases, and the system was allowed to reach stable conditions. At this point, the region of interest was imaged at a resolution of 1.5 μm to generate grayscale images of the fluid occupancy in the pore space. More details about the fluids and experimental apparatus and procedure can be found elsewhere [16].

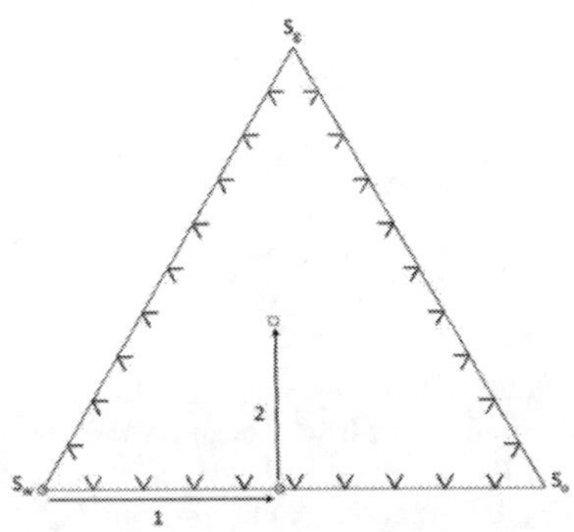


**Fig.1.** Three-phase saturation path of the core-flooding experiment.

## 3. Image analysis

The raw grayscale images acquired during the core-flooding experiment were processed using the Avizo software [16]. The images were first filtered using the non-local means filter to improve the signal-to-noise ratio, combined with the corresponding pore map, and subsequently segmented using the interactive thresholding method to generate the pore-scale fluid occupancy maps, i.e., images with unique labels assigned to the gas, oil, aqueous, and solid phases. We then analyzed these images using specific techniques, as mentioned in later sections, to obtain pore-by-pore fluid occupancy data, investigate the volume, connectivity, and morphology of the oil and gas clusters, and characterize the in-situ capillary pressures.

### 3.1. Pore-by-pore fluid occupancy

Pore-scale fluid occupancy maps can be used to calculate the saturation of each phase in various ranges of pore sizes. Previously, Scanziani et al. [22] proposed a technique to superimpose spheres with specific sizes – representing the pore elements [23]– on the wet images. The dominant phase in each pore is then assigned to the corresponding sphere, and the results are aggregated to calculate the saturation of each phase in various pore-size ranges. In this study, however, we employed a different technique that uses images of the separated pores instead of the spheres extracted using the maximal ball technique. The separated pores were generated using a specific module in the Avizo software that computed the watershed lines of the pore map to identify the local dilations and constrictions in the pore space. The image of the separated pores and the pore-scale fluid occupancy map were then used as input to an in-house MATLAB code, which superimposed the two images and quantified the pore-by-pore fluid occupancy (i.e., volume/saturation of each phase in the pores). The input images were partitioned into sub-volumes, and the computations were performed on multiple CPU cores to account for the large size of the data. Finally, the results were recombined for the entire image. This methodology is assumed to be more accurate since it preserves the pore space geometry, and hence the fluids occupying pore corners and crevices are included in the pore occupancy calculations. An example of the segmented wet image and the separated pore map are shown in Figure (2).

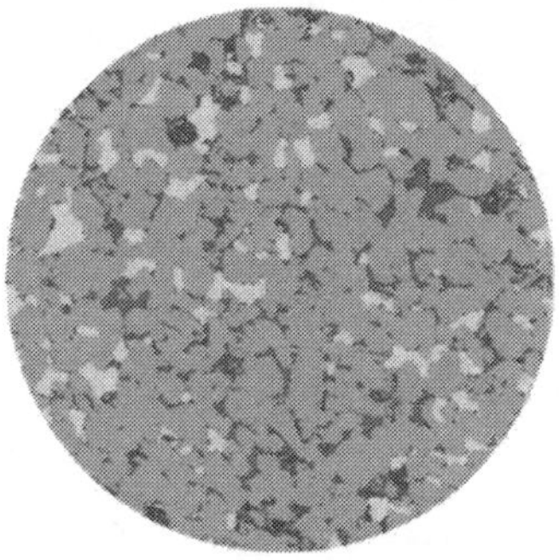

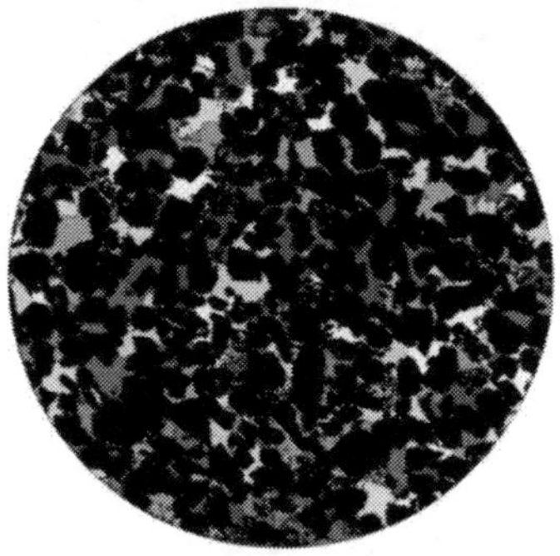

**Fig.2.** Two dimensional cross-sectional views of (top) the segmented wet image during the gas injection process, and (bottom) the separated pore map with different labels assigned to the individual pore elements.

### 3.2. Oil and gas cluster analysis

Each of the fluid phases in the pore space is formed of numerous connected and disconnected clusters of various sizes that can reside in the center of a single pore or span over several pore elements. The study of the connectivity and morphology of these clusters allows one to gain better insights into the multiphase phenomena within the pore space. To conduct this type of analysis, we generated separate images of the oil and gas phases and analyzed each set to identify the size of the individual clusters forming that phase. We also computed the Euler number for each phase, a measure of the connectivity, given by the formula: $\chi = \beta_0 - \beta_1 + \beta_2$, where $\beta_0$ is the number of disconnected objects, $\beta_1$ is the number of loops and $\beta_2$ is the number of holes. Furthermore, these data sets were used to quantify the contribution of clusters with various sizes to the total volume of the respective phase and investigated the relationship between the surface area (A) and volume (V) of the oil or gas globules. In addition, we computed the Euler characteristic for the oil and gas phases, which is indicative of the fluid's connectivity [24].

### 3.3. In-situ capillary pressure characterization

We also characterized the in-situ capillary pressure using the interfacial curvature analysis method described by Zankoor et al. [18]. To this end, we first extracted the free interfaces between each pair of fluids (i.e., oil-water, gas-oil, and gas-water) from the segmented images using the Marching cubes algorithm. The extracted meshes were then smoothed to reduce the impact of discretization on curvature analysis using an implicit fairing method [25, 26]. Afterward, the local curvature values were computed at each point (vertex) on each interface, and a representative average value was found to estimate the local capillary pressure between each fluid pair using the Young-Laplace equation:

$$P_{C_{ij}} = \sigma_{ij} K_T \tag{1}$$

where $P_{C_{ij}}$ is the capillary pressure in Pa, $\sigma_{ij}$ is the interfacial tension between phases $i$ and $j$ in N/m, and $K_T$ is the average total curvature of the interface in m$^{-1}$ (i.e., summation of the principal curvatures). To obtain an average value of the total curvature, we use arithmetic averaging of the per-vertex curvatures. An alternative approach is to use an area-weighted average, where each per-vertex curvature value is given a weight based on the area of the surrounding mesh elements, which, however, requires further investigation which is beyond the scope of the current study. It should be noted that the vertices at the peripheral rings of the fluid-fluid interfaces (i.e., in contact with the solid surfaces) were excluded from the curvature analysis since these regions introduced significant uncertainties to the measurements. Further details regarding the methodology used for the extraction and smoothing of the fluid-fluid interfaces and the sensitivity analysis on the parameters used in the analysis can be found elsewhere [18].

## 4. Results and discussion

### 4.1. Pore-by-pore fluid occupancy

We used the method discussed in section (3.1) to generate the following pore occupancy data from the segmented wet images of the gas injection process:

(i) Size distribution of pore elements that are mostly saturated with gas, oil, or water phases (see Figure 3.a). To this end, we first categorize the pores into three groups based on the dominant occupying fluid (i.e., gas-, oil-, or water-filled) and then generate the pore size distribution for each of these groups.
(ii) Fluid saturation in pores with different sizes (see Figure 3.b).
(iii) The ratio of cumulative fluid volume to the total volume (i.e., cumulative saturation) for each phase as a function of pore size (see Figure 3.c).

Figure 3.a illustrates that each of the three phases occupied a wide range of pore sizes. Nevertheless, it is observed that most of the water-filled elements are of the

 https://doi.org/10.1051/e3sconf/202336601006

smallest pores (smaller than 4.0 E+05 μm³), while the distribution of the oil- and gas-filled pores shift toward more intermediate- and large-size pores, respectively. This pore occupancy preference agrees with the wettability order of this rock-fluid system, with water, oil, and gas being the most wetting, intermediate wetting, and nonwetting phases, respectively [27].

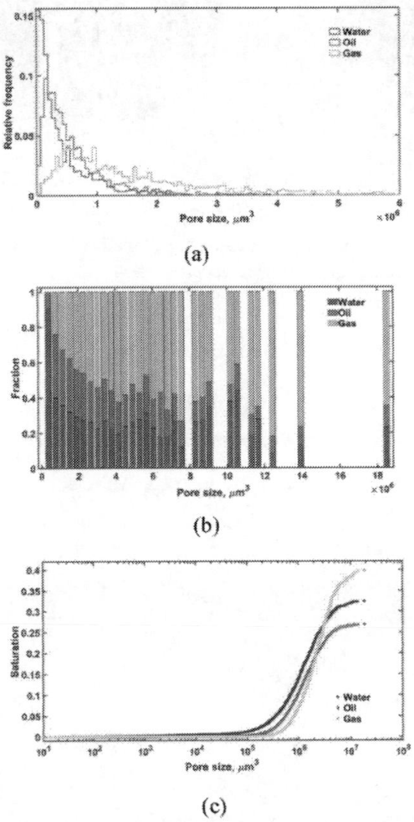


**Fig. 3.** Pore-scale fluid occupancy data during the secondary gas injection process: (a) the size distribution of pores mostly occupied by each phase; (b) fluid saturation in various pore-size ranges; and (c) the cumulative saturation of each phase as a function of the pore size.

Likewise, Figure 3.b shows that water saturation is steadily higher than the oil and gas saturation in the smallest pore elements. This behavior occurred because the threshold capillary pressures for gas-to-water or oil-to-water displacement were the highest in the smallest pores, and hence the oil and gas phases had a smaller chance to invade those pore elements. In the intermediate-size ranges, however, a significant portion of the pore space is occupied by each of the three phases while declining trends for water and oil saturations are observed as the pore size increases.

This observation is expected since it was more favorable for the gas (the most nonwetting) phase to invade the large pores. It is worth mentioning that the rise in water saturation in some of the intermediate-size pores (i.e., the range of 8.0 E+06 – 1.2 E+07 μm³) might be attributed to that the invading phases (oil and gas) did not have access to the entrance of all pore elements in this range, especially at areas with high clay content. As a result, some of the pores stayed water-filled during the drainage processes and contributed to the observed water saturation.

The observed saturation trend versus pore size is also reflected in Figure 3.c. This figure indicates that the water saturation built up earlier than the other phases (i.e., in smaller pores), whereas the gas saturation was increased significantly only when the contribution from the largest pores was added to the cumulative saturation. As seen, the final water, oil, and gas saturations in the FOV reached 0.32, 0.26, and 0.40, respectively.

### 4.2 Oil and gas cluster analysis

As mentioned previously, the oil and gas clusters after the gas injection process were extracted, and their size and distributions were investigated. Figure 4 shows the overall distribution of these clusters for each of the fluid phases. This figure demonstrates that the water phase was well connected (the connected blue cluster) over the FOV, whereas for the oil and gas phases, a significant number of globules with considerable sizes were disconnected and could be potentially trapped in the pore space. The connectivity of these clusters was further studied by quantifying the contribution of gas/oil clusters with various sizes to the total volume of gas/oil in the FOV. As shown in Figure 5, the disconnected gas and oil globules formed approximately 40% and 34% of the total gas and oil saturations, respectively. It should be noted that some of these clusters might be incorrectly considered disconnected. For instance, some of these clusters were truncated at the borders of the FOV and might be connected outside of it. Further analysis of the images revealed that the Euler numbers for the gas and oil phases were 3747 and 2442, respectively.

**Fig. 4.** Three-dimensional visualization of the (from left to right) gas, oil, and water clusters after the gas injection process. The blue clusters are connected from the bottom to the top of FOV, whereas the other colors represent disconnected clusters.

For the oil phase, it is expected to see disconnected oil clusters since the oil phase in this study was non-spreading in the presence of gas and water phases, and hence it may form lenses enclosed by the other phases if certain local capillary pressure conditions are established in the system [28]. For the gas phase, however, these observations are particularly interesting since (i) the continuous flow of gas was maintained during this process, and hence, one would expect to observe a well-connected network of gas clusters from the inlet to the outlet of the FOV, and (ii) the gas injection was a drainage event during which the trapping of the invading phase was less favorable compared to an imbibition process. Trapping of the most nonwetting phase during the secondary gas injection process can be explained based on some pore-scale displacement events such as Haines jump [29, 30, 20, 11] and Roof snap-off [31, 32]. During a drainage process, the pressure of the invading phase gradually builds up at the pore restrictions. Once the capillary pressure exceeds the threshold capillary pressure of a neighboring pore element, the nonwetting phase invades into that pore with an abrupt movement (jump) of the fluid interface and a transient pressure response. The pore-filling event causes the fluid to be withdrawn from high-pressure locations to supply the invasion front in the low-pressure regions. As a result, and especially in three-phase systems, some clusters may be disconnected from the bulk of the invading phase and be hydraulically trapped in the pore space. Moreover, when the gas phase moves from a pore restriction (i.e., throat) to a pore body, the local capillary pressure can fluctuate significantly. This process may lead to swelling of the wetting phase layer and trigger local Roof snap-off events at the restriction that can trap the gas phase. In addition to Haines jump and Roof snap-off, it is also hypothesized that double or multiple displacement chains (e.g., gas-to-oil-to-gas) [6, 9], and the ripening/anti-ripening phenomena [33] [34] could also contribute to the configuration of the gas phase in the porous medium. During the anti-ripening phenomenon, the gas phase occupying natural porous media may diffuse from large clusters into smaller ones, unlike in bulk mediums, since the former can be at higher local capillary pressures due to the confinements of the pore space.

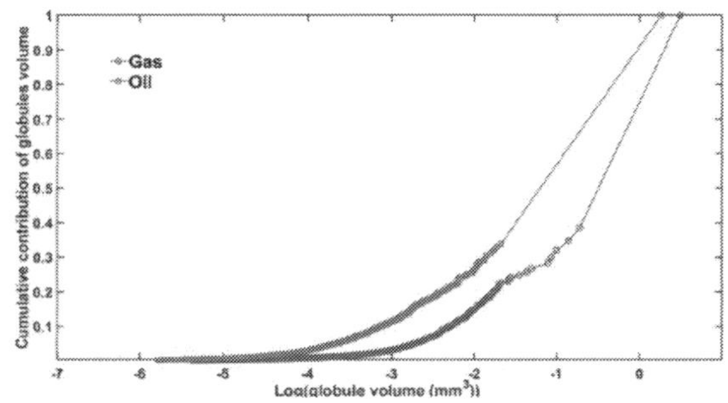


**Fig. 5.** The cumulative contribution of gas/oil ganglia with various sizes to the total gas/oil volume.

We have also investigated the morphology of the gas ganglia by quantifying the relationship between the volume and surface area of these objects. These parameters can be related through a power-law correlation of $A \alpha V^b$ [35]. The correlation exponent $b$ is equal to 0.66 for a perfect sphere and increases as the morphologies of the ganglia become more tortuous or elongated. The result of such a comparison for the gas clusters in this study is shown in Figure 5. The power-law exponent for the gas phase was found – using the least-squares method – to be 0.79, which was consistent with the values reported in literature studies [15].

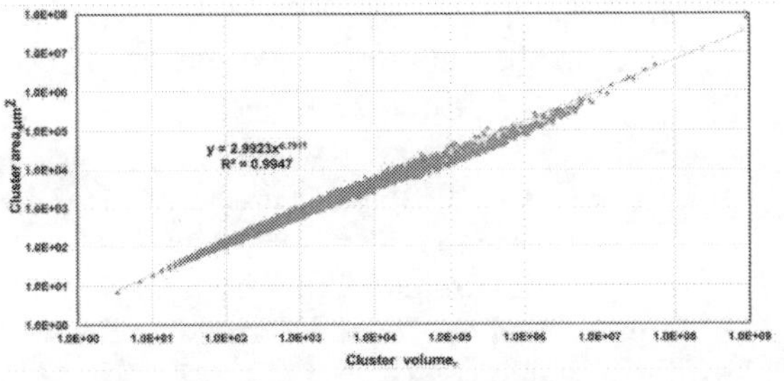


**Fig. 6.** The surface area versus volume of the gas clusters during the secondary gas injection experiment.

### 4.3 In-situ capillary pressure

To quantify the in-situ capillary pressure using the methodology described in section 3.3, we partitioned the segmented image into 18 sub-volumes with equal sizes of 620×620×850 voxels. For each sub-volume, the interfacial curvature of the gas-oil and oil-water interfaces were estimated, and the representative average values were used to calculate the respective capillary pressures [11]. The results are presented in Figure 7. The computed capillary pressures for each pair of fluids show some degrees of variation across the FOV. This variation is more significant for the gas-oil capillary pressures which is probably attributed to the fact that a significant portion of the gas phase existed in the form of disconnected globules, as discussed in section 4.2., which may have possessed different local capillary pressures [36].

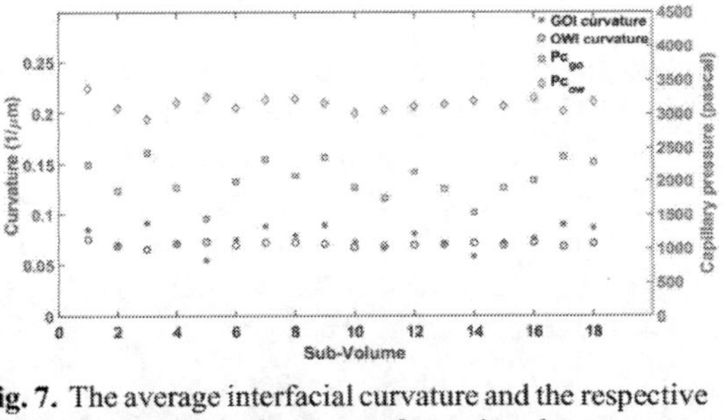


**Fig. 7.** The average interfacial curvature and the respective capillary pressures in the twenty-four sub-volumes.

The average oil-water capillary pressures were always greater than the gas-oil values, even though this is not the case for the average oil-water interfacial curvatures. This is attributed to the locations of the interfaces, as shown in Figure 8, which demonstrate that the oil-to-water displacements mostly took place at narrower capillary elements compared to the gas-to-oil events. Furthermore, the oil-water IFT is significantly higher than the gas-oil IFT ($\sigma_{ow}$ = 45.16 mN/m > $\sigma_{go}$ = 26.61 mN/m), while the receding contact angle is lower in the case of gas-oil interfaces [16] (which raises their curvatures). The combined effect of these three factors is that the threshold capillary pressures of the oil-to-water events would have presumably been higher than the gas-to-oil events. In other words, the gas-to-oil displacement would take place at lower capillary pressures (i.e., $P_g$-$P_o$) and in larger pores compared to the water-to-oil displacements.

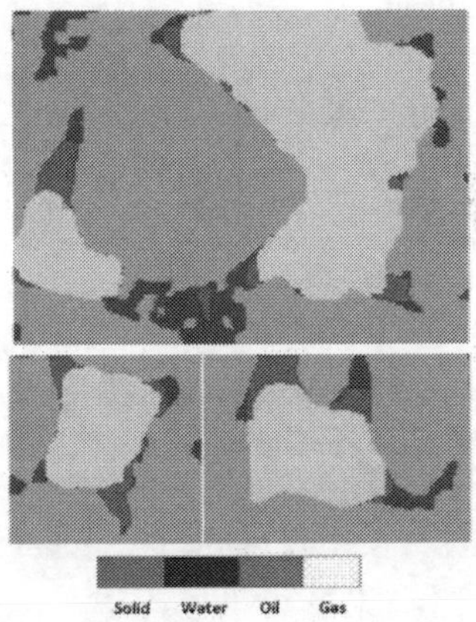


**Fig. 8.** Examples of three-phase interfaces during the secondary gas injection process (rock, water, oil, and gas are shown in grey, blue, red, and yellow respectively).

## 5. Conclusions

We characterize the pore-scale flow behavior during a steady-state gas injection process, where the three phases were injected simultaneously into a miniature Berea sandstone core sample. We used specific techniques to obtain pore-by-pore fluid occupancy data, analyze oil and gas clusters, and characterize the in-situ capillary pressure. In agreement with the wettability order of the system, the gas, oil, and water phases mostly occupied the largest, intermediate, and smallest pore elements of the system, respectively. Interestingly, we observed a significant amount of the invading gas phase trapped in the form of disconnected globules during this drainage process. The presence of these disconnected clusters was attributed to special displacement events such as the Haines jump and Roof snap-off. Moreover, the gas diffusion due to the anti-ripening phenomenon was also believed to contribute to the trapping of the gas phase. Finally, the oil-water local capillary pressures were found to be higher than the gas-oil ones. This observation was attributed to the fact that the oil-to-water displacement took place in smaller pore elements in which higher threshold capillary pressures need to be overcome.

We gratefully acknowledge the financial support of Thermo Fisher Scientific, Hess Corporation, and the University of Wyoming.

## References

[1] A. Mohamed, M. Khishvand and M. Piri, "A pore-scale experimental investigation of process-dependent capillary desaturation," *Adv. in Water Res.,* vol. 144, 2020.

[2] Y. Gong and M. Piri, "Pore-to-Core Upscaling of Solute Transport Under Steady-State Two-Phase Flow Conditions Using Dynamic Pore Network Modeling Approach," *Transp. Porous Med.,* vol. 135, 2020.

[3] O. Vizika and J. M. Lombard, "Wettability and spreading: Two key parameters in oil recovery with three-phase gravity drainage," *SPE Reservoir Eng.,* vol. 11, 1996.

[4] A. H. Alizadeh and M. Piri, "Three-phase flow in porous media: A review of experimental studies on relative permeability," *Reviews of Geophysics,* vol. 52, 2014.

[5] P. Øren, J. Billiotte and W. Pinczewski, "Mobilization of Waterflood Residual Oil by Gas Injection for Water-Wet Conditions," *SPE Formation Eval.,* vol. 7, 1992.

[6] P. Øren and W. Pinczewski, "Fluid Distribution and Pore-Scale Displacement Mechanisms in Drainage Dominated Three-Phase Flow," *Trans. in Porous Med.,* vol. 20, 1995.

[7] A. Keller, M. Blunt and A. Roberts, "Micromodel Observation of the Role of Oil Layers in Three-Phase Flow," *Trans. in Porous Med.,* vol. 26, 1997.

[8] M. Sohrabi, D. Tehrani, A. Danesh and G. Henderson, "Visualization of Oil Recovery by Water-Alternating-Gas Injection Using High-Pressure Micromodels," *SPE J.,* vol. 9, 2004.

[9] M. van Dijke and K. Sorbie, "Pore-scale modelling of three-phase flow in mixed-wet porous media: multiple displacement chains," *J. of Pet. Sci.and Eng.,* vol. 39, 2003.

[10] M. Piri and M. J. Blunt, "Three-dimensional mixed-wet random pore-scale network modeling of two- and three-phase flow in porous media. I. Model description," *Phys. Rev. E,* vol. 71, 2005.

[11] A. Zankoor, M. Khishvand, A. Mohamed, R. Wang and M. Piri, "In-situ capillary pressure and wettability in natural porous media: Multi-scale experimentation and automated characterization using x-ray images," *Journal of Colloid and Interface Science,* 2021.

[12] R. T. Armstrong, C. Sun and et al., "Multiscale Characterization of Wettability in Porous Media," *Transport in Porous Media,* 2021.

[13] R. T. Armstrong, M. L. Porter and D. Wildenschild, "Linking pore-scale interfacial curvature to column-scale capillary pressure," *Advances in Water Resources,* vol. 46, 2012.

[14] S. Iglauer, A. Paluszny and M. Blunt, "Simultaneous oil recovery and residual gas storage: A pore-level analysis using in situ X-ray micro-tomography," *Fuel,* vol. 103, 2013.

[15] S. Iglauer, T. Rahman, M. Sarmadivaleh, A. Al-Hinai, M. A. Fernø and M. Lebedev, "Influence of Wettability on Residual Gas Trapping and Enhanced Oil Recovery in Three-Phase Flow: A Pore-Scale Analysis by Use of Microcomputed Tomography," *SPE J.,* vol. 21, 2016.

[16] M. Khishvand, A. Alizadeh and M. Piri, "In-situ characterization of wettability and pore-scale displacements during two- and three-phase flow in natural porous media," *Adv. in Water Res.,* 2016.

[17] Z. Qin, M. Arshadi and M. Piri, "Micro-scale experimental investigations of multiphase flow in oil-wet carbonates. I. In situ wettability and low-salinity waterflooding," *Fuel,* vol. 257, 2019.

[18] M. J. Sabti, A. H. Alizadeh and M. Piri, "In-situ investigation of the impact of spreading on matrix-fracture interactions during three-phase flow in fractured porous media," *Adv. in Water Res.,* vol. 131, 2019.

[19] A. Alhosani, A. Scanziani, Q. Lin, Z. Pan, B. Bijeljic and M. J. Blunt, "In situ pore-scale analysis of oil recovery during three-phase near-miscible CO2 injection in a water-wet carbonate rock," *Adv. in Water Res.,* vol. 134, 2019.

[20] A. Alhosani, A. Scanziani, Q. Lin, A. Selem, Z. Pan, M. J. Blunt and B. Bijeljic, "Three-phase flow displacement dynamics and Haines jumps in a hydrophobic porous medium," *The Royal Society,* 2020.

[21] A. Scanziani, K. Singh, H. Menke, B. Bijeljic and M. J. Blunt, "Dynamics of enhanced gas trapping applied to CO2 storage in the presence of oil using synchrotron X-ray micro tomography," *Applied Energy,* vol. 259, 2020.

[22] A. Scanziani, K. Singh, T. Bultreys, B. Bijeljic and M. J. Blunt, "In situ characterization of immiscible three-phase flow at the pore scale for a water-wet carbonate rock," *Adv. in Water Res.,* vol. 121, 2018.

[23] A. Q. Raeini, B. Bijeljic and M. J. Blunt, "Generalized network modeling: Network extraction as a coarse-scale discretization of the void space of porous media," *Phys. Rev. E,* vol. 96, 2017.

[24] J. E. McClure, R. T. Armstrong, M. A. Berrill, S. Schlüter, S. Berg, W. G. Gray and C. T. Miller, "Geometric state function for two-fluid flow in porous media," *Phys. Rev. Fluids,* vol. 3, 2018.

[25] M. Desbrun, M. Meyer, P. Schröder and A. H. Barr, "Implicit Fairing of Irregular Meshes Using Diffusion and Curvature Flow," *ACM Press/Addison-Wesley Publishing Co.,* 1999.

[26] D. Kroon, "Smooth Triangulated Mesh," *Matlab central file exchange,* 2021.

[27] A. Alhosani, B. Bijeljic and M. J. Blunt, "Pore-Scale Imaging and Analysis of Wettability Order, Trapping and Displacement in Three-Phase Flow in Porous Media with Various Wettabilities," *Transport in Porous Media,* 2021.

[28] M. Piri, "Pore-Scale Modelling of Three-Phase Flow," *Imperial College,* 2003.

[29] W. Haines, "Studies in the physical properties of soil. V. The hysteresis effect in capillary properties, and the modes of moisture distribution associated therewith.," *The J. of Agric. Sci.,* vol. 20, 1930.

[30] S. Berg and e. al., "Real-time 3D imaging of Haines jumps in porous media flow," *Proc. Natl. Acad. Sci.,* vol. 110, 2013.

[31] J. Roof, "Snap-off of oil droplets in water-wet pores," *Soc. Petrol. Eng. J.* , vol. 10, 1970.

[32] A. Kovscek, G.-Q. Tang and C. Radke, "Verification of Roof snap off as a foam-generation mechanism in porous media at steady state," *Colloids and Surfaces A: Phys. and Eng. Aspects,* vol. 302, 2007.

[33] K. Xu, R. Bonnecaze and M. Balhoff, "Egalitarianism among Bubbles in Porous Media: An Ostwald Ripening Derived Anticoarsening Phenomenon," *Phys. Rev. Lett.,* vol. 119, 2017.

[34] S. Berg, Y. Gao and e. al., "Determination of Critical Gas Saturation by Micro-CT," *Petrophysics,* vol. 61, 2020.

[35] C. J. Landry, Z. T. Karpyn and M. Piri, "Pore-scale analysis of trapped immiscible fluid structures and fluid interfacial areas in oil-wet and water-wet bead packs," *Geofluids,* vol. 11, 2011.

[36] M. Andrew, B. Bijeljic and M. J. Blunt, "Pore-by-pore capillary pressure measurements using X-ray microtomography at reservoir conditions: Curvature, snap-off, and remobilization of residual CO2," *Water Resources Research,* vol. 50, 2014.

# Dissolution behaviour in carbonate reservoirs during WAG injection: A preliminary experimental study

*Mohamed* Khather[1], *Matthew* Myers[2], *Ali* Saeedi[1], and *Ausama* Giwelli[2,*]

[1]Department of Petroleum Engineering, Curtin University, Kensington, WA, 6151, Australia
[2]CSIRO Energy, Kensington, WA, 6151, Australia

**Abstract.** In this study, a core flooding experiment using a water-alternating-gas (WAG) injection was conducted to evaluate its impact on the petrophysical properties of an initially oil-saturated heterogeneous carbonate core sample. Carbon dioxide ($CO_2$) and synthetic formation brine were injected (0.5 pore volume $CO_2$ alternating with 0.5 pore volume brine) alternately following establishment of waterflooding residual oil saturation under reservoir conditions. Gas porosity, gas permeability, NMR (nuclear magnetic resonance) $T_2$ measurements, and X-ray CT scanning were conducted pre- and post-core flooding. The results show that $CO_2$-WAG injection resulted in substantial additional oil recovery (~30 %) under the applied experimental conditions. The results also show an increase in the permeability of the tested sample from 1.5 to 16 mD, which could be attributed to mineral dissolution. X-ray CT imaging shows signs of excessive mineral dissolution and formation of wormhole structures. It is believed that dissolution within the tested core plug caused the WAG fluids to follow the newly wormhole (causing them to enlarge further), and consequently bypassing many parts of the sample. Therefore, despite a significant increase in oil recovery, a large amount of oil is still left behind.

## 1 Introduction

With a pressing need to address climate change, carbon dioxide ($CO_2$) sequestration in hydrocarbon reservoirs (either depleted or producing) is increasingly being considered for many reasons including its large storage capacity (i.e. scalability) compared to other emissions mitigation strategies. Furthermore, initially, the cost of carbon capture and storage (CCS) could be offset by the added value of enhanced hydrocarbon recovery which is a by-product of storing $CO_2$ in producing reservoirs. Carbonate reservoirs represent about fifty percent of producing oil and gas reservoirs around the world [1]. However, $CO_2$ injection into carbonate reservoirs for storage/enhanced oil recovery (EOR) is more challenging due to their extreme heterogeneous nature (compared to comparable sandstone reservoirs) and dominant composition of highly reactive minerals (i.e. calcite and dolomite which are readily reactive with carbonated brine) in their rock formations. With the presence of highly reactive minerals in their composition, carbonate rocks may undergo appreciable alterations to their properties to a much larger extent compared with sandstone [2-4]. A combination of mineral dissolution, mineral precipitation, mechanical/physical compaction and asphaltene precipitation during $CO_2$-EOR injection scheme have been identified as the dominant mechanisms that occur [5-10].

Water-alternating-gas (WAG) injection as an EOR technique was introduced to improve macroscopic sweep efficiency in gas injection processes [11]. Furthermore, this technique is often more economical by lowering the gas volume required to be injected into the reservoir [12]. Each $CO_2$-EOR method (e.g. continuous $CO_2$ injection, carbonated water injection, water-alternating $CO_2$ injection, and cyclic $CO_2$ injection) has its own merits and disadvantage [13, 14].

Majority of the WAG research studies have focused on the field applications, pilot tests, coreflooding experiments, and simulation studies [15-18]. For instance, Caudle et al [15] and Kulkarni et al [16] have demonstrated that a high recovery of up to 90% could be achieved during coreflooding experiments. Also, WAG technique has been successfully applied in the North Sea fields such as Gullfaks, Stafjord, South Brae, Snorre, and Oseberg Ost [19].

In this study, we aim to evaluate the impact of $CO_2$-WAG on the internal pore structure of a carbonate reservoir core sample using nuclear magnetic resonance (NMR) and X-ray CT analysis. The objective is to determine whether the negative effects from $CO_2$-brine induced processes are reduced relative to other $CO_2$ injection processes while maintaining effective additional oil recovery. Crude oil and formation brine were used in the coreflooding experiment. Gas permeability, gas porosity, NMR-$T_2$ analysis and X-ray CT scan techniques were utilised to characterise the selected core sample before and after coreflooding test.

## 2 Material and Testing Method

A core flooding experiment was conducted on a typical sample collected from a Middle East carbonate reservoir. Dry gas permeability and porosity were 1.5 mD and 12.7%, respectively. The plug consisted (as measured by power X-ray) of calcite (82.2%), dolomite (13.1%), quartz (0.5%), halite (0.6%), albite (1%) and Ankerite (2.2%). The core plug was cleaned prior to measurement, and it had a length and diameter of 63 mm and 35.5 mm, respectively (see Fig. *1*).

Nuclear magnetic resonance (NMR) $T_2$ measurements at ambient condition were conducted on a brine saturated core sample, using Oxford-GIT Instruments Geospec 2 Plus Analyzer. Furthermore, X-ray CT imaging was used to evaluate mineral heterogeneity inside the core sample, as well as core-scale/macroscopic changes to some extent pore-scale features (e.g. wormhole formation) due to the flooding procedure. In this study, all X-ray CT scans (pre- and post-test) were performed at room temperature and atmospheric pressure using an X-ray energy beam of 140 kV and current of 1500 mA. A helical acquisition mode (pitch at 350 μm) was used to enable the reconstruction of 3D X-ray images with a voxel size of about 110 μm × 110 μm × 400 μm (512 x 512 pixels).

* Corresponding author: Ausama.giwelli@csiro.au

© The Authors, published by EDP Sciences. This is an open access article distributed under the terms of the Creative Commons Attribution License 4.0 (http://creativecommons.org/licenses/by/4.0/).

https://doi.org/10.1051/e3sconf/202336601007

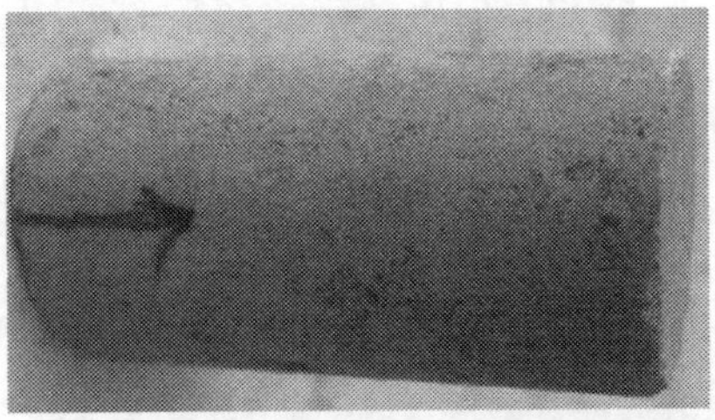

**Fig. 1.** Reservoir core sample used in this study.

Brine composition and test conditions were representative of reservoir conditions from which the core was acquired. The experiment was carried with a pore pressure of 14 MPa, temperature of 65 °C and overburden pressure of 27 MPa. Synthetic formation brine composition used in this experiment is shown in Table 1 and is based on the data available for the selected field. Crude oil used in the experiment was also obtained from the same field in the Middle East. The crude oil has a 38°API gavity and chemical compositions as presented in Table 2. Also, asphaltene content measurement (ASTM D6560) was conducted, and the result shows the crude has 0.075 wt% asphaltene composition.

**Table 1.** Concentration of major ions used to prepare the synthetic brine.

| Major ions | mg/L |
|---|---|
| $Na^+$ | 11870 |
| $K^+$ | 259 |
| $Ca^{2+}$ | 2368 |
| $Mg^{2+}$ | 452 |
| $Cl^-$ | 24046 |

**Table 2.** Composition of the crude oil used in this study.

| Component | Mol % | Component | Mol % |
|---|---|---|---|
| C2 | 0.210 | C18 | 2.687 |
| C3 | 1.155 | C19 | 1.983 |
| iC4 | 0.543 | C20 | 1.995 |
| nC4 | 2.640 | C21 | 1.768 |
| iC5 | 1.773 | C22 | 1.566 |
| nC5 | 3.467 | C23 | 1.375 |
| C6 | 6.253 | C24 | 1.214 |
| C7 | 7.056 | C25 | 1.056 |
| C8 | 8.211 | C26 | 0.984 |
| C9 | 7.563 | C27 | 0.876 |
| C10 | 7.120 | C28 | 0.834 |
| C11 | 5.835 | C29 | 0.784 |
| C12 | 4.953 | C30 | 0.687 |
| C13 | 4.587 | C31 | 0.625 |
| C14 | 4.025 | C32 | 0.579 |
| C15 | 3.984 | C33 | 0.493 |
| C16 | 3.160 | C34 | 0.461 |
| C17 | 3.258 | C35 | 0.426 |

## 2.1 Core flooding experimental procedure

Flooding with syntheric formation brine was conducted until residual oil saturation was established; this was followed by water alternating gas ($CO_2$-WAG) flooding using, the above mentioned, synthetic formation brine and pure $CO_2$. The following steps were taken in a chronological order:

1. Load the cleaned core sample into the core-holder. The core sample was wrapped into a composite sleeve consisting of a layer of FEP heat shrink, a layer of aluminium foil and a conventional Viton rubber sleeve before being inserted into a standard biaxial core-holder. More information about this step can be found in [3, 20].

2. A low overburden pressure was applied slowly on the sample to eject any trapped air. A vacuum pump was then connected to the core-holder for about 12 hours.

3. The core sample was then aged/saturated for 4 weeks in crude oil, shown in Table 2, under reservoir conditions.

4. The core holder was then connected to the flooding system, shown in Fig. *2*, and synthetic brine injection started at a low flow rate while monitoring the pore pressure. The flow rate was then adjusted to 0.5 mL/min throughout the experiment. Brine injection continued until no more oil was produced, indicating establishment of residual oil saturation. Produced liquid was collected in small, graduated tubes to keep track of the oil recovery profile against time accurately.

5. After 24 hours under reservoir conditions, $CO_2$-WAG (pure $CO_2$ and synthetic brine) injection was then started at a 1:1 ratio, half pore volume $CO_2$ alternating with half pore volume brine (for a total of ten cycles were used). At the end of this stage, WAG produced no further crude oil, allowing the injected fluids a relatively long time to interact with the rock sample and promote alteration to its petrophysical properties (if any).

6. At the completion of the experiment, the confining and pore pressures were released gradually, and the core sample was removed, cleaned as per [21], and then dried in the oven at 60° C. Post-test characterisation (NMR, gas porosity, permeability, and medical X-ray CT scan) were then carried out.

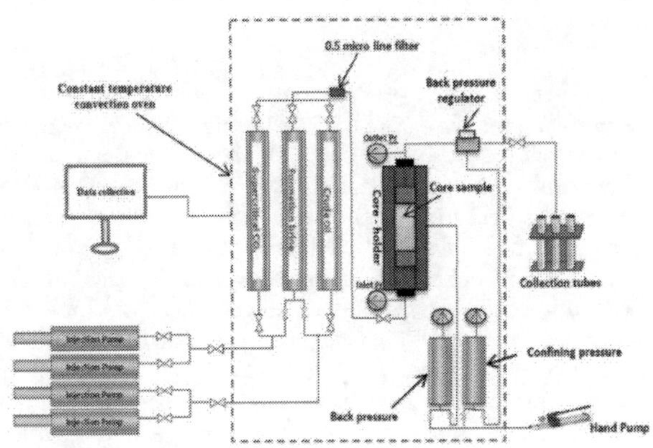


**Fig. 2.** Schematic diagram of core flooding apparatus used in this study [20].

## 3 Results and Discussion

Fig. **3** shows oil recovery factor (RF%) obtained throughout the experiment. The RF is defined as the recoverable amount of oil that initially placed inside the core plug. As stated above, the core sample was aged in crude oil under reservoir condition and then flooded by the synthetic formation brine to displace oil. Afterward, $CO_2$-WAG cycles, total of 10 cycles, was injected. An early water breakthrough was observed at about ~0.17 PV during brine injection/flooding corresponding to 24 % oil recovery (see Fig. **3**). At the end of the brine flooding (~1.2 PV), more oil was produced resulting in approximately 40 % recovery. It should be mentioned that no bump flow was performed during this stage.

The low recovery result obtained during brine injection agrees with [22]. The probable cause appears to be as a result of the complex internal pore structure of the rock sample, and/or wettability effects as most carbonate reservoirs are classified as oil wet [23]. The oil recovery during $CO_2$-WAG was improved to about 51 % at ~5 PV and then increasing to 71 % at ~6 PV. The result revealed that more oil was able to produce during $CO_2$- WAG process, but did not eventually attain any higher recovery (Fig. **3**). Unfortunately, upstream pressure transducer had a technical issue during the experiment and thus, we decided not to show the differential pressure.

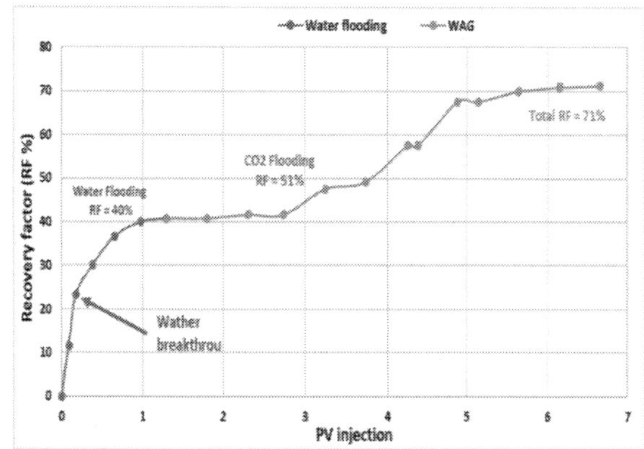


**Fig. 3.** Oil recovery percentage versus injected pore volume for both brine and $CO_2$-WAG stages.

Gas permeability results (pre- vs post-flood) showed significant increase (from 1.5 to 16 mD). Such a change can be attributed to the effect of the mineral dissolution mechanism (a wormhole formation) during $CO_2$-WAG injection. This hypothesis was supported by core effluents analyses, which was collected during the WAG flooding (see Table 3). Geochemistry result shows the concentration of calcium in the first cycle was higher than the base brine composition, which indicates an increase in dissolution. However, as more WAG cycles were injected, the concentration decreased throughout the experiment. In other words, the dissolution of calcium from the calcite dominate core plug mainly occurred in the first few cycles (<5 cycles). Then, the contact/reactivity with the $Ca^{2+}$ in the mineral solid phase is reduced and as a result, less $Ca^{2+}$ was mobilised. On the other hand, magnesium was present in the effluent but at small concentrations and there was, almost, no dissolution throughout the experiment. Gas porosity, on the other hand, shows a slight decrease of about ~ 5% reduction (12.7 to 12.1%). Such observation may have been caused by compaction mechanism (exacerbated by mineral dissolution physically weakening the core sample) caused by the overburden pressure applied during the experiment [10].

CT imaging was used to qualitatively analyse possible spatial changes of the CT attenuation profile/bulk density (within the CT scan resolution > 0.1 mm) along the plug length (as shown in Figure 4). Also a direct threshold segmenntation of the CT histogram was used to visulaise low density areas (pores) in 3D images (Figure 5) to illustrate possible simultaneous mechanisms (e.g. dissolution, precipitation and/or both), which might occur during the experiment. As mention above, x-ray CT images were generated before and after flooding at ambient conditions, and the sample was cleaned and dried in oven. The CT profile (Fig. *4*) along the sample length shows a slight decrease in CT values in comparison with pre-flood state. The value difference between the pre-and post-WAG is small and within the error bar sensitivity of the machine at ± 5 HU. However, 3D x-ray images support mineral dissolution at the inlet face of the plug (a light green colour in Fig. *5*), which has led to wormhole that visible in the images. The reaction extend almost to the entire length of the core sample.

We also calculated Peclet ($P_e$) and Damkohler ($D_a$) numbers of the dissolution and deposition process [24-29]. The Peclet number is the ratio of convection speed to characteristic diffusive velocity, while the Damkohler number is the ratio of the reaction to the mass transport rate. These dimensionless numbers provide a useful means of combining the physical and chemical processes that control dissolution and deformation regimes (e.g., [24, 27]). The Peclet and Damkohler numbers were estimated as $P_e = n/LD$ and $D_a = A_r k_r L/\phi v C_{eq}$; where $v$ is the interstitial velocity, $L$ is the characteristic length, $D$ is the molecular diffusion coefficient ($8\times10^{-10}$ m/sec), $A_r$ is the reactive surface area, $k_r$ is the reaction rate coefficient ($1\times10^{-4}$ cf. [30]), $\phi$ is the porosity and $C_{eq}$ is the average calcium concentration. At the core length, the tested sample yield high $P_e$ and $D_a$ values characteristic of $5.22\times10^3$ and $7.26\times10^3$, respectively.

The $P_e$ number is >> 1, indicating that the advective transport is dominant over diffusion at the core scale. The fact that $D_a$ >> 1 also suggests an extensive dissolution occurred at the beginning of the flow system [31]. Based on $P_e$-$D_a$ diagram [26], it could be concluded that the tendency of reaction is dominant wormhole regimes (i.e., permeability increases greatly due to the dissolution process). This result is consistent with the XCT images shown in Fig. *5*.

Although, there was a 30% increase in oil production during $CO_2$-WAG cycles, we believe the dissolution patterns, wormhole in this case, has negatively influenced the total amount of oil that could be recovered from the tested sample (with substanital amounts of oil remaining unrecovered). Apparently, the wormhole has created a preferential flow path for the injected fluids ($CO_2$ and brine) to bypass parts/islands of the oil inside the pores under the experiment conditions. Brine-$CO_2$ had a sufficient time to interact at different locations (flowline, inside the rock sample) during WAG, which resulted in accelerated mineral dissolution.

Asphaltene precipitation and to some extent mineral precipitation were expected to occur during the $CO_2$-WAG cycles. However, it seems that the effects of these precipitation mechanisms on permeability have been masked and suppressed by more dominant mineral dissolution (and wormhole formation).

**Table 3.** Calcium and magnesium concentration in the core effluents collected during WAG injection.

| WAG Cycle number | Brine effluent concentration | |
|---|---|---|
| | calcium ($Ca^{2+}$) mg/L | magnesium ($Mg^{2+}$) mg/L |
| 1 | 2500 | 250 |
| 5 | 2100 | 210 |
| 8 | 2000 | 190 |
| 10 | 2000 | 200 |

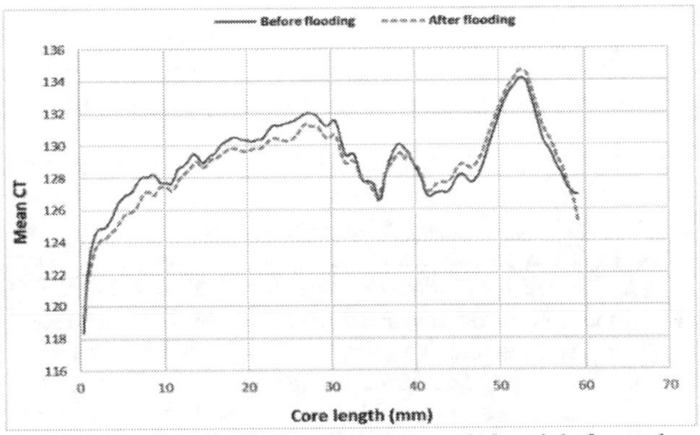


**Fig. 4.** CT number distribution along the sample length before and after WAG injection. The CT values show signs of dissolution in the first half of the sample, then turns to precipitation toward the end of the sample.

Fig. *6* shows the cumulative and incremental NMR-$T_2$ spectra of pre- and post-test measurements. The NMR-$T_2$ relaxation time was used to calculate the connected porosity filled by brine and demonstrate (qualitatively) the pore size distribution. The result shows a shift towards smaller values, meaning that the pore sizes/porosity became smaller and a slight shift in its pore size distribution towards smaller pore sizes (i.e., a reduction in the $T_2$ relaxation times). However, the NMR-$T_2$ distribution only examines pore space that is filled with water. This is consistent with gas porosity reduction (~5%) discussed above. Liteanu et al [10] showed $CO_2$ injection can increase mechanical compaction in carbonate by 7 orders of magnitude. However, in order to clarify how the mechanical strength changes during the experiment, we intend to implement fibre optic sensing technique [32, 33] to measure sample deformation. The NMR-$T_2$ distribution was not able show larger pores (Figure 6) corresponding to the wormholes, evident in the x-ray CT imaging. This is because water cannot be held inside the pore during ambient NMR-$T_2$ measurement, as the wormhole created an easy flow path for water to get out of the sample while NMR-$T_2$ measurement is conduced. Similar behaviour was observed in [34].

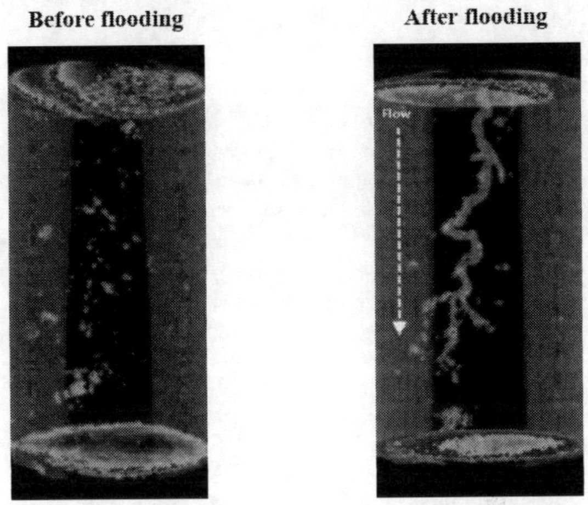


**Fig. 5.** 3D x-ray CT images for the tested core sample before and after flooding. The images show some low-density areas (corresponding to a light green colour) occurred due to the creation of wormholes along the length of the sample because of dissolution of carbonate minerals.

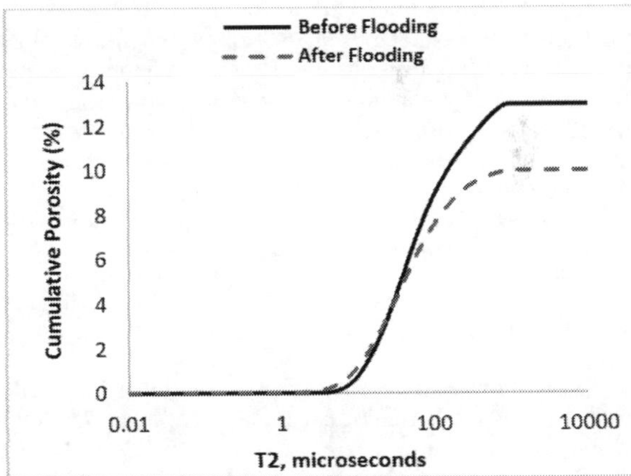

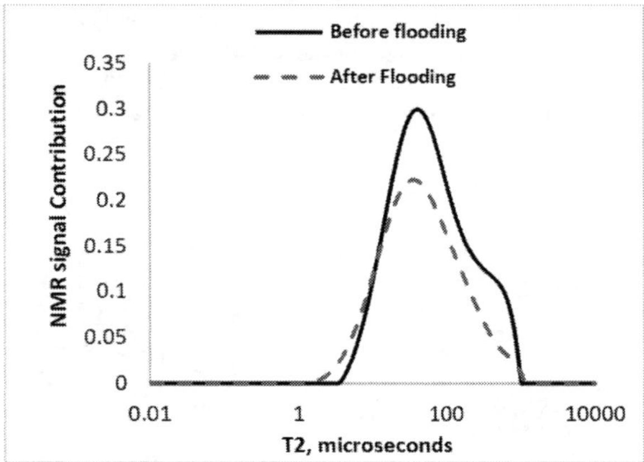


**Fig. 6.** Change in the cumulative porosity (top) and incremental NMR-$T_2$ distribution (bottom) of the tested sample, before and after flooding. NMR $T_2$ distribution shifted slightly to the left, indicating that the sample's pore sizes have become smaller after flooding.

## 4 Conclusions

While this study has only used one core sample, it is expected that within the context of this study that $CO_2$-WAG injection into carbonate rocks would improve sweep efficiency, to some extent, and it would also lead to minerals dissolution. In this particular case, fluid-rock interaction was significate and has eventually created wormhole formation across the entire length of the sample. With a high permeability channelling exists along the core sample, large amount of oil that is trapped in lower permeability portions of the rock was not recovered. All of these considerations have significant implications in terms of deployment in the field. Thus, we suggest more experimental studies should be considered to investigate the combined mechanisms to maximise oil recovery factor. Further research should be done to examine the effect of $CO_2$-WAG injection with the aid of other chemical additives such as surfactants, polymers, nanoparticles into the carbonate reservoirs.

## References

[1] SJ Mazzullo. Overview of porosity evolution in carbonate reservoirs. Kansas Geological Society Bulletin, **79**, 20-8 (2004)

[2] M Seyyedi, HKB Mahmud, M Verrall, A Giwelli, L Esteban, M Ghasemiziarani, B Clennell. Pore structure changes occur during $CO_2$ injection into carbonate reservoirs. Sci Rep, **10**, 3624 (2020)

[3] A Giwelli, MZ Kashim, MB Clennell, L Esteban, RRP Noble, C White, S Vialle, M Ghasemiziarani, M Myers, S. Md Shah Salwani. *$CO_2$-brine injectivity tests in high $CO_2$ content carbonate field, Sarawak basin, offshore East Malaysia.* In proceedings of: The 2018 International Symposium of the Society of Core Analysts, Trondheim, Norway, (2018)

[4] M. Khather, A. Saeedi, R. Rezaee, R.R.P. Noble, D. Gray. Experimental investigation of changes in petrophysical properties during $CO_2$ injection into dolomite-rich rocks. International Journal of Greenhouse Gas Control, **59**, 74-90 (2017)

[5] P. Mahzari, AP. Jones, EH. Oelkers. An integrated evaluation of enhanced oil recovery and geochemical processes for carbonated water injection in carbonate rocks. J Petrol Sci Eng, **181**, (2019)

[6] ME. Amin, AY. Zekri, R. Almehaideb, H. Al-Attar. *Optimization of $CO_2$ WAG processes in a selected carbonate reservoir-An experimental Approach. SPE-161782-MS.* In proceedings of: Abu Dhabi International Petroleum Conference and Exhibition, Abu Dhabi, UAE, (2012)

[7] M. Motealleh, R. Kharrat, A. Hashemi. An Experimental investigation of water-alternating-$CO_2$ coreflooding in a carbonate oil reservoir in different initial core conditions. Energy Sources, Part A: Recovery Utilization and Environmental Effects, **35**, 1187-96 (2013)

[8] IM. Mohamed, J. He, HA. Nasr-El-Din. Experimental analysis of $CO_2$ injection on permeability of vuggy carbonate aquifers. J Energ Resour-Asme, **135**, (2013)

[9] AY. Zekri, SA. Shedid, RA. Almehaideb. Experimental investigations of variations in petrophysical rock properties due to carbon dioxide flooding in oil heterogeneous low permeability carbonate reservoirs. J Pet Explor Prod Te, **3**, 265-77 (2013)

[10] E. Liteanu, CJ. Spiers, CJ. Peach, AN. Obdam. *Effect of $CO_2$ injection on compaction of carbonate rocks.* In proceedings of: American Geophysical Union, Fall Meeting 2006, (2006)

[11] S Touray. *Effect of water alternating gas injection on ultimate oil recovery.* Masters of Engineering, Petroleum Engineering, Dalhousie University, (2013)

[12] G.J. Pariani, K.A. McColloch, S.L. Warden, D.R. Edens. An approach to optimize economics in a West Texas $CO_2$ flood. J Pet Technol (SPE-22022-PA), **44**, 984-1025 (1992)

[13] D.M. Malcolm, S.M. Frailey, A.S. Lawal. *New approach to $CO_2$ flood: Soak alternating gas (SPE-70023-MS).* In proceedings of: The SPE Permian Basin Oil and Gas Recovery Conference, Midland, Texas, (2001)

[14] Xg. Xu and A. Saeedi. Evaluation and optimization study on a hybrid EOR technique named as chemical-alternating-foam floods. Oil Gas Sci Technol Rev IFP Energies nouvelles, **72**, (2017)

[15] BH. Caudle and AB. Dyes. Improving miscible displacement by gas-water injection. T Am I Min Met Eng, **213**, 281-4 (1958)

[16] MM. Kulkarni and DN. Rao. Experimental investigation of miscible and immiscible Water-Alternating-Gas (WAG) process performance. J Petrol Sci Eng, **48**, 1-20 (2005)

[17] JR. Christensen, EH. Stenby, A. Skauge. Review of WAG field experience. Spe Reserv Eval Eng, **4**, 97-106 (2001)

[18] E. Fernandez Righi, J. Royo, P. Gentil, R. Castelo, A. Del Monte, S. Bosco. *Experimental study of tertiary immiscible WAG injection. SPE-89360-MS.* In proceedings of: The SPE/DOE Symposium on Improved Oil Recovery, Tulsa, Oklahoma, (2004)

[19] S. Afzali, N. Rezaei, S. Zendehboudi. A comprehensive review on enhanced oil recovery by Water Alternating Gas (WAG) injection. Fuel, **227**, 218-46 (2018)

[20] M. Khather, A. Saeedi, R. Rezaee, R.R.P. Noble. Experimental evaluation of carbonated brine-limestone

interactions under reservoir conditions-emphasis on the effect of core scale heterogeneities. International Journal of Greenhouse Gas Control, **68**, 128-45 (2018)

[21] C McPhee, J Reed, I Zubizarreta. *Core Analysis: A Best Practice Guide*. Elsevier (2015)

[22] M. Khather, A. Saeedi, M.B. Myers, M. Verrall. An experimental study for carbonate reservoirs on the impact of $CO_2$-EOR on petrophysics and oil recovery. Fuel, **235**, 1019-38 (2019)

[23] L.E. Treiber and W.W. Owens. A laboratory evaluation of the wettability of fifty oil-producing reservoirs (SPE-3526-PA). SPE Journal, **12**, 531-40 (1972)

[24] CN Fredd and HS Fogler. Influence of transport and reaction on wormhole formation in porous media. AIChE Journal, **44**, 1933-49 (1998)

[25] CN Fredd and HS Fogler. Optimum conditions for wormhole formation in carbonate porous media: Influence of transport and reaction. SPE Journal, **4**, 196-205 (1999)

[26] F Golfier, C Zarcone, B Bazin, R Lenormand, D Lasseux, M Quintard. On the ability of a Darcy-scale model to capture wormhole formation during the dissolution of a porous medium. J Fluid Mech, **457**, 213-54 (2002)

[27] ML Hoefner and HS Fogler. Pore evolution and channel formation during flow and reaction in porous media. AIChE Journal **34(1)**, (1998)

[28] S Vialle, J Dvorkin, G Mavko. Implications of pore microgeometry heterogeneity for the movement and chemical reactivity of $CO_2$ in carbonates. Geophysics, **78**, L69-L86 (2013)

[29] S Vialle, S Contraires, B Zinzsner, JB Clavaud, K Mahiouz, P Zuddas, M Zamora. Percolation of $CO_2$-rich fluids in a limestone sample: Evolution of hydraulic, electrical, chemical, and structural properties. J Geophys Res-Sol Ea, **119**, 2828-47 (2014)

[30] RS Arvidson, IE Ertan, JE Amonette, A Luttge. Variation in calcite dissolution rates: A fundamental problem? Geochimica Et Cosmochimica Acta, **67**, 1623-34 (2003)

[31] CI Steefel and K Maher. Fluid-rock interaction: A reactive transport approach. Rev Mineral Geochem, **70**, 485-532 (2009)

[32] Y. Kovalyshen, Banks S., A. Giwelli. *Measurement of rock strain using Fiber Bragg Grating sensors*. In proceedings of: The 52$^{nd}$ US Rock Mechanics/Geomechanics Symposium, Seattle, Washington, USA, (2018)

[33] B. da Silva Falcão, L. Esteban, A. Giwelli, Y. Kovalyshen, S. Banks, A. Al-Yaseri, A. Keshavarz, S. Iglauer. *Monitoring fluid migration using in-situ nuclear magnetic resonance core flooding system integrated with fiber optic sensors: A proof of concept*. In proceedings of: The 35$^{th}$ International Symposium of the Society of Core Analysts (SCA Annual Symposium), Online Event, (2021)

[34] M. Khather, A. Saeedi, MB. Myers, A. Giwelli. Effects of $CO_2$-saturated brine on the injectivity and integrity of Chalk reservoirs. Transport Porous Med, **135**, 735-51 (2020)

# Monitoring fluid migration using in-situ nuclear magnetic resonance core flooding system integrated with fiber optic sensors: A proof of concept

*Bruno* da Silva Falcão[1,2,3,*], *Lionel* Esteban[2], *Ausama* Giwelli[2], *Yevhen* Kovalyshen[2], *Stephen* Banks[2], *Ahmed* Al-Yaseri[1,4], *Alireza* Keshavarz[1], and *Stefan* Iglauer[1]

[1]School of Engineering, Edith Cowan University, Joondalup, WA 6027, Australia
[2]CSIRO Energy Business Unit, Kensington, WA 6151, Australia
[3]Currently at Fluid Science and Resources Division, Department of Chemical Engineering, University of Western Australia, Crawley, WA 6009, Australia
[4]Currently at Center of Integrative Petroleum Research (CIPR), College of Petroleum Engineering & Geosciences, King Fahd University of Petroleum & Minerals, Saudi Arabia, Dhahran 31261, Saudi Arabia

**Abstract.** In-situ nuclear magnetic resonance (NMR) core flooding system has enabled researchers to monitor several rock properties such as porosity, pore size distribution, and fluid saturation along the tested samples with high resolutions and under reservoir conditions. However, spatially resolved rock strength/mechanical property alteration coupled to fluid migration/substitution remains poorly characterized. To this end, Fiber Bragg Grating (FBG) multiplex sensors were integrated with NMR core flooding system to monitor rock strength changes, or generally speaking, to observe hydro-mechanical-chemical coupling mechanisms during core flooding tests. In this study, we present a novel approach on how to conduct core flooding experiments, while simultaneously monitoring NMR and FBG strain response of the tested limestone plug. The NMR cell was modified to integrate FBG technology without impeding the NMR signal and core flooding high pressure/temperature capacity. A high spatial resolution optical fiber was attached onto the sample radial surface. The results show the successful association of NMR and FBG sensors to track any change at each stage of brine injection. The FBG is capable of measuring the rock strain variations induced by rock-fluid interactions during brine injection, allowing it to capture the fluid front location along with the sample and at a faster rate than the NMR.

## 1 Introduction

Nuclear magnetic resonance (NMR) core flooding system became an important tool for reservoir rock characterizations [1-3]. It enables the evaluation of several rock properties such as porosity, pore size distribution, and fluid saturation to name a few with high resolution and under reservoir conditions [4-6].

Likewise, the evaluation of rock mechanical properties also plays an important role in reservoir characterization, especially for risk assessment and economic production sustainability. Measuring rock strength changes is essential under reservoir conditions, even more when the rock experiences potential mineralogy alteration during fluids displacements, which can lead to rock deformations, faults reactivation, reservoir leakage pathways [7], and the like. Conventionally, rock strain deformations are measured by linear variable differential transformers (LVDTs) and electrical resistance strain gauges (ESG) [8-10]. However, those conventional sensors are incompatible with the NMR core flooding system as the small confining space and the magnet in-place of the NMR probe prohibit the insertion of such metallic sensors.

Optical fibers were introduced in the 1960s with the sole purpose of transmitting light, which later had a positive impact on telecommunication systems. However, due to the change in the transmitted light properties by the nearby environment, optical fibers were considered a good option for sensing applications [11] Fiber optic sensors (FOS) have been applied in a vast variety of industries from aerospace [12], medicine [13], civil engineering [14] to petroleum [15]. FOS can be classified into two main categories: distributed and discrete. Distributed FOS is based on Raman, Rayleigh, or Brillouin scattering mechanisms and measure changes all along the fiber's length. Discrete FOS, however, such as fiber Bragg grating and Fabry-Perot, provide information at specific locations [14, 16]. Fiber Bragg Grating (FBG) sensors can be multiplexed enabling them to measure changes at different locations using a single fiber being considered as a quasi-distributed sensor [14].

FBG sensor is a very promising deformation measurement tool acting as an alternative to the conventional sensors [8-10, 17] with less sensors-sample setup time, more efficient, and cost-effective. It is capable of measuring strain, temperature, pressure, among other parameters. FBG sensors also offer the advantages of being small size, flexible, low weight, high sensitivity, not affected by electromagnetic interference, resistant to corrosion and water, resilient to high-pressures and high-temperatures, and multiplexing capability [8, 10].

In recent years, the application of FBG sensors to monitor rock strain changes in real-time at laboratory scale has expanded and it seems to be a good alternative to LVDTs or ESG but it is still in the exploration stage. Sun et al. [8] proposed the use of multichannel FBG sensor arrays to monitor rock strain. The authors measured both radial and axial strains on the lateral area of a sandstone sample undergoing a uniaxial loading. Sun et al. [9] also employed

* Corresponding author: bruno.falcao@msn.com

© The Authors, published by EDP Sciences. This is an open access article distributed under the terms of the Creative Commons Attribution License 4.0 (http://creativecommons.org/licenses/by/4.0/).

multichannel FBG sensors to detect strain responses caused by water and $CO_2$ gas injection during core flooding experiments through a low permeability sandstone. Kovalyshen et al. [10] reported the applicability of FBG sensors to measure strain variations of a limestone sample under confining pressure. Zhang et al. [17] tracked the $CO_2$/brine displacement front in a rock sample composed of both reservoir and caprock formations. The fluid displacement was monitored by measuring the strain using the distributed fiber optic strain sensing (DFOSS) technique. X-ray computed tomography imaging was used to estimate the fluid saturation.

In this paper, we present a novel approach on how to conduct core flooding experiments associating NMR core flooding system with FBG sensing technology. To our knowledge, this is the first NMR core flooding experiment that is capable of measuring rock strain deformations at reservoir conditions. An optical fiber with 8 FBG sensors was radially placed around the rock sample surface to monitor the rock strain deformations induced by brine injection while under reservoir pressure condition. The NMR core analyzer was used to monitor the fluid displacement and saturation along the sample throughout the experiment.

## 2 Material and methods

### 2.1 Principle of FBG sensing

A Fiber Bragg Grating (FBG) sensor is an optical deflector or filter of specific wavelength responses. Its functioning relies on the optical fiber core's refractive index being periodically altered by UV light. When a light source is emitted into the FBG sensor, it reflects part of the energy in a limited wavelength band around its peak wavelength and transmits all other wavelengths. The reflected wavelength is known as Bragg wavelength $\lambda_B$ and it is defined by Equation (1) [8-10, 18, 19].

$$\lambda_B = 2n_{eff}\Lambda \tag{1}$$

where $n_{eff}$ is the FBG core's effective refraction index and $\Lambda$ is the grating spacing. Modifications on $n_{eff}$ or $\Lambda$ values, or even a combination of both, will represent any changes that occur in the physical properties such as strain and temperature [8-10, 18, 19].

The shift in the Bragg wavelength caused by changes in the strain and temperature is given by Equation (2).

$$\Delta\lambda_B = 2\left(\Lambda\frac{\partial n_{eff}}{\partial l} + n_{eff}\frac{\partial \Lambda}{\partial l}\right)\Delta l + 2\left(\Lambda\frac{\partial n_{eff}}{\partial T} + n_{eff}\frac{\partial \Lambda}{\partial T}\right)\Delta T \tag{2}$$

The first term in Equation (2) represents the change in the Bragg wavelength due to the strain effect, while the second term describes the effect of temperature [18, 19].

For the FBG used in this study, the wavelength sensitivity coefficient related to strain and temperature at room conditions, according to the manufacturer, are 0.776 pm/με, 8.53 pm/°C, and 0.0023 pm/°C. The optical fiber was manufactured by FBGS (FBGS Technologies GmbH), with 125 μm cladding diameter and Ormocer coating. Each single optical fiber has 8 FBG sensors of low bend loss and every sensor is 8 mm long. The eight gratings cover an extent of 315 mm ranging from 1530 to 1565 nm wavelength, with a difference of 5 nm between the gratings to avoid overlap.

### 2.2 Nuclear magnet resonance (NMR)

When an external magnetic field is applied, protons $^1H$ (abundant in reservoir fluids) behave like spinning magnets due to their magnetic moment and angular momentum [4, 6]. Rotational proton procession produces signals that decay with time also known as proton relaxation that can be measured by a Nuclear Magnetic Resonance spectrometer (NMR) [4, 6].

The NMR signal decay **M**(t) is described by Equations (3)–(5) from two mechanisms: NMR signal can be measured along the applied magnetic field corresponding to proton polarization or longitudinal relaxation $T_1$; while the proton decay post-polarization and then tipped transversally to the magnetic field corresponds to the transversal relaxation $T_2$ [4, 20, 21].

$$M_x(t) = \left[M_x(0)\cos(\omega t) - M_y(0)\cos(\omega t)\right]e^{(-t/T_2)} \tag{3}$$

$$M_y(t) = \left[M_x(0)\cos(\omega t) - M_y(0)\cos(\omega t)\right]e^{(-t/T_2)} \tag{4}$$

$$M_z(t) = M_{eq} + \left[M_z(0) - M_{eq}\right]e^{(-t/T_1)} \tag{5}$$

Generally speaking, on very porous rock materials, $T_1$ is sensitive to the environment and fluids interactions while $T_2$ is more sensitive to surface relaxation and diffusion mechanisms. With negligible fluids interactions, both $T_1$ and $T_2$ provide pore size distribution information, though $T_1$ measurement is more time consuming than $T_2$ [6].

A 2.37 MHz Geospec2 NMR core analyzer and a P5 overburden NMR probe from Oxford Instruments were used to obtain the brine saturation profile along the sample.

### 2.3 Sample preparation and characterization

A Savonnières limestone core (Fig. 1), a heterogeneous rock composed of 99% of calcite [22-24], was used in this study and cut along the bedding plane. The yellow dashed lines in Fig. 1 highlight a centimeter-thick bedding composed of high-density/low porosity structure that pervades the middle of the sample. This tight bedding is better illustrated in Fig. 2, which shows a 3D volume rendering from X-ray CT images viewed in two angles color-labeled by density intensity from single threshold method on the X-ray CT attenuation histogram where magenta color corresponds to a highly porous area while yellow color marks much denser and tighter material. The images illustrate that the tight bedding is actually composed of alternating thin-sub layers of high- and low-density materials with some thin dense layers propagating toward one side of the sample. The plug can then be divided into three structural areas along the sample from left to right in Fig. 2: (1) 1.5 cm thick highly porous layer, (2) 1.5 cm thick tight layer, and (3) < 1 cm thin highly porous layer. Table 1 shows the location of the sensors on the sample with respect to the described structural layers: (i) sensors facing the section with the highest void fraction with sensors 1, 4, 5, and 7; (ii) sensors facing the section with the lowest void fraction with sensors 3 and 6; (iii) sensors located perpendicular to the beddings with sensors 2 and 8. The dimension of the tested plug is shown in Fig. 1. The sample was initially dried at 60 °C under air vacuum for 24 hours before attaching FBG wire and then place the assembly in the NMR core flooding system. Further details about the rock properties can be found in the open literature [22, 23, 25].

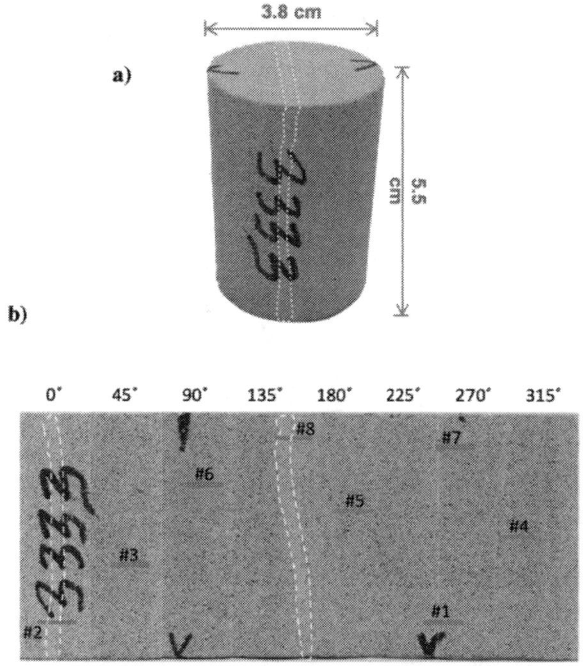


Fig. 1. (a) Savonnières limestone sample used in this experiment. (b) Position of FBG sensors on the sample after unwrapping the plug surface. The dashed yellow area indicates the position of a bedding of higher density/lower porosity.

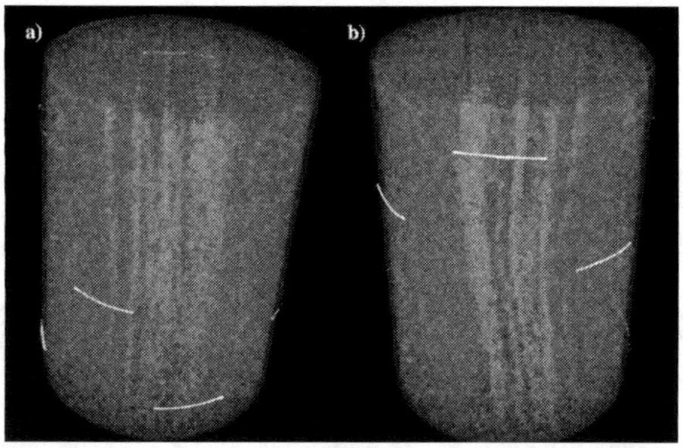


Fig. 2. 3D volume rendering from X-ray CT images of Savonnières limestone sample. The color magenta represents the pores of the rock sample, the yellow represents the existing beddings of higher density in the middle of the sample, and the white (enhanced) indicates the sensors' location. (a) Front view of the sample showing the sensors 3, 2, 4, and 1 at the bottom (from right to left). (b) back view of the sample showing the sensors 5, 8, and 6 (from right to left).

Table 1. Position of the FBG sensors related to the void fractions and beddings.

| Category | Sensor | Description |
|---|---|---|
| (i) | 1, 4, 5, 7 | Facing the section with the highest void fraction. |
| (ii) | 3, 6 | Facing the section with the lowest void fraction. |
| (iii) | 2, 8 | Perpendicular to the beddings. |

The optical fiber was spirally attached onto the sample surface using Loctite Super Glue All Plastics and cured for 24 h. This glue adequately couples the optical fiber with the rock surface, without modifying the rock stiffness properties [10]. The position of the sensors is shown in Fig. 1 and Table 2. For the core flooding test, the FBG wire was inserted through the pore fluid line using a modified end-platen, specifically designed for this experiment. A feedthrough component was used to prevent any leakage while the system was pressurized. After the curing process, the whole assembly was covered by a Teflon heat shrink jacket and a Viton sleeve and then placed inside the NMR P5 probe.

The optical fiber was connected to a FAZT I4 optical interrogator (FAZ Technology) through an LC/APC connector at the lead-in. The optical interrogator is capable of detecting wavelengths ranging from 1529 to 1568.2 nm, which limits the total number of sensors to eight due to the 5 nm grating difference.

Table 2. Position of FBG sensors: the distance was measured from the bottom of the sample to the center of the FBG sensor.

| Sensor | Distance (mm) |
|---|---|
| 1 | 7.8 |
| 2 | 8.2 |
| 3 | 20.9 |
| 4 | 26.9 |
| 5 | 32.6 |
| 6 | 38.1 |
| 7 | 46.7 |
| 8 | 48.4 |

### 2.4 Testing procedure

The P5 probe assembly was inserted into the NMR spectrometer and a constant hydrostatic confining pressure of 1000 psi was applied at a rate of 50 psi/min using Fluorinert as confining fluid. After stabilization of the confining pressure, dopped brine with 5 g/L $NiCl_2$ was continuously injected at 0.1 cc/min from the bottom of the sample.

During the imbibition/saturation stage, the outlet of the system was open to the atmosphere such as the FBG sensors can accurately and slowly track the brine front by recording the changes in the strain values. A 200 psi backpressure was then applied, to monitor the effect of pore pressure change on the FBG sensors during brine injection at 0.1 cc/min.

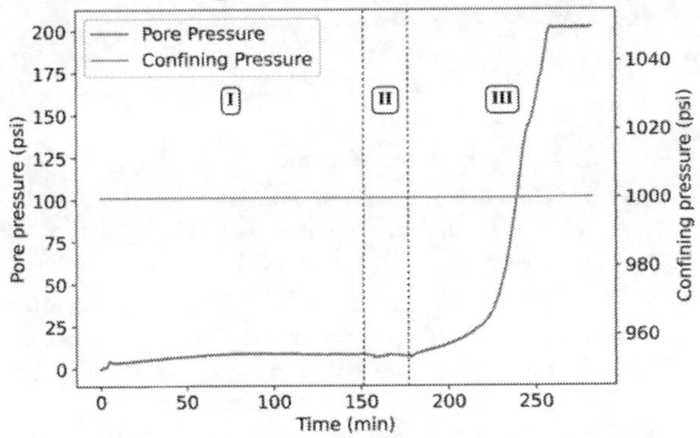


Fig. 3 shows the confining and pore pressure data during the entire experiment. Section I refers to the brine injection while the outlet was open to the atmosphere. Section II represents the period during which the injection was stopped to connect the backpressure regulator. Section III shows the pore pressure behavior after the backpressure was activated and set at 200 psi.

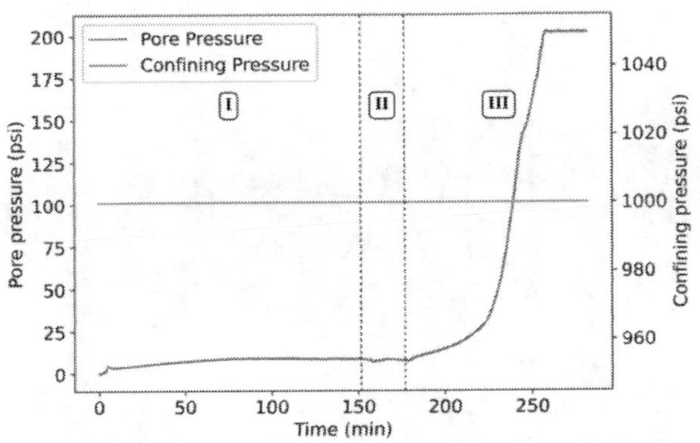


Fig. 4. Pore pressure and confining pressure monitoring with time during imbibition exercise on Savonnières sample under 1000 psi confining pressure: I – brine injection at 0.1 cc/min while outlet was open to the atmosphere; II –injection stopped to connect the backpressure; III – backpressure activated and set to 200 psi.

Three Teledyne ISCO syringe pumps were used to control the confining pressure, pore pressure, and backpressure. All pressure values were recorded using in-house LabView code. The entire experiment was performed at a constant temperature of 22 °C in order to avoid any temperature interference on the strain measurement. To obtain that the NMR system is connected to external chiller throughout the experiment.

### 2.5 Data acquisition and processing

During the experiment, all the measured NMR data were collected and computed automatically via Green Imaging Technologies software [26, 27]. Water content profile along the sample used the following settings:

Recycle delay: 3750 ms
Tau: 500 μs
Gradient duration: 300 μs
Filter width: 125 kHz
Resolution: 64

A data acquisition system, using LabView, was designed to continually log the data from the syringe pumps every second. The optical interrogator systems enable the streaming of the data over an ethernet connection. The FBG data acquisition and graphical display software ran on a PC via an ethernet connection (0.1 Gbit/s) at a sampling rate of 1000 Hz. An in-house algorithm was developed to process the raw data and generate all the relevant figures.

## 3 Results and discussion

### 3.1 Confining pressure and strain

In the first stage of the experiment, the objective was to analyze the relationship between confining pressure and rock strain changes in dry state. Initially, the confining pressure was increased from 20 psi to 1000 psi at a rate of 50 psi/min as shown in

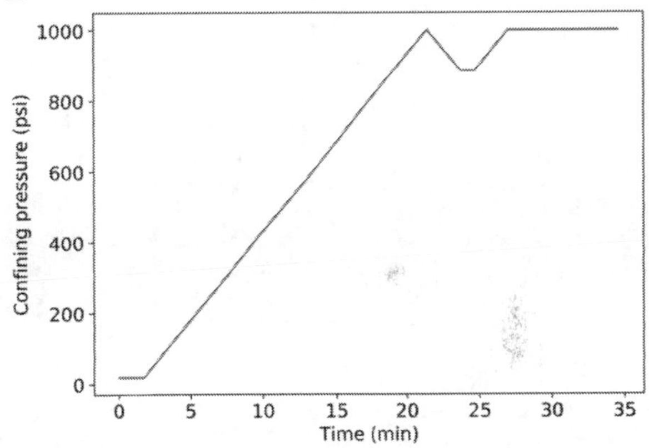


Fig. 5. After the confining pressure reached 1000 psi, it was temporarily decreased to 885 psi (after ~ 22 min) and kept constant for a few minutes. FBG strain changes caused by the increase in the confining pressure are shown in Fig. 5, which demonstrates the good agreement between the changes in the confining pressure and the rock strain responses. Fig. 5 highlights some different sensor behaviors while the confining pressure increases or decreases. Though most of the sensor responses can be explained from the structural heterogeneities observed in Fig. 1 and Fig. 2, sensors 2 and 3 showed similar responses to the changes in the confining pressure while being within distinct structural areas (Fig. 1). This was also observed for sensors 5 and 8. As depicted in Fig. 9, the strain values of all sensors decreased with the increase in the confining pressure. As a result, the temporary changes in the confining pressure demonstrated the fast and accurate response of the FBG sensors. The final strain values at a constant confining pressure of 1000 psi ranged from -250 to -350 με.

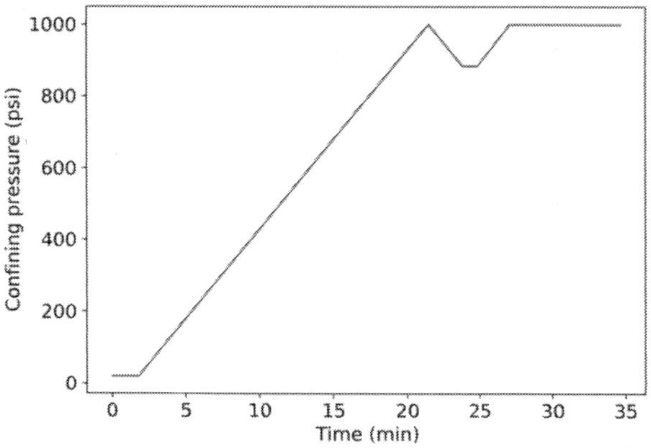


Fig. 6. Confining pressure profile prior to the brine injection.

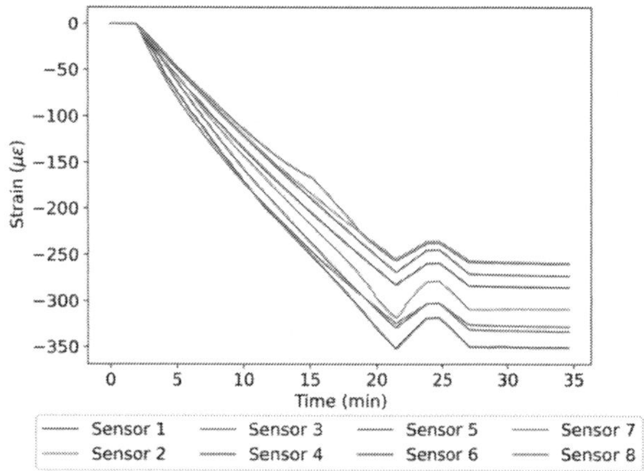


Fig. 7. Corresponding strain changes on the dry rock during the confining pressure build-up, measured by FBG sensors.

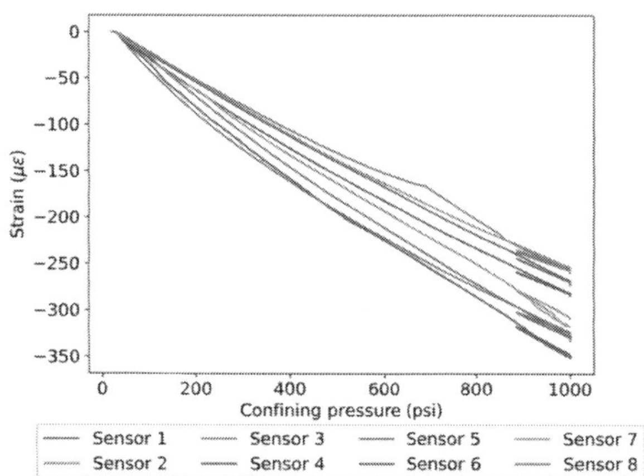


Fig. 8. Measurements of strain at varying confining pressures for each sensor, measured during the build-up of the confining pressure on dry rock.

### 3.2 Brine front monitoring

In order to monitor the fluid migration with FBG sensors, we kept the outlet pore line opens to the atmosphere, section I in

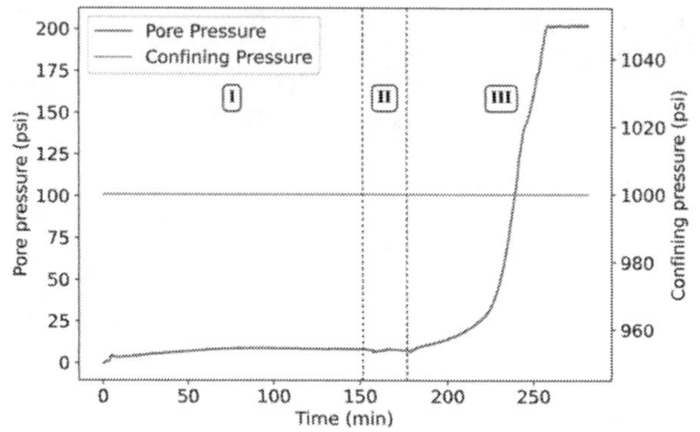


Fig. 9. This brine injection continues until breakthrough occurred, once the brine came out of the sample. Pore pressure was as low as 9 psi, at a 0.1 cc/min brine flow rate.

Fig. 10 illustrates that the FBG sensors effectively monitored the brine front during the injection process. It is possible to identify the location of the brine front in the rock sample at a specific time by monitoring the rock strain changes. The strain changes corresponded well to the fluid migration, and all sensors exhibited an increase in the strain when the fluid reached the sensor location. The differential strain kept increasing to values between 2 and 3 με, until the brine breakthrough.

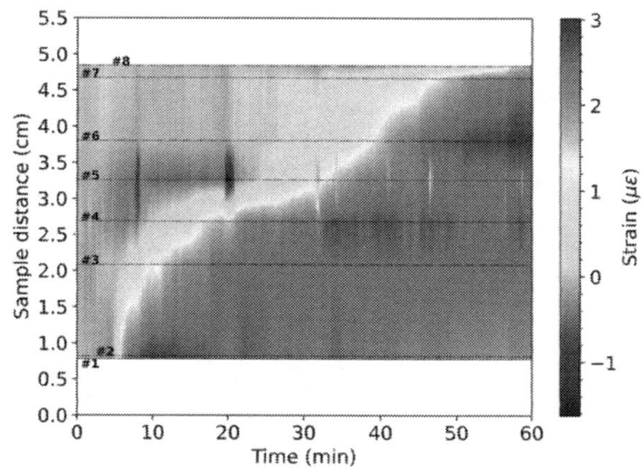


Fig. 10. Mapping of the strain changes throughout the sample until the brine breakthrough. The locations of the sensors are labeled on the figure in dashed lines. Note that the differential strain data were linearly interpolated between the sensors. Information about strain changes before sensor 1 and after sensor 8 is unavailable.

The saturation profiles measured by NMR are shown in Fig. 11. The NMR saturation profile provides information about the migration of the fluid front and the corresponding volume distribution of the fluid within the sample. The results from both FBG and NMR are in good agreement. For instance, it can be estimated that the brine front reached sensor 4 after approximately 10 minutes. However, while the FBG interrogator was recording 1000 points/second, the NMR core analyzer took approximately 2 minutes to acquire each saturation profile measurement. Therefore, the combination of FBG sensors with NMR analysis allows a more accurate time tracking of the fluid front migration during core flooding experiments. It is worth mention that after about 60 mins of injection, the sample did not show uniform saturation distribution. This could be due to air bubble at the top of the sample not fully removed yet or permeability impediment during imbibition.

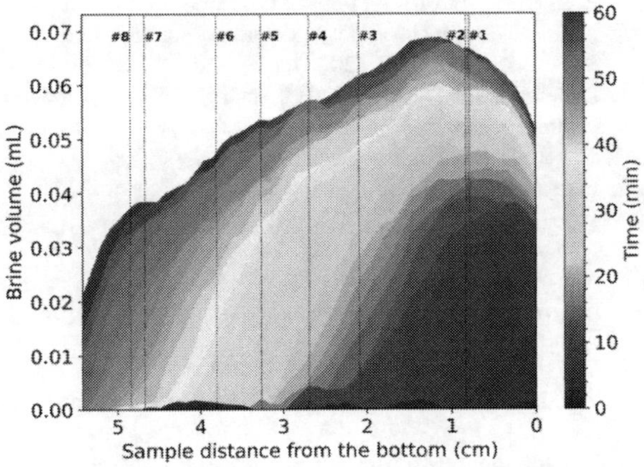


Fig. 11. NMR saturation profiles along the length of the sample.

### 3.3 Pore pressure effect

After the brine breakthrough occurred, a backpressure regulator was applied. A backpressure regulator at 200 psi was used with the purpose to analyze the effect of the pore pressure on the FBG sensors. After the activation of the backpressure, the pore pressure started to increase, while keeping brine injected at 0.1 cc/min, as shown in

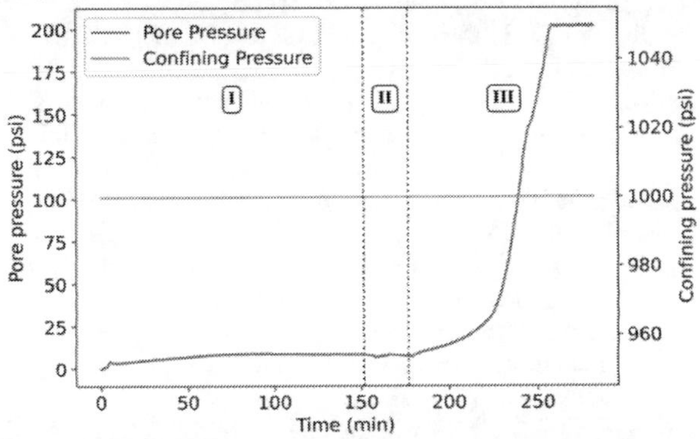


Fig. 12 (section III).

Fig. 13 shows the FBG strain changes from each sensor during the entire experiment. Comparing sections I and III in Fig. 13, it is observed that the increase in the pore pressure caused higher changes in the strain values. While the strain changes in section I were between 2 and 3 με, the strain changes increased to values between 30 and 50 με at the end of section III. This change can be explained by Fig. 14, which shows that an increase in the pore pressure causes an increase in the strain measurements of all sensors.

According to Fig. 13 and Fig. 14, each sensor shows a different strain behavior while the pore pressure is increased, mainly due to rock heterogeneity and internal pore structure (see Fig. 1 and Fig. 2). For instance, sensors 4 and 5, glued on the permeable region of the rock, see Table 1, showed the highest strain value/deformation among the sensors. On the other hand, sensors 2 and 3, glued on/near the bedding area, see Table 1, shows the least strain value. This is a clear indication that all the FBG sensors were very sensitive to detect the internal pore fluid distribution of the sample. As pore pressure is gradually increased inside the rock, brine navigating its way through more permeable section, the pore pressure distribution is not uniform anymore, and accordingly, the net overburden pressure acting on the sample surface is not uniform. It is worth noting that, we anticipated sensor 8 would behave similarly to sensor 2 (see Table 1), but surprisingly the sensor 8 showed a higher deformation. This could be due to test platen effects that caused more damage around the sensor area.

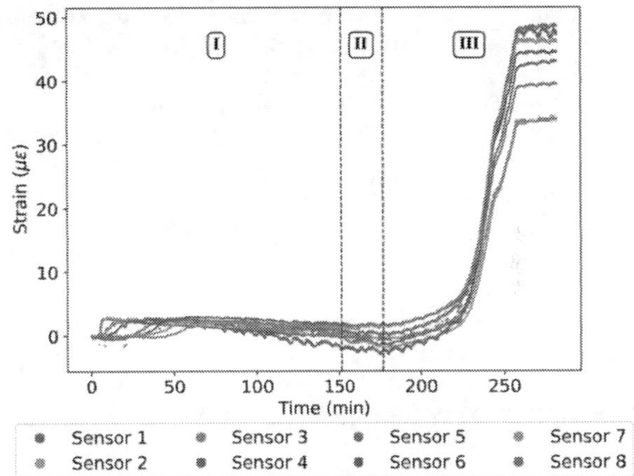


Fig. 13. Strain changes, measured by FBG sensors, during the entire experiment: I – brine injection at 0.1 cc/min while the outlet section was open to the atmosphere; II – injection paused to connect the backpressure; III – backpressure activated and set to 200 psi.

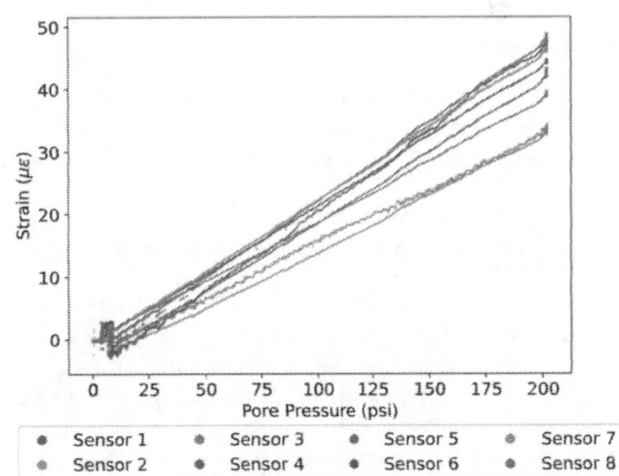


Fig. 14. Differential strain measurements at varying pore pressures for each sensor, measured during the pore pressure build-up.

## 4 Conclusion

In this study, a new method is presented to perform NMR core flooding experiments associated with fiber optic sensing technology. FBG array of 8 sensors were employed to monitor rock strain induced by the brine injection. We demonstrated that the FBG sensors responded quickly and accurately to the changes in the confining pressure, pore pressure and also were able to track the fluid migration through the Savonnières sample. In addition, we showed that the response of the FBG sensors was related to the sample's structure. Sensors facing the high void fraction sector of the sample were more sensitive to both confining and pore pressure changes than the sensors facing the lower void fraction sector (i.e., denser material).

Prior to brine injection, the confining pressure was increased to 1000 psi and kept constant throughout the experiment stages. The FBG measured changes in the rock strain with high accuracy caused by the increase in the confining pressure, showing an inversely proportional behavior. During the first stage of the brine injection, the outlet was kept open to the atmosphere to capture the fluid front migration. The FBG sensors demonstrated the capability of monitoring the fluid front migration, determining its location along the sample, at a faster rate and higher resolution than the NMR system. After the brine breakthrough, a backpressure regulator was activated to analyze the effect of pore pressure build-up on the FBG strain measurements.

In summary, this pilot experiment proves to be useful to monitor rock strain changes during NMR core flooding stages. Future studies are planned to assess the behavior of different rock types, injection fluids, including $CO_2$.

The first author acknowledges Edith Cowan University's financial support. We also acknowledge the financial support provided by CSIRO Energy Business Unit through a Strategic Research Fund.

## References

1. M. Li, D. Xiao, M. Shakerian, A. Afrough, F. Goora, F. Marica, L. Romero-Zerón, and B. Balcom, International Symposium of the Society of Core Analysts, Snowmass, Colorado, 21-26 (2016)
2. T. Hiller, G. Hoder, A. Amann-Hildenbrand, N. Klitzsch, and N. Schleifer, E3S Web of Conferences, 03005 (2020)
3. S. Hua, Y. Li, and Q. Di, E3S Web of Conferences, (2020)
4. G. R. Coates, L. Xiao, and M. G. Prammer, *NMR logging: principles and applications* (Haliburton Energy Services Houston, 1999)
5. J. Mitchell, T. C. Chandrasekera, D. J. Holland, L. F. Gladden, and E. J. Fordham, Phys. Rep., **526**, 165-225 (2013)
6. C. McPhee, J. Reed, and I. Zubizarreta, *Core analysis: a best practice guide* (Elsevier, 2015)
7. J. Rutqvist, and C.-F. Tsang, Environ. Geol., **42**, 296-305 (2002)
8. Y. Sun, Q. Li, D. Yang, C. Fan, and A. Sun, Eng. Geol., **213**, 1-10 (2016)
9. Y. Sun, Q. Li, and C. Fan, Int. J. Greenh. Gas Control., **60**, 186-198 (2017)
10. Y. Kovalyshen, S. Banks, and A. Giwelli, 52nd US Rock Mechanics/Geomechanics Symposium, (2018)
11. X. Qiao, Z. Shao, W. Bao, and Q. Rong, Sensors (Basel), **17**, (2017)
12. P. F. Díaz-Maroto, A. Fernández-López, J. García-Alonso, M. Iglesias, and A. Güemes, Thin-Walled Struct., **132**, 375-384 (2018)
13. J. Heijmans, L. Cheng, and F. Wieringa, 4th European Conference of the International Federation for Medical and Biological Engineering, 2330-2334 (2009)
14. H.-N. Li, D.-S. Li, and G.-B. Song, Eng. Struct, **26**, 1647-1657 (2004)
15. C. S. Baldwin, Fiber Optic Sensors and Applications XII, 94800D (2015)
16. B. Glisic, and Y. Yao, Struct. Health Monit., **11**, 696-711 (2012)
17. Y. Zhang, Z. Xue, H. Park, J. Q. Shi, T. Kiyama, X. Lei, Y. Sun, and Y. Liang, Water Resour. Res., **55**, 856-867 (2019)
18. L. Grattan, and B. Meggitt, *Optical fiber sensor technology: advanced applications-Bragg gratings and distributed sensors* (Springer Science & Business Media, 2013)
19. C. E. Campanella, A. Cuccovillo, C. Campanella, A. Yurt, and V. M. N. Passaro, Sensors (Basel), **18**, (2018)
20. I. B. Trindade, and R. O. Louro: 'Introduction to biomolecular nuclear magnetic resonance and metals': 'Practical Approaches to Biological Inorganic Chemistry' (2020), pp. 155-199
21. P. R. Connolly, J. Sarout, J. Dautriat, E. F. May, and M. L. Johns, Petrophysics-The SPWLA Journal of Formation Evaluation and Reservoir Description, **61**, 151-161 (2020)
22. M. Seyyedi, H. K. B. Mahmud, M. Verrall, A. Giwelli, L. Esteban, M. Ghasemiziarani, and B. Clennell, Sci Rep, **10**, 3624 (2020)
23. A. Raza, R. Gholami, and M. Sarmadivaleh, AAPG Bull., **103**, 83-96 (2020)
24. Y. Zhang, M. Lebedev, M. Sarmadivaleh, A. Barifcani, and S. Iglauer, SPE asia pacific oil & gas conference and exhibition, (2016)
25. M. Lebedev, M. E. J. Wilson, and V. Mikhaltsevitch, Geophys. Prospect., **62**, 1253-1265 (2014)
26. Geo-Spec 2-53 User Manual, Version 1.8, Oxford Instruments
27. GIT Systems and LithoMetrix User Manual, Revision 1.9, Green Imaging Technologies

# Core cleaning and wettability restoration – selecting appropriate method

*Raheleh* Farokhpoor[1], *Lars* Sundal[1], *Anne* Skjærstein[1], *Albert* Hebing[2], *Xiangmin* Zhang[2], and *Lucian* Pirlea[2]
[1]Lundin-Energy, Norway, [2]PanTerra Geoconsultants, The Netherlands

**Abstract.** The core cleaning approaches aim to remove native crude oil, mud filtrates and evaporated salts which leads to strongly water-wet cores. In the standard core cleaning method, the solvent(s) injection continues until no more oil is observed in effluent. This is generally confirmed by visual examination, rather than by analyzing the composition of effluent, simply because it is expensive.

In this work, we selected two preserved rock sections from an oil reservoir in North Sea. 1.5-inch diameter core plugs were cleaned by two different flush cleaning programs (so-called Program-1 and 2). In Program-1, samples were flushed first with 75% toluene + 25% methanol and then followed by alternate of toluene-chloroform-methanol until the samples were deemed cleaned from visual examination. In Program-2, samples were flushed by alternating toluene-tetrahydrofuran-chloroform for 8 cycles. Oil effluents from Program-2 were collected for composition GC analyses.

The efficiencies of two flush cleaning programs were compared by measuring the wettability of the samples by combined Amott-USBM wettability method. The USBM index shows samples after cleaning program-2 are more water wet. Samples were then aged in crude oil at reservoir conditions for 4 weeks. Drainage and imbibition capillary pressure using centrifuge and porous plate and the end-point liquid permeabilities were then measured on restored samples. End-point properties after aging process, showed wettability being restored to intermediate water-wet as is expected from field production history.

This study shows that standard cleaning does not fully clean these samples. Meanwhile, the elaborate Program-2 is often impractical, takes too long, is expensive and limited by laboratory capacity.

## 1. Introduction

Not every feature in core is geological! No matter how careful the attempt to collect and preserve core is, it can be affected by the process of collection, like drilling induced fractures, discoloration by drilling fluids and cutting marks.

A big challenge in SCAL analysis is restoring wettability of core material to a representative reservoir wettability. Wettability determines fluid distribution and fluid flow of wetting and non-wetting phases and is greatly affected by wettability restoration protocol. A standard protocol consists of cleaning the samples to a strong water-wet state, establishing representative initial water saturation and then, aging the samples with dead crude oil at reservoir conditions [1], [2]. The last step is believed to reproduce the core wettability to a representative reservoir state. Inadequate cleaning at the first step may results in a weak water-wet state and disrupt the final wettability state of the core.

The motivation of this present work comes from an earlier extensive SCAL study on preserved core samples from a sandstone reservoir in the Norwegian Continental Shelf (NCS). Large variations in fluid flow properties from different types of experiments was observed. Experiments included steady-state relative permeability, imbibition capillary pressure from centrifuge and porous plate and combined Amott-USBM wettability tests. These contradictory results led us to re-examine the wettability restoration procedure and ultimately the cleaning program. Cleaning is often performed using solvent in Soxhlet extraction or flushing in a core holder at controlled temperature. A typical solvent system used is combination of toluene and methanol, or chloroform or azeotropic solvents mixture to increase efficiency. But we question if this approach works equally efficient for all types of conventional oil.

In this study, the typical cleaning protocol is compared with a more effective program which tends to remove any trace of hydrocarbon in the cores and it was inspired by a relevant study performed on sandstone cores from North

© The Authors, published by EDP Sciences. This is an open access article distributed under the terms of the Creative Commons Attribution License 4.0 (http://creativecommons.org/licenses/by/4.0/).

 https://doi.org/10.1051/e3sconf/202336601009

Sea field [3]. This was followed by compositional analysis of effluents after each cleaning cycles. Afterwards, wettability of the clean samples was determined by Amott-USBM method to undermine the effectiveness of the two programs. Later, the most water-wet samples were aged by dead crude oil at reservoir conditions. And then, imbibition capillary pressure, endpoint saturation and relative permeabilities were measured and compared with not-aged samples.

## 2. Core material

Two preserved core materials (10 cm in diameter and 20 cm in length), from oil and waters zones, were selected for this study. X-ray CT is used to visualize the internal structure of whole core pieces and plugs and are shown in Figure 1. Heterogeneities, fractures, mud invasion and other factors that affect the local density will show up on a CT-image. The 35mm diameter core plug samples are drilled parallel to apparent bedding with a synthetic formation water (SFW). From preserved samples (PS), a subsample was taken for petrographic analysis (Thin section, SEM and XRD Whole Rock and Clay Fraction) and high pressure mercury injection (MICP) on cleaned and dry samples.

The samples are characterized as fine- to medium-grained subarkoses mainly containing quartz grains with abundant K-Feldspar grains. Primary porosity is moderately interconnected and authigenic illite locally fills intergranular pores. Clusters of kaolinite and mixed layer illite-smectite was also observed, see Figure 2. Whole rock and clay characterization was done by XRD analysis on end-trims and are given in Table 1. From MICP experiments, fairly similar pore throat size distribution was derived for both samples. As it can be observed in Figure 3, pore throat size is predominately unimodal with mean hydraulic radius of 11-15 μm.

Final check was to make sure the preserved cores were not damaged by drilling mud which is water-based mud in this case. A large thin section (10 x 4 cm) from a whole core was taken with the short side parallel to the core axis and the long axis parallel to bedding. High resolution images (Figure 4) shows mud particles are only present in thin layers at the outermost edges of the core and have not invaded the core deeper than 3 or 4 layers of sand-sized grains. Therefore, it was concluded the inner part of the core is unaffected by mud particle.

Figure 1: Photo of core section and CT scans of preserved samples (PS).

Table 1: XRD analysis of whole rock and clay fraction composition

| | Whole Rock Composition (%BW) | | | | | | | Clay Fraction (%BW) | | | |
|---|---|---|---|---|---|---|---|---|---|---|---|
| Sample | Quartz | Orthoclase | Kaolinite/ Chlorite | Halite | Illite | Dolomite | Ankerite | Illite | Illite/ Smectite | Kaolinite | Chlorite |
| PS 1 | 60 | 32 | 4 | tr | 1 | - | 2 | 20 | 12 | 54 | 14 |
| PS 2 | 62 | 28 | 6 | 1 | 2 | 1 | - | 10 | 2 | 88 | - |

Figure 2: Images of thin section plates and SEM analysis (to the right). Grain dissolution resulted in mouldic pores (MP) and oversized pores and some places is filled by deformed kaolinite (kao) or patches of dolomite cement (dol). Grain surfaces exhibit continuous coatings illite and illite-smectite (il). Q: quartz, Fsp: feldspar, Cl: chlorite, IP: Intercrystalline pores, SP: secondary porosity.

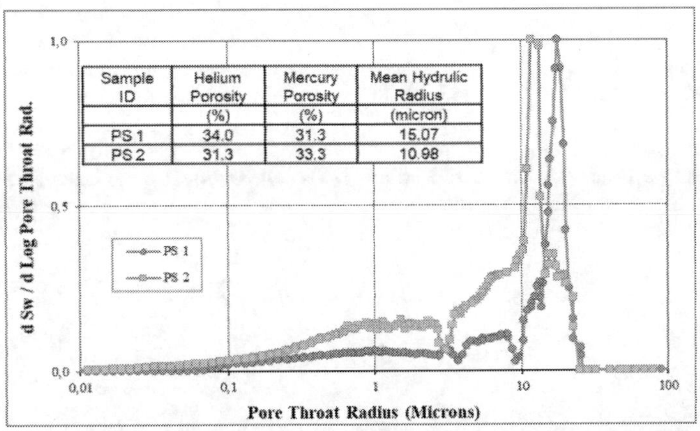


| Sample ID | Helium Porosity (%) | Mercury Porosity (%) | Mean Hydrulic Radius (micron) |
|---|---|---|---|
| PS 1 | 34.0 | 31.3 | 15.07 |
| PS 2 | 31.3 | 33.3 | 10.98 |

Figure 3: Pore throat size distribution from MICP experiment.

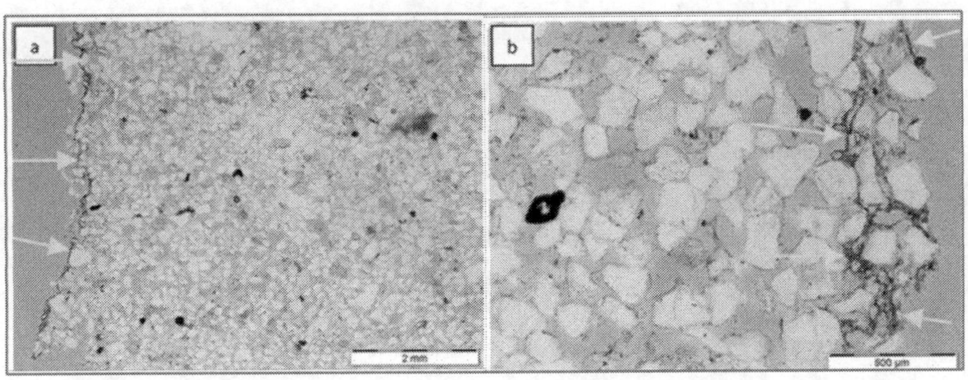


Figure 4: Thin section photomicrographs illustrating the occurrence of mud in the core sample.

## 3. Core cleaning

In this section the comparative cleaning results of 4 core plug samples are described. Cleaning Program-1 is a standard protocol and cleaning Program-2, is the more elaborate (aggressive) cleaning method. The flush cleaning in both Program-1 and 2 were performed at 60 °C and 20 bars confining pressure with the flow rate of 0.3 ml/min. The individual sample was jacked in a Viton sleeve and loaded into hydrostatic core holder. At controlled temperature, solvent was injected via an automated flushing cleaning system equipped with three KNAUER pumps. Program-1 and 2 are described below:

*Program 1:*

1. Started with low rate (0.5 ml/min) flush cleaning of mixture of 75% toluene and 25% methanol.
2. Then followed by alternate injection of toluene-chloroform-methanol until effluent was clear.
3. For each sample about 251 ml (~16 pore volumes) solvent was flushed through the sample
4. Absolute permeability to methanol in both directions was measured before samples were sent to hot oven drying and is given in Table 2.

*Program 2:*

1. The plugs first were flushed with two pore volumes (PV) of toluene to remove the bulk oil.
2. Flushing by 2 PV of tetrahydrofuran (THF) to remove remaining oil and water and effluent was collected (see Figure 5).
3. The oil composition in effluent was measured up to $C_{42+}$ utilizing a high-resolution capillary gas chromatography (GC) and an FID detector.
4. Plugs were then flushed by 2 PV of chloroform to desorb THF in step 2.
5. Steps 1 to 4 were repeated for 8 cycles (~ 42 PV) and effluent after each THF flooding (a total of 8 effluents) were analyzed by GC.
6. At the end, plugs were flushed by methanol to remove polar solvents, salts and water.
7. Absolute permeability to methanol in both directions was measured before samples were sent to hot oven drying.

Table 2: petrophysical properties of core plugs

| Sample | Helium Porosity | Klinkenberg Permeability | Methanol Permeability | |
|---|---|---|---|---|
| | | | Forward | backward |
| Number | % | (mD) | (mD) | (mD) |
| 2 - PS 1 | 32.9 | 1644 | 1143 | 1117 |
| 4 - PS1 | 32.3 | 1374 | 1137 | 958.8 |
| 9 - PS2 | 31.3 | 1373 | 1102 | 967.4 |
| 10 - PS2 | 31.0 | 1122 | 632.2 | 600.0 |

During cleaning Program-2, each subsequent cleaning cycle (methanol/THF/chloroform) removed more oil. The remaining oil fractions became heavier with each flushing, as the molecular weight of crude changed from 225 g/mol at first cycle to 550 g/mol at cycle 8 (see Figure 6), which corresponds to fractions heavier than $C_{42}$. This is explained by the fact that the relatively light fractions are flushed first and the residual oil composition shifting towards a higher C number. This can be also noticed in the shapes of the chromatograms in Figure 7, where there is a clear reduction in lighter fraction and a shifting of the whole spectrum towards the heavy end (after the retention time of 50 on the graph).

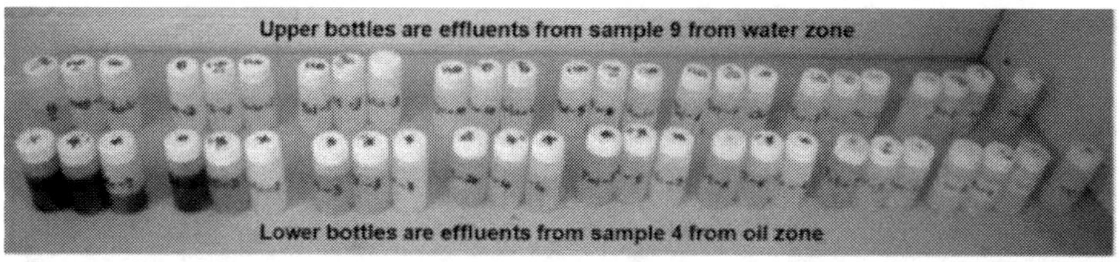


Figure 5: Effluents collected from core plugs after cleaning Program-2. Stack of three bottles representing different cycles.

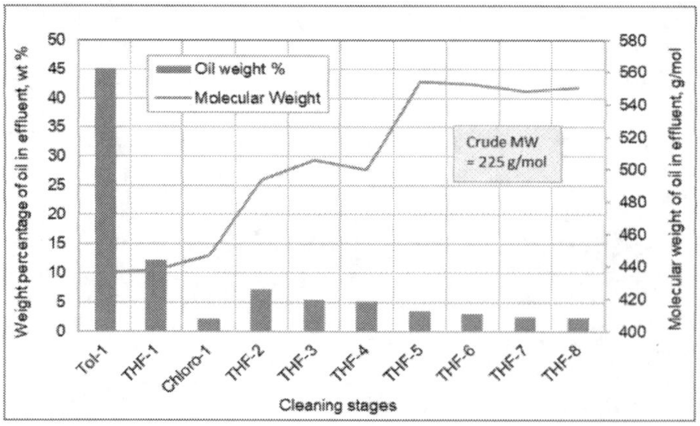


Figure 6: Evolution of oil fraction wt % and molecular weight of oil fraction during the cleaning stages.

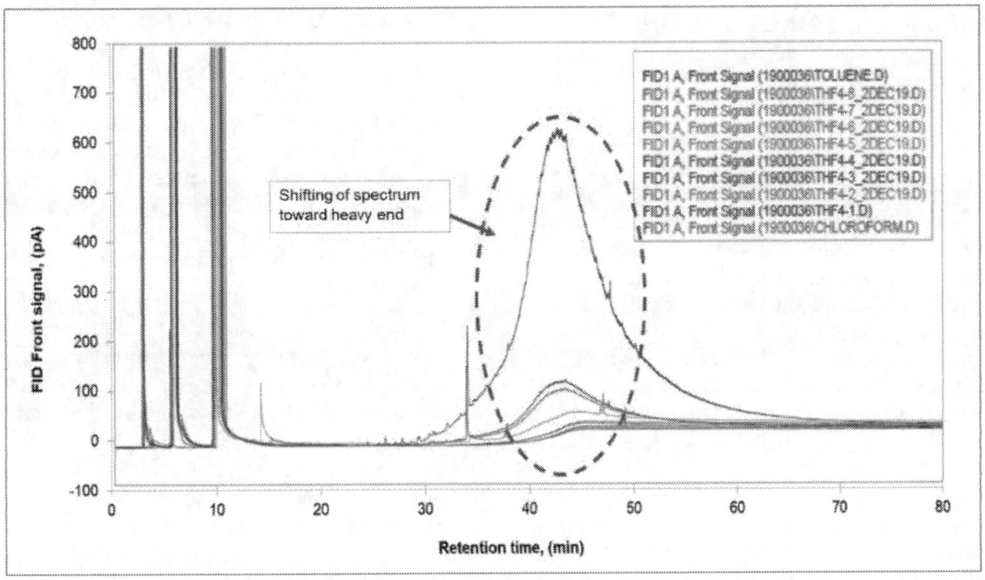


Figure 7: FID front signal over time showing overlapped chromatograms of effluents. First peak is for THF, second for Chloroform and third for toluene. The last peaks are for oil heavy component as it is shifted toward longer retention time from first effluent to eighth effluent.

## 4. Wettability analysis on clean cores

Two cleaned samples from each of the two different cleaning programs were used for combined Amott and USBM wettability analyses [4], [5], [6], [7]. The aim was to compare the effect of cleaning methods on wettability.

The clean and dry samples were first fully saturated with synthetic formation brine. Each sample was first placed in an Amott cell for two weeks filled with mineral oil. The target water saturations (~20-25%) was achieved by a centrifuge oil-brine drainage process. The samples were then placed in glass Amott cells filled with synthetic formation brine and spontaneous imbibition of brine was monitored before the samples were loaded into the centrifuge for forced imbibition. After establishment of $S_{or}$, the samples were loaded into Amott cells where spontaneous and forced imbibition of oil was measured.

Based on the same experimental data, two types of wettability indices are calculated, the USBM wettability index (W) and Amott-Harvey Index and are listed in Table 3. All four samples are water-wet based on wettability index. It was observed that water was much easier to spontaneously imbibe into samples than Soltrol oil. This is a direct indication that water is the wetting fluid. Although the Amott-Harvey index does not show much difference in wettability of the four samples, the USBM index seems to show that sample 4 and 9, cleaned by program-2, are more water wet compared to sample 2 and 10, cleaned by Program-1.

Table 3: The Amott-Harvey and USBM indexes after combined Amott-USBM Wettability test at ambient conditions

| Sample Number | Amott- Harvey Index | USBM Index | Cleaning Program |
|---|---|---|---|
| 2-PS 1 | 0.54 | 0.29 | Program-1 |
| 4-PS 1 | 0.47 | 0.75 | Program-2 |
| 9-PS 2 | 0.52 | 0.52 | Program-2 |
| 10-PS 2 | 0.57 | 0.18 | Program-1 |

Effective permeability for oil at irreducible water saturation ($S_{wi}$) and for water at residual oil saturation ($S_{or}$) were also measured. The core was mounted in a core holder with 20 bar net overburden pressure, measurements were performed at room temperature with four different flow rates. The results are summarized in Table 4. The oil permeability at irreducible water saturation ($K_O$ @ $S_{wi}$) is close to the Klinkenberg permeability and significantly higher than water permeability at residual oil saturation ($K_W$ @ $S_{or}$). These end-points also support that the samples are water-wet and are consistent with combined Amott-USBM wettability test.

Table 4: The end-point oil and water saturations and permeabilities

| End Point data from Amott-USBM | | | | | |
|---|---|---|---|---|---|
| Sample Number | $S_{wi}$ (frac) | $K_W@S_{or}$ (mD) | $S_{or}$ (frac) | $K_O@S_{wi}$ (mD) | $S_{wi}$ (frac) |
| 2-PS 1 | 0.200 | 83.6 | 0.298 | 1116 | 0.217 |
| 4-PS 1 | 0.200 | 359.3 | 0.194 | 1131 | 0.236 |
| 9-PS 2 | 0.250 | 85.6 | 0.296 | 1135 | 0.289 |
| 10-PS 2 | 0.250 | 130.6 | 0.255 | 921.7 | 0.284 |

## 5. Capillary pressure and endpoint permeability after wettability restoration

For further investigation, four more core plugs were selected from the two preserved samples and then they were flush cleaned by Program-2. The clean and dry samples were first fully saturated with synthetic formation brine. The samples were spun in centrifuge to bring to the initial (target) water saturation ($S_{wi}$) by injecting crude oil. Samples were then aged in crude dead oil at reservoir conditions (80 °C and 190 bars) for 4 weeks. Properties of formation water and reservoir oil are given in Table 5. Afterwards, the crude oil in the sample was replaced first by decaline and then by Soltrol oil. Flooding the core plugs by decaline is to avoid any chemical reaction between crude oil and Soltrol. The oil permeability at initial water saturation ($S_{wi}$) was measured by the end of the flushing by Soltrol oil. Oil-brine drainage and imbibition capillary pressures were then measured by two different methods, multi-speed centrifuge and porous plate.

Table 5: Reservoir fluid properties

| Parameter | Unit | Value |
|---|---|---|
| Viscosity of reservoir oil | cp | 0.52 |
| Density of stock tank oil | kg/Sm³ | 855 |
| Density of reservoir oil | kg/Sm³ | 700 |
| Density of formation water | kg/Sm³ | 1050 |

### 5.1 Capillary pressure by centrifuge:

After measurement of $K_O$ @ $S_{wi}$, the samples were weighed and loaded into an ultracentrifuge in (brine) imbibition mode. The volume of displaced oil was monitored on consecutive days at each capillary pressure to ensure samples had reached stable saturations. The samples were stored for few days to allow the fluid to redistribute across the sample before measuring effective permeability for water at residual oil saturation ($K_W$ @ $S_{or}$).

The samples were loaded again into an ultracentrifuge in a drainage mode where brine was displaced by mineral oil (secondary drainage process). End-face fluid saturation at each capillary pressure step was modelled with Hassler-Brunner method. Capillary pressure at the final stage of centrifuge and corresponding bound numbers for imbibition and drainage are given in Table 6. Afterwards, the effective permeability for oil at irreducible water saturation ($K_O$ @ $S_{wi}$) was measured. The results are presented in Table 7 and Figure 8.

Table 6: Capillary pressure and corresponding bound number used at the last stage of multi-speed centrifuge

| Sample | Imbibition | | Secondary Drainage | |
|---|---|---|---|---|
| | End-Pc | Bound No. | End-Pc | Bound No. |
| Number | (Pci) | [-] | (Pci) | [-] |
| 8 - PS 1 | -62.6 | -2.4E-04 | 50.3 | 1.9E-04 |
| 14 - PS 2 | -62.7 | -5.5E-04 | 50.4 | 4.4E-04 |

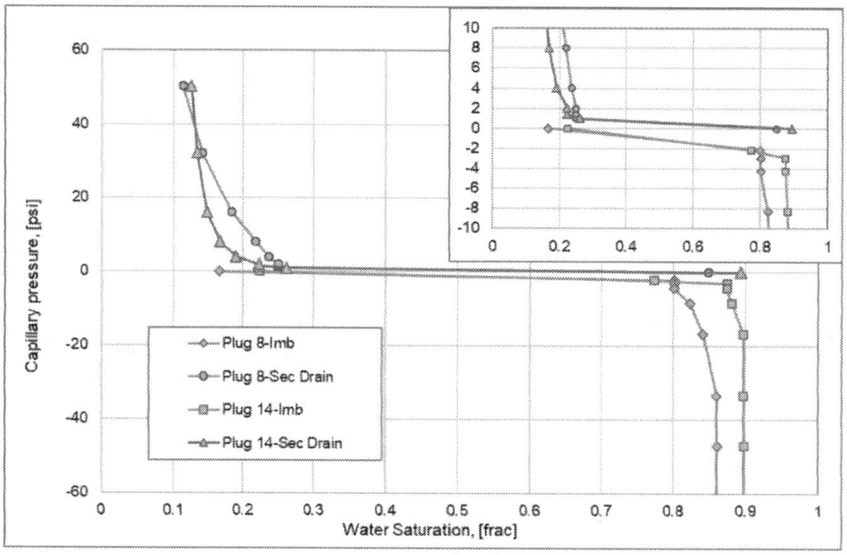


Figure 8: Oil-brine imbibition and secondary drainage capillary pressure measured by centrifuge on aged samples.

Table 7: Endpoint saturation and permeabilities after multi-speed centrifuge test

| Routine Analysis | | | Primary Drainage | | Imbibition | | Secondary Drainage | |
|---|---|---|---|---|---|---|---|---|
| Sample Number | Porosity (%) | Klink Perm (mD) | $S_{wi}$ (ave) (%Vp) | $K_O$ @ $S_{wi}$ (mD) | $S_{or}$ (ave) (%Vp) | $K_W$ @ $S_{or}$ (mD) | $S_{wi}$ (ave) (% Vp) | $K_O$ @ $S_{wi}$ (mD) |
| 8 - PS 1 | 33.1 | 1264 | 0.166 | 717.3 | 0.152 | 586.4 | 0.163 | 841.1 |
| 14 - PS 2 | 33.4 | 2875 | 0.223 | 1505 | 0.106 | 2065 | 0.146 | 2179 |

## 5.2 Capillary pressure by Porous Plate:

The other two samples were utilized in Pc-RI experiments where capillary pressure was determined by the porous plate method. Samples saturated with Soltrol at $S_{wi}$, were mounted in core holders with a net overburden pressure of 115 bar. SFW at different pressure stages was injected from bottom. The top-end of the core is in capillary contact with an oil-wet membrane (porous plate) and it is only permeable to oil under pressure differentials less than the pore entry pressure of the membrane. Multi-step imbibition capillary at different saturations was measured, and then porous plate was removed and water permeability at residual oil saturation ($K_W$ @ $S_{or}$) was measured. Afterwards, porous plate was replaced by a water-wet ceramic plate and Soltrol was injected to the core. Once equilibrium was achieved (this took few weeks), the pressure is increased, and the process repeated until a full curve of secondary drainage capillary pressure and resistivity index was obtained. Oil permeability at irreducible water saturation ($K_O$ @ $S_{wi}$) was then measured and is given in Table 8.

Imbibition and secondary drainage capillary pressure curves is shown in Figure 9. Despite differences in petrophysical properties of sister plugs, for similar capillary pressure, generally lower residual oil saturation (after imbibition) and irreducible water saturation (after secondary drainage) was achieved by centrifuge experiment (see Table 7 and Table 8).

Table 8: Endpoint saturations and permeabilities after imbibition and secondary drainage porous plate test

| Routine Analysis | | | Primary Drainage | Imbibition | | Secondary Drainage | |
|---|---|---|---|---|---|---|---|
| Sample | Porosity | Klink Perm | $S_{wi}$ (ave) | $S_{or}$ (ave) | $K_W$ @ $S_{or}$ | $S_{wi}$ (ave) | $K_O$ @ $S_{wi}$ |
| Number | (%) | (mD) | (%Vp) | (%Vp) | (mD) | (% Vp) | (mD) |
| 7 - PS 1 | 34.1 | 2002 | 0.232 | 0.201 | 330.7 | 0.248 | 431.4 |
| 13 - PS 2 | 34.1 | 1045 | 0.242 | 0.209 | 350.1 | 0.180 | 1079 |

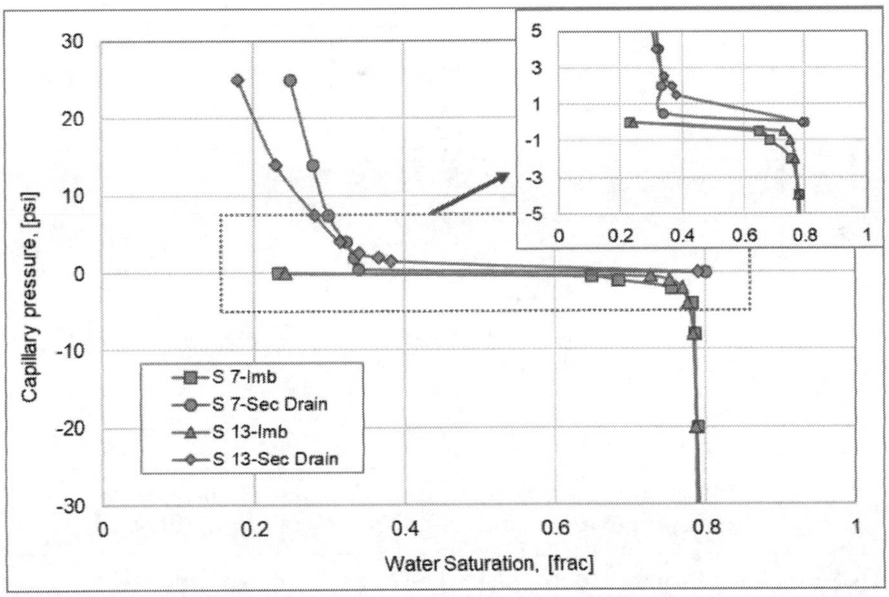


**Figure 9:** Oil-brine imbibition and secondary drainage capillary pressure measured by porous plate on aged samples.

## 6. Discussion

Four core samples (two sister plugs) were flushed-cleaned by two different cleaning methods; standard method (Program-1) and more aggressive method (Program-2). After cleaning, wettability was confirmed with combined Amott-USBM method as water-wet. The Amott-Harvey index showed similar wettability for the four samples, while the USBM index exhibit samples cleaned by Program-2 are more water wet compared to sample cleaned by Program-1.

For further investigation, 4 more core samples (two sister plugs) from similar preserved sample were flushed cleaned by Program-2. Then, the core samples were aged with crude oil at reservoir conditions to restore the in-situ wettability. Afterwards, to examine the core wettability state in terms of fluid flow, oil-brine imbibition and secondary drainage capillary pressure and end-point saturations and relative permeabilities were measured with two different approaches; multi-speed centrifuge and porous plate.

Imbibition capillary pressure measured on aged and not aged samples from preserved sample 1 and 2 are compared in Figure 10 and Figure 11. Oil-brine imbibition capillary pressure was measured using centrifuge on plugs 2 and 4 from PS 1 and plugs 9 and 10 from PS 2, right after cleaning and without aging process. The plugs are strongly water-wet (as shown earlier in Table 3) and it explained the large increase in water saturation by spontaneous water imbibition. On the other hand, for plugs 7 and 8 from PS 1 and plugs 13 and 14 from PS 2 which are aged, almost no spontaneous water imbibition was observed. The wettability of these plugs is quantified as intermediate water-wet, although less than 1 psi capillary pressure was required to make a large shift in water saturation.

Considering the PS 1 is from oil zone and PS 2 is from water zone, hence the original wettability for these preserved samples was significantly different. Meanwhile, the cleaning and aging process has resulted in similar wettability behavior for both samples as it is shown in Figure 10 and Figure 11. Therefore, we believe the current rock wettability is entirely dependent on crude oil properties and interaction between rock and fluid. Since the aging time (4-weeks) is not comparable with geological time that porous media were exposed to oil, the current wettability might be underestimated toward oil-wet system.

End-point properties including residual oil saturation ($S_{or}$) and water permeability ($K_{rw}$ @ $S_{or}$) at the end of water-flooding for aged and not-aged samples are compared Table 9 and Figure 12. For not-aged samples, $K_{rw}$ @ $S_{or}$ do not correlate with cleaning programs. Low $K_{rw}$ @ $S_{or}$ is simply due to strong water-wet system and it varies with core structures. For aged samples, residual oil saturation at the end of water flooding is lower and subsequently water relative permeability is higher reflecting big shift in end-points properties as wettability changes towards intermediate water-wet system.

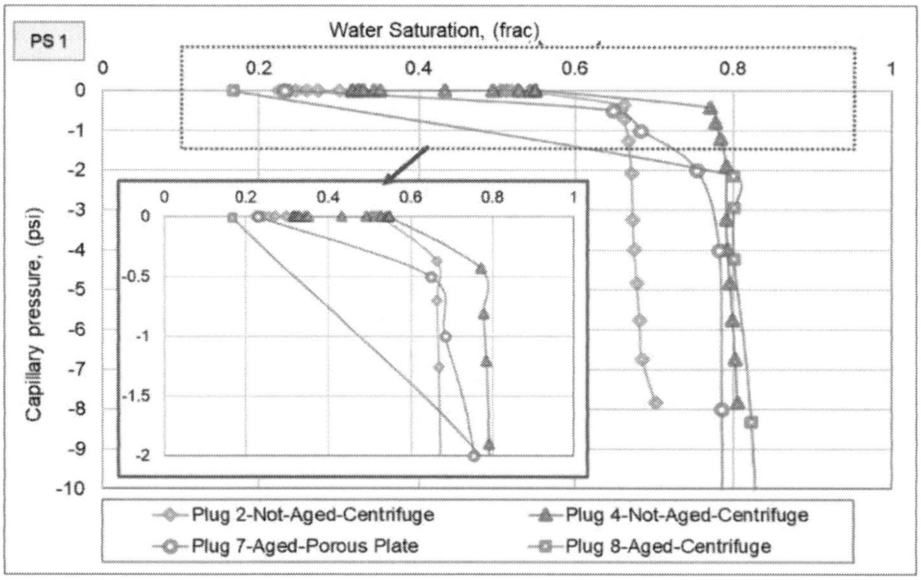


Figure 10: Oil-brine imbibition capillary pressure on sister-plugs from PS 1. Plug 2 (yellow line) is after cleaning program 1, plug 4 (blue triangle) is after cleaning program 2. Plug 7 (green circle) and plug 8 (orange square) are after cleaning program 2 and aging.

Table 9: initial and residual oil saturation and water relative permeability at end of water flooding

| Sample No. | Aging [-] | Cleaning Program | He Poro. (% of Vb) | Klink Perm (mD) | $S_{oi}$ (frac) | $S_{or}$ (frac) | $K_{rw}$ @ $S_{or}$ (frac) |
|---|---|---|---|---|---|---|---|
| 2-PS 1 | × | Program 1 | 32.9 | 1644 | 0.800 | 0.298 | 0.07 |
| 4-PS 1 | × | Program 2 | 32.3 | 1374 | 0.800 | 0.194 | 0.32 |
| 7-PS 1 | ✓ | Program 2 | 34.1 | 2002 | 0.768 | 0.201 | 0.77 |
| 8-PS 1 | ✓ | Program 2 | 33.1 | 1264 | 0.834 | 0.152 | 1.00 |
| 9-PS 2 | × | Program 2 | 31.3 | 1373 | 0.750 | 0.296 | 0.08 |
| 10-PS 2 | × | Program 1 | 31.0 | 1122 | 0.750 | 0.255 | 0.14 |
| 13-PS 2 | ✓ | Program 2 | 34.1 | 1045 | 0.758 | 0.209 | 0.32 |
| 14-PS 2 | ✓ | Program 2 | 33.4 | 2875 | 0.777 | 0.106 | 0.95 |

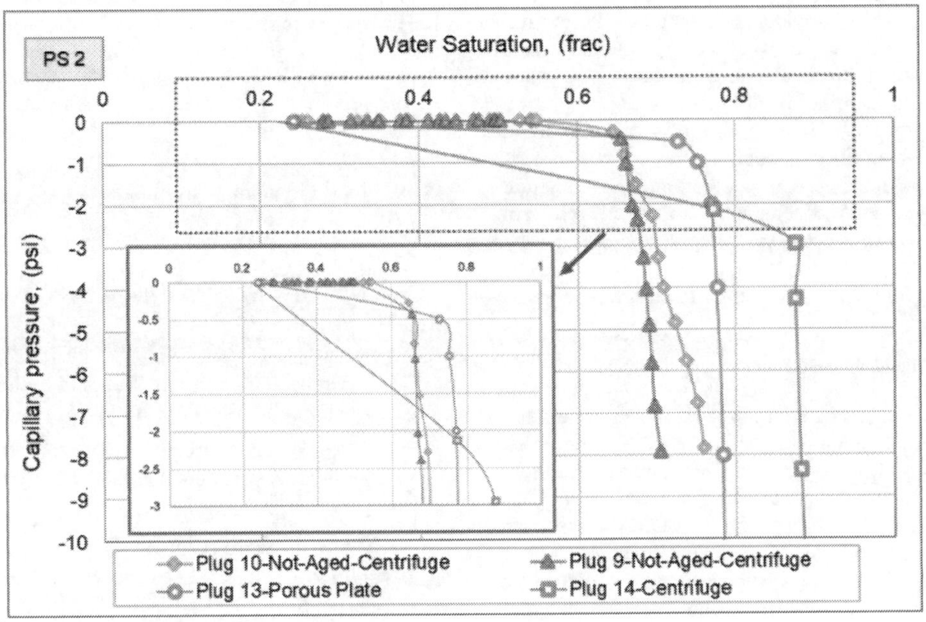


Figure 11: Oil-brine imbibition capillary pressure on sister-plugs from PS 2. Plug 10 (yellow line) is after cleaning program 1, plug 9 (blue triangle) is after cleaning program 2. Plug 13 (green circle) and plug 14 (orange square) are after cleaning program 2 and aging.

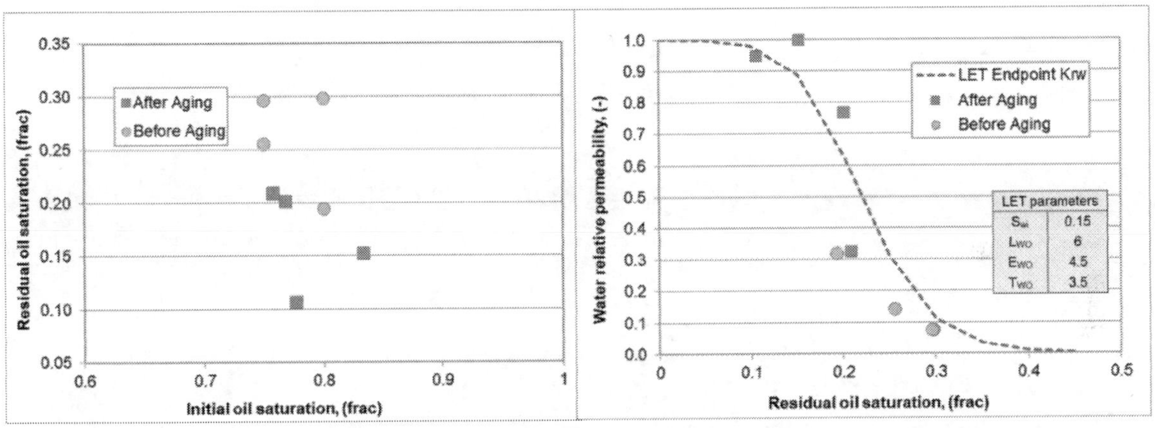


Figure 12: Residual oil saturation and water relative permeability at end of water flooding on aged and not aged samples.

## 7. Conclusion

In the presented work, effect of two cleaning approaches is studied in detail. Core plugs from two preserved samples with similar petrography and petrophysical properties, but one from oil zone and the other from water zone, were flush cleaned by two programs; Program-1 and Program-2. Gas chromatography analysis on effluents produced from the core plugs showed the presence of heavy hydrocarbon (fractions heavier than $C_{42}$) even after more than 40 PV solvent injection.

Wettability of the core plugs were then tested using combined Amott-USBM wettability method. The Amott-Harvey index does not show much different in wettability of the samples. But, the USBM index showed plugs cleaned by the Program-2 are more water-wet compared to those cleaned by Program-1. Spontaneous water imbibition and endpoint properties on clean samples before aging reflects the strong water-wet behavior but did not correlate with cleaning programs. Afterwards, two more plugs from each preserved sample were drilled and cleaned by program-2 and then aged at reservoir conditions. Oil-brine imbibition capillary pressure, endpoint saturation and relative permeability were measured by centrifuge and porous plate. Almost no spontaneous water imbibition was observed on these plugs and endpoint properties were greatly shifted towards intermediate water-wet system. Although the original wettability of the samples was not the same (one from oil zone and the other from water zone), the core treatment procedure resulted in almost similar water-oil flow properties.

Based on our observations, cleaning Program-2 removed more oil, especially heavy hydrocarbons, from the core than typical cleaning protocol so called Program-1. On the other hand, the elaborate Program-2 is often impractical, takes too long, is expensive and limited by laboratory capacity. We believe proper cleaning approach is specific to certain rock and fluid and should be derived separately for an oil field. An effective cleaning method can simply be confirmed by performing gas chromatography compositional analysis of the final effluent at the end of core cleaning.

## Acknowledgements

This study is a part of ongoing SCAL (special core analysis) project on core plugs from an exploration well from Edvard Grieg oil field. Therefore, the authors would like to thank Lundin-Energy Norway and our partners, Wintershall Dea Norge and OMV Norge, for funding this project and for allowing to publish these data. We would also give special thanks to PanTerra Geoconsultants for performing the experiments and providing excellent advices.

## References

[1] C. McPhee, J. Reed and I. Zubizarreta, *Core Analysis: a Best Practice Guide*, Amsterdam: Elsevier, 2015.

[2] E. Donaldson, R. Thomas and P. Lorenz, "Wettability Determination and Its Effect on Recovery Efficiency," **no. 13-20, 1969**.

[3] G. Hirasaki, J. Rohan, S. Dubey and H. Niko, "Wettability Evaluation During Restored-State Core Analysis," in SPE Annual Technical Conference and Exhibition, New Orleans, Louisiana, 1990.

[4] E. Amott, "Observations Relating to the Wettability of Porous Rock," **Vols. 216, (156-162)**, 1956.

[5] W. Anderson, "Wettability literature survey - Part 2: Wettability measurement," **vol. 38:12**, 1986.

[6] W. Anderson, "Wettability Literature Survey - Part 4: Effects of Wettability on Capillary Pressure," **no. 1283-1300**, 1987.

[7] W. Anderson, "Wettability Literature Survey - Part 5: Effects of Wettability on Relative Permeability," **no. 1453-1468**, 1987.

# Comparing Centrifuge, Steady-State and Semi-Dynamic Methods for Relative Permeability and Capillary Pressure Determination: New Insights

*Fabrice* Pairoys[1,*], *Cyril* Caubit[1], *Micah* Alexander[2], and *Javier* Ramos[2]

[1]TOTAL CSTJF, Pau, France
[2]Schlumberger Reservoir Laboratories, Houston, United States

**Abstract.** Relative permeability and capillary pressure are essential parameters for understanding multiphase flow in porous media and scenarios of production in oil or gas reservoirs. There are several experimental methods for determining the relative permeability curves: unsteady-state (USS), steady-state (SS), and semi-dynamic (SD) methods. Each method has advantages and weaknesses. Although the USS approach leads to fast data results, the interpretation neglects the capillary pressure effects and provides a limited amount of data points obtained after breakthrough. The SS method is time consuming but enables covering a wider range of saturation with data points if the test is well designed. The SD method may be more time consuming than the SS method but provides both relative permeability to the injected phase and capillary pressure. The relative permeability to the produced phase is then determined by numerical means. The main objective of this study was to compare the water-oil relative permeability curves obtained from the steady-state and semi-dynamic methods performed at reservoir conditions with live fluids. Carbonate core plugs of same rock type and same properties were selected for this experimental program. The samples were brought to the same irreducible water saturation at a constant brine-oil capillary pressure using a centrifuge before being dynamically aged with live oil. In addition to monitoring the average saturation using material balance (MB), a linear X-ray scanner was used for in-situ saturation monitoring (ISSM) along the core samples. The oil relative permeability from the SD method was simulated with fixed water relative permeability and capillary pressure by history-matching the oil production and the differential pressure signal. Two additional centrifuge tests on twin plug were performed in order to measure imbibition capillary pressure and oil relative permeability at pseudo-reservoir conditions. This comparative study shows that the SD method provides similar capillary pressure and oil relative permeability curves to those obtained by centrifuge methods. Even if all Kr curves are in an acceptable envelop, some differences are observed between SD and SS Kr curves: several investigative leads are given to explain this discrepancy. It is also shown that a better saturation method needs to be implemented, especially when dealing with heterogeneous rocks. While a more robust ISSM method is being tested at TOTAL, the results presented in this paper are very encouraging.

## 1 Introduction

Relative permeability and capillary pressure are important petrophysical parameters for interpreting fluid flow in reservoirs and for calibrating appropriate reservoir simulation models. They are obtained in the laboratory, separately in most of the cases. The only method able to dynamically measure both parameters on a same test is the semi-dynamic method. This method was first implemented by [1]. By recirculating the produced fluid at constant pressure at the outlet of rock sample, the capillary pressure Pc at the core inlet is directly equal to the differential pressure dP, assuming the outlet capillary pressure is nil. The injected fluid relative permeability $Kr_{inj}$ can be obtained from the slope of the injected fluid rate Qi plotted against the inlet Pc (or dP). The inlet injected phase analytical saturation given by [1] has a similar form to that of Welge equation. With this method, relationships for relative permeability of the injected phase and capillary pressure are obtained.

Using the semi-dynamic approach, [2] developed a method to provide both positive and negative Pc for drainage and imbibition cycles at ambient conditions, using an ultrasonic method for monitoring the local saturation. Later, the same method was validated and used by [3] to build an integrated petrophysical tool able to determine full Pc curves, resistivity index and end-point relative permeability at reservoir conditions. [4] used the technique for measuring full Pc curves with live fluids at reservoir conditions. Relative permeabilities were derived by history matching of the transient pressures and saturation profiles acquired by X-ray attenuation method. [5] studied the variability in Pc curves due to local

* Corresponding author: fabrice.pairoys@total.com

© The Authors, published by EDP Sciences. This is an open access article distributed under the terms of the Creative Commons Attribution License 4.0 (http://creativecommons.org/licenses/by/4.0/).

heterogeneity using the semi-dynamic method with x-ray in situ saturation monitoring. [6] and [7] used the method at reservoir conditions with x-ray saturation monitoring for estimating oil recovery from transition zones. More recently, [8] developed a full petrophysical rock characterization workflow similar to [3] but included the full relative permeability curve, the produced phase relative permeability being derived from the electrical properties or using history match process. Few other studies can be found in the literature, there is not one comparing the semi-dynamic capillary pressure Pc and relative permeability Kr to those obtained from the standard methods, centrifuge and steady-state methods. The aim of the study is to compare data results from all these methods, performed at true reservoir conditions with live fluids (except for centrifuge tests). A 1D linear x-ray scanner was used to monitor the saturation profiles obtained at capillary equilibrium. It is shown that a more robust in situ saturation monitoring method is required to avoid erroneous interpretation, especially for heterogeneous rocks.

## 2 Background

The steady-state method is well known and accepted in the industry for determining relative permeability, it is not really the case for the semi-dynamic method despite being investigated since the '90s. A description of the method is found in [1]. The main idea is to recirculate the produced fluid at constant pressure at the outlet of the rock while injecting the displacing fluid at several constant rates at the inlet until steady-state condition (no more variation of saturation and differential pressure). In the following, consider an imbibition cycle, water displacing oil.
The method allows obtaining the relative permeability of the injected water phase, Krw, the analytical water saturation, Sw, and the capillary pressure, Pc, at the inlet of the core. It is a unique way of determining both Kr and Pc during a single coreflooding test.

A set of pressure drops, final average saturations and flow rates are required inputs for the method which consists of three steps:

1- The capillary pressure at the inlet simply equals the pressure drop (with oil gradient pressure and outlet Pc equal to 0):

$$Pc = dP = P_i(inlet) - P_o(outlet) \quad (1)$$

2- The relative permeability of water at the inlet of the sample is calculated using Darcy's law:

$$Krw = \frac{\mu_w L}{k_{eo}(Swi) A} \frac{dQ_w}{dPc} \quad (2)$$

The relative permeability to oil Kro, with fixed relative permeability to water and capillary pressure, is then determined by numerical simulation and history match of the oil production volume, Vo, and differential pressure dP

3- The oil saturation at the inlet is calculated by differentiating the product of steady-state water Darcy velocity and average oil saturation, leading to the following simplified equation:

$$So = <So> + Q_w \frac{d<So>}{dQ_w} \quad (3)$$

and

$$Sw = 1 - So = 1 - \left(<So> + Q_w \frac{d<So>}{dQ_w}\right) \quad (4)$$

Equation 4 can be substituted with a direct measurement of the inlet saturation using in situ saturation monitoring such as x-ray method.

## 3 Rocks and Fluids

In this study, a tight carbonate rock was selected. Several core plugs were extracted from a same whole core to perform composite stack coreflooding experiments, semi-dynamic and steady-state tests, and also centrifuge tests, with the aim of comparing the methods for Kr and Pc characterization. The core plug selection was based on the degree of heterogeneity of the samples, using CT scan images. All fractured core plugs were discarded.

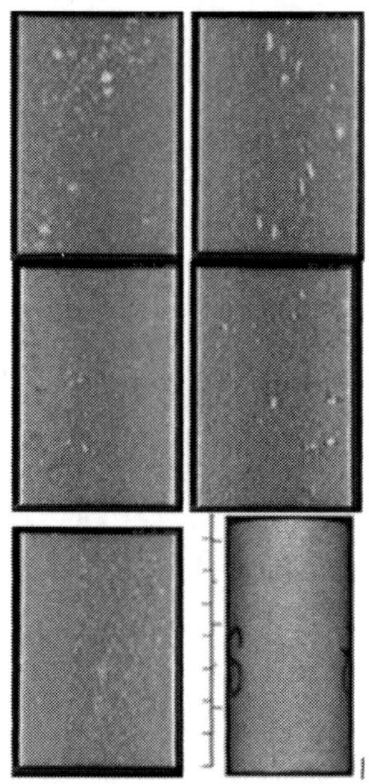

**Fig. 1.** CT scan images of samples S1/S2 (SDM), S3/S4 (SS), S5 (Centrifuge) and 1 core plug photography of sample S1

There were no miscible tracer tests programmed in this study. However, it is highly recommended to perform these tests to discard the rocks with too high longitudinal heterogeneity (especially for the unsteady-state experiments) but also to compare the degree of heterogeneity from one rock to another one.

XRD was performed on the end-trims to confirm the selected core plugs had the same mineralogy. The clay-free samples represent limestones composed of 90-95% calcite with traces of anhydrite, potassium feldspar and quartz. Before measuring the porosity and permeability, the core plugs were batch-cleaned in Soxhlet using a chloroform/methanol azeotropic mixture and then dried in a convection oven at 116°C. The routine properties were measured at 2000 psi of net confining stress (NCS), representative of the NCS applied during the coreflooding and centrifuge tests. Table 1 proves that the samples have similar properties.

**Table 1.** Routine core properties

| Plug Id | Test Type | He ϕ (%) | Kg (mD) | Kw (mD) |
|---|---|---|---|---|
| S1 | SD | 12.7 | 1.35 | 0.75 |
| S2 | SD | 12.1 | 1.25 | 0.51 |
| S3 | SS | 15.6 | 2.09 | 0.92 |
| S4 | SS | 14.7 | 1.21 | 0.63 |
| S5 | Pc centri | 16.1 | 1.78 | 0.66 |

The core plugs were individually saturated with a 220 kppm synthetic formation brine before being desaturated with dead oil. A recombined oil was also prepared to replace the dead oil before performing the coreflooding experiments with live fluids. Fluid properties at various experimental conditions are listed in Table 2:

**Table 2.** Fluid properties

| | Pp (psig) | T (°C) | $\rho_w$ (g/cc) | $\rho_o$ (g/cc) | $\mu_w$ (cP) | $\mu_o$ (cP) |
|---|---|---|---|---|---|---|
| SD&SS | 3000 | 100 | x | x | 0.48 | 3.17 |
| Centrifuge | 0 | 70 | 1.12 | 0.85 | 0.66 | 5.45 |

Both brine and oil were filtered through a 0.1 μm filter to limit potential plugging issues during the tests (based on MICP pore throat size distribution).

Figure 2 shows pore throat radius going down to 0.05μm, highlighting the necessity of the 0.1μm fluid filtering process. The distribution looks unimodal but spread along the pore throat radius range. All selected samples have similar pore throat size distributions.

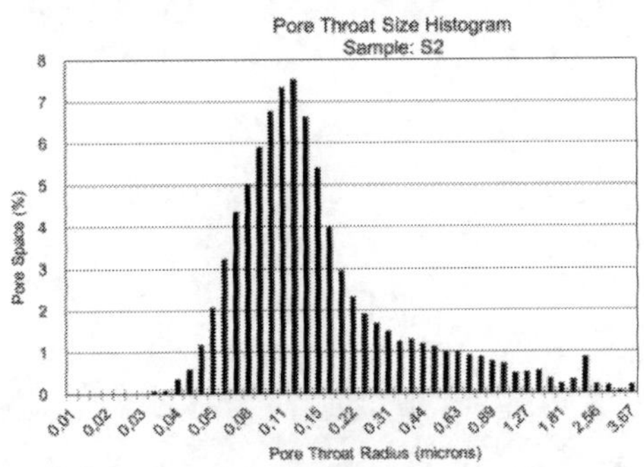


**Fig. 2.** MICP pore throat size distribution

Figure 3 represents an example of thin section (TS) and Scanning Electron Microscopy (SEM) images from S1 end-trims, confirming a certain degree of heterogeneity.

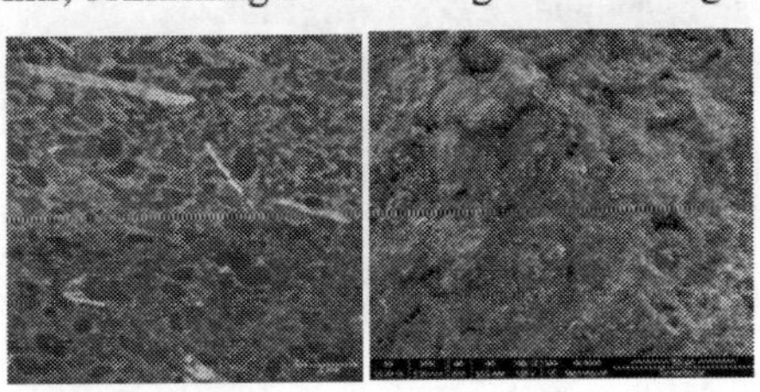

**Fig. 3.** TS and SEM images

## 4 Experimental Setups, Conditions and Methodology

The same coreflooding system was used to perform the SD and SS tests (Figure 4):

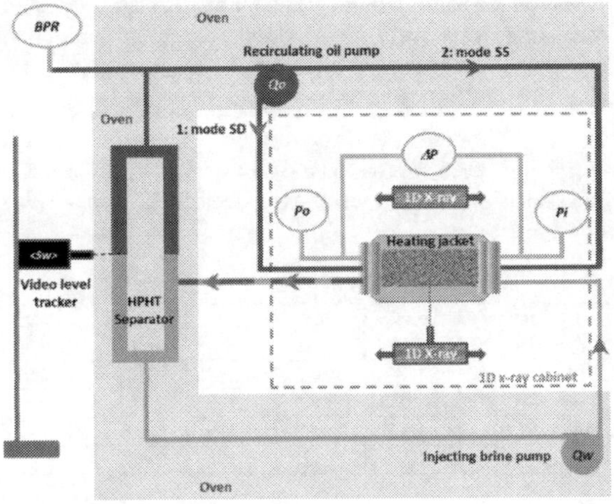


**Fig. 4.** Schematic of the HPHT coreflooding system

The coreflooding system is composed of a triple-pump system per phase -brine and oil- for injection and recirculation (mode 1 for SD method, mode 2 for SS method), an x-ray transparent core holder with heating jacket in x-ray protective cabinet, a 1D linear x-ray system for measuring saturation profiles, one High Pressure High Temperature (HPHT) video level tracker for material balance monitoring, three differential pressure sensors of different ranges with automatic switches, two -inlet and outlet- pressure transducers, a back pressure regulator with single pump pore pressure control, and one single pump controlling the confining pressure. All measurements are automatically recorded during the tests.

A centrifuge with ability to perform the tests under confining pressure was used to determine multispeed centrifuge Pc and single-step Kro tests. Note it is not possible to perform live oil experiments in a centrifuge (no pore pressure, limited temperature). In the centrifuge tests, NCS of 2000 psi was applied (Table 2).

After the 100% brine saturation under vacuum and applied hydrostatic pressure of 5000 psi, the samples were centrifuged at a constant capillary pressure Pc with dead oil up to irreducible water saturation Swi. To reduce the capillary end-effects, they were flipped to flatten the saturation profile during a second centrifuge step. The samples were then loaded in the x-ray transparent core holders (two composite stacks of two samples) and brought to the reservoir conditions (100 °C of temperature, 3000 psi of pore pressure Pp and 5000 psi of confining pressure Pconf. Samples with the lowest permeability were placed at the outlet of each composite stack. Stack 1 was composed of samples S1 and S2 for the SD test. Stack 2 was composed of samples S3 and S4 for the SS test. Sample S5 was loaded in a Hassler core holder and dynamically aged with dead oil for four weeks. S5 was then measured for effective permeability to oil at

https://doi.org/10.1051/e3sconf/202336601010

irreducible water saturation, then loaded in a centrifuge core holder for the forced imbibition cycle (without spontaneous imbibition). After cleaning and measuring porosity and permeability properties to ensure the rock was not altered during the centrifuge multistep test and cleaning, the same sample was later aged again and tested for single-step centrifuge desaturation in order to calculate the relative permeability to oil.

Four weeks of dynamic aging was performed to restore the rock wettability. As for S5, the effective oil permeability Keo at Swi on the two stacks (stack 1 for SD and stack 2 for SS tests) was measured prior water flooding.

**Table 3.** Initial conditions

| Test Type | Sample Id | Oil Type | Keo (mD) | Swi (frac.) | X-Swi (frac.) |
|---|---|---|---|---|---|
| SD | S1/S2 | Live | 0.387 | 0.223 | 0.224 |
| SS | S3/S4 | Live | 0.385 | 0.193 | 0.188 |
| Centri Pc | S5 | Dead | 0.216 | 0.197 | x |
| Centri Kro | S5 | Dead | 0.196 | 0.203 | x |

The design of the SD and SS Kr corefllooding tests and multistep Pc and single-step Kro centrifuge tests was performed using the core analysis software CYDAR, using a first guess of Kr/Pc anticipating a slightly water-wet behaviour based on previous works.

## 5 Experimental Results

<u>SD results on S1/S2</u>:

Figure 5 represents the differential pressure and the oil production versus time acquired during the SD experiment:

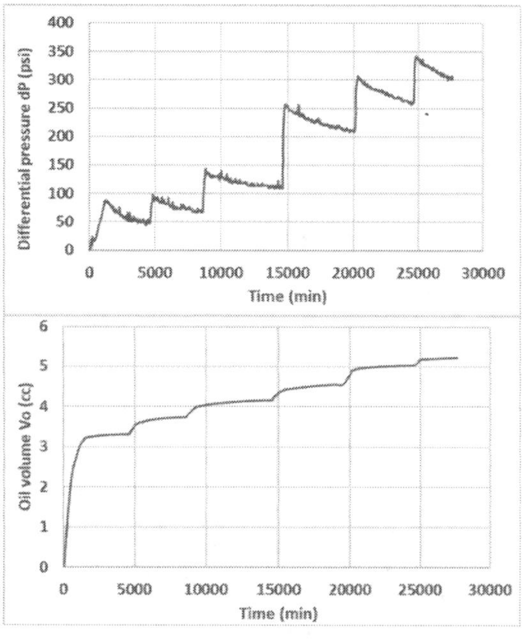


**Fig. 5.** dP and produced Vo versus time during the SD waterflood

The stabilization criteria were difficult to reach: if the stabilization criterion for oil production (<0.5% PV change in 24hr) was obtained, the dP stabilization was not fully satisfying, leading to a potential slight shift of the resulting Pc curve. Concerning the saturation profiles, the applied stabilization criterion was 10 overlapping and consecutive scans. This criterion was satisfied.

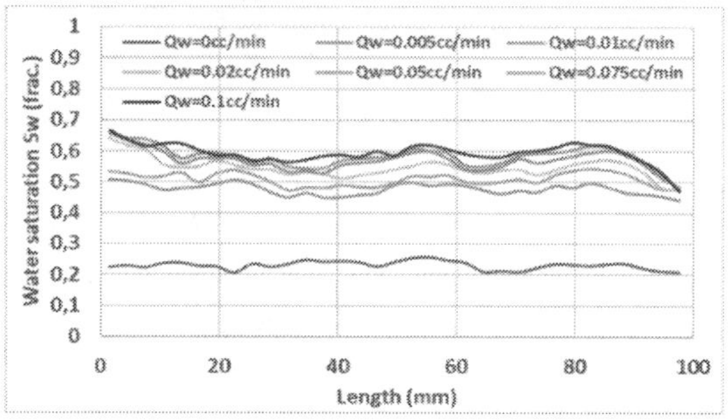


**Fig. 6.** Saturation profiles at equilibrium

To reduce the error on the counting, a minimum of 10,000 counts was successfully targeted, leading to less than 1% error on the number of transmitted photons (calculation via Poisson's law): it was achieved by applying a counting time per point of 5 seconds, at the specific applied energy.

The experimental design and specification limitations did not allow to obtain saturation profiles well spread on the full saturation range (Figure 6). The average of the water saturation profiles <$Sw_{x\text{-}ray}$> obtained from x-ray attenuation technique and average water saturation <$Sw_{sep}$> obtained from material balance using the separator at each stabilized step were found to be very close (<3 saturation units, s.u.), as shown on Figure 7:

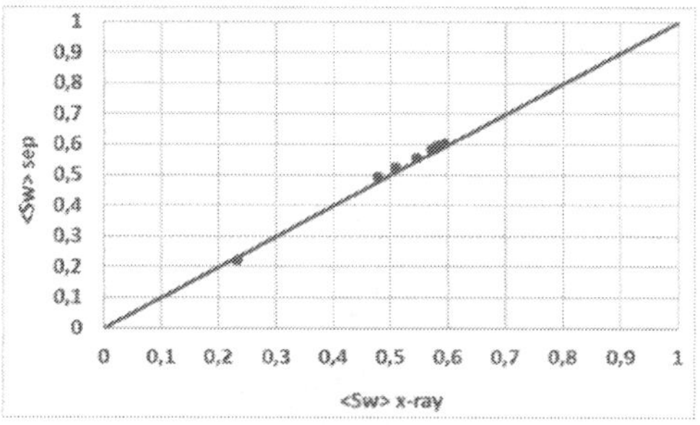


**Fig. 7.** Comparison between separator <Sw> and x-ray <Sw> during the SD waterflood

Figure 7 shows the robustness of both methods to acquire averaged saturations. On the other hand, higher discrepancy was observed between the analytical inlet local Sw calculation using Equation 4 and inlet local Sw value from the linear 1D x-ray method:

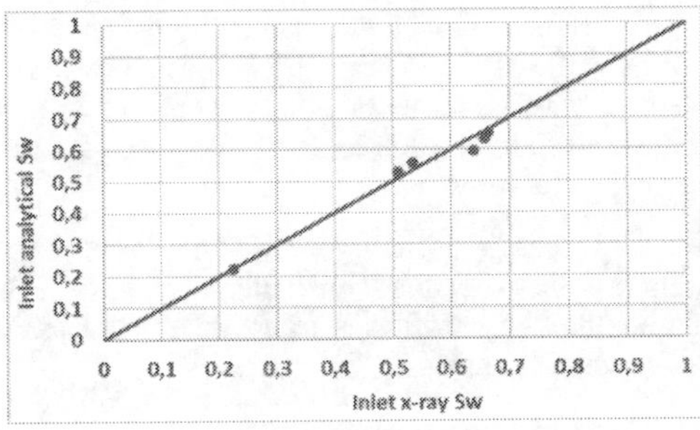


**Fig. 8.** Comparison between inlet local x-ray and analytical saturations at each equilibrium step

Although the comparative result looks acceptable, a single point of x-ray measurement does not allow to accurately capture the local saturation at the inlet section due to the degree of heterogeneity, the difference going up to 5 s.u. It is highly recommended for future tests to acquire more points on the full inlet section, requiring at least a 2D x-ray acquisition.

Nevertheless, the saturation profiles can help validate the numerical simulations and provide qualitative information on the rock wettability state.

Figure 9 represents the SD relative permeability curves on Cartesian and semi-log scale:

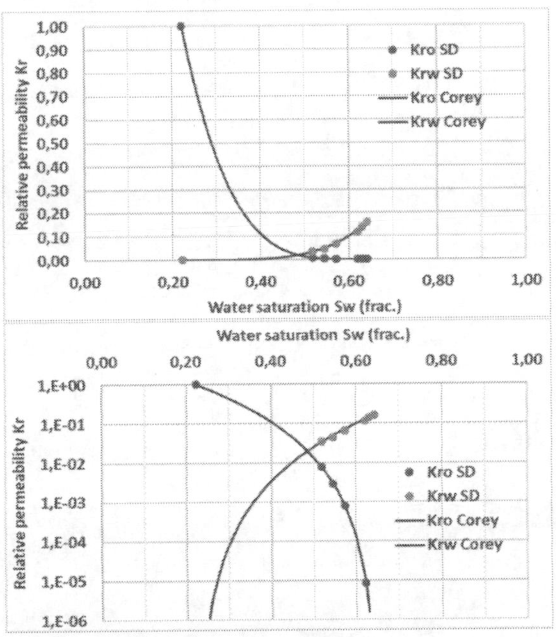


**Fig. 9.** SD Kr curves (Cartesian and semi-log plots)

Krw was directly obtained using Equation 2 while Kro was obtained by history matching oil production and differential pressure. The Sw values in Figure 9 are the ones analytically calculated using Equation 4, with <So> calculated using the produced oil volume from the separator. In the simulations, Pc was also entered using measured data points, fitted with a log(beta) function allowing positioning of the saturation Sw at Pc=0: this fitting function was optimized during the history-matching process.

The experimental program could not fully satisfy the CYDAR design due to too low water rate required to obtain Kr data points well spread over the saturation range, as shown in Figure 9. The minimum water rate applied during the test was 0.005 cc/min (lowest limit), the maximum rate was 0.1 cc/min (due to dP limitation).

To best fit the Kr data points, a Corey model was used. The values of the water and oil Corey exponents are:

Nw=4.5
No=4.0
Krw Max=0.16
Swi=0.224
Sor=0.346

The capillary pressure curve was directly obtained from the differential pressure values at each equilibrium, as explained in the background section. Figure 10 represents the measured differential pressure or capillary pressure versus Sw, with the fitting log(Beta) function used for the numerical simulations.

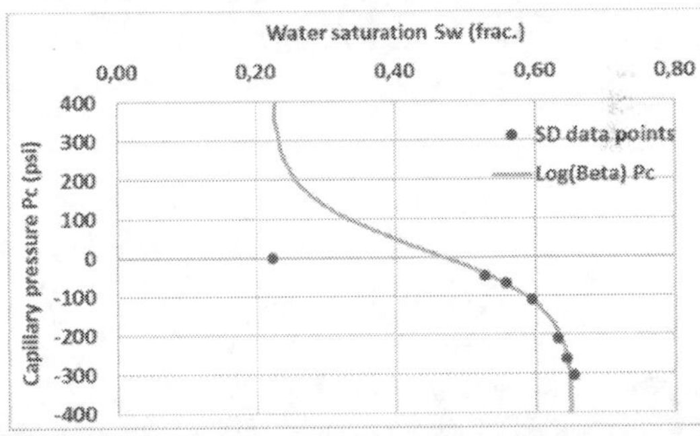


**Fig. 10.** SD Pc data points and optimized fitted Pc curve using a log(Beta) function

Unfortunately, the positive part of the Pc curve was not acquired for this project. A first attempt of fitting function going from Pc=0 at Swi value and passing through all data points using a single hyperbola function did not allow to match the oil production and differential pressure well. A log(Beta) function was used to fit the measured data points and to successfully history match the oil production and differential pressure. The data point at Pc=0 can be at a value of Sw varying from Swi to Sor. The saturation profiles were used to provide an estimation of this value (Sw=0.47 at Pc=0). The results of the numerical simulations and history matching are presented later.

The log(Beta) function is a 3 input-parameter function [Po, β, Sw(Pc=0)] in CYDAR, as defined in Equation 5:

$$Pc = cP_m \ln \frac{(1 - Sw^{*B})}{Sw^{*B}} - b \tag{5}$$

With Pm a pressure coefficient to control the magnitude of the Pc curve, Sw* the reduced saturation (Sw*=[Sw-Swi]/[1-Swi-Sor]), B coefficient to control the asymmetry of the function, b, a function dependent on the water saturation at Pc=0, and c parameter calculated as a function of β to impose a slope equal to Po at the middle of the Pc curve (Sw*=0.5). In the following simulations, the parameters Po and β were kept constant for both SD and SS numerical simulations: only the value of Sw(Pc=0)

https://doi.org/10.1051/e3sconf/202336601010

was changed, according to the observation on the saturation profiles of the SD and SS tests.

**Table 4.** Parameters of the log(Beta) Pc function

| Tests | Po (bars) | β | Sw(Pc=0) |
|---|---|---|---|
| SD | 20 | 1 | 0.47 |
| SS | 20 | 1 | 0.57 |

SS results on S3/S4:

Figure 11 represents both dP and produced Vo versus time:

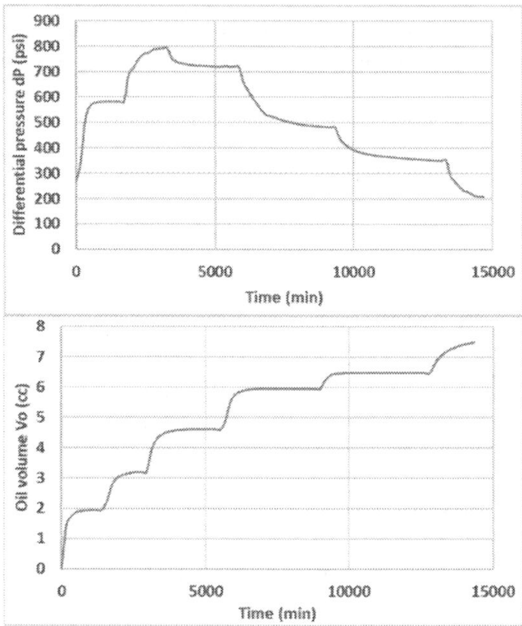


**Fig. 11.** dP and produced Vo versus time during the SS waterflood

For this test, the stabilization criteria were all obtained.

Concerning the saturation monitoring (Figure 12), the saturation profiles were noisier than the ones from the SD test: despite the choice of similar or "twin" samples and same experimental protocol, rocks may have different degree of heterogeneity.

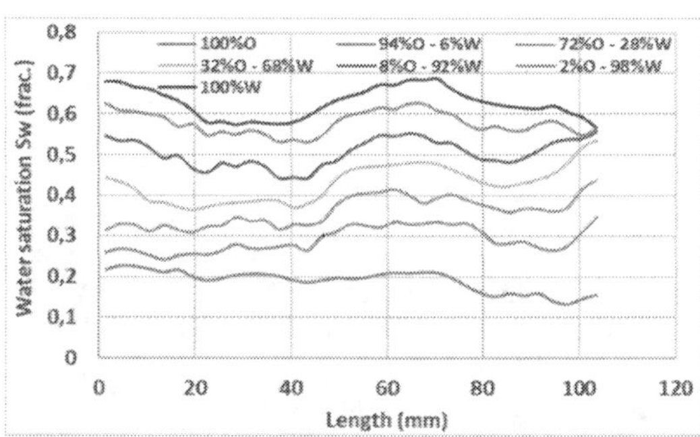


**Fig. 12.** Saturation profiles at equilibrium

Test parameters were well designed, with data points well spread over the saturation range. The outlet Sw seems to converge towards a value of 0.57. this information helps in choosing the imbibition log(beta) Pc curve for further numerical simulations.

Again, the comparison between the average water saturation <Sw> obtained by material balance and by averaging the x-ray profiles is acceptable, with less than 3 s.u. difference (Figure 13):

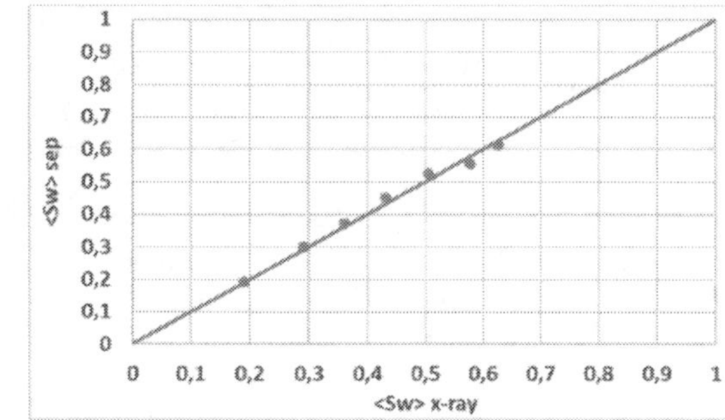


**Fig. 13.** Comparison between $<Sw_{sep}>$ and $<Sw_{x\text{-}ray}>$ during the SS waterflood

Figure 14 represents the non-interpreted SS relative permeability curves on Cartesian and semi-log scale:

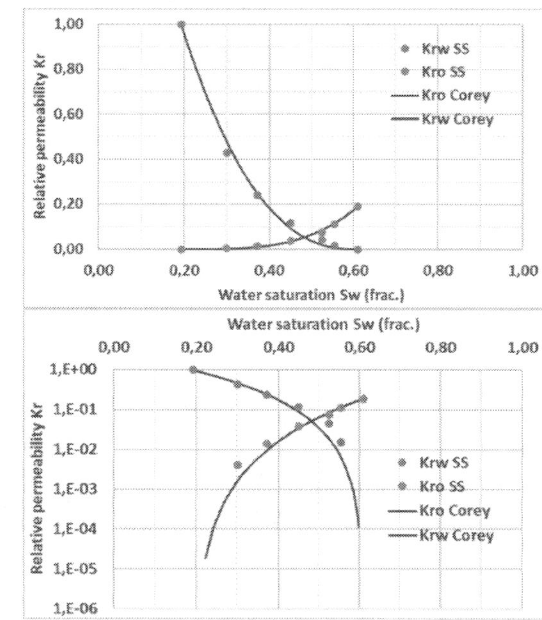


**Fig. 14.** SS Kr curves (Cartesian and semi-log plots)

The saturation Sw points are those calculated by material balance. Similar curve is obtained by using the average saturation profiles. As for the SD test, a Corey model was used to best fit the data points. The values of the water and oil Corey exponents are:

Nw=3.5
No=2.5
Krw Max=0.191
Swi=0.193
Sor=0.389

 https://doi.org/10.1051/e3sconf/202336601010

Pc centrifuge results on S5:

A forced imbibition multistep centrifuge test was performed to measure the capillary pressure Pc of sample S5. Unfortunately, there was no spontaneous imbibition performed before starting the forced imbibition. This step would have been interesting to compare with the value of SD Sw at Pc=0.

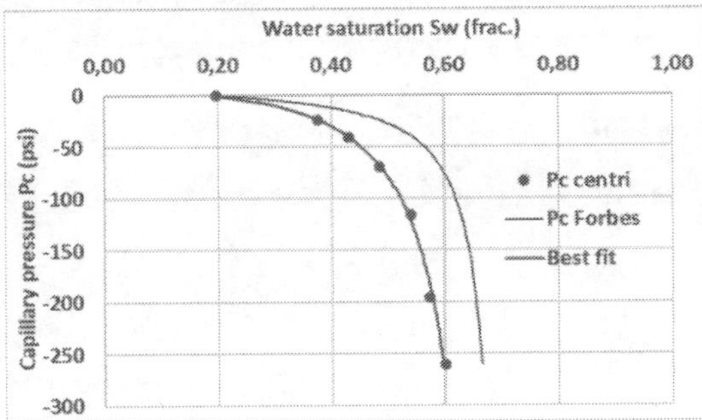


**Fig. 15.** Capillary pressure Pc (experimental and local Forbes Pc)

A modified hyperbolic function was first used to fit the experimental data points before calculating the local inlet saturation using Forbes' approach.

The same sample S5 was cleaned, brine-saturated, brought to Swi and run for single-step centrifuge (10000rpm) with dead oil to determine Kro. Note that intermediate porosity and permeability measurements were performed to ensure good sample integrity.

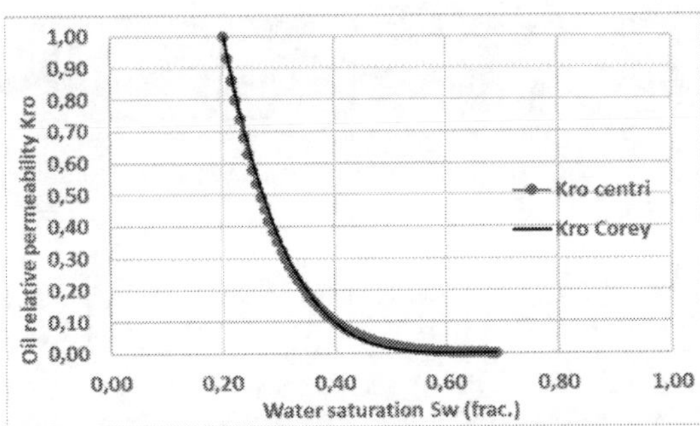


**Fig. 16.** Relative permeability to oil Kro from single-step centrifuge

Collected data were analysed using techniques published in [9] and [10], taking into account the correction for ramp up, capillary hold up, and mobility effects. The resulting oil relative permeability was then fitted with a Corey model. The No exponent was found to be equal to 4.0.

**Interpretation: Simulations and History Matching**

**SD Method:**

The results of the history-match process on the semi-dynamic waterflood are shown in Figure 17:

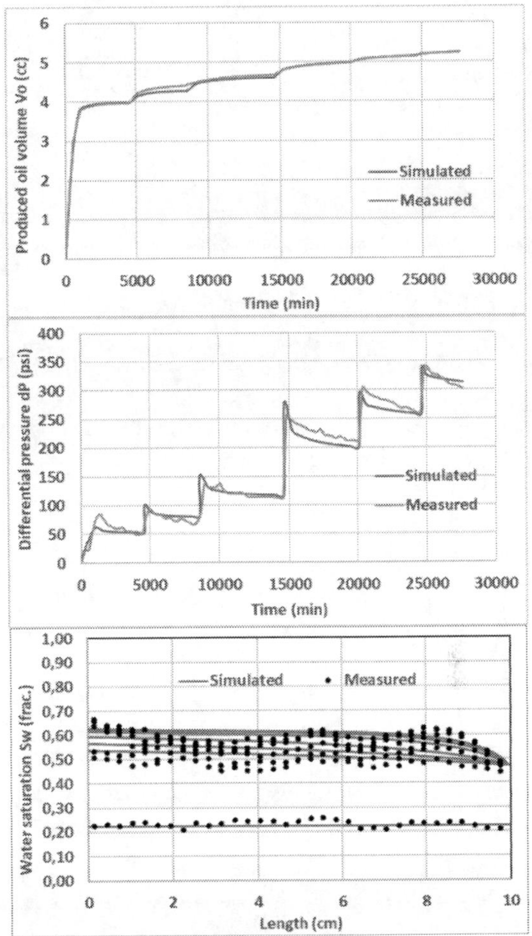


**Fig. 17.** Comparisons between measured properties and numerical simulations for the SD waterflood

Figure 17 represents the results of the produced oil Vo, differential pressure and saturation profiles history matches by tuning the Kro parameter during the simulations. As a matter of fact, local and inlet Krw and Pc are directly determined during the experiment. Pc parameters from log(beta) function were also tuned to fit the experimental data points and to find an optimal value of Sw(Pc=0) for improving the history-match results.

The history match appears acceptable for production Vo and differential pressure, but it is less obvious for the saturation profiles. This is mainly the result of the 1D x-ray limitation and the degree of heterogeneity rock. As explained before, using one single point of x-ray acquisition through the diameter at the core inlet does not represent the average water saturation through the inlet corresponding section.

In order to better history-match the production and differential pressure, a modified Corey or L.E.T. function may have been preferred. But in the frame of comparing the SD and SS methods, keeping consistency/coherence and limiting the number of matching parameters, it was decided to keep a Corey model for Kr and provide the best match possible. The resulting Corey parameters were:

Nw=4.5
No=4.0
Krw Max=0.16
Swi=0.224
Sor=0.346

**SS Method:**

In the SS simulations, the SD local Pc was taken as first guess input, preserving the same parameter values discussed earlier but with the value of Sw(Pc=0) adjusted by using saturation profile observations.

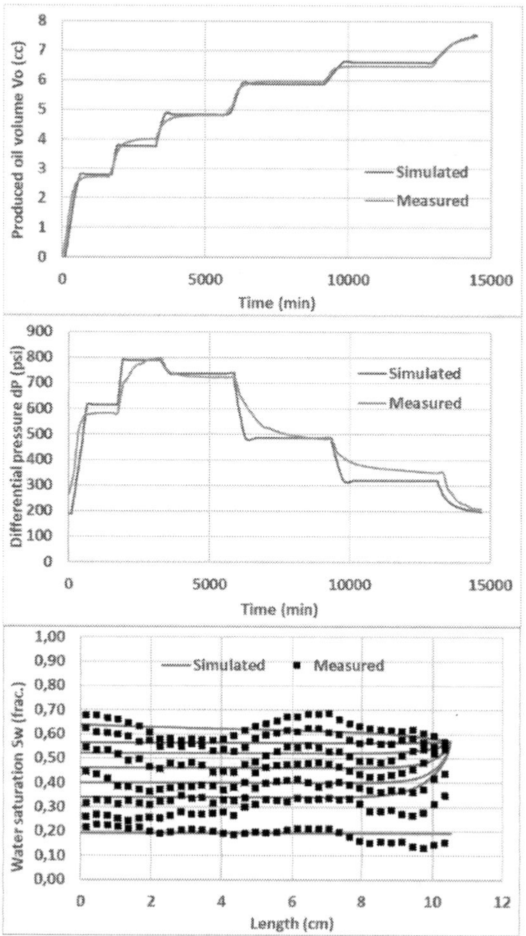


**Fig. 18.** Comparisons between measured properties and numerical simulations for the SS waterflood

As for the SD numerical simulations and history-match, it was difficult to improve the quality of the production and differential pressure history-match results using a simple Corey model. As explained before, even if better results may have been obtained using a modified Corey or L.E.T. function, it was decided to keep consistency/coherence and to limit the number of matching parameters by using a simple Corey model for Kr. The resulting Corey parameters were:

Nw: 3.5
No: 3.0
Krw Max=0.220
Swi=0.193
Sor=0.330

Figure 19 represents both uninterpreted and interpreted SS Kr and Pc curves:

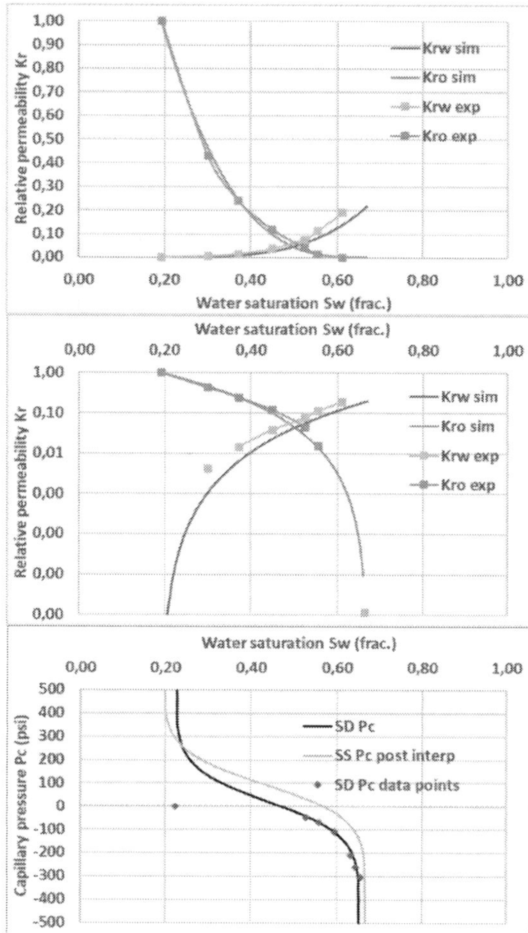


**Fig. 19.** Pre- and post-interpretation Kr and Pc curves for the SS waterflood

Note that Forbes Pc curves as shown in Figure 15 were tested for the two SD and SS waterfloods. An acceptable history-match of dP and Vo were obtained using newly interpreted Kr curves. The observations were:

- The newly interpreted Kr curves were found to be very close to the Kr curves obtained with log(Beta) Pc
- The Forbes Pc function does not mimic the outlet Sw, as shown in saturation profiles in Figure 16 and Figure 17 where a log(Beta) Pc function was used with an appropriate value of Sw(Pc=0).

**Centrifuge Pc Simulation and History-Matching:**

The local centrifuge capillary pressure was obtained using the Forbes' approach (Figure 15). The Kr curve was built with a Corey model, with No equal to 4, as for the single-step test. A first guess of water Corey exponent Nw equal to 2 and Krw Max of 0.2 were tested before running the history-match optimization. The best history match was finally obtained with the below Kr and Pc curves:

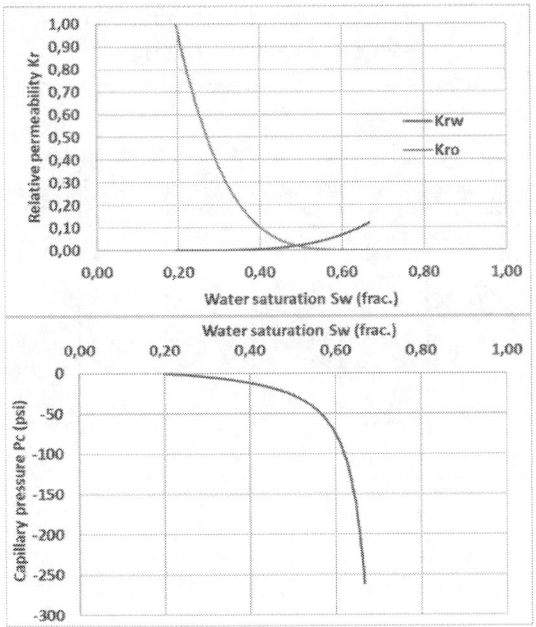


**Fig. 20.** Interpreted Kr curve for the Pc multistep test with the associated Forbes' Pc curve

Nw=3.8
No=4.0
Krw Max=0.12
Swi=0.197
Sor=0.334

The quality of the history-match is presented in Figure 21:

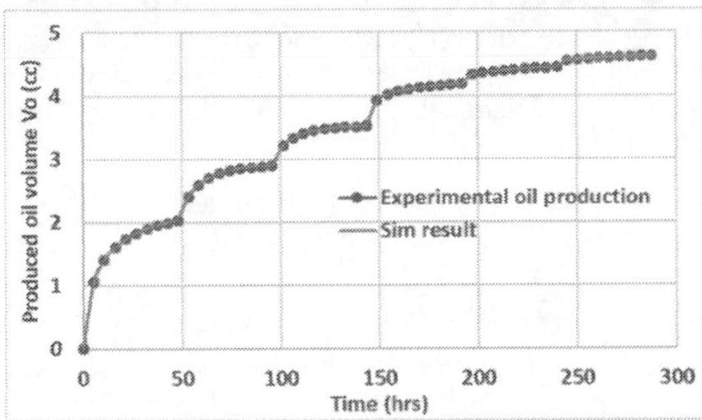


**Fig. 21.** Interpreted Kr curve for the Pc multistep test with the associated Forbes' Pc curve

### Centrifuge Kro Simulation and History-Match:

An history-match of the single-step centrifuge oil production was performed. The Kro was taken as calculated via Hirasaki's method (Figure 16). The same Forbes Pc curve as is in Figure 15 was used as input. An initial Krw guess with Corey exponent Nw equal to 4 was attempted first. Krw Max, Sor and Nw were then tuned for the history-match. Considering the Pc and Kro parameters are well determined, the resulting Kr and Pc curves are shown in Figure 22:

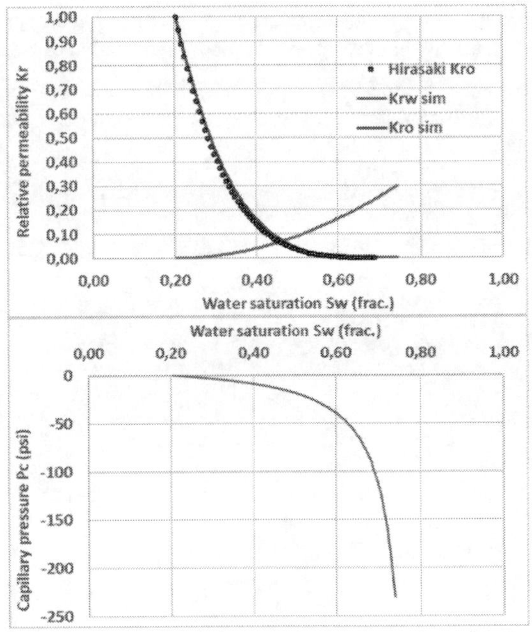


**Fig. 22.** Optimized Kr curve for single-step centrifuge with fixed Forbes Pc

To history-match the oil production Vo, a Corey coefficient of 2 was taken for Nw.

Nw=2
No=4
Krw Max=0.30
Swi=0.203
Sor=0.257

The resulting history-match is shown in Figure 23:

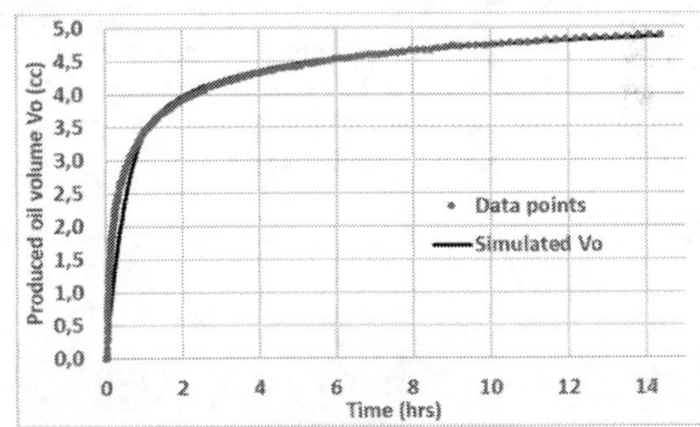


**Fig. 23.** Measured and simulated Vo for the single-step Kr centrifuge test

It is difficult to history-match the early oil production well, probably because no spontaneous displacement was performed before the forced imbibition cycle.

## Kr and Pc Comparisons

### Kr Comparison:

For comparing the petrophysical parameters Kr and Pc from different experiments but on the same rock type, it is generally recommended to plot the results using the same value of Swi (the averaged value) and preserving the recovery information. In this study, the Swi values were found to be very close. It was decided to directly compare the results without rescaling them to a same value of Swi.

Reminder: for the SD and SS tests, the material balance method was preferred to the x-ray method to interpret the data.

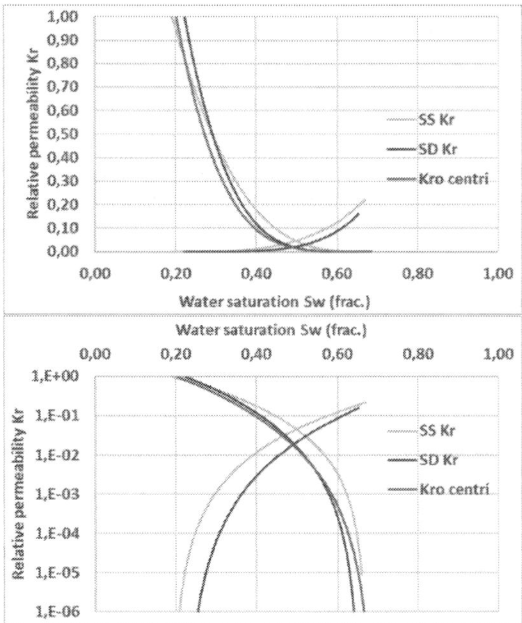


**Fig. 24.** Comparison of SD, SS and centrifuge relative permeability curves

Figure 24 shows that centrifuge and SD Kro curves overlap nicely while SS and SD Kr curves on the Cartesian plot look quite similar. But the semi-log graph better highlights the differences. All can see that SS Kr shows higher oil and water mobility at same saturation compared to the SD. It may be due to the heterogeneity or wettability differences. Fraction flow curve helps in comparing the SD and SS Kr curves:

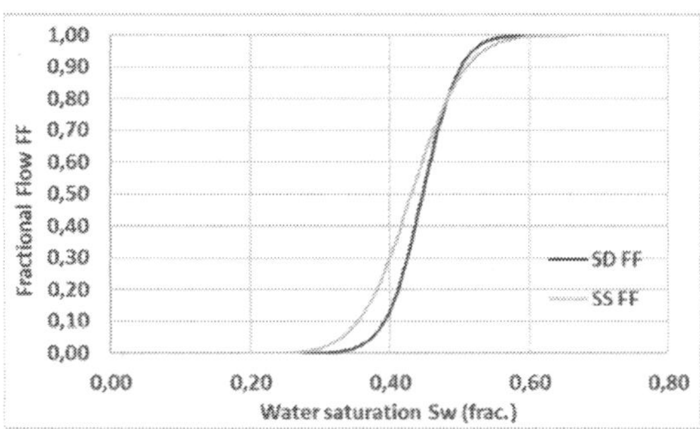


**Fig. 25.** Comparison of relative permeability ratios

Figure 25 shows that the SD case is more optimistic from Swi to Sw~50% while SS case is more optimistic from Sw~50% to residual oil saturation Sor. The difference can be explained by the difference in degree of heterogeneity, the lack of equilibrium during the SD test, as shown on the dP signal in Figure 17, possible effect of the capillary contact between the samples. Wettability difference may also explain the difference despite the tests were run at same conditions, with same fluids, at same Swi and with initial uniform fluid distribution for both stacks.

The Kr Corey exponents and saturation ranges are not so different as presented in Table 5:

**Table 5.** Final Kr Corey parameters and saturation range

| Test | Krw Max | Swi frac. | Sor frac. | Nw | No |
|---|---|---|---|---|---|
| SD | 0.16 | 0.223 | 0.346 | 4.5 | 4.0 |
| SS | 0.22 | 0.193 | 0.330 | 3.5 | 3.0 |
| Pc centri | 0.12 | 0.197 | 0.334 | 3.8 | 4.0 |
| Kr centri | 0.30 | 0.203 | 0.257 | 2.0 | 4.0 |

Wettability could have also explained the Kr difference, but the parameters in Table 5 highlight an intermediate-wet state of the rocks (typically with $3<Nw<5$ and $3<No<6$), also confirmed by the positions of the Kr cross-points.

Even if it is not observed on the saturation profiles (Figure 6 and Figure 12), a poor capillary contact could lead to non-uniform fluid distribution and so, lead to a difference in flow behaviour. The use of composite stack may not be a good way to perform the waterflooding tests.

Finally, the use of Pc curve for the SS optimization process/history-match, fixed with same SD Pc parameters may also explain the discrepancy. If porous plate or centrifuge Pc could have been individually performed on the composite core plugs prior the SS test, an averaging process (J-function…) would have been necessary to build an averaged Pc curve, introducing again an uncertainty.

It is concluded that having both Kr and Pc measured during similar test on a non-composite stack (to avoid capillary contact issue), is the most acceptable way of avoiding erroneous interpretation.

Even if it is not an objective of the study, the SD and SS Kr curves were compared to the interpreted Kr curves from the single and multistep centrifuge tests:

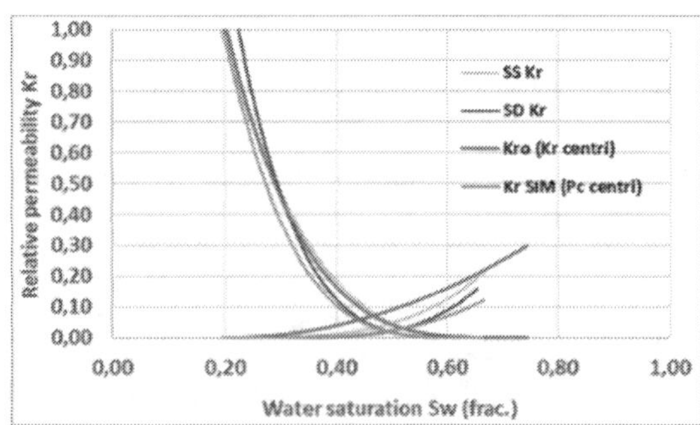

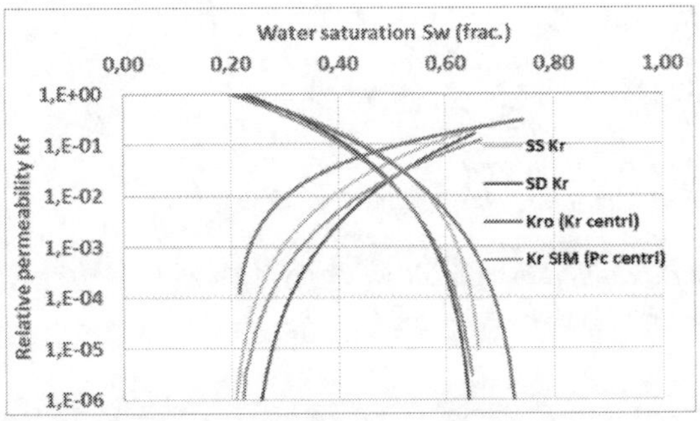


**Fig. 26.** All interpreted Kr curves from corecoflooding and centrifuge tests

The Kr curves obtained from the different methods are not so different, except the one from the single-step Kr centrifuge test, probably due to the small amount of signal/data to be history-matched and re-use of the multistep Pc sample after cleaning it again, restored wettability being maybe different. Other curves are in the same range of saturation, close Kr cross-points, close Krw Max and close Kr curvatures, showing a certain consistency in the obtained results. In reservoir engineering, it is not unusual to observe such Kr envelops when dealing with a same rock type.

**Pc Comparison:**

Both the local Pc from centrifuge obtained using the Forbes approach and local Pc directly obtained from SD method are plotted together:

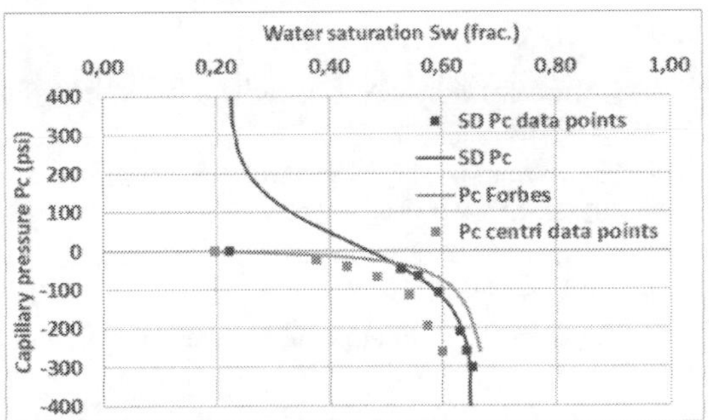


**Fig. 27.** SD and SS capillary pressure Pc curves

Despite using oil at different conditions (dead oil for centrifuge – live oil for SD), Figure 27 shows an acceptable similarity between the SD and Forbes Pc curves. It is very encouraging to notice that the SD method provides similar local Pc to the Forbes method using centrifuge. Again, the SD Sw was analytically calculated and not obtained using the 1D x-ray method. Also, note that there was no spontaneous imbibition performed before running the forced centrifuge and SD imbibition tests. It could have helped in validating the SD Pc, especially the Sw value at Pc=0.

Figure 28 shows all measured and interpreted Pc curves. The interpreted negative Pc parts are all close together, proving a certain consistency of the results and adequate interpretation workflow. The positive part of the Pc curve was unfortunately not measured. It could have been done using the semi-dynamic method.

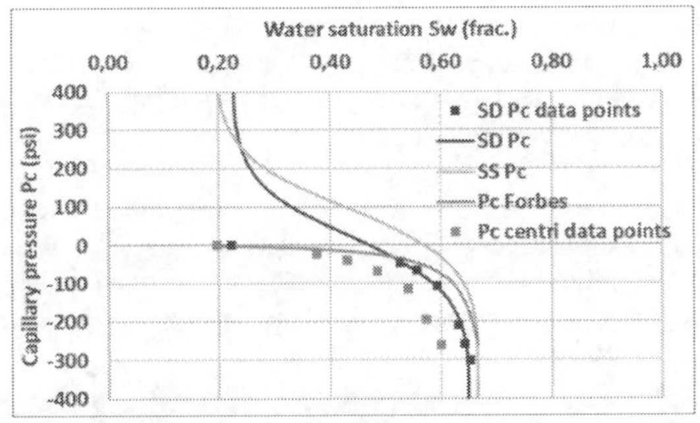


**Fig. 28.** Capillary pressure Pc curves

Based on the capillary pressure curves, it is observed that the SS stack behaves as a more water-wet rock than the SD stack, which is consistent with the observation on the fractional flow (Figure 25). Further investigations must be done to better understand the potential wettability change during the forced imbibition using different methods.

## 6 Conclusions

In this study, several methods for measuring relative permeability and capillary pressure were tested in order to compare the data results and validate the SD method as a unique experiment able to provide both petrophysical parameters during a same test, on same core, with live fluids at real reservoir conditions. It was concluded that:

- The SD Pc data points were found to be similar to the local analytical Forbes Pc obtained during the multistep centrifuge, despite aging and testing with two different oils (dead versus live). The interpreted SD Pc curve using the log(Beta) function passing through the data points is also forced to pass on the true value of Sw(Pc=0) observed on the saturation profiles. Unfortunately, neither the positive part of SD Pc nor the spontaneous imbibition displacement prior centrifuge test were acquired in this study

- The SD Kr was quite different from the SS one. Assuming that the wettability was identical in both tests, the difference may be explained by the rock heterogeneity. The interpreted SS Kr curve was obtained using SD Pc curve (and using centrifuge Pc curve, not presented here), while SD Kr and Pc curves are directly measured during the same test on same composite stack. This is a real advantage to avoid erroneous interpretation

- The simulated SD Kro was consistent with centrifuge Kro. This observation is very encouraging, validating the workflow for the SD history-matching process

- Large rocks should be preferred to composite stacks to avoid potential capillary contact issue

- The quality of the simulated and interpreted Kr curves from the centrifuge tests (multistep and single step) are more questionable because they are based on oil production history match only

If the results of the study are building confidence in the semi-dynamic approach to determine both relative permeability and capillary pressure, some experimental improvements are still required:

- 2D or even 3D x-ray method should be preferred to the 1D or analytical means to measure the inlet water saturation, especially for heterogeneous rocks. A new x-ray setup is being tested in TOTAL [11]. This system allows having enough contrast between water and oil without additional dopants, reducing the risk of a wettability change during the waterfloods

- Pc curve, including the positive part, must be measured with the SD method to determine the full Pc and to better quantify the value of Sw(Pc=0). This can easily be done with the current SD setup, by reducing the oil flow rate and recirculating water at the outlet before injecting water

- The conclusions of this study would need to be confirmed by performing more tests and ensuring the repeatability of the observed results

- Wettability alteration, from the initial restoration while recirculating oil at the core outlet during the semi-dynamic test or while co-injecting oil during the steady-state test, both in a closed loop, requires more attention and investigation

- Finally, a large project for multiple history-matching process is ongoing at TOTAL with the aim of finding Kr and Pc curves via simultaneous assisted history-match on data collected from several experiments [12]

The authors would like to thank TOTAL and Schlumberger for permission to publish this work.

## Nomenclature

*A: section area, in* $cm^2$
*CT: Computed Tomography*
*dP: differential pressure, in psi*
*HPHT: High Pressure High Temperature*
*ISSM: in situ saturation monitoring*
*Kg: apparent gas permeability, in mD*
*Kw: water permeability, in mD*
*Keo: oil relative permeability, in mD*
*Kr: relative permeability*
*Krw: relative permeability to water*
*Kro: relative permeability to oil*
*Krw Max: maximum water relative permeability*
*MB: Material Balance*
*MICP: mercury injection capillary pressure*
*NCS: net confining stress, in psi*
*Nw: water Corey exponent*
*No: oil Corey exponent*
*Pc: capillary pressure, in psi*
*Pp: pore pressure, in psi*
*Pconf: confining pressure, in psi*
*Pi: inlet pressure, in psi*
*Po: outlet pressure, in psi*
*Qw: water flow rate, in cc/min*
*Qo: oil flow rate, in cc/min*
*ROS: Remaining Oil Saturation*
*SD: Semi-Dynamic method*
*SS: Steady-State method*
*SEM: Scanning Electron Microscopy*
*So: oil saturation, in v/v or %*
*Sor: residual oil saturation, in v/v or %*
*<So>: average of oil saturation, in v/v or %*
*Sw: water saturation, in v/v or %*
*Swi: irreducible water saturation, in v/v or %*
*<Sw>: average of water saturation, in v/v or %*
$\langle Sw_{sep} \rangle$*: average of water saturation using separator, in v/v or %*
$\langle Sw_{x\text{-}ray} \rangle$*: average of water saturation using x-ray attenuation method, in v/v or %*
*S*: normalized water saturation, in v/v or %*
*T°C: temperature, in °C*
*TS: Thin Section*
*USS: Unsteady-State method*
*Vo: produced oil volume*
*XRD: X-Ray Diffraction*
*X-Swi: irreducible water saturation measured with x-ray attenuation method, in v/v or %*

$\phi_{He}$*: helium porosity, in v/v or %*
$\rho_w$*: water density, in g/cc*
$\rho_o$*: oil density, in g/cc*
$\mu_w$*: water viscosity, in cP*
$\mu_o$*: oil viscosity, in cP*

## References

1. T.S. Ramakrishnan, A. Cappiello, Chem. Eng. Sci., **46** (1991)
2. R. Lenormand, A. Eisenzimmer, Society of Core Analysts, SCA1993-22 (1993)
3. R. Lenormand, P. Schmitz, Society of Core Analysts, SCA1997-17 (1997)
4. J-M. Lombard, P. Egermann, R. Lenormand, Society of Core Analysts, SCA2002-09 (2002)
5. J-M. Lombard *et al.*, Society of Core Analysts, SCA2004-32 (2004)
6. D.R. Maloney, Society of Core Analysts, SCA2006-05 (2006)
7. M.C. Spearing *et al.*, ADIPEC, SPE-171892 (2014)
8. F. Pairoys, Society of Core Analysts, SCA2014-05 (2014)
9. G.J. Hirasaki *et al.*, SPE Reservoir Engineering, SPE-25290 (1992)
10. G.J. Hirasaki *et al.*, SPE Advanced Technology Series, **3**, SPE-24879 (1995)
11. G. Puyou *et al.*, Society of Core Analysts, SCA2017-37 (2017)
12. A.E. Sylte, Society of Core Analysts, SCA2004-17 (2004)

# Of rats and rocks: using pre-clinical PET imaging facilities in core analysis

*Bergit Brattekås[1], Martin A. Fernø[1], Malin Haugen[1], Tore Føyen[1], Marianne Steinsbø[1], Arne Graue[1], Njål Brekke[2], Tom Christian Holm Adamsen[2], Cecilie Brekke Rygh[3]* and *Heidi Espedal[4]*

[1]Dept. of Physics and Technology, University of Bergen, Norway
[2]Center for Nuclear Medicine - PET, Department of Radiology, Haukeland University Hospital, Norway
[3]Department of Radiology, Haukeland University Hospital, Norway
[4]Molecular Imaging Center, Department of Biomedicine, University of Bergen, Norway

**Abstract.** Positron emission tomography (PET) is routinely used for medical imaging; a current surge in published geoscientific research utilizing this modality also infer increasing interest for *in-situ* PET imaging in core analysis. Excellent signal to noise ratio coupled with high temporal and spatial resolution suggest that PET might become the new method-of-choice for core analysis. Obstacles related to production, transfer and handling of radioactive fluids and gases must, however, be dealt with for PET to become a widely used core scale imaging technique. This paper describes an ongoing, true multidisciplinary collaboration, where pre-clinical PET imaging facilities are routinely used in core analysis to investigate dynamic fluid flow at high pressure conditions. We detail challenges and opportunities related to porous media research in established pre-clinical laboratory facilities designed for small-animal imaging, and demonstrate the significant potential of PET imaging in core scale analysis in a context related to long-term porous media carbon storage.

Explicit imaging of several fluid phases is possible by PET imaging using a range of readily available radiotracers. Relevant radiotracers to carbon storage in porous media are e.g. the carbon radioisotope $^{11}C$ and water-soluble tracer $^{18}F$. These are both short-lived tracers (20 - 110 min) and must be used in high doses of radiation, which present challenges related to safe transfer and handling. Although there are several obstacles to conduct advanced core analysis in hospital imaging facilities (some of which are detailed in this paper), significant advantages include trained personnel on-site to operate a local cyclotron, procedures in place to ensure safe and efficient transfer of short-lived radiopharmaceuticals from the cyclotron, and advanced image analysis capabilities available. Cyclotrons are widely available worldwide (currently more than 1200 operating cyclotrons), often located in close proximity to medical and pre-clinical imaging facilities and academic institutions. Similar collaborations may therefore also be possible elsewhere, reducing the need for allocated geophysical PET-scanners and lowering the threshold for routinely using PET imaging in core analysis.

## 1 Background

### 1.1 A quick introduction to Positron Emission Tomography imaging

Positron emission tomography (PET) imaging utilizes positron-emitting radionuclides to gain insight into fluid flow and accumulation. These radionuclides are rich in protons and decay by beta plus decay; the spontaneous physical phenomenon where an emitted positron annihilates with an electron and consequently emits a pair of specific energy photons (511 keV) in opposite directions. PET scanners can determine annihilation events with high spatial and temporal accuracy; by detecting gamma rays (photons) occurring simultaneously on opposite sides of an object. The number, position and size of PET detector elements, as well as the distance between the object and detectors, strongly influence the spatial resolution of PET imaging [1]. PET is established as a diagnostic tool in medicine, where the radioactively labelled "object" is often a human (in clinical imaging) or animal (in pre-clinical research) but has also recently been demonstrated as a highly useful tool for non-medical purposes: flow visualization in porous rocks [2-15]. For practical purposes the beta plus decay is insensitive to pressure and temperature [16]. PET imaging is, hence, a suitable tool to visualize fluid flow and accumulation within core plugs placed in confinement vessels under high pressure and temperature conditions; e.g. to investigate porous media $CO_2$ storage at relevant storage conditions.

© The Authors, published by EDP Sciences. This is an open access article distributed under the terms of the Creative Commons Attribution License 4.0 (http://creativecommons.org/licenses/by/4.0/).

PET imaging has several advantages, which in sum renders PET a strong competitor to other imaging methods initially used for diagnostic purposes, e.g. computed tomography (CT) and magnetic resonance imaging (MRI). Advantages include 1) high spatial resolutions, 2) high temporal resolutions (i.e. possibility to investigate fast processes at high flow rates), 3) excellent signal-to-noise ratio (i.e. a small amount of radioactivity is sufficient to produce high-quality images; without altering the properties of the labelled fluid), 4) explicit detection of radioactively labelled fluid only (i.e. fluid flow paths are imaged without the porous medium influencing image quality). For a general introduction to PET imaging in water resources research we recommend Zahasky and Benson (2019) [21].

During core scale imaging by CT and MRI, high spatial resolution may be achieved by decreasing temporal resolution: a trade-off which must be set within each experiment. Micro-CT for example enables micrometer visualization of the pore space, but the time scale for acquiring 3D images is several hours. PET imaging does not suffer from this tradeoff: photon emissions are continuously recorded during imaging, and spatial and temporal resolutions are set during post-processing. 3D images may, hence, be achieved during short time spans (seconds), while maintaining a high spatial resolution. A simple schematic of imaging methods and resolution relations is shown in **Fig. 1**.

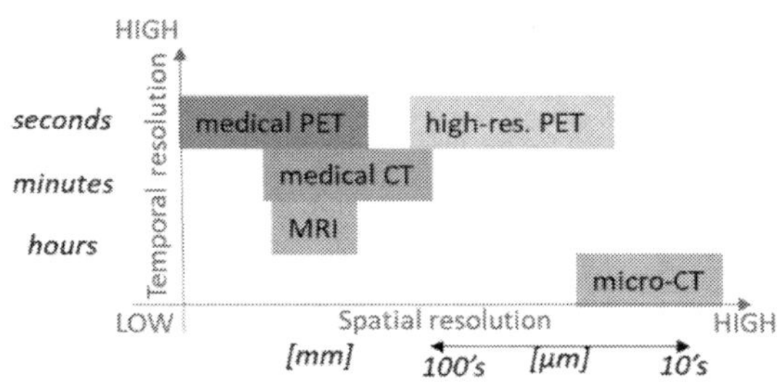


**Fig 1**. Imaging technologies used for 3D *in-situ* visualization of core material and core floods. Spatial resolution may be high for MRI and CT imaging, although that typically increases the scan time. PET images are post-processed; and maintain high spatial resolution regardless of temporal resolution preferences.

### 1.2 Existing opportunities

Although some research groups have procured or built their own PET imaging systems (e.g. [17-20]), this is not a realistic solution for the average core analyst. Several necessary specifications and requirements must be fulfilled and a number of challenges (both perceived and real, recently described by Zahasky *et al.* [21]) dealt with to perform core analysis with PET imaging. Most obstacles relate to the production, transportation, safe use and handling of radioisotopes, especially short-lived radioisotopes, that are often used for geoscientific purposes. These obstacles are manageable, but requires significant funds and space for necessary equipment and personnel. The threshold for PET-imaging utilizing private facilities therefore remains high, and other options may be explored to make PET-imaging available for a broader range of users.

Hospitals world-wide are actively applying PET imaging for diagnostic purposes [22, 23], and clinical PET-scanners have become widespread during the last few decades. Obstacles related to production, transportation and safe handling of radioisotopes have therefore already been handled to routinely perform PET-imaging in hospitals. Clinical scanners are intended for human use, and are normally reserved for patients during operating hours. The time frame available for non-essential flow experiments is therefore limited. Access to operational IT infrastructure connected to clinical scanners is also restricted, to prevent insight into sensitive patient information. Further limitations related to cleanliness (operating potentially messy equipment in a sterile lab) and spatial resolution (limited to 2-5mm, because gantry and detector element sizes are constructed for human use [24]) in sum prevents clinical PET-scanners to be widely utilized in core analysis, although occasional experiments are feasible, and have been reported [2, 25, 26].

Another option also exists within many hospital research facilities: high-resolution PET-scanners used for pre-clinical imaging of small animals. This paper highlights the opportunity in using pre-clinical PET-imaging facilities in core analysis, and describes an ongoing collaboration that has successfully utilized this opportunity. Pre-clinical PET shares the advantages of clinical imaging in terms of radioisotope production, transportation and handling albeit with weight and size limitations not normally encountered using clinical scanners. A higher spatial resolution, however, results from the smaller size (due to decreased distance between scanned object and detector elements, and smaller detector element size, see e.g. Bailey *et al.* [1]), and the high-resolution PET-CT scanner used in this work features voxel sizes of 0.3-0.4 mm. Issues related to experimental duration may be largely overcome and the pre-clinical scanners available for longer continuous time spans, depending on the number and frequency of parallel animal studies.

**Table 1** summarizes the obstacles associated with PET imaging in private facilities, and using existing clinical and pre-clinical imaging facilities. The table illustrates that several obstacles are overcome using pre-clinical PET imaging, but some remain; e.g. it may be a challenge using software and settings intended for soft-tissue small animals in studies on sedimentary rocks. Tailor-made software is an advantage reserved for geo-scientifically oriented, private PET-imaging facilities. Additionally, access requirements may vary between facilities; the pre-clinical PET-scanner used in this work is a part of the Molecular Imaging Centre (MIC) (Department of Biomedicine, University of Bergen, Norway) and is located at the PET-centre at Haukeland University Hospital.

 https://doi.org/10.1051/e3sconf/202336601011

**Table 1 :** General obstacles in PET-imaging may be overcome within medical imaging facilities- although new obstacles occur. An overview of obstacles for each option (private, clinical or pre-clinical PET-imaging) is shown below.

| Obstacles | Present | Partly handled | Handled | |
|---|---|---|---|---|
| Production of radioisotopes | PET (general) | | Pre-clinical PET | Clinical PET |
| Transportation of radioisotopes | PET (general) | | Pre-clinical PET | Clinical PET |
| Safe handling of radioisotopes | PET (general) | | Pre-clinical PET | Clinical PET |
| Experiment duration | Clinical PET | Pre-clinical PET | | PET (general) |
| Access for personell | Clinical PET | Pre-clinical PET | | PET (general) |
| IT access | Clinical PET | Pre-clinical PET | | PET (general) |
| Software modification | Clinical PET | Pre-clinical PET | | PET (general) |
| Cleanliness | Clinical PET | Pre-clinical PET | | PET (general) |
| Spatial resolution | Clinical PET | PET (general) | | Pre-clinical PET |
| Weight and size of laboratory equipment | Pre-clinical PET | Clinical PET | | PET (general) |

MIC was started in 2003 and is an open access national (Normolim[a]) and international/European (Euro-BioImaging[b]) level, providing access to advanced small animal imaging to the research community. Obstacles related to researcher access have therefore been minor. MIC maintains instrumentation and has highly qualified scientific and technical personnel to facilitate a variety of research (usually in the field of biomedicine). Their experience in interacting across research groups have also made the leap from physiology to flow physics research relatively short. The existence of this operational open core facility is obviously a significant advantage in using pre-clinical PET-imaging for core analysis.

### 1.3 Availability of pre-clinical PET scanners

The pre-clinical PET-CT scanner used in this work is a nanoScan® PET-CT from Mediso, who is one of the main suppliers of pre-clinical PET scanners in addition to Bruker, MR Solutions and Siemens. Bruker reports 60-70 delivered pre-clinical PET scanners world-wide. Mediso reports 61 operational PET-CT and 30 PET-MRI scanners. MR Solutions do not wish to provide information about the exact number of scanners delivered by their company, but features more than ten customers on their web home page. Siemens no longer produce pre-clinical scanners, but some may still be in operation. Tracers used for animal studies are usually short-lived, and the PET-scanners are therefore often located in close proximity to cyclotrons.

### 1.4 Availability of cyclotrons

Medical cyclotrons are used to produce isotopes used in diagnostic imaging. More than 1200 cyclotrons exist world-wide, according to the International Atomic Energy Agency (IAEA) [27], and many are localized in close proximity to research institutions and/or hospitals that are actively applying PET-imaging as a diagnostic tool.

A searchable and detailed overview of cyclotrons may be found at: https://nucleus.iaea.org/sites/accelerators/Pages/Cyclotron.aspx, providing location, affiliation, make and model. The cyclotron connected to the current research facility is shown in **Fig. 2**.

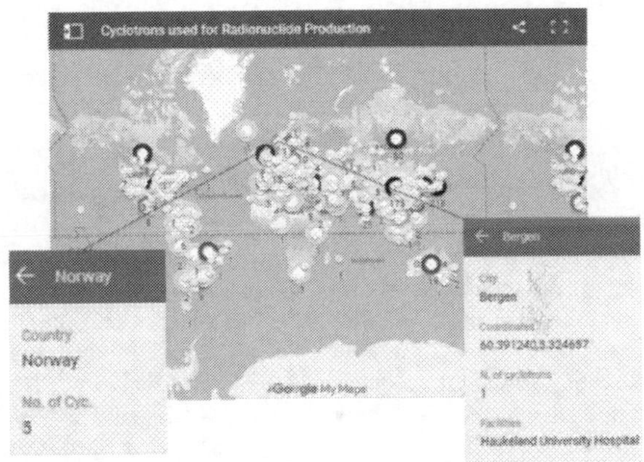


**Fig. 2.** World-wide map of cyclotrons from International Atomic Energy Agency (IAEA), with country and city inserts. Modified screenshot from [27].

## 2 Overcoming the obstacles

We have used the MIC pre-clinical PET-CT scanner since 2013 and have developed a procedure to routinely utilize shared pre-clinical imaging facilities for core analysis. Our experience is described, and attempted generalized, in the following; to highlight the possibilities of similar collaboration elsewhere.

---

[a] *Normolim: NORwegian MOLecular IMaging infrastructure. More information available on normolim.w.uib.no*
[b] *See eurobioimaging.eu for more information*

### 2.1 Production of radioactive tracers

Medical imaging often utilizes short-lived radionuclides, produced at local cyclotrons. Radionuclides are attached to larger molecules before imaging. $^{18}$F (Fluorine-18) e.g. often substitutes the normal hydroxyl group in glucose to make radiotracer $^{18}$F-FDG (Fludeoxyglucose). The accelerated uptake of this glucose analogue by cancer cells enables visualization of cancerous tissue within the human body. $^{18}$F-FDG is also the most used radiotracer in fluid flow visualization for geoscientific purposes [21], due to its miscibility in water coupled with PET sensitivity. Minute amounts of $^{18}$F-FDG is needed to procure accurate, high-quality images, and can be added to water without notably altering its properties. E.g. in the MIC research facility the $^{18}$F-FDG concentration is high; at around 20 GBq in 24 mL of water directly after production. We typically receive <1 mL of this solution to label the desired aqueous solution. A variety of aqueous solutions have so far been labelled by $^{18}$F-FDG, including: brine, polymer solution and surfactant solution. The small volume of tracer compared to the aqueous solution volume ensures that macroscopic fluid properties are unaltered during PET imaging. This is one of the main advantages of PET-imaging compared to e.g. attenuation methods (exemplified by CT), where density contrasts are necessary to capture flow dynamics. $^{18}$F-FDG is routinely produced at the PET-centre (Haukeland University Hospital) to facilitate medical imaging, and trained personnel are available on a daily basis. Daily production of $^{18}$F-FDG for clinical use reduces the need for dedicated $^{18}$F-FDG production for pre-clinical imaging. The half-life of $^{18}$F-FDG is short, at $t_{1/2}$ = 109 minutes, and it is delivered in vials as a liquid.

$^{11}$C (Carbon-11) can be attached to Carbon dioxide ($CO_2$) to make radiotracer $^{11}CO_2$, suitable for explicit visualization of $CO_2$ flow in porous rocks. $^{11}CO_2$ has a half-life of $t_{1/2}$ = 20 minutes. The short half-life requires higher doses of radiation to be handled in flow experiments, because the radiation (which will, in turn, produce the PET signal) swiftly decays. The preferred state of radiotracer used for $CO_2$ flow experiments is gaseous- which also makes transfer more challenging (see **Section 2.2**).

The challenges related to production of radioisotopes are very well handled by trained personnel at the PET-centre. The end user- in this case core analysts aiming to use PET to visualize fluid flow- should therefore not perceive the production of radiotracers as a large obstacle. Basic understanding of radiation is, however, necessary to accurately design and implement flow experiments.

### 2.2 Safe transfer of radioactive tracers

Safe transportation of short-lived radiotracers varies with the type and state of radiotracer. This far in our work, $^{18}$F has been used as $^{18}$F-FDG mixed in water, delivered in vials within lead containers. The containers are taken into the imaging lab through a wall hatchway (**Fig. 3**). $^{18}$F-FDG radioactivity is measured and mixed with aqueous solution within the imaging lab. The obstacles related to transportation of liquid radiotracer at ambient conditions between labs in this facility are minor.

Transfer of gaseous $^{11}CO_2$ is more challenging and requires an introduction to the layout of the interconnected labs (**Fig. 3**). $^{11}CO_2$ is produced locally and transferred from the cyclotron through a hot cell (shielded nuclear radiation containment chamber) located at the chemistry lab. A Swagelok nylon tubing is connected to the hot cell in one end to prepare for gas transfer, stretched across the floor and through the wall hatchway to the control room. The other end of the tubing is connected to a receiving single-piston pump located within the PET-CT imaging lab. Two operators are needed for safe transfer: one operator present by the cyclotron, to transfer $^{11}CO_2$ to the hot cell and monitor radiation levels within the receiving laboratories, and one operator monitoring and controlling the receiving pump. During gas transfer the radiation levels within both the chemistry and imaging lab are high, and both rooms should be empty. The receiving pump (**Fig. 4**) should therefore be remote controlled to avoid operator exposure to high levels of radiation.

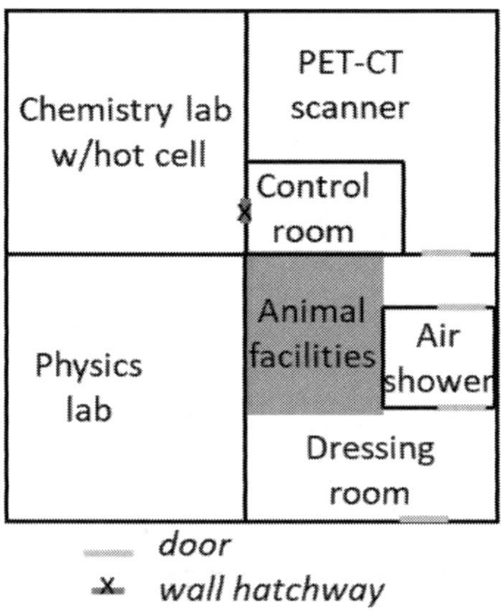


**Fig. 3.** Schematic layout of relevant labs at MIC/PET-centre.

$^{11}CO_2$ is sent to the hot cell by the cyclotron operator (normally a physicist or radiochemist employed at the PET-centre) and progresses directly through the tubing and into the pump, which should be empty and set to receive. Depending on $CO_2$ injection volume, rate, conditions and desired radiation dose, several consecutive volumes of $^{11}CO_2$ may be transferred. The system is thereafter flushed by non-radioactive gas ($N_2$) to displace $^{11}CO_2$ from the tubing. When laboratory radiation levels have decreased sufficiently, the pump is isolated from the hot cell and connected to the experimental set-up (Fig. 4), by activating a set of valves. We use pressure- and remote-controlled valves for this purpose. Manual valves are also mounted on the pumps as a precaution. The received $^{11}CO_2$ is mixed and swiftly pressurized to the experimental conditions by transferring pre-pressurized $CO_2$ from a second pump (**Fig. 4**). Although obstacles remain and the risk for failure is higher when transferring gas (e.g. gas leakage could force temporary shut-down of several labs, with implications for hospital capacities), safe transfer has repeatedly been demonstrated, which enable visualization experiments of high scientific value (**Section 3**).

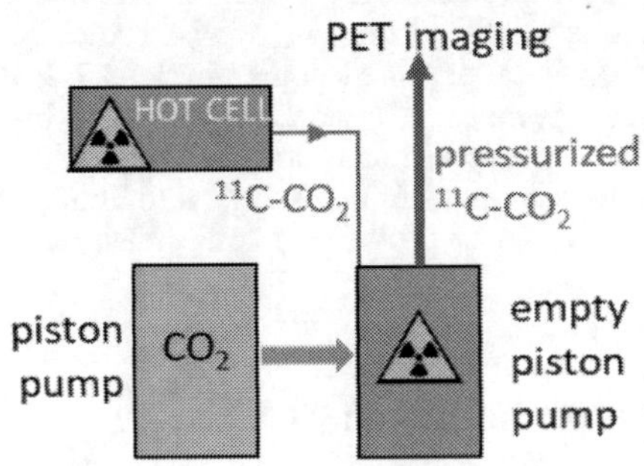


**Fig. 4.** Safe transfer of radioactive gas using two single-piston pumps. The pumps are connected to each other, the experimental setup and the hot cell via remote controlled valves.

### 2.3 Safe use and handling of radioactivity

During experimental work with radiotracers the well-known principles of radiation safety apply: ALARA, "as low as reasonably achievable". The basic protective measures are to increase distance (from the radiation source), decrease time (close to the radiation source) and use proper shielding (between radiation source and personnel). Shielding is important both in terms of health, safety and environmental guidelines for workers (HSE) and to improve image quality, by avoiding radiation from the experimental setup into the PET scanner (increase in background radiation leads to higher level of noise in images). Portable lead walls have been used for shielding towards users in the control room and in adjacent imaging labs. During $^{11}CO_2$ injection radiation levels are initially high, and shielding walls have also been placed between the experimental setup and the PET-CT scanner, or the $CO_2$ pumps have been shielded by placing custom made lead rings around the pistons.

An additional important human safety precaution during core analysis with PET is to avoid leakages of radioactive fluids from the experimental setup. Hence; all parts of the setup, especially pumps, valves and line connections should be checked thoroughly at the experimental conditions before radioactive fluids are introduced. Performing baseline tests with non-radioactive fluids is recommended. At ambient conditions, where radioactively labelled liquid is used at low radiation dose, manual control of pumps and effluent production is possible. $^{18}$F-FDG mixed in aqueous solution may be produced into shielded graded cylinders, flasks or beakers where the radiation can decay, and is thereafter discharged according to local regulations. For more complicated experiments, for example injecting radioactive gas at high pressure conditions, installation of remote-controlled computers and pressure control valves is highly recommended. Hence; injection rates, direction of flow and fluid production may be monitored remotely and operator exposure to radiation minimized. Radioactive $^{11}CO_2$ should not be produced into air, and is collected in a piston accumulator at the production end of the core holder. The lower end of the accumulator is pressurized to the experimental conditions by mineral oil or distilled water, which is produced through a back-pressure regulator as produced effluents accumulate above the piston. After decay, the gas is vented into air. Image acquisitions may be remote controlled from an office outside the preparation room. The PET signal is typically recorded in time spans of 20 min – 2 hours per acquisition, and the signal is post-processed into three-dimensional images. The temporal resolution in each image frame is flexible and set during post-processing.

### 2.4 Weight and size of laboratory equipment

Safe transfer and use of radioactively labelled fluids for PET imaging is only possible by choosing suitable experimental equipment. Experimental conditions, such as pressure and temperature, and possibilities, such as flow rates and directions, directly relates to the different mechanical parts of an experimental setup, and their inherent limitations. Inevitably: operational limits of equipment often relate to their weight and size.

Clinical scanners are intended for human use, hence the gantry opening diameter is larger and can fit large and heavy equipment. Because the PET detector elements are also larger and positioned further away from the imaged object, the spatial resolution will, however, decrease. High-resolution (pre-clinical) PET scanners tend to have smaller gantry openings, which improves spatial resolution but also limits the possible size of an imaged object. In our case, the size of equipment that can be moved into the PET-scanner (i.e. core and core holder) is limited by 1) the gantry opening diameter (16 cm), 2) the PET field of view (FOV) of 120*120*97mm in the x,y,z directions, and 3) a mechanical arm which moves the equipment into the FOV and has inherent weight limitations (**Fig. 5**).

Several experiments have been performed using epoxy-coated cores, where the core circumference is cast in epoxy and the end faces are fitted with POM (polyoxymethylene) end pieces that are attached to nylon fittings and lines [28-33]. The most influential factors in terms of weight are then the core material and saturating fluids. Use of epoxy allows flexibility in terms of core shape and defined no-flow boundaries, and small cores may be aligned either horizontally or vertically within the scanner. Experimental investigations of subsurface carbon sequestration, however, often require relevant conditions, i.e. higher pressures, and it is necessary to use a core holder and apply overburden pressure to the core.

https://doi.org/10.1051/e3sconf/202336601011

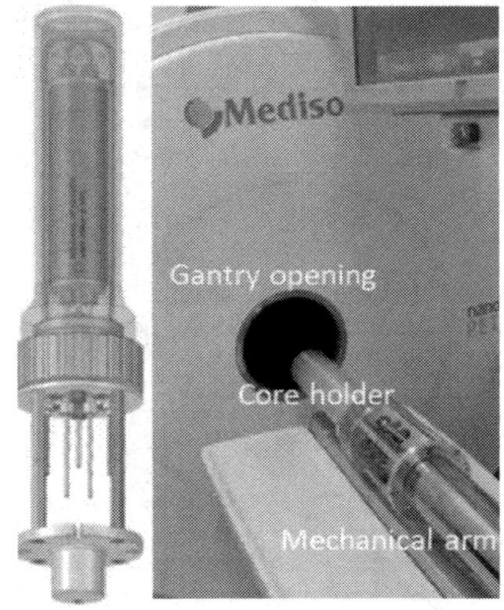


**Fig. 5.** A custom-made core holder by RS systems © (Trondheim, Norway) is installed on a mechanical arm for imaging with high-resolution PET.

The core holder is then the single most influential factor in terms of weight, and the pressurizing fluid, fittings, valves, lines, core and saturating fluid will also contribute to the total weight. We use a custom-made, aluminium core holder (**Fig. 5**) for our setup (©RS systems, Trondheim, Norway). The core holder is attached directly to the mechanical arm, and can fit cores of 1.5'' diameter and 10cm length. The core holder can withstand confinement pressure up to 200Bar. With thicker aluminium walls, higher pressures may be achieved-albeit with consequences for the total weight of the equipment.

The mechanical arm moves the core horizontally for positioning within the CT and PET FOV. The core is always horizontally aligned when using the core holder. Epoxy coated cores rest on plastic animal beds attached to the mechanical arm and can also be oriented at an angle.

In the combined Mediso PET-CT the CT scanner is located further into the scanner, and the PET scanner ring is located closer to the gantry opening. The two modalities are run in sequence and simultaneous images from the two modalities cannot be acquired. The mechanical arm moves the core between modalities and is controlled by the operator through the scanner software. CT is initially acquired to position the core. The mechanical arm thereafter moves the core into the defined position for PET acquisition. Lines connecting the core holder to the experimental setup therefore needs to be flexible and allow for horizontal movement during experiments. We have successfully used both 1/8'' peek and 1/16'' stainless steel lines, due to their low weight, flexibility and ability to withstand high pressures.

Another obstacle in terms of size and weight during pre-clinical imaging is that all equipment must pass through the air shower illustrated in **Fig. 3**. The purpose of the air shower is to remove animal allergens from researchers going out of the imaging lab. Air showers are proven effective to remove allergens from hair and clothes, to prevent spreading to the surrounding facilities. Because the imaging lab and surrounding facilities were intended for animal studies and users, this procedure must be followed also during core analysis (it is not possible to physically bypass the air shower). We have solved this for our use by mounting all laboratory equipment (pumps, accumulators, core holder, pressure transducers etc) on a portable trolley, that fits within the air shower (width 60cm, length 100cm). The trolley is wheeled into the imaging lab through the air shower before imaging experiments, and is stored in the physics lab (**Fig. 3**) in between experiments. In lab facilities shared by several research disciplines, compact and easily transportable experimental setups are necessary, as the time available for imaging may be short (down to few days).

## 3 Opportunities in CCUS

Permanent storage of $CO_2$ in deep sedimentary reservoirs is being developed worldwide, to curb atmospheric carbon emissions. Demonstrations of safe storage on the field scale must be supported by experimental data on smaller scales, to fully understand how $CO_2$ storage influences, and is influenced by, the pore space. PET imaging has significant potential to capture dynamics during $CO_2$-flow in porous media, both implicitly (labelling the aqueous phase) and explicitly (radioactively labelling the $CO_2$). This section summarizes the opportunities and demonstrated potential for PET in CCUS (Carbon Capture Utilization and Storage) research. For an overview of previous application of PET imaging in water resources research, we refer to Zahasky *et al.* [21].

An example experimental setup is demonstrated in **Fig. 6**: the setup is designed to be flexible and can easily be modified for the purpose of the experiment, e.g. to facilitate circulation of fluids at the inlet end face.

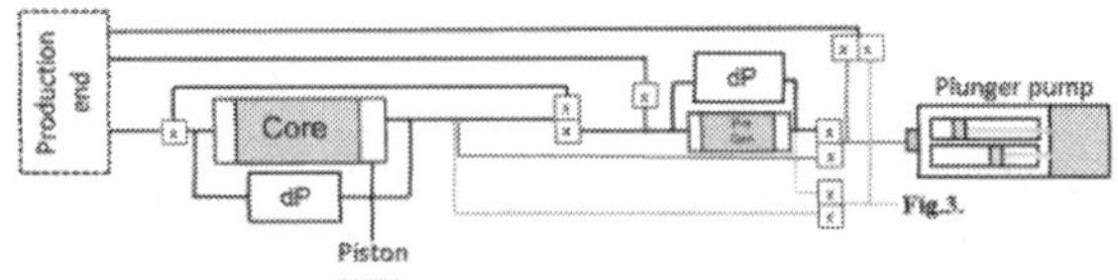


**Fig. 6.** Example of experimental setup used for PET imaging of CO2-foam. The setup is very flexible, and can with few adjustments be used for a variety of experiments. "x" marks pressure control valves and "dP" marks pressure measurements. CO2 only flows in orange lines and only the aqueous phase in blue lines. The production end varies with radioactively labelled phase, as described in **Section 2.3.**

### 3.1 $CO_2$ storage

*In-situ* experimental observations of forced $CO_2$ displacement into porous media is necessary to explain the influence of heterogeneities on $CO_2$ saturation development and capillary trapping: to accurately predict $CO_2$-brine migration in carbon sequestration reservoirs. Fernø *et al.* [2] initially demonstrated decoupled PET-CT imaging for $CO_2$ storage applications. They injected $CO_2$ into water-saturated sandstone at 90 Bar pressure to measure decoupled saturation information from each method. $CO_2$ saturation development was determined implicitly by quantifying the decrease in signal from $^{22}$Na-labelled (half-life of 2.6 years) water, using a clinical PET-CT scanner with lower spatial resolution. Exhaustive sensitivity studies comparing PET and CT imaging were performed by Pini *et al.* [7] and Zahasky and Benson [3], applying a previously derived analytical method [6, 34] to extract capillary pressure and relative permeability functions on the sub-core scale. Zahasky and Benson [3] used high-resolution PET imaging to measure single- and multiphase transport properties in heterogeneous sandstone during gas injections: Nitrogen constituted the gas phase and the pressure was approximately 1.4 Bar, i.e. the experimental conditions were less relevant for CCUS, but the data analysis provided a solid foundation for using PET in further quantification experiments. Zahasky and Benson [35], [36] also used high-resolution PET to investigate $CO_2$-flow and trapping in heterogeneous sandstone cores at pressures ranging from 1-8 Bar. $CO_2$ displacement was implicitly imaged by injecting pulses of $^{18}$F-FDG labelled water, or co-injecting water and Nitrogen. PET-imaging was again repeatedly and successfully used to assess flow parameters on the sub-core scale. Brattekås and Haugen [37] recently demonstrated explicit and dynamic imaging of $CO_2$ during flow through a low-permeable chalk core at elevated pressure conditions. High-resolution PET imaging was used to visualize $CO_2$, labelled explicitly by $^{11}$C-$CO_2$ and injected into a chalk core at 50 Bar pressure (**Fig. 7**). The procedure for $CO_2$ radiolabelling and injection was similar to the procedure detailed in Section 2; and accurate quantification of spatial phase saturations were achieved during both brine and $CO_2$ injections, at the [mm] scale. Visualization revealed heterogeneous displacement patterns, as expected during gas-brine displacements at less-than-ideal mobility ratio. The heterogeneities were attributed to a sub-core variation in permeability, which could be determined from PET displacement data at the millimetre scale.

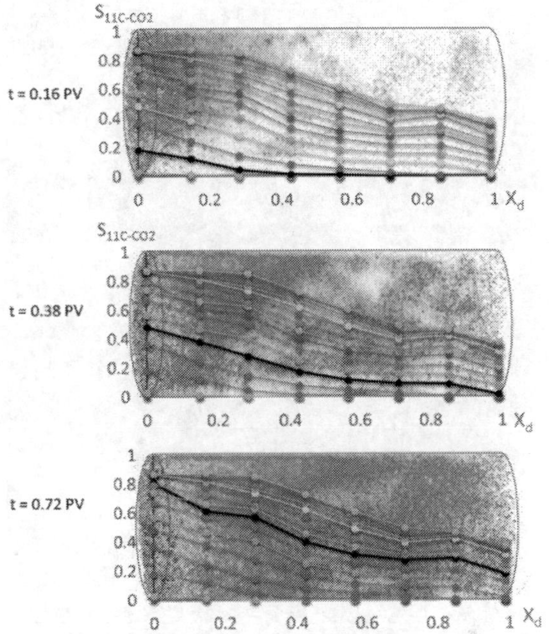


**Fig. 7.**[c] Dynamic saturation development during $^{11}$C-CO2 injection into a chalk core, demonstrated by three timesteps. Qualitative 3D images of core saturation are overlaid by quantitative 1D profiles derived from PET. The red colour in 3D images represents the highest PET signal intensity, followed by green and blue. White areas do not contain $^{11}$C-$CO_2$. The variations in the colour map hence indicate large spatial variations in $CO_2$ saturation. The experiment and results are detailed in [37].

These recent publications illustrate that PET imaging can (and should) be utilized to investigate $CO_2$-brine displacements at a range of conditions and flow rates, where insight into local saturations and flow patterns allow for accurate determination of flow functions. Few publications are yet available where $CO_2$ injection is captured with high-resolution PET at high pressure conditions; but may be provided as necessary by broader access to PET imaging technology.

Spontaneous imbibition is another important mechanism during $CO_2$ sequestration, and may control the redistribution and remobilization of fluids after $CO_2$ injection. High-resolution PET imaging has in several instances been used to image the imbibing, radioactively labelled, water front [28-30]. The results confirmed that PET is an excellent tool to quantify dynamics during spontaneous imbibition, **Fig. 8**.

[c] A modified version of this figure was first published in *Explicit tracking of CO2-flow at the core scale using Micro-Positron Emission Tomography*. Journal of Natural Gas Science and Engineering, 2020. 77.

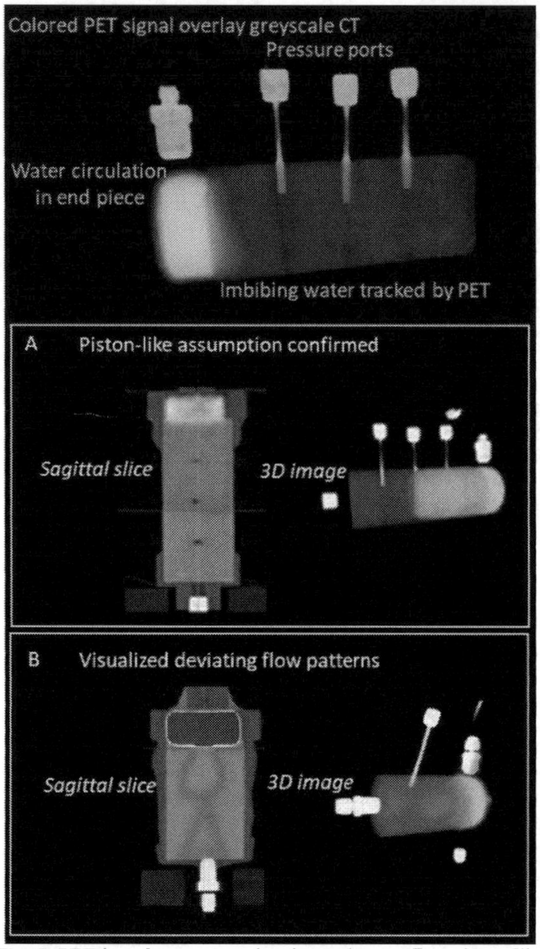


**Fig. 8.** PET imaging was previously used to confirm piston-like behaviour during spontaneous imbibition [28], and to visualize and explain behaviour deviating from general assumptions [33].

PET imaging can confirm the assumption of piston-like behaviour necessary to extract relative permeability and capillary pressure from spontaneous imbibition data [28], or provide insight to displacement patterns deviating from the general assumptions, where extraction of flow functions would be inaccurate [33]. The referenced works were simple in terms of experimental conditions and equipment: using epoxy-coated cores at ambient conditions. With minor adjustments, however, the high-pressure experimental setup in **Fig. 6.** can be modified for high-pressure spontaneous imbibition at relevant $CO_2$ storage conditions, applying e.g. the method proposed in Zahasky and Benson [38]. Relevant saturation functions for CCUS can then be extracted from the data set when important assumptions are confirmed by imaging.

### 3.2 $CO_2$ utilization ($CO_2$-EOR)

Combined $CO_2$ storage and enhanced oil recovery (EOR) can lower the carbon footprint of produced oil, while improving the recovery factor in many reservoirs. PET imaging can provide insight into complex flow systems used for EOR applications; including the flow of complex fluids such as polymers [39], polymer gels [32, 40-42] and foams [10] in heterogeneous and fractured porous media.

Displacement of oil by $CO_2$ can be imaged by PET similarly to $CO_2$-brine displacements (**Section 3.1**) using $^{11}C$ to explicitly label $CO_2$. Water and $CO_2$ flow may be explicitly imaged by PET in the same pore network, but each fluid must then be allowed to decay before injection of the other- i.e. only one fluid phase may be imaged at a time. Multimodal imaging by PET-CT [2] or PET-MRI [26] allows for explicit and simultaneous imaging of up to three fluid phases within the same porous medium, with high potential benefits for core scale research.

Foam represents an opportunity to reduce $CO_2$ mobility, to improve the displacement of oil. Foam is a dispersion of gas separated by liquid lamella, where both foam constituents may often be imaged by PET: aqueous surfactant solution labelled by $^{18}F$-FDG or injected $CO_2$-foam gas phase labelled by $^{11}CO_2$. Brattekås *et al.* [10] used PET imaging to quantify foam flow in natural fracture networks, where an impermeable matrix ensured fluid flow in the fracture network only. The spatial resolution of PET is too low to facilitate direct observation of foam structure (bubble size etc), but is an efficient tool to establish foam distribution within fractures. PET may also be used to establish improved sweep efficiency within adjacent (permeable) matrix.

Polymer gel is often used for water shutoff purposes, and may also have significant potential in $CO_2$-EOR applications. Polymer gel reduce fracture conductivity, without influencing matrix properties; hence, polymer gel placed in fractures improve sweep efficiency by diverting subsequently injected fluids into the matrix. Previous studies showed that the efficiency of polymer gel in EOR is heavily dependent on core wettability, the properties of the fracture-filling gel, and the properties of the chase fluid [43]. $CO_2$ has several inherent benefits as a chase fluid, such as miscibility with matrix oil. Recent studies demonstrated that PET imaging is an efficient method to establish gel behaviour and matrix sweep during chase floods. Brattekås *et al.* [32] used PET imaging and $^{18}F$-FDG labelled water to gain insight into dynamics during chase waterflooding of gel-filled fractures. The water injection rate was initially low, and gel rupture was controllably measured: the pattern of gel rupture was found to be heterogeneous within the fracture, and its width could not be derived from global measurements of flow rate and differential pressure. Further, Brattekås and Seright [31] and Brattekås and Seright [42] used PET imaging to establish how the rupture channels through gel evolved when the salinity of the injected water-phase was reduced. They proved that gel swelled when contacted by low-salinity water and consequently improved fracture blocking efficiency. The coupled dynamics of gel-swelling with continuous water injection improved matrix displacement in fractured cores. The studies used $^{18}F$-FDG mixed in high-salinity or low-salinity water to explicitly image water flow through the gel-filled

 https://doi.org/10.1051/e3sconf/202336601011

fractures and adjacent matrix. The gel was first ruptured using high-salinity water (w/[18]F-FDG), which was flushed out or decayed before low-salinity water w/[18]F-FDG was imaged. Because of the short half-life of [18]F-FDG, tracing of several water composition was possible within the duration of the experiments; i.e. hours to days. PET imaging was a significant advantage in assessing the behaviour of different water phases within gel-filled fractures: where other imaging methods would require chemical alterations of the water phase(s) or gel (e.g. using $D_2O$ instead of $H_2O$ for MR imaging, or adding sodium iodide for CT imaging) to distinguish the behaviour of the injected phase from gel or previously injected water. The PET signal is not influenced by injected fluids that have decayed, hence the labelled phase is tracked during injection without disturbances from the surroundings.

### 3.3 $CO_2$ storage security

PET imaging has significant potential in establishing seal security for $CO_2$ storage applications. Seal ability to prevent leakage of buoyant $CO_2$ is important in field scale applications; and demonstration of safe $CO_2$ storage is crucial to implement widespread CCUS. Fernø *et al.* [25] used a clinical PET-CT scanner to visualize $CO_2$ flow in tight shale and demonstrated the excellent sensitivity of PET imaging compared to CT in tight, low-porous and -permeable rocks, **Fig. 9**; where a very low volume of $^{11}C$-$CO_2$ was sufficient for PET quantification of saturation. The uneven fluid displacement within shale, visible by PET, could be tied to structural heterogeneities, visible by CT.

PET imaging could be further used to quantify $CO_2$ capillary entry pressure in seals by modifying established methods (previously summarized in e.g. [44]) for imaging. The established experimental methods (without imaging) applies pressure to $CO_2$ in contact with the core end face, either stepwise (step-by-step approach) or continuously (continuous injection approach), while pressure and production is monitored and logged. The capillary entry pressure may be calculated from pressure and production data when the cores are assumed to be homogeneous. The same approach may be followed during imaging, applying $^{11}CO_2$ pressure at the inlet end face of the core (**Fig. 6**) and monitoring gas entry into the core by PET, and the capillary entry pressure may be extracted by connecting the imaged gas entry to pressure data. The high temporal resolution of PET ensures that the entry pressure is measured at the correct time, and the influence of core heterogeneities is accounted for rather than neglected. Longer-term experiments to assess seal wettability or $CO_2$ diffusion, which are also important in terms of seal security [45], are more challenging and requires radiotracers with longer half-lives to be assessed by PET.

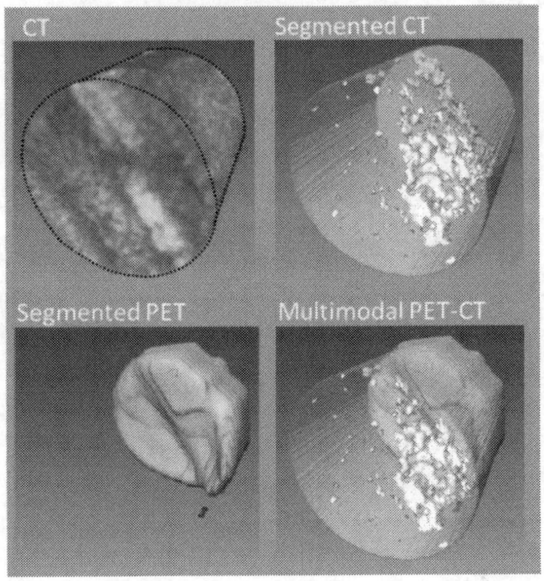


**Fig. 9**[d]. PET imaging captures the inflow of 11CO2 in tight shale. The nonuniform displacement front can be tied to structure, visible by CT. For in-depth discussion and results of the experiment showed in the figure we refer to [25].

$CO_2$ storage is most commonly developed in sandstone formations, but carbonates also play an important role in the development and widespread implementation of CCUS: both due to the world-wide abundancy of saline aquifers within carbonate formations, and as candidates for $CO_2$-EOR with combined storage. Acidification of formation brine during $CO_2$ injection in carbonates may cause dissolution of the rock material and reactive flow patterns are developed that significantly influence fluid flow and, inevitably, carbonate $CO_2$ storage security. PET imaging enables experimental investigations of reactive transport dynamics at the core scale. Dynamic quantification of dissolution (rate and degree) at the core scale may further support upscaling to larger scales. **Fig. 10** shows preliminary images from an ongoing dissolution study. The CT modality shows structural changes in the core resulting from dissolution. The PET modality shows changes in the preferred fluid flow pattern, consistent with the development of wormholes. Carbonate dissolution and formation of wormholes was previously observed using CT [46, 47]. Multi-modal PET-CT imaging represents a new addition to the experimental portfolio available to study dissolution; by simultaneously connecting the flow of labelled fluids to the structure through which it flows, **Fig. 10**. The unique contribution of PET imaging to dissolution studies is, hence, to non-invasively, dynamically and explicitly study the development of fluid flow patterns. Flow patterns may be tied to minor or major changes in structure by CT scans.

[d] A modified version of this figure was first published in *Description of $CO_2$ EOR and Storage in Tight Shale Formations by Explicit $CO_2$ Flow Visualization,* Geophysical Research Letters. 2015.

CT scans may be performed either dynamic, pre- and/or post- flow experiment imaging. Pre- and post-imaging of micro-CT may than also be applied to visualize structural changes at a much higher spatial resolution.

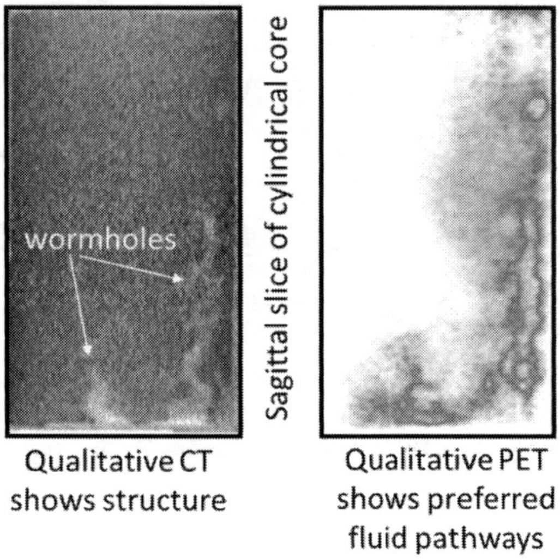


**Fig. 10:** Dissolution of carbonate during $CO_2$ or carbonated water injection is often visualized using CT, but PET imaging may be an important contribution. While CT (left) portrays structural changes during dissolution, PET determine dynamic changes in the preferred flow paths within the core.

*This work was supported by Norwegian Research Council under project 280341.The authors acknowledge Kristine Fasmer and Kjartan Foldnes for valuable help during radiotracer production and transfer, and PET-imaging. Thank you to Torunn Veien and Olav Parelius Folkvord for providing image data from ongoing dissolution studies. Pre-clinical imaging is routinely performed at the Molecular Imaging Center, Dept. of Biomedicine, University of Bergen. Personnel and equipment associated with the PET-centre at Haukeland University Hospital contributes greatly to the collaboration. The mechanical workshop at the Dept. of Physics and Technology, University of Bergen tailored experimental equipment to our needs, for which the authors are thankful.*

## References

1. Bailey, D.L., et al., eds. *Positron Emission Tomography*. Basic Sciences. 2005, Springer.
2. Fernø, M.A., et al., *Combined positron emission tomography and computed tomography to visualize and quantify fluid flow in sedimentary rocks*. Water Resources Research, 2015.
3. Zahasky, C. and S.M. Benson, *Micro-positron emission tomography for measuring sub-core scale single and multiphase transport parameters in porous media*. Advances in Water Resources, 2018. **115**: p. 1-16.
4. Krevor, S.C.M., et al., *Capillary heterogeneity trapping of CO2 in a sandstone rock at reservoir conditions*. Geophysical Research Letters, 2011. **38**(15): p. L15401.
5. Krevor, S.C.M., et al., *Relative permeability and trapping of CO2 and water in sandstone rocks at reservoir conditions*. Water Resources Research, 2012. **48**(2): p. W02532.
6. Pini, R., S.C.M. Krevor, and S.M. Benson, *Capillary pressure and heterogeneity for the CO2/water system in sandstone rocks at reservoir conditions*. Advances in Water Resources, 2012. **38**(0): p. 48-59.
7. Pini, R., et al., *Quantifying solute spreading and mixing in reservoir rocks using 3-D PET imaging*. J. Fluid Mech., 2016. **796**: p. 558-589.
8. Brattekås, B., et al., *New Insight to Wormhole Formation in Polymer Gel during Water Chasefloods using Positron Emission Tomography (PET)*, in *SPE Bergen One Day Seminar*. 2016: Bergen, Norway.
9. Dechsiri, C., et al., *Positron emission tomography applied to fluidization engineering*. Can J Chem Eng, 2005. **83**(1): p. 88-96.
10. Brattekås, B., et al., *Foam Flow and Mobility Control in Natural Fracture Networks*. Transport in Porous Media, 2020. **131**(1): p. 157-174.
11. Khalili, A., A.J. Basu, and U. Pietrzyk, *Flow visualization in porous media via Positron Emission Tomography*. Phys Fluids, 1998. **10**(4): p. 1031-1033.
12. Kulenkampff, J., et al., *Evaluation of positron-emission-tomography for visualization of migration processes in geomaterials*. Phys Chem Earth, 2008. **33**((14-16)): p. 937-942.
13. Maucec, M., et al., *Imaging of Fluid Mobility in Fractured Cores Using Time-lapse Positron Emission Tomography*, in *SPE Annual Technical Conference and Exhibition*. 2013, Society of Petroleum Engineers: New Orleans, LA, USA.
14. Maguire, R.P., et al., *Positron emission tomography of large rock samples using a multiring PET instrument*. Ieee T Nucl Sci, 1997. **44**(1): p. 26-30.
15. Degueldre, C., et al., *Porosity and pathway determination in crystalline rock by positron emission tomography and neutron radiography*. Earth Planet Sc Lett, 1996. **140**((1-4)): p. 213-225.
16. Emery, G.T., *Perturbation of Nuclear Decay Rates*. Annual Review of Nuclear Science, 1972. **22**(1): p. 165-202.
17. Haugan, A. *A Low-Cost PET System for Use in Flow Experiments of Porous Media*. in *SPE Annual Technical Conference and Exhibition*. 2000. Dallas, Texas.
18. Kulenkampff', J., et al., *Evaluation of positron-emission-tomography for visualisation of migration processes in geomaterials*. Physics and Chemistry of the Earth, 2008. **33**(14-16): p. 937-942.
19. Richter, M., et al., *Positron emission tomography for modelling of geochemical transport processes in clay*. Radiochimica Acta, 2005. **93**: p. 643-651.
20. Gründig, M., et al., *Tomographic radiotracer studies of the spatial distribution of heterogeneous geochemical transport processes*. Applied Geochemistry, 2007. **11**: p. 2334-2343.
21. Zahasky, C., et al., *Positron emission tomography in water resources and subsurface energy resources engineering research*. Advances in Water Resources, 2019. **127**(May 2019): p. 39-52.
22. Rohren, E.M., T.G. Turkington, and R.E. Coleman, *Clinical Applications of PET in Oncology*. Radiology, 2004. **231**(2): p. 305-332.
23. Anand, S.S., H. Singh, and A.K. Dash, *Clinical Applications of PET and PET-CT*. Medical journal, Armed Forces India, 2009. **65**(4): p. 353-358.

24. Bailey, D.L., et al., *Positron Emission Tomography*, ed. M.N. Maisley. 2005: Springer.
25. Fernø, M.A., et al., *Flow visualization of CO2 in tight shale formations at reservoir conditions.* Geophysical Research Letters, 2015. **42**: p. 7414-7419.
26. Brattekås, B., et al., *Unlocking multimodal PET-MR synergies for geoscience.* Advances in Water Resources, 2020. **142**: p. 103641.
27. Agency, T.I.A.E. https://nucleus.iaea.org/sites/accelerators/Pages/Cyclotron.aspx. 2020.
28. Fernø, M.A., et al., *Quick and Affordable SCAL: Spontaneous Core Analysis*, in *the International Symposium of the Society of Core Analysts*. 2015: St.John's, Newfoundland&Labrador, Canada.
29. Fernø, M.A., et al., *Spontaneous Imbibition Revisited: A New Method to Determine Relative Permeability and Capillary Pressure by Inclusion of the Capillary Backpressure*, in *EAGE 18th European Symposium on Improved Oil Recovery*. 2015: Dresden, Germany.
30. Ruth, D., et al. *Matching Experimental Saturation Profiles by Numerical Simulation of Combined Co-/Counter-Current Spontaneous Imbibition.* in *International Symposium of the Society of Core Analysts*. 2016. Snowmass, CO, USA.
31. Brattekås, B. and R.S. Seright, *Implications for improved polymer gel conformance control during low-salinity chase-floods in fractured carbonates.* Journal of Petroleum Science and Engineering, 2018. **163**: p. 661-670.
32. Brattekås, B., et al., *New Insight Into Wormhole Formation in Polymer Gel During Water Chase Floods With Positron Emission Tomography.* SPE Journal, 2017. **22**(01): p. 32-40.
33. Føyen, T.L., M.A. Fernø, and B. Brattekås, *The Effects of Nonuniform Wettability and Heterogeneity on Induction Time and Onset of Spontaneous Imbibition.* SPE Journal, 2019. **24**(03): p. 1192-1200.
34. Pini, R. and S.M. Benson, *Simultaneous determination of capillary pressure and relative permeability curves from core-flooding experiments with various fluid pairs.* Water resources research, 2013. **49**(6): p. 3516-3530.
35. Zahasky, C. and S.M. Benson. *Phase saturation validation and tracer transport quantification using microPET in a heterogeneous sandstone core.* in *International Symposium of the Society of Core Analysts*. 2016. Snow Mass, Colorado, USA: Society of Core Analysts.
36. Zahasky, C. and S.M. Benson, *Using micro-positron emission tomography to quantify single and multiphase flow in heterogeneous reservoirs.* Energy Procedia, 2017. **114**: p. 5070-5082.
37. Brattekås, B. and M. Haugen, *Explicit tracking of CO2-flow at the core scale using micro-Positron Emission Tomography.* Journal of Natural Gas Science and Engineering, 2020. **77**.
38. Zahasky, C. and S.M. Benson, *Spatial and temporal quanitification of spontaneous imbibition.* Geophysical Research Letters, 2019.
39. Sandnes, M.F., *Wetting Stability of Aged Limestone in the Presence of HPAM Polymer*, in *Department of Physics and Technology*. 2020, University of Bergen: Bergen Open Research Archive.
40. Brattekås, B. and M.A. Fernø, *New Insight from Visualization of Mobility Control for Enhanced Oil Recovery Using Polymer Gels and Foams*, in *Chemical Enhanced Oil Recovery (cEOR) - a Practical Overview*, L. Romero-Zèron, Editor. 2016, InTech.
41. Brattekås, B. and R.S. Seright, *Implications for improved polymer gel conformance control during low-salinity chase-floods in fractured carbonates.* Journal of Petroleum Science and Engineering, 2017.
42. Brattekås, B. and R. Seright, *The Mechanism for Improved Polymer Gel Blocking During Low-Salinity Waterfloods, Investigated Using Positron Emission Tomography Imaging.* Transport in Porous Media, 2020. **133**(1): p. 119-138.
43. Brattekås, B., et al., *Fracture Mobility Control by Polymer Gel- Integrated EOR in Fractured, Oil-Wet Carbonate Rocks*, in *EAGE Annual Conference & Exhibition incorporating SPE Europec*. 2013: London, UK.
44. Busch, A. and N. Müller, *Determining CO2 /brine relative permeability and capillary threshold pressures for reservoir rocks and caprocks: Recommendations for development of standard laboratory protocols.* Energy Procedia, 2011. **4**: p. 6053-6060.
45. Iglauer, S., C.H. Pentland, and A. Busch, *CO2 wettability of seal and reservoir rocks and the implications for carbon geo-sequestration.* Water Resources Research, 2015. **51**(1): p. 729-774.
46. Menke, H.P., et al., *Dynamic Three-Dimensional Pore-Scale Imaging of Reaction in a Carbonate at Reservoir conditions.* Environmental science & technology, 2015. **49**: p. 4407-4414.
47. Ott, H., et al., *Core-flood experiment for transport of reactive fluids in rocks.* Review of Scientific Instruments, 2012. **83**(8): p. 084501.

# Integrated workflow for rotary sidewall cores orientation: best practices and examples from planning to execution

*Giorgio* Volonté[1], *Antonella* Bersani[1], *Roberto* Berto[1], *Riccardo* Cerri[1,*] *and Fabio* Pinelli[1]

[1]Eni S.p.A., 20097 San Donato Milanese, Italy

**Abstract.** Over the last five years, the acquisition of rotary sidewall cores has become increasingly important in the O&G business. It is a matter of fact that the current tendency is to prefer them as a replacement to bottom hole cores, essentially for rig time related cost saving. This fact leads to force their use also for those types of characterisation usually reserved to larger size samples so, by consequence, some specific issues associated to their small size and to the lack of knowledge of *in situ* orientation have to be faced. For several applications such as geomechanics (whose experimental analysis require vertical plugs), petrophysics (permeability anisotropy, natural fracture orientation) and sedimentology, the original orientation, or its *a posteriori* identification, is mandatory in order to correctly measure properties relevant for reservoir management and studies. The full orientation of rotary sidewall cores requires to acquire or compute three couples of spatial information: high-side and low-side, wellbore end and formation end, trend and plunge of the samples. A dedicated workflow to *a posteriori* reconstruction of the orientation of large size rotary sidewall cores has been developed by means of the integration of multiple scale imaging techniques: images from logs and sidewall core are integrated through commercial software usually adopted for dip analysis. This approach has been successfully applied to several sidewall cores gathered in different geological environments. Currently, new coring and conveyance technologies allowing a predetermined sidewall cores orientation are emerging, increasing the reliability of the orientation workflow. Therefore, different scenarios of log planning strategy, operational solutions and post-coring analysis can be expected, depending on the confidence in the geological-structural interpretation and on the complexity of the technical environment. In all cases, it is extremely important to plan specific data acquisition and handling operations from wellsite to laboratory; including formalized minimum requirements and dedicated procedures, in order to maximize the quality of the subsequent laboratory analyses.

## 1 Introduction

The recovery of bottom hole cores (BHC) has always been a significant component of worldwide drilling costs. Over recent years, however, the general trend has become to acquire an increasing number of large size rotary sidewall cores (RSWC) instead, trying to find a balance between reducing costs and ensuring the acquisition of enough material for laboratory analyses. Unfortunately, RSWCs are not currently suitable for several types of characterization usually reserved for larger size samples, since they recover a much lower amount of material than BHCs and their original *in situ* orientation is unknown. The first drawback makes the overall quality of the acquired RSWCs extremely important (in order to maximize the data obtainable from most types of experimental analyses). Moreover, several applications require also the knowledge, or the *a posteriori* identification, of the original orientation of the recovered plug. For example, in petroleum-related experimental geomechanics, this requirement arises from the need to simulate in the laboratory the stress and deformation boundary conditions observable in a reservoir; similarly, the sedimentology can base its studies on oriented thin sections [1]. Furthermore, reservoir characterization often requires oriented permeability measurements [1] or the evaluation of natural fracture distribution also in terms of their spatial orientation [2, 3, 4]. A complete reconstruction of the *in situ* orientation of a RSWC requires the determination of the following parameters (see Fig. 1 below):

- trend and plunge, which are function of the borehole orientation (hole azimuth and deviation) and of the RSWC azimuth;
- wellbore end and formation end, corresponding to the two opposite sides of the RSWC facing the wellbore and the formation respectively;
- high side and low side, which are the geometrical intersection of the RSWC with the vertical plane.

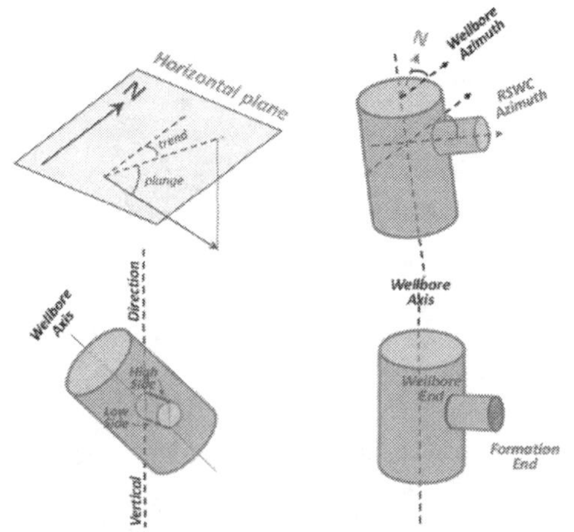


**Fig. 1.** Reference nomenclature for RSWC goniometry.

[*] Corresponding author: riccardo.cerri@eni.com

© The Authors, published by EDP Sciences. This is an open access article distributed under the terms of the Creative Commons Attribution License 4.0 (http://creativecommons.org/licenses/by/4.0/).

Some of the above parameters, such as the borehole orientation, are generally known or acquired during the operation of coring and logging. The azimuth of the RSWC is not usually recorded during coring operations, unless specific conveyance methods (i.e. tool taxi) are adopted or a dedicated image log is run after coring. Wellbore and formation ends are generally easily identifiable since the most common commercial tools for RSWC acquisition store the samples in the core container with the same side-up (Fig. 2).

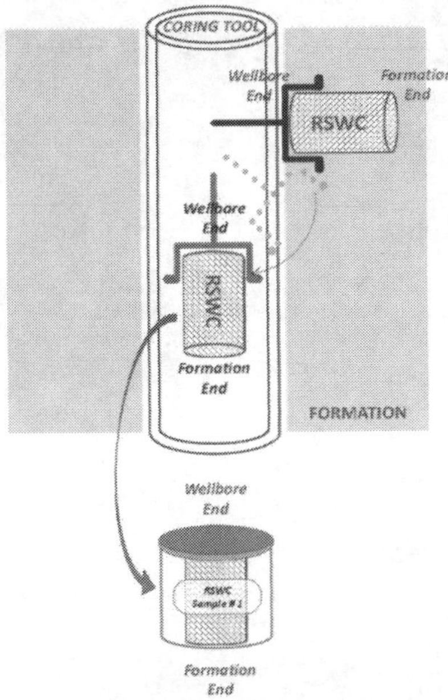


**Fig. 2.** Recovery of a RSWC.

When RSWC quality is high, this information can also be inferred by inspecting the RSWC: the wellbore end is the borehole wall and it is concave (and usually shows a more polished surface, Fig. 3a), while the formation end can show an indentation generated by the breaking of the RSWC (Fig. 3b) [2, 3].

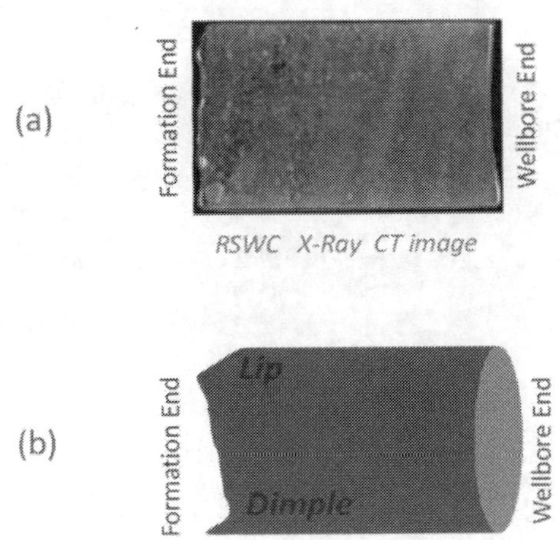


**Fig. 3.** RSWC characteristics helpful for orientation.

The high/low sides of the RSWC are never recorded during coring. In some cases, where the RSWC quality is particularly good and recovery procedures are known, at least the direction in which the RSWC has been broken by tilting, which corresponds to wellbore axis, can be inferred from the first inspection of the core (lip/dimple in Fig 3b and Fig. 6) [2, 3].

Therefore, to provide an accurate *a posteriori* reconstruction of the parameters describing the *in situ* orientation of RSWCs, a workflow mainly based on laboratory X-Ray CT scans and wellbore log images has been developed by integrating all the available wellsite and laboratory data.

Regardless of the adopted procedure, the *a posteriori* orientation of a RSWC is not a straightforward task and, to maximize the probability of success of this operation, it is extremely important to accurately track all the required data and to preserve the quality of the RSWC along all its lifetime. So, all the recovery, data acquisition and handling operations from wellsite to laboratory have to be specifically planned. In the following paragraphs, the identified minimum operative requirements will be formalized and dedicated procedures will be suggested in order to maximize the accuracy of the orientation process and, by consequence, the quality of the subsequent laboratory analyses. Indeed, orientation is particularly important for some petrophysical properties that strongly depend on the measurement direction (e.g. to validate permeability in anisotropic geological environments).

## 2 Minimum requirements and procedures for coring

### 2.1 Job planning and execution

Whenever the orientation of the RSWC is required for the subsequent studies, careful planning of the acquisition operations must be put in place and all relevant information and best practices gathered in a dedicated coring protocol to guarantee the reliability of the dataset required by the orientation procedure. This paragraph dictates the specific requirements that must be followed throughout the coring operations, namely: a) purpose of core acquisition and formation characteristics, b) selection of core samples, c) log acquisition and correlation plan, d) rotary sidewall coring operational procedures, e) wellsite handling of core samples.

a) The objectives of the acquisition must clearly state that the core orientation is requested based on the core analysis purposes. Furthermore, the lithological and sedimentological formation characteristics significant to the acquisition have to be highlighted. The Uniaxial Compressive Strength (UCS) computation from sonic $\Delta t$ (using the appropriate algorithm calibrated on nearby wells or analogues) has to be provided in order to properly plan the operations.

b) The criteria for selecting number and depth of the sampling points must be defined taking into account the expected quality in terms of size and shape, always in connection with the requirements for the laboratory analyses. The sequence and priority of sampling points have to be considered; extra cores must be planned when targeting thin beds or bed boundaries, since depth accuracy issues are emphasized.

c) The log suite suggested to support the orientation of the RSWC would better include two image log runs, before and after the coring operation.

- The image log acquired before coring is aimed at defining the formation texture and structure; therefore high resolution image tools are needed where complex and not homogeneous geological environments are expected in order to correctly plan the RSWC coring points; a full coverage image log is not mandatory. The complete logging suite will also allow basic lithological and petrophysical characterization.
- The image log run after coring has the main target to define the correct depth of sampling points and, moreover, the azimuth of the RSWC. Thus, a full coverage view of the borehole wall is needed, since it ensures the location of the holes left by cutting the RSWC. High resolution image logs are not necessary, provided that the quality is enough to clearly identify the coring points.

The two previous targets can be reached by combining image logging while drilling (LWD) and/or wireline (WL) in the safest and most cost effective ways:

- Image log acquisition before the RSWC performed in LWD mode, using either low resolution (e.g. density azimuthal imaging) or high resolution (e.g. resistivity) tools depending on the geological environment complexity; then a WL image acquisition after coring, even with low resolution tools (e.g. ultrasonic).
- WL image log acquisition before and after coring, choosing high resolution tools for the first run when the geology is complex and low resolution otherwise.
- Regardless of whether the image log acquisition before coring is planned in LWD or WL mode, it can be provided in LWD mode after coring if a wiper trip is needed.

d) Depending on the available tool, dedicated coring and breaking procedures to maximize the RSWC recovery and quality must be selected on the based on the formation characteristics and the well environment (mud type, overbalance, solids content etc.). Recommendations must be detailed and strictly followed for approaching each station, by reducing the speed and according to the well depth, to minimize both the cable stretch and creep; finally, a dedicated tripping out procedure must be pursued according to the formation petrophysical and lithological characteristics and well environment conditions.

e) All wellsite surface procedures are aimed at minimizing damage during handling and shipping to the laboratories. Particular care is needed to preserve the RSWC order of collection within the container and its formation end/wellbore end orientation. The documentation must include for future reference a close-up picture of each core in the container before its removal (Fig. 4) This is crucial when no core separators are used due to tool design and/or well deviation [5].

Care is recommended when wiping off the samples and packing them in the transportable jars with the wellbore end facing up. Additional wrapping of the sample is required with plastic or aluminium foil on which red and black parallel lines should be marked to preserve the correct sequence and top/bottom ends of the samples (red on the left, facing the core from wellbore end towards formation end) (Fig. 5). Shock adsorbing materials to stabilize the RSWC inside the containers are always suggested to preserve the sample integrity.

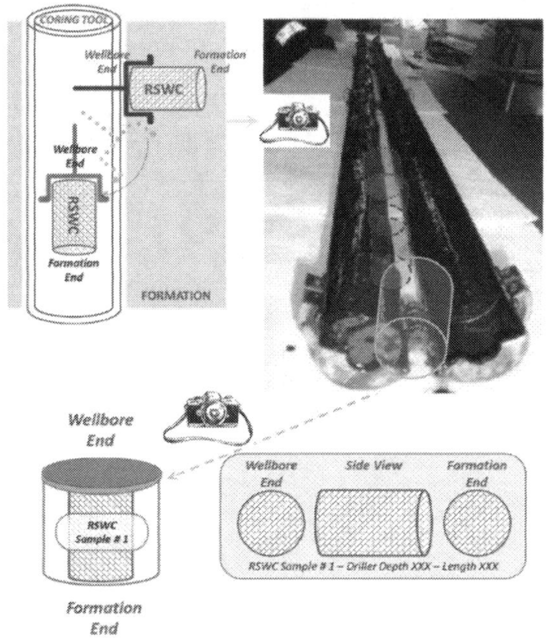


**Fig. 4.** Suggested pictures to be acquired on site.

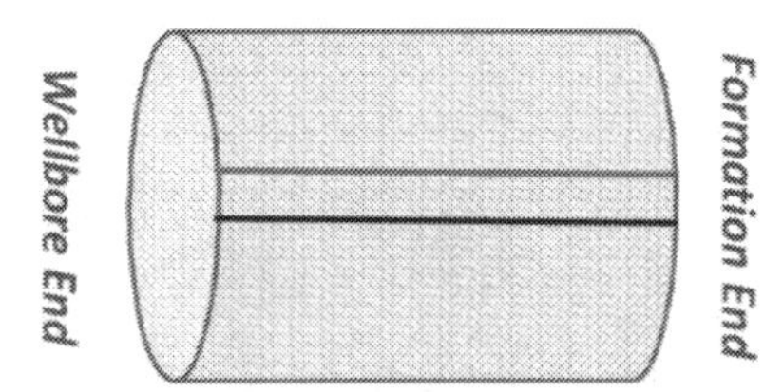


**Fig. 5.** Marking of RSWC packing to track formation/wellbore ends.

## 2.2 Laboratory processing of samples

The procedures to be observed when receiving and handling sidewall cores are analogous to those adopted for bottom hole cores: it is mandatory to treat them with great care in order to avoid mechanical shock; to protect them from extreme temperatures, humidity and dehydration and to limit as much as possible the use of fresh water to clean them. Labelling and marking require even greater attention. In fact, it is extremely important to preserve the proper sequence and core orientation ensuring that individual samples are not out of order or turned upside down.

More in detail, specific actions have to be performed in terms of: a) wellsite data retrieval; b) visual inspection of the core and RSWC labelling and marking; c) RSWC photos; d) X-Ray CT scan.

a) Generally speaking, all the coring-related well site data should be retrieved. In particular, the following cannot be neglected: coring parameters (weight on bit, rate of penetration, coring time, etc.); sample description, included its quality (intact, fractured, broken, crumbled, etc.); effective length and diameter of each recovered sample; trace of the wellbore/formation ends of the sample, according to the established conventions (see Fig. 5).

b) It is mandatory to preserve the correct sequence of the samples and their orientation, properly labelling and marking each RSWC. This implies the validation of the collected field data through a visual inspection of the core and, when possible, the data integration through the determination of the direction in which each RSWC has been broken by tilting, which corresponds to wellbore axis (Fig. 6).

The red and black parallel lines drawn at wellsite on the external surface of the wrapping must be transferred onto the RSWC sample. Moreover, to provide a further reference for the subsequent operations, it is recommended to draw the double line along the upper side with respect to the determined tilting direction, if previously identified.

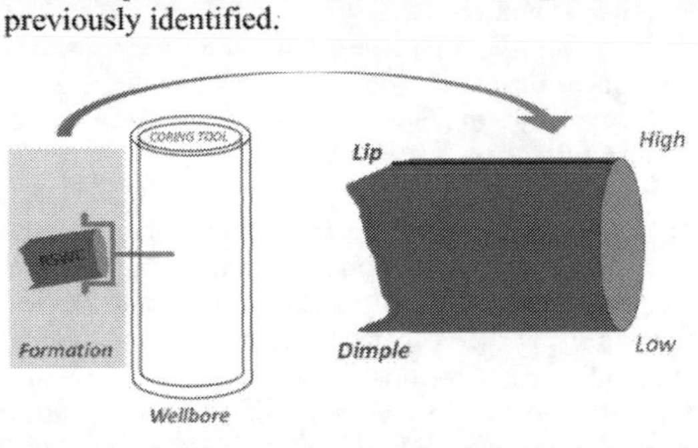


**Fig. 6.** Marking of RSWC to track the identified formation and wellbore ends

c) A complete set of RSWC photos (Fig. 7) should be acquired. Wellbore end, formation end and side views must be photographed, taking care to correctly mark the references. A comparison with the wellsite photos is also recommended to confirm the RSWC ends identification (see Fig. 4).

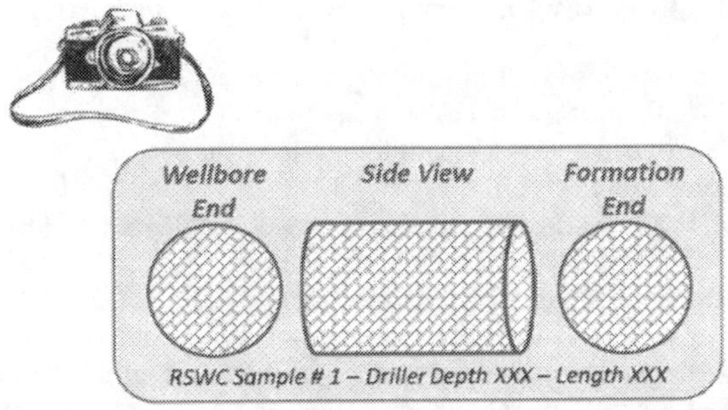


**Fig. 7.** Suggested views for RSWC photos at the laboratory.

d) A 3D X-Ray Computed Tomography of the sidewall core must be acquired at the maximum feasible resolution (the typical resolution of a medical CT scanner, 250 µm, is assumed satisfactory in most of the cases) A 2D 360° unrolled image must be extracted from the acquired volume, by following the maximum possible circumference, compatibly with the surface roughness. This picture is the main input for the goniometry procedure described in section 3. Moreover, it is mandatory that the position of the starting point of the unrolled images be recorded and recoverable on the sample. This starting point position (left border of the 2D image) should be measured as the angle from the line representing the side corresponding to the tilting direction, if its estimate has been possible. A corresponding reference notch or sign must be created on the RSWC.

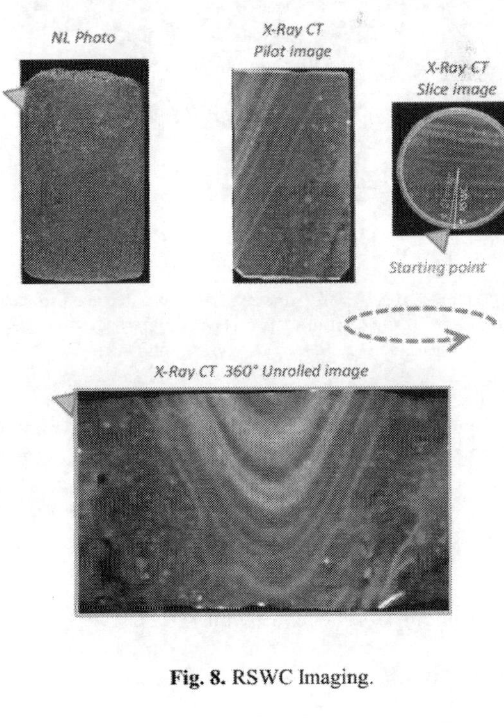


**Fig. 8.** RSWC Imaging.

## 3 The sidewall core goniometry procedure

The proposed "Eni SideWall Core Goniometry" (e-SWCG) procedure has been developed in order to provide an exhaustive methodology for the *a posteriori* reconstruction of the original *in situ* orientation of RSWCs. As described in Fig. 1, the determination of three main variables is required: wellbore/formation ends, RSWC azimuth and high/low sides.
The e-SWCG workflow is based on the identification of geological features (layering, fractures, stylolites, etc.) that are recognizable both at the laboratory scale on the X-Ray CT image of a RSWC and at the well scale on the image logs. Since these two scales are very different, to properly apply this procedure, the analysed large size RSWC must honor a diameter as close as possible to the nominal one (1.5") and a minimum length around 1.5". By consequence, lithologies without features detectable at both image scales of resolution or smaller diameter RSWC are not suitable for this approach.
The workflow starts with the 3D X-Ray CT acquisition of the analysed RSWC. Then, a 360° unrolled image is extracted from the volume; this unrolled "sampling" must be obtained at the maximum possible radius able to avoid the external surface roughness (Fig. 9).
The unrolled X-Ray CT image is converted into a digital borehole image format, uploaded onto an image analysis software platform and treated in the same way as an image log with its own deviation coordinates: inclination and azimuth. The top of the unrolled image is assumed to be coincident with the wellbore end of the RSWC.

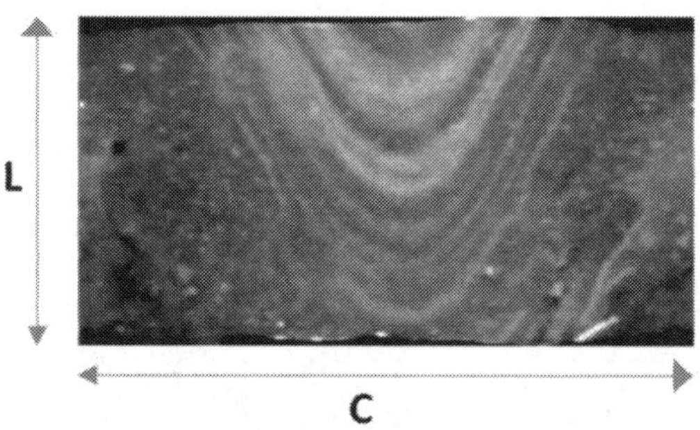


**Fig. 9.** Unrolled X-Ray CT image of a RSWC, where L is the length and C is the circumference. The actual image diameter is 1.34" (see field case of paragraph 4.2).

Since the commercial software platforms for log analysis usually manage the data loading according to the borehole depth and the sidewall core is oriented at 90° with respect to the well axial direction, the unrolled image has to be loaded assuming the following rules to preserve its correct aspect ratio:

- Top: Core wireline depth – 0.5L;
- Bottom: Core wireline depth + 0.5L;

where L is the length of the RSWC.

Bedding planes, fractures or other geological features must be identified and manually interpreted on both RSWC and log images.
The geological feature visible on the image log is oriented by picking the sinusoid and the corresponding tadpole will be generated (see 4th track on Fig. 11).
This tadpole, which represents a plane specifically oriented in space (true dip), has to be replicated on the RSWC image. The amplitude and offset (apparent dip) of the sinusoid drawn on the core image depend on the three unknown orientation variables.
To obtain the original orientation of the RSWC, the three variables must be tuned until the geological feature visible on the RSWC image matches exactly the sinusoid. The final orientation is best achieved step by step with a trial and error procedure. If the orientation variables are all unknown, it is suggested to start the procedure from the four base cases shown on Fig. 10 and then fine tuning the variables starting from the case that provides the best match.
The e-SWCG procedure allows the unique determination of the vertical direction along the RSWC, which is an essential input for the subsequent studies. On the other hand, an ambiguity of 180° persists on azimuth and high/low sides: in fact, two opposite choices of azimuth and wellbore/formation end of the RSWC can both honor the image log evidence. This ambiguity can be removed, as previously outlined, by providing additional information through one of the following actions:
- by observing the azimuth on an image log acquired after the coring, or
- by adopting coring tools allowing the user to define in advance the azimuth, or
- by identifying the high side from the RSWC topology, but only in vertical wells, or
- by recognizing the wellbore end by inspecting the physical RSWC, the RSWC pictures or the RSWC CT Scan images.

Once the match is achieved on the images, the final step consists of physically orienting the sample.The obtained inclination of the unrolled image coincides with the true inclination of the RSWC and the lateral shift (expressed as angle) of the unrolled image corresponds to the rotation to be applied to the RSWC initially positioned with the reference line up. This information should be provided to core analysts in order to reproduce the true orientation of the RSWC in laboratory before going ahead with analysis.

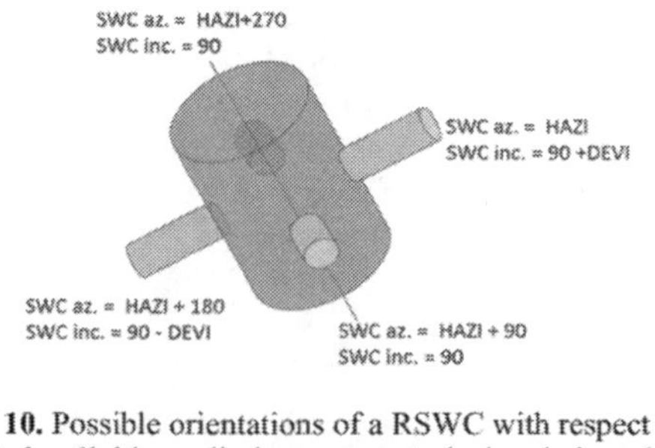


**Fig. 10.** Possible orientations of a RSWC with respect to a deviated well: blue cylinder represents the borehole and the 4 orange cylinders represent the RSWC in 4 base cases of the

https://doi.org/10.1051/e3sconf/202336601012

possible orientations: 2 cases on the plane of the borehole direction and another 2 orthogonal to the previous ones.

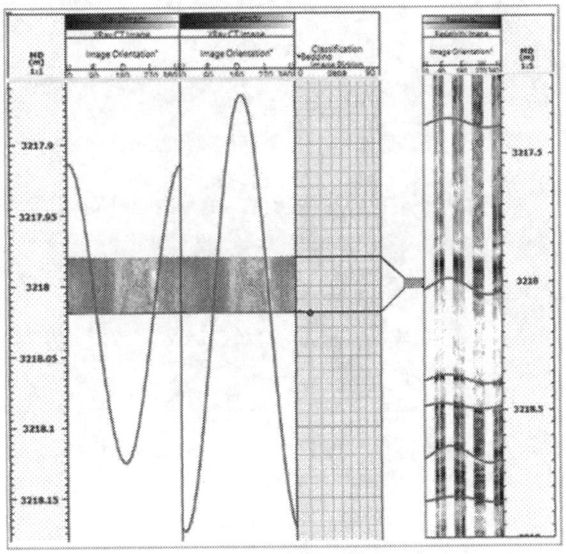

**Fig. 11.** Green sinusoids represent the bed boundary picked on the image log but show different attitudes because the orientation of the RSWC images are offset by 90° from each other.

## 4 Case histories

The two field applications of the e-SWCG procedure presented in this section are related to wells that are very different in terms of lithology, borehole environment and, more importantly, *a priori* knowledge of the RSWC orientation. In the first case, all the variables were unknown, while in the second the RSWC azimuth was predefined using a specific conveyance method (tool taxi). Both case histories successfully provided the correct parameters to orient the physical rock samples.

### 4.1 Carbonate reservoir

The first case is related to a sub-vertical deepwater well crossing Cretaceous to Miocene carbonate shelf and reef build-ups. WL resistivity images at very high vertical resolution were acquired in a 12.25" borehole with water based mud (Fig. 13). On the RSWC, a lamination pattern is visible at low angle with respect to the core axis (Fig. 14). On the image logs, many bed boundaries are well recognisable at the surrounding depth of the SWC, the sequence being well laminated and the borehole imaging at very high resolution (Fig. 13). The bed boundary closest to the RSWC depth has an attitude of 10° dipping toward W. The exercise of orientation of this RSWC is to match the lamination of the RSWC with the sinusoid from the image log interpretation.

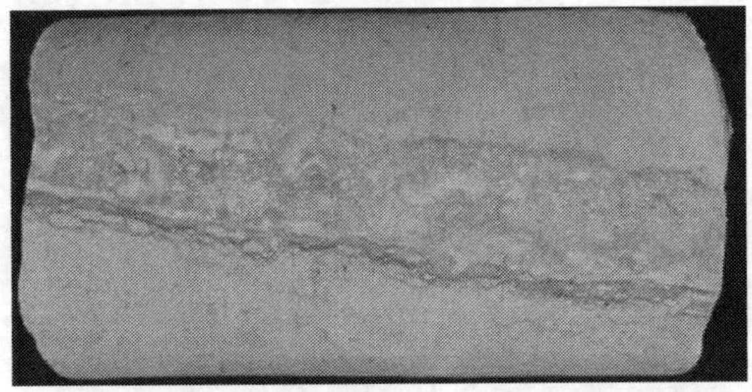

**Fig. 12.** Side view in white light photo of carbonate RSWC (wellbore end on the right side).

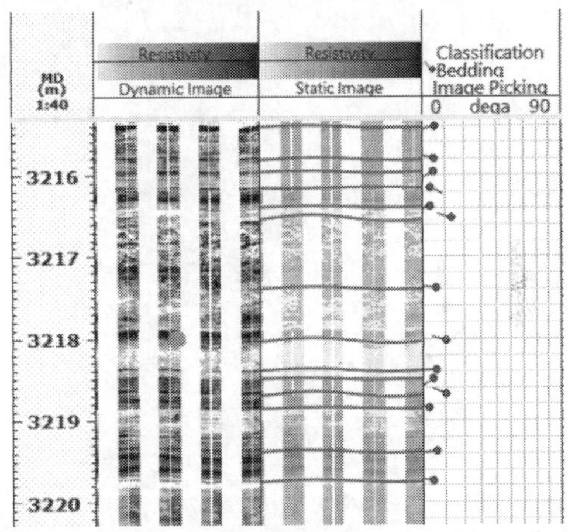


**Fig. 13.** Image logs and picking in the interval surrounding the carbonate RSWC.

For this RSWC, all three orientation variables, defined in paragraph 1, are unknown because the RSWC azimuth was neither measured nor imposed a priori. For this reason, the search for the match was started by trying the 4 RSWC azimuths corresponding to the 4 base cases of Fig.10 and selecting the case with the closest match.
Then, the one in which the best-fitting feature has an orientation as close as possible to the sinusoid was selected. To improve the match quality, fine tuning was also performed by applying a further rotation of 25° to the RSWC image and the final match was obtained (Fig. 14).

 https://doi.org/10.1051/e3sconf/202336601012

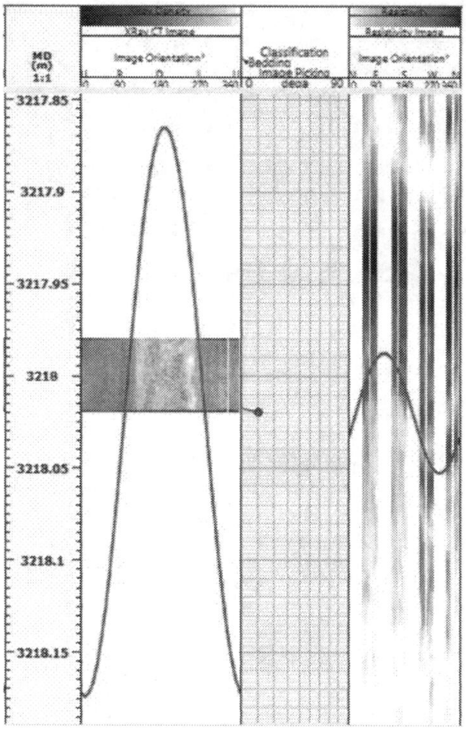


**Fig. 14.** Final match of carbonate sample.

## 4.2 Clastic reservoir with conveyance tool

The second case was an offshore 40° slanted well drilled through a Pliocene deltaic to turbiditic siliciclastic sequence where several RSWCs were cut inside the sandstones. The 12.25" phase in the reservoir section was drilled using oil based mud and the available image log was Density LWD with low vertical resolution. One of the RSWCs was taken in a portion of rock where laminations are well visible at the image log scale. On visual inspection, the core sample is apparently homogeneous without sedimentary laminae (Fig. 15), whereas the X-Ray CT shows that it is actually well laminated, with the layering displaying a high angle with respect to the RSWC axis (Fig. 9). Fig. 16 shows the LWD density image including the bedding interpretation. At the RSWC depth, the bedding inclination is between 10° to 20°. dipping toward NNW.

**Fig. 15.** Side view in white light photo of siliciclastic RSWC (wellbore end on the left side) .

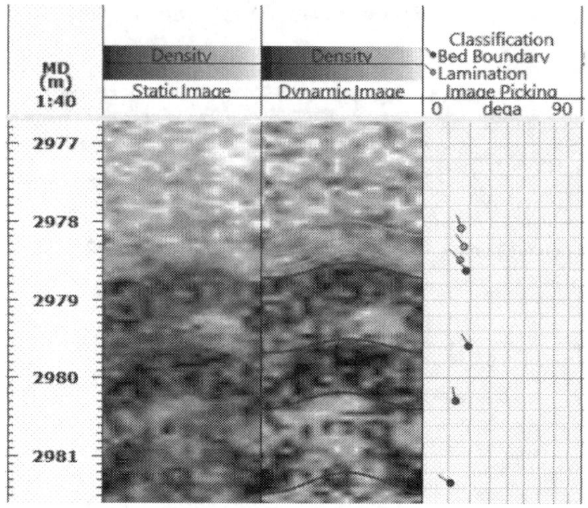


**Fig. 16.** Image logs and picking in the interval surrounding the siliciclastic RSWC.

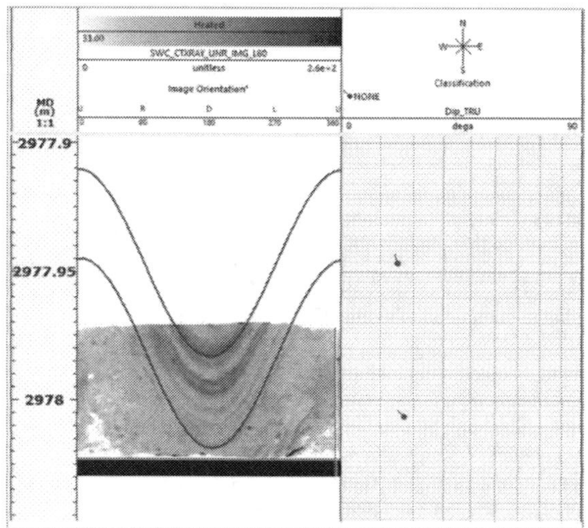


**Fig. 17.** Final match of clastic sample.

The RSWC azimuth and inclination are known because the coring tool was run with the tool taxi and the RSWC was taken in the lower side of the borehole. The wellbore end information was preserved during the wellsite recovery from the core barrel. The only unknown variables are the high and low sides of the RSWC. The match between sinusoids and laminations was achieved by choosing the right angular offset of the RSWC, which means a 10° lateral shift on the RSWC image. The position of the reference line can be used to orient the real sample (Fig. 17).

## 5 Conclusions

The original *in situ* orientation of RSWC is a fundamental input for obtaining reliable results from geological studies

https://doi.org/10.1051/e3sconf/202336601012

and for the proper execution of experimental analyses. The e-SWCG procedure presented in this paper can deliver a complete *a posteriori* goniometry of the RSWC, provided that an appropriate core and log acquisition program has been planned, as well as correct handling operations at wellsite in order to obtain and accurately track all the data needed for the orientation procedure. Moreover, the overall quality of the acquired RSWC is extremely important to maximize the data that can be recovered through experimental analyses and studies from an inherently more limited amount of material with respect to BHC. The presented case studies highlight the possibility of applying the orientation procedure in different geological contexts. Particularly, the case history on carbonates shows the reliability of the procedure even with limited data. On the other hand, the example related to a clastic reservoir emphasizes the advantages of using coring tools providing greater control on RSWC positioning.

The authors gratefully thank Eni S.p.A. for the authorization to publish this work, and colleagues Mauro Rossi for his valuable collaboration in X-Ray CT image acquisition/processing and Anna Maria Lyne for the manuscript review.

## References

1. D.K.T. Hansen, M.B. Enderlin, "Petrophysical significance of oriented sidewall cores for improved reservoir characterization", in Worthington, P.P., and Longeron, D., eds., *Advances in core evaluation II - reservoir appraisal*, p. 135-144 (Gordon and Breach Science Publishers, 1991)
2. B. A. Hardage, E. Doherty, S.E. Laubach, F. T. Hentz, "Secondary Natural Gas Recovery in the Appalachian Basin: Application of Advanced Technologies in a Field Demonstration Site, Henderson Dome, Western Pennsylvania", PhD Thesis Bureau of Economic Geology The University of Texas at Austin Austin, TX (1998).
3. S.E. Laubach, E. Doherty, "Oriented Drilled Sidewall Cores for Natural Fracture Evaluation", *SPE Annual Technical Conference and Exhibition*, (SPE56801, 1999).
4. J.F.W. Gale, L.A Gomez, "Late opening-mode fractures in karst-brecciated dolostones of the Lower Ordovician Ellenburger Group, west Texas: Recognition, characterization, and implications for fluid flow", AAPG Bullettin, **91**, 7, p. 1005-1023 (2007).
5. V. Torlov, C. Bonavides, A. Belowi, "Data Driven Assessment of Rotary Sidewall Coring Performance", *SPE Annual Technical Conference and Exhibition*, (SPE187107-MS, 2017).

# Evolution of electrical resistivity and NMR during early-stage maturation of an organic-rich chalk

*Eva G.* Vinegar[1,*], *Yoav O.* Rosenberg[2], *Scott V.* Nguyen[3] , *Larry W.* Lake[1], and *Harold J.* Vinegar[4]

[1]The University of Texas at Austin
[2]Geological Survey of Israel
[3]17TeraWatts
[4]Vinegar Technologies, LLC

**Abstract.** This paper studies the changes in the electrical resistivity of an organic-rich source rock chalk containing a Type-IIs kerogen during early-stage maturation of the kerogen. As bitumen is generated during maturation, the chalk changes from water wet to a complex mixed-wet pore system, with the micritic pore space remaining water wet while the bitumen expelled into the macro pores converts some pores to oil wet. This mixed-wet system results in non-Archie electrical behavior with pronounced curvature of the I-$S_w$ curve. We find that this curvature is well fit by connectivity theory. The connectivity theory parameters are determined both on native state core and from artificial laboratory pyrolysis which generates bitumen. NMR is used to distinguish the brine in micritic, water-wet pore spaces from that in the intergranular, mixed-wet pore spaces, from which the observed connectivity parameters may be independently estimated. Comparison between electrical resistivity and dielectric logs show that the connectivity parameters match the laboratory results over a calcareous section more than 300 m thick.

## Introduction

The Late Cretaceous Ghareb-Mishash formation in Israel provides a unique opportunity to study the evolution of electrical resistivity and NMR responses to organic maturation in a source rock chalk. This chalk has an extremely high TOC up to 15 wt%. In the coastal plain containing the Shefela basin, the organic-rich chalk has experienced shallow burial and the kerogen is immature. In contrast, in the Golan basin in northern Israel, the same Ghareb-Mishash formation has been buried much deeper and experienced a higher geothermal gradient, leading to a higher level of kerogen maturation to bitumen. Using core and log data from both basins, we have been able to follow the change in the electrical resistivity and NMR responses from immature to early-stage organic maturation.

The early-stage mature, kerogen-rich Late Cretaceous formation in the Golan basin is characterized by large amounts of bitumen and light hydrocarbons which affect the connectivity of the fluids within the rock and thus the electrical resistivity and NMR responses. Increasing burial of these organic-rich chalks affects connectivity by compaction (reduction in porosity) and by organic maturation of the kerogen (hydrocarbon generation and wettability changes).

Electrical resistivity, dielectric and NMR data are used to aid the interpretation of petrophysical properties of reservoir rocks, such as porosity, pore size, fluid type, fluid saturation, tortuosity, and permeability [1, 2].

Improvements to the models and interpretations used to evaluate the petrophysical properties have been ongoing with the development of unconventional resource plays. For example, extremely high values of electrical resistivity are common in mixed-wet micritic calcites and are difficult to model using Archie's resistivity model which is applicable to conventional reservoir rocks. Cementation and saturation exponents are about 2 for conventional rocks, but often vary significantly for vuggy rocks, shaly sands, and mixed or oil-wet systems. Furthermore, the Archie exponents selected may not fit well for a range of water saturations. Several resistivity models have been developed to better fit the electrical resistivity in different types of rock systems, but few fit the curvature of the resistivity index with water saturation [3-6]. A more general form of Archie's resistivity model based on connectivity theory provides a more consistent fit with relatively few parameters [5,6].

In agreement with connectivity theory, the electrical resistivity in the Ghareb-Mishash formations shows a sharp increase in I-Sw curvature for early-stage thermal maturation. The change from Archie-type behavior in the Shefela basin to I-Sw curvature in the Golan basin is due to the generation and expulsion of bitumen from the kerogen, which blocks many of the pore throats in the dual-porosity system and creates mixed wet or oil-wet surfaces in the larger, intergranular pore spaces.

* Corresponding author: evavinegar@gmail.com

© The Authors, published by EDP Sciences. This is an open access article distributed under the terms of the Creative Commons Attribution License 4.0 (http://creativecommons.org/licenses/by/4.0/).

We first present a brief review of connectivity theory. We then show how the resistivity index changes from Archie-type behavior to non-Archie behavior once the organic-rich chalk is artificially matured. We show how bitumen content increases with depth and maturation in the Ghareb basin. As bitumen is generated during maturation, the chalk changes from water wet to a complex mixed-wet pore system, with the micritic pore space remaining water wet while the bitumen expelled into the macro pores converts some pores to oil wet. This mixed-wet system results in non-Archie electrical behavior with pronounced curvature of the I-$S_w$. We show that this curvature is well fit by connectivity theory. The connectivity theory parameters are then determined on native state core and found to match those found during artificial laboratory pyrolysis. NMR is used to distinguish the brine in micritic, water-wet pore spaces from that in the intergranular, mixed-wet pore spaces, from which the observed connectivity parameters may be independently estimated. Finally, we compare electrical resistivity and dielectric logs and show that the connectivity parameters match the laboratory results over a 350 m calcareous section.

## Electrical resistivity and connectivity theory

The petroleum industry has long used Archie's equation [7] to estimate the hydrocarbon saturation of reservoirs using resistivity and porosity logs. Archie's equation is:

$$R_t = \frac{R_w}{\phi^m S_w^n} \tag{1}$$

where $R_t$ is the total resistivity of the rock, $R_w$ is the resistivity of the brine, $\phi$ is the porosity, $S_w$ is the water saturation, m is the cementation exponent, and n is the saturation exponent.

The sensitivity of oil reserve estimates to the value of these exponents has been critical for reservoir evaluation and project decisions. For many reservoirs, m is about 2, but can be much higher for vuggy rocks. n is usually about 2 in clean, water-wet sandstones, less than 2 in shaly sands, and much more than 2 in mixed-wet and oil-wet systems. Heterogeneous systems further make it difficult to fit these exponents to a dataset.

Deviation from the Archie straight line values on a resistivity index versus water saturation log-log plot is a common issue in assessing many reservoir rocks. The Archie saturation exponent needed to match data on this plot changes significantly with water saturation [6, 8]. Several methods were developed to represent the saturation in non-Archie rocks. Montaron [5,6] proposed a more generalized version of Archie's equation based on percolation theory:

$$R_t = R_w \left[\frac{1-P_c}{P-P_c}\right]^{\mu} \tag{2}$$

where $R_t$ is the electrical resistivity of the system, $R_w$ is the electrical resistivity of the conductive portion, P is the probability that a portion of the rock will be conductive (water volume fraction), Pc is the percolation threshold (Montaron calls $P_c$ a water connectivity correction index, $\chi_w$), and μ is the conductivity exponent (~2). The water volume fraction is the water saturation times the porosity of the rock, $S_w\varphi$, giving:

$$R_t = R_w \left[\frac{1-\chi_w}{S_w\phi-\chi_w}\right]^{\mu} \tag{3}$$

This connectivity equation elegantly accounts for the curvature of the resistivity index with saturation in some systems. We note that when the percolation threshold is 0, the connectivity equation reduces to Archie's equation if m=n. Montaron further derived this connectivity equation for a variety of cases using a modified CRIM mixing law [9,10]. We find this particularly applicable to the organic-rich Ghareb-Mishash chalk.

As described in Vinegar et al. [11], **Fig. 1** shows a core sample from the Shefela containing immature organic matter that was artificially matured and its resistivity index versus brine saturation. $T_{max}$ is 421°C for the post-pyrolysis sample and the porosity is 50 pu. The blue circles represent measurements taken post-pyrolysis (dry pyrolysis), and the red circles represent measurements taken pre-pyrolysis. An Archie equation fits the pre-pyrolysis data with a saturation exponent n=1.93. The model based on connectivity theory fits the post-pyrolysis data with $\chi_w = 0.03$ and $\mu = 2$. The results show curvature beginning to develop at lower values of water saturation.

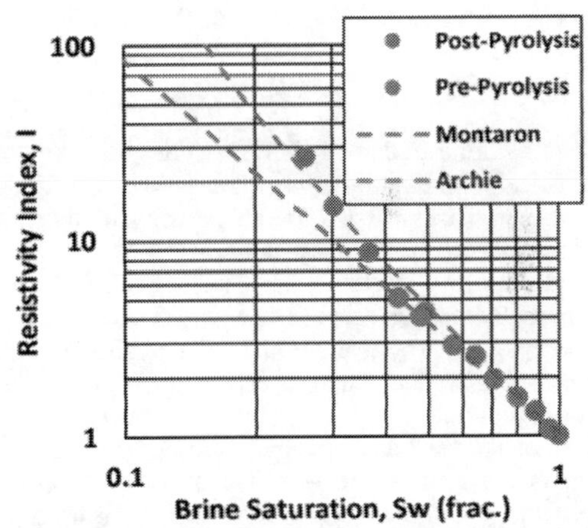


**Fig. 1**: Resistivity index versus brine saturation for an immature core sample from the Shefela basin pre- and post-pyrolysis [11].

We can show the deviation of the electrical resistivity from Archie's equation using connectivity theory [6]. The water connectivity correction index primarily controls the curvature of the resistivity index versus water saturation plot. There are two main contributors to the water connectivity correction index: one due to the micritic pore space and one due to the oil-wet pore space.

The water-filled pores of the micritic space are highly connected, and the conductivity exponent of micritic grains is often taken to be 1.5 as a result (as Montaron points out, 1.5 is the conductivity exponent for a packing of spherical grains). If the micritic space contributes to the electrical resistivity, then the electrical resistivity would be slightly less resistive than if one were to consider only larger, intergranular pore spaces. A non-organic, water-wet carbonate containing a dual micro and meso-macropore system would consist of three phases: micritic grains, water in the intergranular pore space, and solid grains. The micritic grains are treated as an independent unit with its micropores Montaron [5,6] derived the equation for the water connectivity correction index for a rock containing micritic grains:

$$\chi_w = x_m\left(\phi_m - \phi_m^{\mu_m/\mu}\right) \quad (4)$$

where $\chi_w$ is the water connectivity correction index, $x_m$ is the volume fraction of the rock containing micritic grains, $\phi_m$ is the porosity of the micritic portion of the rock, $\mu_m$ is the conductivity exponent of the micritic portion of the rock, and $\mu$ is the conductivity exponent of the intergranular portion of the rock.

Assuming that $x_m$ is around 0.25, $\mu_m$ is 1.5, and $\mu$ is 2, $\chi_w$ can be strongly negative, as shown in **Fig. 2a**. Negative values of $\chi_w$ suggest that compared to Archie, the resistivity index would be much smaller, especially with decreasing water saturation. If the system is fully water-saturated, however, in many cases the resistivity index will be close to that predicted by Archie.

If we neglect the micritic space and consider a system where the larger pore spaces are mixed-wet, with some water-wet pores and some oil-wet pores, then the water connectivity correction index as derived by Montaron [6] is:

$$\chi_w = x_o S_{co} \phi_M \quad (5)$$

where $x_o$ is the fraction of pores that are oil-wet, $S_{co}$ is the critical oil saturation at which the conductive fluids no longer percolate, and $\phi_M$ is the porosity of the macro pores. $\chi_w$ is positive and grows linearly.

In the early-stage mature case, where the bitumen has partially blocked the pore throats of the micritic space, $\chi_w = x_o S_{co} \phi_M$ is positive and the resistivity index compared to Archie is higher, especially with decreasing water saturation.

For the case in which there are oil-wet spaces and there is connectivity between the brine in the micritic space and that in the larger pore spaces, $\chi_w$ is the sum of the two terms:

$$\chi_w = x_m\left(\phi_m - \phi_m^{\frac{\mu_m}{\mu}}\right) + x_o S_{co} \phi_M \quad (6)$$

where the first term is negative, and the second term is positive.

**Fig. 2a** shows the behavior of $\chi_w$ with $x_m$ and $\phi_m$ due to the micritic grains. Over a large range of porosities inside the micritic portion of the rock, $\chi_w$ does not vary significantly, particularly if the bulk volume fraction of micritic grains is low. **Fig. 2b** shows the behavior of $\chi_w$ due to the oil-wet pore spaces. With burial and compaction, the porosity of the oil-wet pore spaces decreases, but the oil-wet fraction of the intergranular pore spaces increases. These factors keep $\chi_w$ more stable along the length of the organic-rich portion of the well.

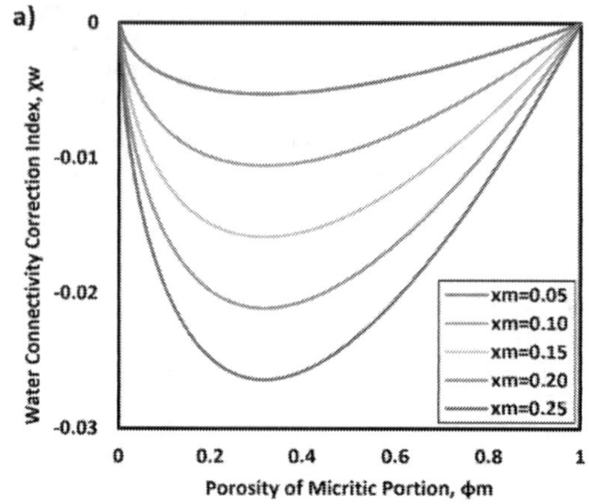


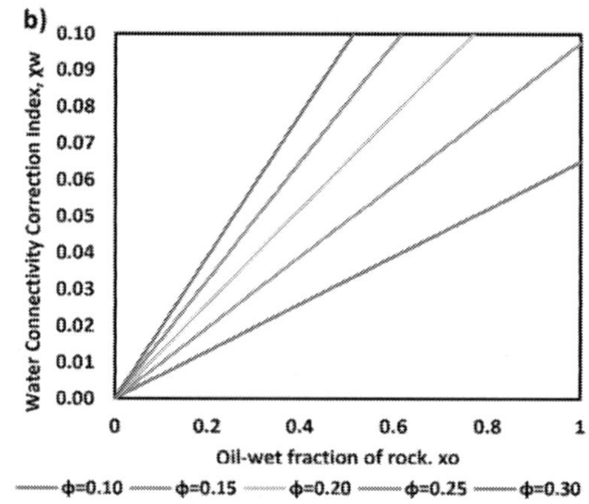


**Fig. 2**: **a)** $\chi_w$ for a micritic portion of the rock in which $\mu_m$ is 1.5 and $\mu$ is 2 in equation (4). **b)** $\chi_w$ for a rock with oil-wet pore spaces in which $S_{co}$ is 0.65 in equation (5). $\phi$ is the porosity of the macro pore space.

## Creating Oil-Wet Spaces with Bitumen Generation

The immature organic-rich chalk from the Shefela basin is water wet. In contrast, the early-stage mature kerogen from the Golan basin has generated bitumen and light hydrocarbons, and thus created oil-wet surfaces. These oil-wet surfaces induce the curvature in the resistivity index. Furthermore, the distribution of the organic matter in the rock cause much of the micritic water-wet pore spaces to be blocked by bitumen.

The SEM images in **Fig. 3** from Vinegar et al. [11] show the distribution of the pore spaces and organic matter. The upper right image is the base SEM image. The circular coccoliths and larger calcite grains are shown in the upper right image in green (calcium mapping using SEM-EDS). The lower left image shows the kerogenous organic matter dispersed between the larger calcite grains. Sulfur mapping was used to identify the organic matter because Type II-s kerogen (S/C ratio above 0.04) is dominant in this rock [12]. The lower right image shows the pore spaces, organic matter, and calcite. The co-deposition of organic matter and calcite minerals allow for bitumen generation to enter the macropore system and block many of the entrances to the micritic pore space. This observation agrees with the change in NMR diffusional coupling from strong to weak as the kerogen generates bitumen [2,11].

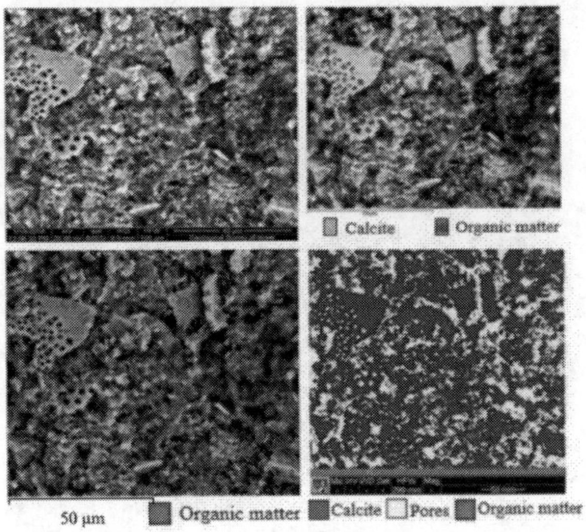


**Fig. 3:** SEM (upper left) and SEM-EDS images showing calcite, organic matter, and pores for the organic-rich chalk [11].

An immature, organic-rich Ghareb rock was artificially matured in a semi-open pyrolysis reactor gradient experiment [13]. The sample was evaluated in a Vinci Technologies Rock-Eval 6 analyzer to determine $T_{max}$ and hydrocarbon content with TOC. Bitumen was obtained from cuttings using a Soxhlet extractor (Vinci Technologies) with dichloromethane and methanol solvent (9:1 V/V). **Fig. 4** shows the bitumen and light hydrocarbon content versus $T_{max}$ and the corresponding windows of hydrocarbon generation. Bitumen generation begins at a $T_{max}$ value of about 417°C.

**Fig. 5** shows the bitumen content (wt%) with depth for the NESS 2 well in the Golan basin. There is significant bitumen content over the 350 meters. This is quite different from the immature organic-rich rock from the Shefela basin, where there is very low bitumen content. The porosity decreases from the upper to lower portion of the well.

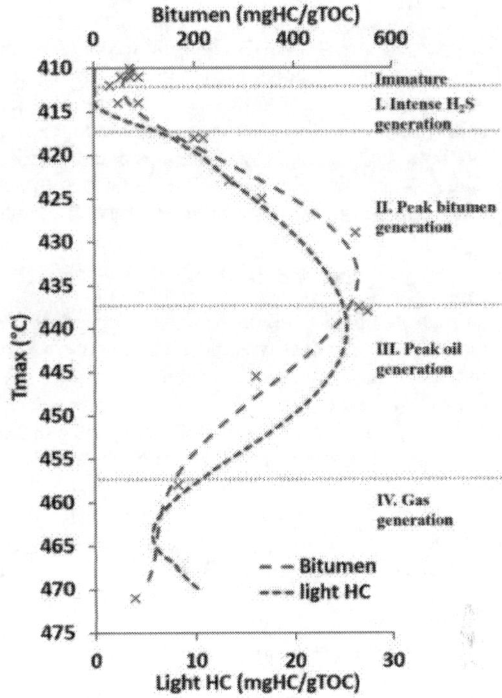


**Fig 4:** Bitumen and light hydrocarbon generation versus $T_{max}$ of an artificially matured organic-rich Ghareb sample from the Shefela basin. The immature stage, hydrogen sulfide, bitumen, oil, and gas windows are marked by the dotted lines [11,13].

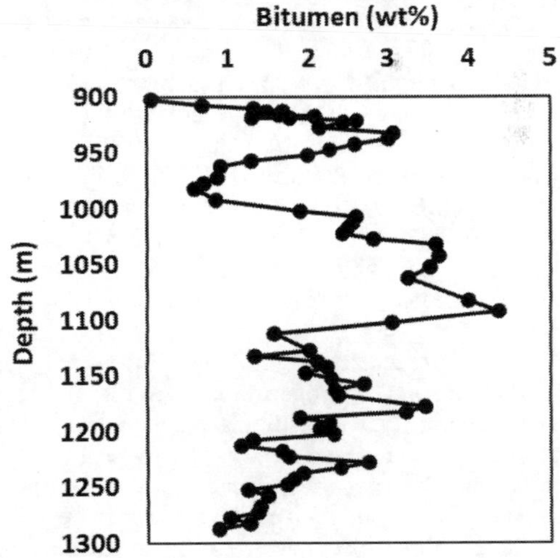


**Fig. 5:** Bitumen (wt%) with depth for the NESS 2 well as determined by Soxhlet extraction.

### $\chi_w$ using NMR core data

We consider the components used for the connectivity theory-based resistivity model in the immature and early-stage mature cases. In the immature stage, the organic-rich chalk contains mostly calcite, immature kerogen, and water in both the macropores and micropores. During early-stage maturation, the kerogen generates bitumen and lighter hydrocarbons. Once bitumen is expelled into the macropores, some of the macropore space becomes oil wet.. We assume that the micropores remain water-wet and contain only water as demonstrated by $D_2O$ diffusion eliminating the entire $H_2O$ signal from the micropores The macropore system is mixed-wet, and for the purposes of our model, we assume that the water is in water-wet macropores, while the light hydrocarbons are in oil-wet pores. Our electrical resistivity model therefore has three classes of pores: water-wet micritic pores, water-wet macropores, and oil-wet macropores.

One method of estimating the values of $\chi_w$ is by using NMR $T_2$ distributions for "as received" core and fully saturated with brine. Downhole, the light hydrocarbons (NGLs) are in the oil-wet pores but the NGLs blow out of the core on the trip to the surface. This provides a method for separating the volume of water-wet and oil-wet macropores.

The laboratory NMR analysis was performed at Rice University as detailed in Chen et al. [2]. An Oxford Instruments GeoSpec2 rock-core analyzer was used with a Larmor frequency of 2.3 MHz to match that of the MREX logging tool.

**Fig. 6** shows the $T_2$ distribution of an early-stage mature Ghareb sample from 920m in the NESS 2 well [2,11]. The region with the shortest T2 values, depicted in the A region of the spectrum (3 ms and shorter), responds to part of the bitumen. Kerogen has relaxation times which are too short to be detected by this NMR spectrometer. Bitumen also has values which are short, but a fraction may be detected if the viscosity is low enough. Region B, which includes $T_2$ values between 3 and 32 ms, is the response due to the water-wet micritic calcite. The signal with the longer relaxation time in region C ($T_2$ greater than 32 ms) responds to the fluids in the macro pore space. The separation of the T2 distribution for water into regions B and C, corresponding to the micritic and macropores, is a result of weak diffusive coupling due to bitumen blocking [11]. The water molecules no longer diffuse easily between the two spaces because bitumen is blocking many of the pore throats. Downhole, region C contains some connate water in the macropore space as well as methane and condensates.

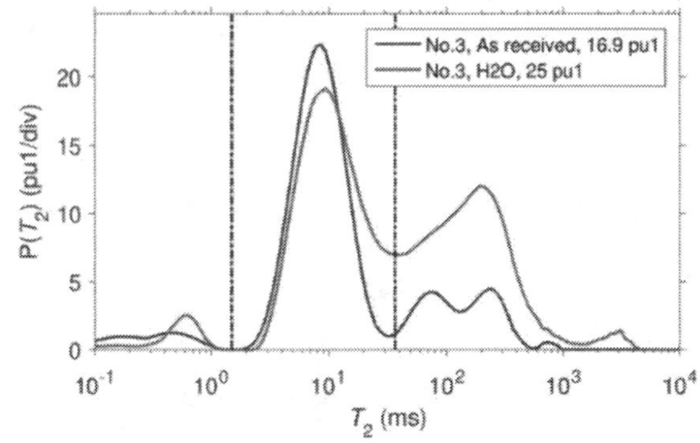


**Fig. 6:** NMR $T_2$ distribution of a core sample from the NESS 2 well at 920m, "as received" and after saturating with brine [2,11].

The light hydrocarbons in this core is taken to be the difference between the porosity in the "as received" core and the porosity when fully brine saturated. The micritic porosity is about 12 pu and the macro porosity of the as received core is about 4.9 pu. When saturated with brine, the macro porosity is 13 pu. The oil-wet macro porosity is taken to be the difference between the brine-saturated total porosity and the "as received" total porosity, or 8.1 pu. The fraction of the oil-wet macropore space is 8.1pu/13pu = 0.62. The bulk volume micritic portion of the rock is 25%, which implies that the porosity of the micritic portion is 48%. $S_{co}$ is about 0.65 [6]. Using $\mu_m = 1.5$ and $\mu = 2$, the value of $\chi_w$ is about 0.028. This value agrees with the 0.03 fit to the resistivity index dataset in **Fig. 1**.

### $\chi_w$ measured from core

The electrical resistivity profile of the organic-rich chalk at these maturity levels shows a significant increase compared to the immature case-more than would be expected for an Archie-type rock because of curvature. **Table 1** shows the Dean-Stark core data for a set of samples from the NESS 5 well in the Golan basin. **Fig. 7a** shows the fit of the Dean-Stark core data to the Archie resistivity model. The fit for the Archie model is best when m=4 and n=3.5; however, the brine resistivity is very low: 0.03 ohm-m. **Fig. 7b** shows the fit of the Dean-Stark core data to the connectivity theory-based resistivity model. The fit when $\chi_w$=0.03 and $\mu$=2 gives a brine resistivity of 1.2 ohm-m, which is the measured brine resistivity in the Upper Ghareb. **Fig. 8** shows the excellent fit between the electrical resistivity using the connectivity theory-based model ($\chi_w = 0.03, \mu = 2,\ R_w = 1.2$ ohm-m) with the Dean Stark core data shown in **Table 1** versus the NESS 5 resistivity well log in the Golan basin.

https://doi.org/10.1051/e3sconf/202336601013

**Table 1:** Dean-Stark core data and resistivity log values for the NESS 5 well from 1050-1056m

| Depth (m) | Phi | $S_w$ | Rt (Ω-m) |
|---|---|---|---|
| 1050.7 | 0.13 | 0.16 | 8370 |
| 1050.9 | 0.12 | 0.31 | 13544 |
| 1051.2 | 0.14 | 0.35 | 7510 |
| 1051.4 | 0.15 | 0.31 | 5545 |
| 1051.8 | 0.13 | 0.38 | 2935 |
| 1052.1 | 0.14 | 0.38 | 2023 |
| 1052.4 | 0.17 | 0.35 | 2023 |
| 1052.7 | 0.16 | 0.47 | 787 |
| 1052.8 | 0.17 | 0.53 | 473 |
| 1054.4 | 0.18 | 0.69 | 79 |
| 1054.8 | 0.20 | 0.73 | 103 |
| 1055.2 | 0.18 | 0.75 | 140 |
| 1055.4 | 0.18 | 0.70 | 167 |
| 1055.7 | 0.16 | 0.69 | 137 |
| 1055.9 | 0.15 | 0.66 | 129 |

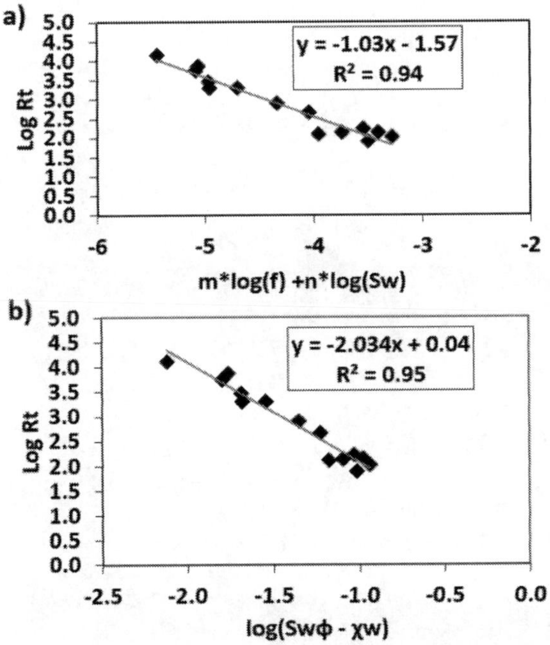


**Fig. 7: a)** Regression to fit Archie parameters for the NESS 5 well from 1050-1056m. m=4 and n=3.5 fit this dataset, but $R_w$ is only 0.03 ohm-m. **b)** Regression to fit connectivity theory parameters for the NESS 5 well from 1050-1056m. $\chi_w = 0.03$ and $\mu = 2$ gives $R_w$=1.2 ohm-m.

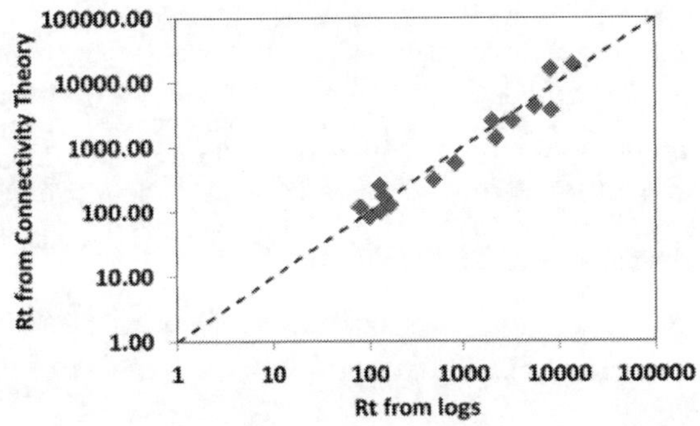


**Fig. 8:** Resistivity from NESS 5 well log from 1050-1056m versus resistivity calculated using the connectivity theory model, with $\chi_w = 0.03$ and $\mu = 2$.

## Comparison between dielectric and resistivity log using connectivity theory

The $S_w\phi$ term in the connectivity theory-based resistivity model (Eq. 3) is directly measured with the dielectric log. One advantage of using the dielectric log in this model is that it is not necessary to use two separate logs to determine $\phi$ and $S_w$ independently.

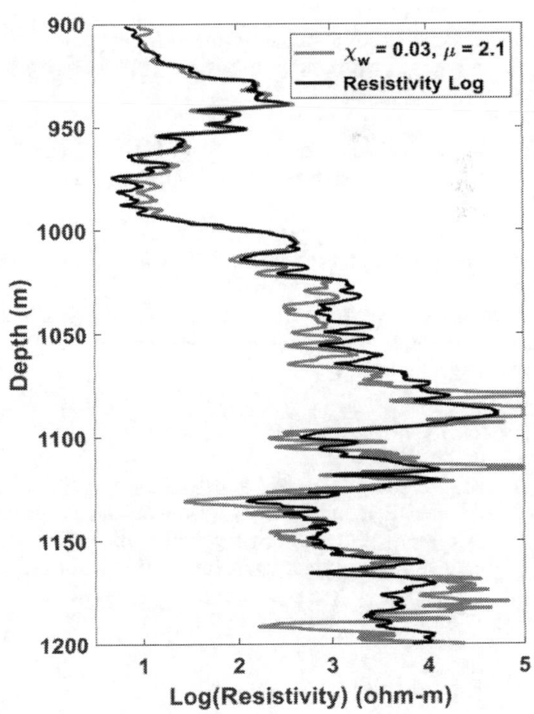


**Fig. 9:** Resistivity log from the NESS 2 well overlaid with the connectivity theory resistivity calculated from Eq. 3 using the dielectric log for $S_w\phi$, $\chi_w = 0.03$, and $\mu = 2.1$.

The NESS 2 resistivity log was overlaid with the connectivity model (Eq. 3) using the dielectric log for $\phi S_w$, $\chi_w = 0.03$, and $\mu = 2.1$ in **Fig. 9**. The brine resistivity $R_w$ is 0.3 ohm-m which matches the $R_w$ using Archie's equation in the organic-free carbonate at the top of the Ghareb. We note there is excellent agreement for over 300 meters of organic-rich rock with resistivity values ranging from less than 10 to over 10,000 ohm-m and porosities from 10 to 35 pu.

## Conclusions

1. The amount of bitumen in macropores increases with organic maturity, creating a mixed-wet pore system with water-wet micritic pores and some oil-wet and some water-wet macropores.

2. The change in wettability causes a curvature in the I-$S_w$ relation, resulting in a non-Archie behavior. NMR measurements on core show the micritic pores remain water-wet even at high bitumen content.

3. Comparing immature and post-pyrolysis resistivity, the change to non-Archie behavior occurs at $T_{max}$ ~ 420°C, where the bitumen content increases rapidly. A connectivity model using $\mu = 2$ and $\chi_w = 0.03$ matches the post-pyrolysis data.

4. Dean Stark core data in NESS 5 shows the connectivity theory-based resistivity model provides a good fit to the I-$S_w$ curvature with the same value of connectivity parameter ($\chi_w = 0.03$).

5. The value of $\chi_w$ can be explained using NMR $T_2$ relaxation data on "as received" and fully brine-saturated core.

6. Comparing the resistivity and dielectric logs in NESS 2, and using the value of $\chi_w = 0.03$ previously measured, we obtain an excellent match over a large calcareous section of ~300 m, covering an extremely wide range in resistivity and porosity.

## Acknowledgements

We thank Weidong Li and Mohammad Chohan from Baker Hughes, Professor George Hirasaki, Professor Philip Singer, Zeliang Chen, and Xinglin Wang from Rice University, and Itay Reznick from the Geological Survey of Israel for their contributions to this paper.

## Symbols and Abbreviations

NMR = Nuclear Magnetic Resonance
$R_t$ = electrical resistivity of the saturated rock
$R_w$ = electrical resistivity of the brine
$\phi$ = porosity
$S_w$ = brine saturation
$m$ = cementation exponent
$n$ = saturation exponent
$P$ = probability that a site will be conductive
$P_c$ = pseudo-percolation threshold
$\chi_w$ = water connectivity correction index
$\mu$ = conductivity exponent
$\mu_m$ = conductivity exponent of the micritic portion of the rock
$x_m$ = bulk volume fraction of rock that contains micritic grains
$\phi_m$ = porosity of the micritic portion of the rock
$x_o$ = fraction of macro pores that are oil-wet
$S_{co}$ = critical oil saturation at which conductive fluids no longer percolate
$\phi_M$ = porosity of the macro pores
$T_{max}$ = temperature of peak S2 response during programmed Rock-Eval pyrolysis
$TOC$ = total organic carbon

## References

1. R.L. Kleinberg, W.E. Kenyon, P.P. Mitra, Journal of Magnetic Resonance, Series A **108**, 206-214 (1994)
2. Z. Chen, P.M. Singer, X. Wang, H.J. Vinegar, S.V. Nguyen, G.J. Hirasaki, Petrophysics **60**, 771-797 (2019)
3. D. Kennedy, Petrophysics **48**, 335-361 (2007)
4. A.P. Garcia, Z. Heidari, A. Rostami, Petrophysics **58**, 454-469 (2017)
5. B. Montaron, *15th SPE Middle East Oil & Gas Show and Conference* (2007)
6. B. Montaron, Petrophysics **50**, 102-115 (2009)
7. G.E. Archie, *SPE 942054-G, Transactions, AIME* **146**, 54-62 (1942).
8. W.G. Anderson, *SPE 13932* (1986)
9. J. Birchak, L. Gardner, J. Hipp, J. Victor, *IEEE Proceedings* **62**, 93-98 (1974)
10. J.G. Berryman, *Mixture Theories for Rock Properties, Rock Physics and Phase Relations- A Handbook of Physical Constants*, 205-228 (1995)
11. E.G. Vinegar, Y.O. Rosenberg, I. Reznick, Y. Gordin, P.M. Singer, X. Wang, Z. Chen, S.V. Nguyen, W. Li, T. Bradley, G.J. Hirasaki, L.W. Lake, S. Feinstein, Y. H. Hatzor, H.J. Vinegar, *SPWLA 61st Annual Logging Symposium* (2020)
12. I. Kutuzov, M.Sc. Thesis. Ben-Gurion of the Negev, Be'er Sheva, Israel (2017)
13. Y. O. Rosenberg, I. J. Reznick, H. J. Vinegar, S. Feinstein, Y. Bartov, Marine and Petroleum Geology (2020)

# Effect of specimen size and shape on the mechanical properties measured on plugs subsampled from sidewall cores

*Giorgio* Volonté[1,*], *Jørn* Stenebraten[2], and *Euripides* Papamichos[2]

[1]Eni S.p.A., 20097 San Donato Milanese, Italy
[2]SINTEF Industry, NO-7465 Trondheim, Norway

**Abstract.** For several disciplines, the increasing acquisition of rotary sidewall cores (RSWC) in the oil and gas industry gives the opportunity for more extensive experimental rock characterization but, on the other hand, the limited amount of recovered material, their shape and orientation are not optimal for geomechanical experiments in the triaxial cell. These drawbacks can be overcome by performing experiments on properly shaped and oriented small diameter plugs subsampled from the RSWCs. These plugs are smaller than conventional ones, so the effects of both specimen shape (i.e., length over diameter L/D ratio) and size (i.e., diameter) on strength, stiffness (i.e., Young's modulus), and dilation (i.e., Poisson's ratio) must be carefully considered. In this work, an experimental study on specimen shape and size effects on rock mechanical properties has been tailored to materials and testing conditions generally encountered in petroleum related geomechanics. Triaxial compression experiments at three confining pressures have been performed on outcrop analogues of typical reservoir rock lithologies, i.e., two sandstones and two limestones, and within a range of shapes and sizes relevant for specimens prepared from RSWCs with emphasis on 10 mm diameter specimens. The strength, stiffness and dilation results were analysed and compared with relevant data and relations available in the literature. Then, analytical equations, useful for the practice of geomechanical experimental studies on RSWCs, have been derived to correlate mechanical properties measured on specimens of any shape or size to the ones obtainable on reference specimens.

## 1 Introduction

Several wellbore and field scale problems in the Oil & Gas industry require dedicated geomechanical experimental characterization that is usually performed on material acquired from 3" to 4" diameter bottom hole cores (BHCs). However, in the recent years, industry is increasingly promoting the partial replacement of bottom hole cores with 1"1/2 diameter rotary sidewall cores (RSWC) since the cost of their acquisition is lower. It should be noticed that RSWCs provide an opportunity to several disciplines to acquire data already from exploratory wells or from non-reservoir facies with a limited additional expense. But, on the other hand, geomechanical testing, e.g. triaxial compression tests, requires the use of multiple plugs sampled at the same depth interval and properly oriented with respect to bedding or in situ stresses. These requirements, easily satisfied by conventional plugs sampled on BHCs, requires that RSWCs are subsampled to plugs whose dimensions are smaller than conventional ones.

Since measured mechanical properties are affected by specimen shape (i.e., length over diameter L/D ratio) and size (i.e., diameter D), in this work an experimental study on their effect on the strength, stiffness (i.e., Young's modulus), and dilation (i.e., Poisson's ratio) of the rock has been tailored to materials and testing conditions generally encountered in petroleum related geomechanics. In the literature several experimental studies investigate these effects and especially the effects on the unconfined compressive strength (UCS) (e.g., [1] and references within). However, most of them are based on larger samples or on lithology not common in the Oil & Gas industry. A summary of these experimental evidence is presented in the following paragraphs.

### 1.1 Effect of specimen size and shape on strength

Generally, the compressive strength of the rock increases with decreasing L/D ratio especially for values lower than 2.5. The reasons for this increase may be related to the impediment of the development of an inclined shear band for failure in short specimens. For such specimens the inclination of the failure plane (shear band) is such that it meets the loading platens and thus it cannot develop freely, while in more slender specimens it meets the side of the specimen and can form a kinematically possible failure mechanism. Another reason is the confinement due to friction at the loading platens which is more dominant in shorter specimens and provides lateral support to the specimen. It should be anyway highlighted that, for small samples, these boundary conditions may dominate the failure, not the intrinsic failure mechanisms. Experimental data for granite specimens presented in [2] are an example of such behaviour. Figure 1 plots various correlations, based on experimental results, that have been proposed in the literature [1,3,4,5] to describe the shape effect on strength. The ASTM suggested equation for the correction of measured strength is also shown [6].

Regarding the size effect, a decrease in strength is observed with increasing specimen size since, statistically, with larger specimens there is a larger probability to include critical flaws in the sample that may lead to specimen failure. Moreover, the strength of a specimen increases as the ratio of specimen size to internal length of the rock, e.g. grain size, decreases. This behaviour has been highlighted by the experimental results reported in [7] and described, e.g., by the following equation proposed in [8]:

$$\frac{UCS}{UCS_{D_{ref}}} = \left(\frac{D_{ref}}{D}\right)^k \tag{1}$$

* Corresponding author: giorgio.volonte@eni.com

© The Authors, published by EDP Sciences. This is an open access article distributed under the terms of the Creative Commons Attribution License 4.0 (http://creativecommons.org/licenses/by/4.0/).

where $D_{ref}$ is the diameter of the reference specimen and k a parameter equal to 0.18 for $D_{ref}$ equal to 50 mm.
However, in [9] it is suggested that the relation presented in [8] is rather valid for hard rocks whereas for carbonates and sandstones a peak strength is observed between 38-54 mm diameter specimens as shown in Figure 2 where the strength decreases in either side of the peak.

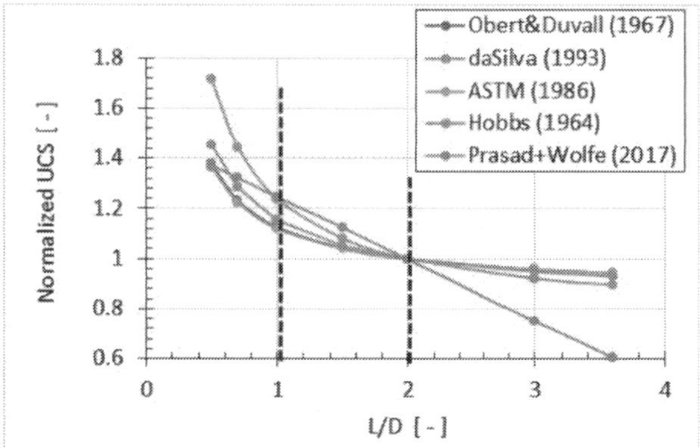


**Fig. 1.** UCS relations with L/D proposed by various investigators. The UCS is normalized with the UCS for L/D = 2. The dotted lines indicate the range of L/D ratios investigated in this study.

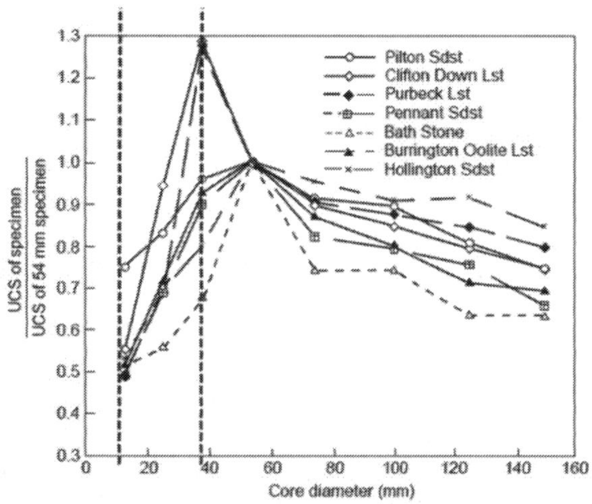


**Fig. 2.** UCS of seven sedimentary rocks measured on samples of eight different diameter [9]. Values are normalized with respect to the strength of the 54 mm diameter sample of the corresponding rock type. The dotted lines indicate the range of diameters investigated in this study.

### 1.2 Effect of specimen size and shape on stiffness

Although the effect of shape and size on stiffness has not been studied as extensively as that on strength, most of the data in the literature indicate that it is relatively small or that there is a significant scatter in the observations. In [10] it is shown that the stiffness modulus is practically independent of the shape. However, both an increase and a decrease of stiffness with increasing L/D ratio have been reported in [11] for kersantite rock and in [12] for Berea sandstone, respectively. In [13] a small, if any, increase in stiffness with increasing L/D in a granite is highlighted. Reference [14] also reports an increase in stiffness with increasing L/D ratio in Red Wildmoor sandstone, but this increase was attributed to the bedding error in the contact between the specimen and the loading platens.
Similarly, the size does not seem to have a large effect on the stiffness of the rock. In [7] and [13] a small effect of the size of granite specimens on stiffness is reported. Reference [13] presents a small increase in stiffness, while [15] reports a larger increase with increasing diameter in limestones. All the experimental results presented are for diameters in the range of 45 – 900 mm, which are larger than the diameters analysed in this study which were in the range of 10-38 mm.

### 1.3 Effect of specimen size and shape on dilation

The effect of shape and size on dilation, i.e. on the tangent Poisson's ratio $\nu_{50}$, has been studied even less than that on stiffness. In [10], from the few available data, a decrease of Poisson's ratio with increasing L/D ratio for granitic rocks is reported. Reference [12] shows a decrease, no effect or increase of Poisson's ratio with increasing L/D for Berea sandstone with no clear effect of the confining stress. In [13] a small, if any, effect on Poisson's ratio is shown for a granite. Similarly, the size does not seem to have a consistent effect on the Poisson's ratio. In [13] variable effects of the size of granite specimens on Poisson's ratio is seen.

## 2 Experimental program

The effect of the specimen shape and size on the triaxial compressive strength (TCS), tangent Young's modulus $E_{50}$ and tangent Poison's ratio $\nu_{50}$ at 50% the peak deviatoric stress was studied experimentally through conventional triaxial compression (CTC) tests at 0.5, 10 and 30 MPa confining stresses. Tests on two sandstones, Carbon Tan (CT) and Saltwash North (SWN), and two limestones, Indiana (IND) and Burlington (BURL), were performed. Table 1 lists the shape and size combinations that were studied experimentally by listing the typical number of tests that were performed. The repetition of tests performed with L/D = 2 specimens allowed to verify the repeatability of the experiments. Some dispersion of the results has been anyway observed (as shown in terms of standard deviation in the plots of the subsequent sections), especially for SWN and IND due to their more accentuated heterogeneity, but this scatter does not hide the observable trends when present. In the cases where tests were repeated also for different L/D ratios, similar repeatability was confirmed.

**Table 1.** Test matrix for the shape (L/D) and size (D) effect on the triaxial compressive strength, the tangent stiffness modulus $E_{50}$, and the tangent Poison's ratio $\nu_{50}$. The table lists the typical number of tests performed for each lithology.

| | ***D* [mm]** | | |
|---|---|---|---|
| ***L/D*** | **10** | **25** | **38** |
| **1** | 1 | - | - |
| **1.5** | 1 | - | - |
| **2** | 3 | 3 | 3 |

Experiments were performed on oven dried specimens (60°C for 2 days). Samples were cored with a rotary core barrel and the end-surfaces were flattened with a surface grinder. The samples were mounted between steel end-pistons and surrounded with a Fluorinated Ethylene Propylene (FEP) heat shrink sleeve. A thin Teflon sheet was placed on each sample-end to reduce end-friction between the sample and end-pistons. Axial deformation was measured with three Linear Variable Differential Transformers (LVDTs), oriented in one plane, 120° apart. Radial deformation was measured with two pairs of strain-gaged cantilevers oriented orthogonally to each other and measuring the diametrical deformation at mid-height of the sample.
The testing was performed in a hydraulic servo-controlled load frame with digital feedback control and data acquisition. Confining pressure was ramped at 36 MPa/h and an axial strain rate of 1E-05 $s^{-1}$ was used for the shearing segment.
The shape effect was studied on 10 mm diameter specimens with L/D equal to 1, 1.5 and 2, while the size effect was studied on specimens with L/D equal to 2 and diameters 10, 25 and 38 mm (Figure 3).

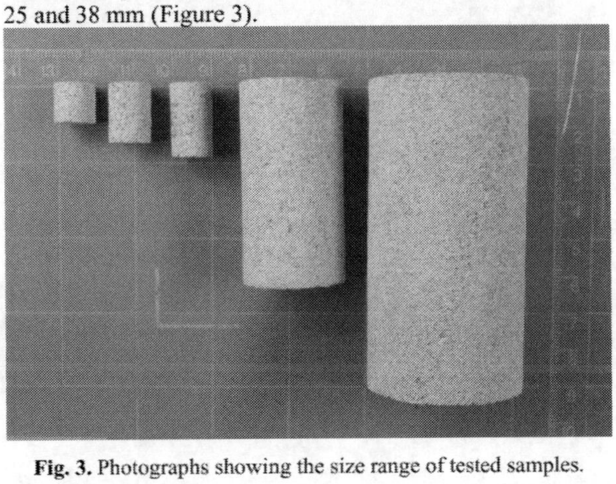

**Fig. 3.** Photographs showing the size range of tested samples.

## 3 Data analysis

The results of the experimental campaign were analysed to derive useful expressions for practical applications that correlate the mechanical properties measured on small specimen retrieved from RSWCs with those that would have been obtained through conventional plugs. For this purpose, a series of elaborations have been performed to reduce, as much as possible, the effect of specimen variability and, subsequently, to highlight or rule out eventual lithology dependent behaviours.
A measure of inherent sample variability can be the oven-dry bulk density of the specimens, since samples with higher density are expected to have a higher strength and stiffness. Indeed, some rocks in the present study showed a systematic difference in mechanical properties for a specific shape and size, that were prepared from blocks of rock with lower or higher density than the remaining samples. This resulted in an apparent lower or higher strength/stiffness, respectively. This effect is especially evident in the Carbon Tan samples whose data showed that the D = 25 mm specimens had systematically lower density and, thus, lower strength and stiffness (Figure 4a).

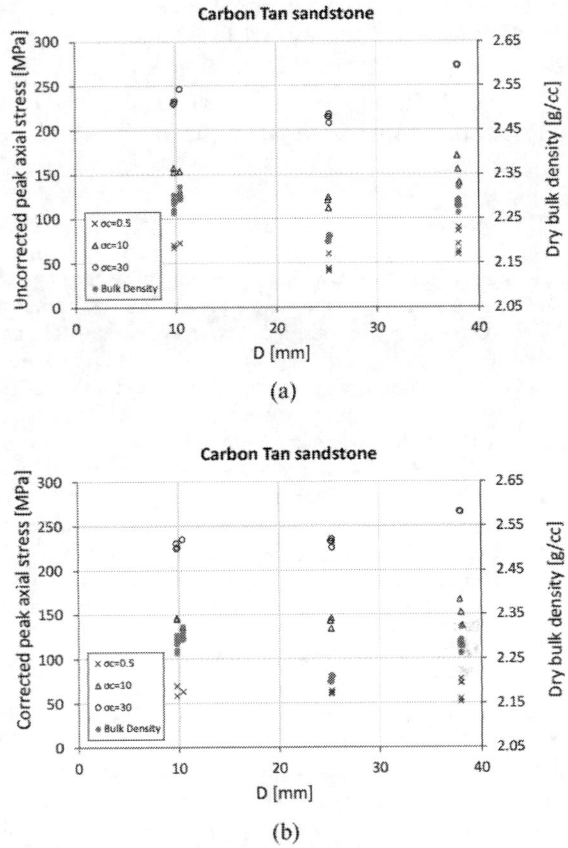


**Fig. 4.** Carbon Tan sandstone. Oven dried density and TCS before (a) and after (b) density correction.

Since this specimen variability influences the size and shape effects, correction correlations were considered necessary for Carbon Tan, Saltwash North, and Indiana. They were derived through the procedure described in the Appendix. An example of the corrected results is reported in Figure 4b for the strength of Carbon Tan sandstone: corrected data are slightly less sparse while trends are more evident and independent of the specimen density. This procedure was not applied to Burlington limestone which showed similar densities between all the samples.
To allow the comparison and the integration of data referring to different lithologies, all the corrected strength, moduli and Poisson's ratio values measured in the experiments at given confining pressure were averaged and normalized with respect to the corresponding average values obtained by testing reference dimension specimens. The reference dimensions have been assumed equal to 10 mm diameter with L/D = 2 for the study of shape effect and 38 mm diameter with L/D = 2 for the analysis of size effect.
Finally, all normalized values of strength, moduli and Poisson's ratio at given confining pressure were averaged across all rocks and integrated to derive correlations with size and shape independent of lithology. The data from the three confining stresses were not combined because the confining stress may play a role in the scale effect as suggested in the literature.

The outcomes of these analyses are described in the following sections where the obtained correlations with size and shape are reported.

## 4 Triaxial compressive strength

### 4.1 Shape effect

The values of normalized compressive strength in Figure 5 show some variability within each lithology at the various confining stresses. In some cases, the strength is higher at lower L/D ratios but in most it is lower. No clear difference can be identified between the four rocks. The plot in Figure 6a indicates that the strength of specimens with L/D = 1.5 and 2 are, for all practical purposes, similar. On the other hand, the strength of the specimens with L/D = 1 seems to be between 5% and 24% lower than the one of specimens with L/D = 1.5 and 2. This finding does not agree with similar findings in the literature which suggest an increase of strength with decreasing L/D ratio. In the range of L/D between 1 to 2, the literature suggests an increase of the strength of specimens with L/D = 1 compared to the one of specimens with L/D = 2 by anything between 12% [6] to 25.5% [1]. This difference can be explained by the small size of the specimens with L/D = 1 used in this study compared to the much larger diameters used in the studies in the literature. In this study the diameter is equal to 10 mm, and thus the specimens are 0.79 $cm^3$ cylinders with dimensions 20 or 30 times their grain size. Due to their relatively small size any damage during specimen coring and polishing may have a considerable effect on the strength.

For the practical applications involving 10 mm diameter specimens, we can combine a formula from literature, e.g., the one suggested by ASTM [6], for L/D > 2 with an equation that accounts for the apparent strength decline due to plug damage for L/D < 2. The combined formula is valid only for 10 mm diameter specimens and may be written as:

$$\frac{TCS_{L/D\_10}}{TCS_{2\_10}} = \begin{cases} 1.007 - e^{-3\frac{L}{D}} & \text{for } \frac{L}{D} \leq 2 \\ 0.88 + \frac{0.24}{\frac{L}{D}} & \text{for } \frac{L}{D} \geq 2 \end{cases} \quad (2)$$

This equation is plotted in Figure 6b together with test data. In this formula and in all the following ones, the first part of the lower index indicates the L/D ratio of the specimen while the second part its diameter in mm. For example, $TCS_{2\_10}$ refers to the TCS of a specimen with L/D = 2 and D = 10 mm. Since, in this study, the shape effect has not been tested in other diameters, any of the formulas in the literature can be used for other diameter values, such as e.g., the following ASTM formula [6]:

$$\frac{TCS_{L/D\_D}}{TCS_{2\_D}} = 0.88 + \frac{0.24}{\frac{L}{D}} \quad \text{for} \quad D \geq 25 \text{ mm} \quad (3)$$

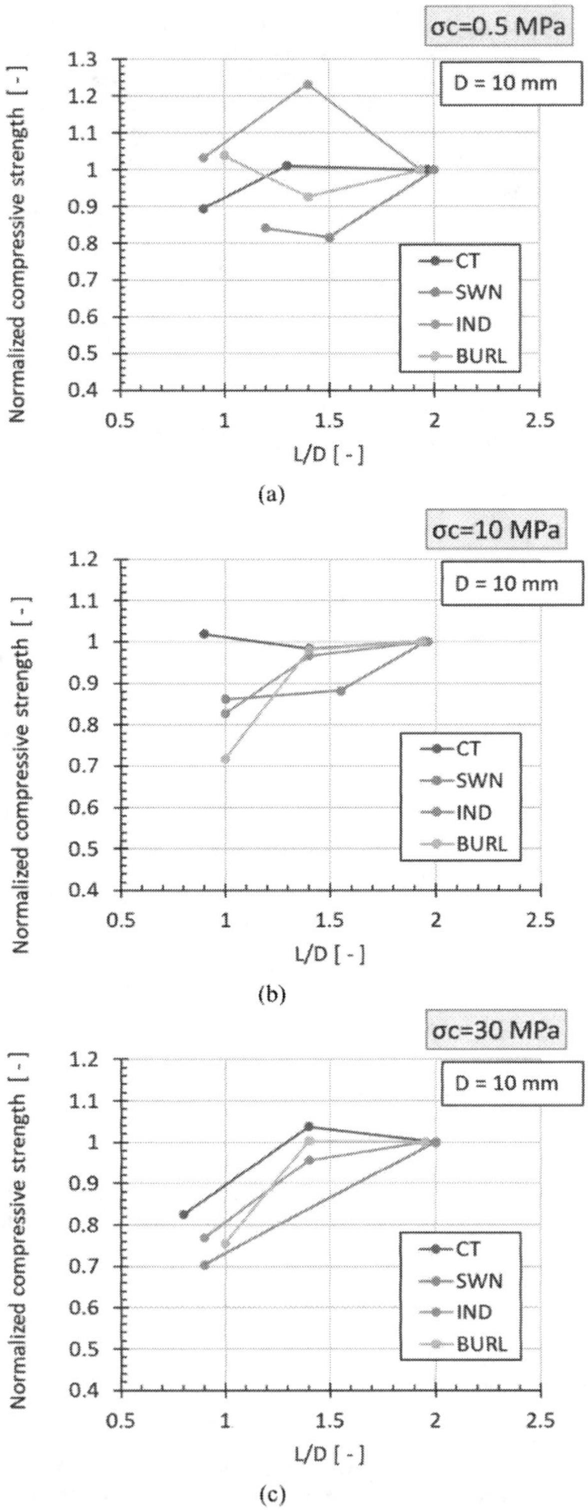


**Fig. 5.** Normalized TCS of all rocks vs. L/D ratio for confining stress $\sigma_c$ = (a) 0.5, (b) 10 and (c) 30 MPa. All samples have diameter equal to 10 mm.

 https://doi.org/10.1051/e3sconf/202336601014

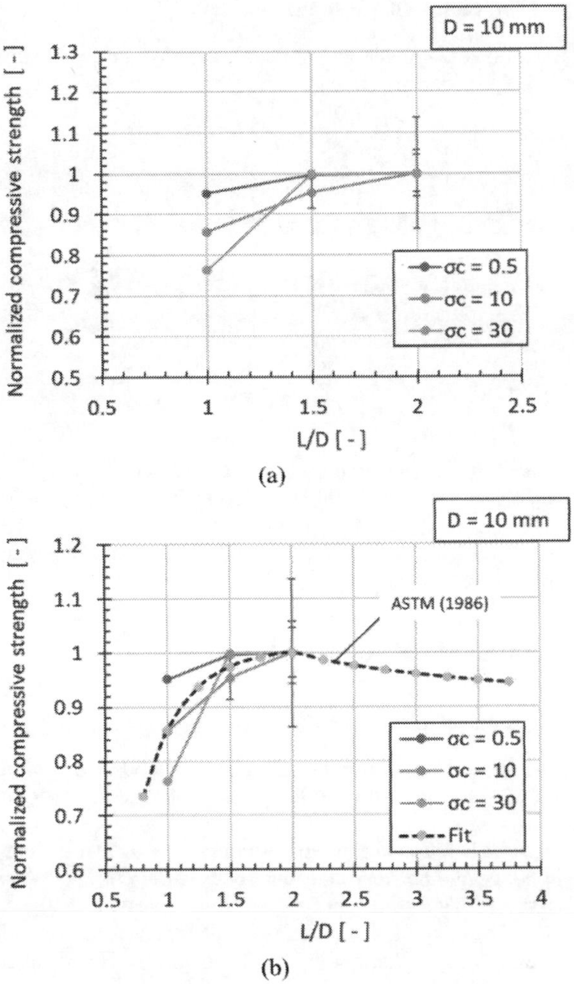


**Fig. 6.** Average normalized TCS of all rocks vs. L/D ratio for confining stress $\sigma_c$ = 0.5, 10, and 30 MPa. All samples have diameter equal to 10 mm. In (a), data with standard deviation are reported while, in (b), they are superimposed with the Eq. (2) approximation.

### 4.2 Size effect

The values of normalized compressive strength in Figure 7 show some variability within each lithology at the various confining stresses. The variability is larger at low confining pressure $\sigma_c$ = 0.5 MPa and lower for $\sigma_c$ = 10 and 30 MPa. This reduction in variability with increasing confinement has been reported also in the literature, e.g., in [16] where earlier works are summarized saying that confining pressure tends to suppress the size effect on failure strength.

No clear difference is noticeable between the four rocks consistent for all confinement stresses. The plot in Figure 8a indicates that the strength decreases with decreasing size for all confining stresses, but the decrease is larger at low confinement, i.e. ca. 21% decrease, and smaller at higher confining stresses, i.e. ca. 7-10%. This decrease for small specimen diameters has also been reported in the literature [9,17] where it is suggested that the peak strength occurs for specimens with diameters between 38 and 54 mm. For larger sizes, the strength decreases again according to an exponential law [7,16,8]. The results in this study seem to corroborate the ones reported in [9] for limestones and sandstones (Figure 2).

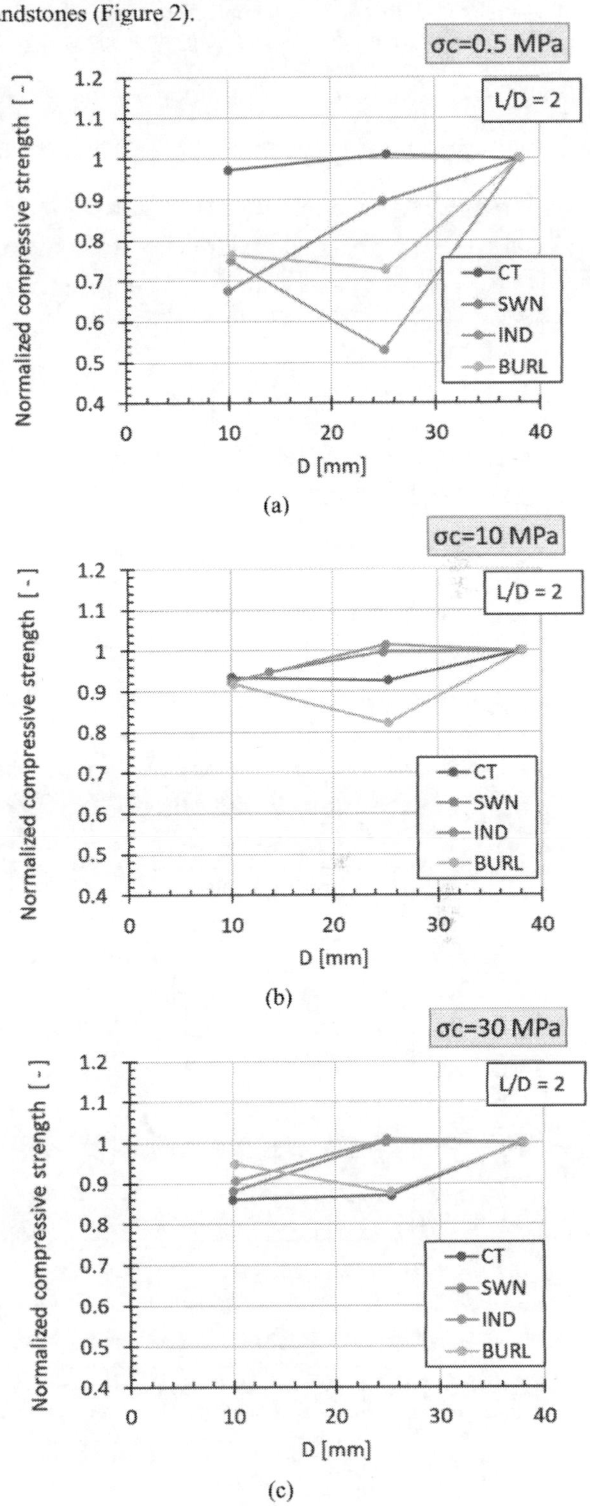


**Fig. 7.** Normalized TCS of all rocks vs. diameter for confining stress $\sigma_c$ = (a) 0.5, (b) 10 and (c) 30 MPa. All samples have L/D ratio equal to 2.

It is thus proposed to combine the exponential decline formula reported in [8] for D > 38 mm with a linear formula where strength decreases with decreasing diameter. The linear relationship assumes a confining stress dependent slope to accommodate the effect of confining stress on the strength reduction. The combined formula may be written as:

$$\frac{TCS_{2\_D}}{TCS_{2\_38}} = \begin{cases} 1 - 0.13\left(1 + 2e^{-\frac{\sigma_c}{3}}\right)\left(1 - \frac{D}{38}\right) & \text{for } D \leq 38 \\ \left(\frac{D}{38}\right)^{-0.18} & \text{for } D \geq 38 \end{cases} \quad (4)$$

where D is in mm. The approximation to the data using Eq. (4) is plotted in Figure 8b together with test data.

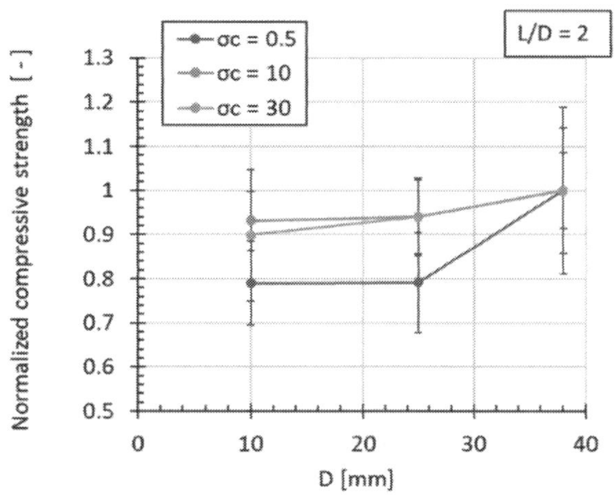


(a)

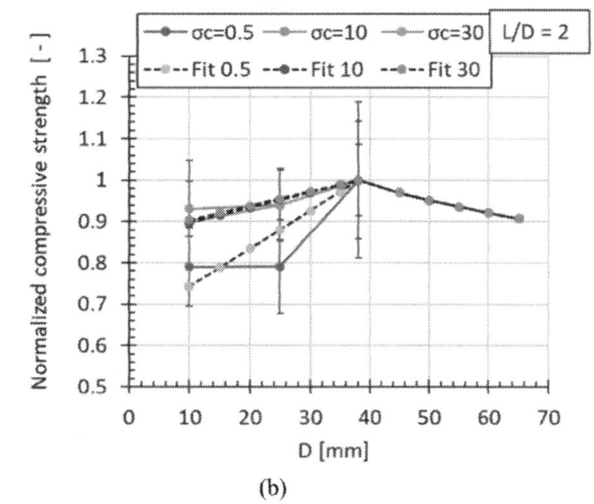


(b)

**Fig. 8.** Average normalized strength of all rocks vs. diameter for confining stress $\sigma_c$ = 0.5, 10, and 30 MPa. All samples have L/D ratio equal to 2. In (a), data with standard deviation are reported while, in (b), they are superimposed with the Eq. (4) approximation.

### 4.3 Combined shape and size effect

Equations (2) or (3) and (4) can be combined to obtain the TCS value of a specimen with arbitrary shape L/D or size D:

$$\frac{TCS_{L/D\_D}}{TCS_{2\_38}} = \begin{cases} \left(0.88 + \frac{0.24}{L/D}\right)\left(\frac{D}{38}\right)^{-0.18} \; for\ D \geq 38 \\ \left(0.88 + \frac{0.24}{L/D}\right)\beta \; for \begin{cases} 20 \leq D < 38 \\ or \\ D \leq 20, L/D > 2 \end{cases} \\ \left(1.007 - 3e^{-3L/D}\right)\beta \; for\ D \leq 20, \frac{L}{D} \leq 2 \end{cases} \quad (5a)$$

where

$$\beta = \left[1 - 0.13\left(1 + 2e^{-\sigma_c/3}\right)\left(1 - \frac{D}{38}\right)\right] \quad (5b)$$

D is in mm and $\sigma_c$ is in megapascals. It should be noted that Eq. (5) is based on both data derived in this study and data available in the literature. In particular, the relations for D >= 38 mm and for L/D > 2 are based on data and expressions from literature.

## 5 Tangent stiffness modulus

### 5.1 Shape effect

The values of the normalized stiffness moduli in Figure 9 show some variability within each lithology at the various confining stresses. In some cases, the stiffness is higher at lower L/D ratios but in most it is lower. No clear difference is noticeable between the four rocks. The plot in Figure 10 suggests that the stiffness modulus increases with increasing L/D. The modulus of the L/D = 1 specimens is between 27% and 58% lower than the L/D = 2 specimens. This difference can be observed also by comparing the stress-strain curves reported in Figure 11 for Saltwash North sandstone. This finding agrees with some results in the literature [11] although other results suggest a stiffness modulus independent of L/D [10], or decreasing with increasing L/D [12], or even both decreasing and increasing with L/D [15]. One of the reasons for the increase of stiffness with increasing L/D ratio is the bedding error associated with the specimen-piston interfaces when the axial strain is calculated from the deformation of the whole specimen [14]. Similarly to what observed for strength, another reason for the lower stiffness modulus of the low L/D ratio specimens is the small size of the specimens with L/D = 1, that allows any damage to have a considerable effect in stiffness as well as in strength.
For practical applications involving 10 mm diameter specimens, experimental data can be regressed with a linear function to model the shape dependency of stiffness modulus as:

$$\frac{E_{50\_L/D\_10}}{E_{50\_2\_10}} = 0.5 + 0.25\frac{L}{D} \quad (6)$$

which is also superimposed to test data in Figure 10. It should be highlighted that in this case data are quite sparse, especially for L/D=1.5. The shape effect has not been tested

https://doi.org/10.1051/e3sconf/202336601014

in other diameters in this study and no formula is available in the literature.

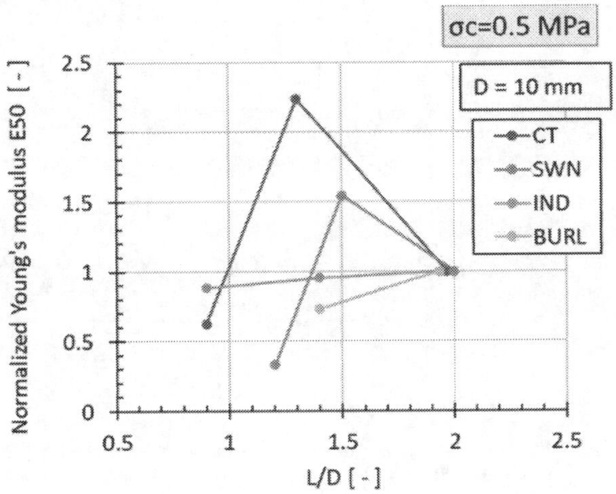


(a)

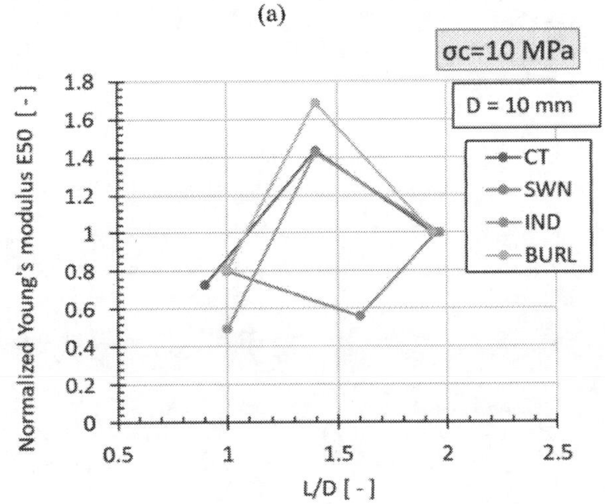


(b)

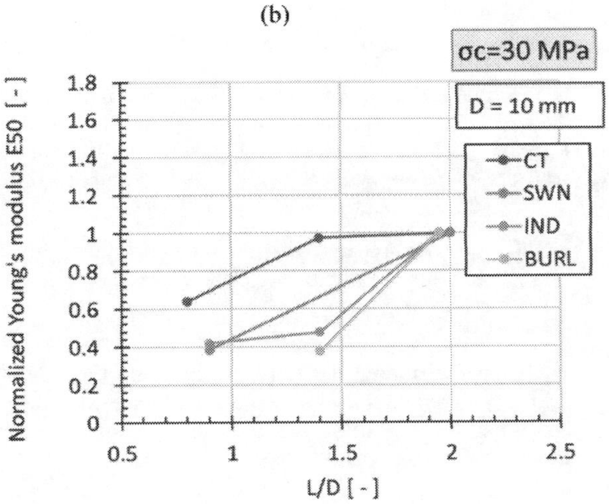


(c)

**Fig. 9.** Normalized $E_{50}$ of all rocks vs. L/D ratio for confining stress $\sigma_c$ = (a) 0.5, (b) 10 and (c) 30 MPa. All samples have diameter equal to 10 mm.

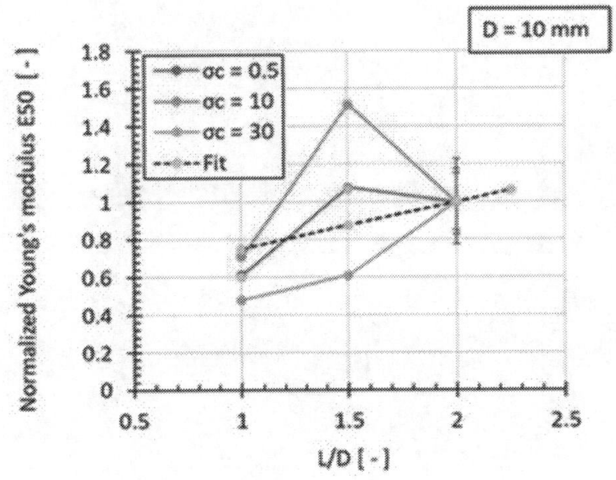


**Fig. 10.** Average normalized $E_{50}$ of all rocks vs. L/D ratio for confining stress $\sigma_c$ = 0.5, 10, and 30 MPa. All samples have diameter D = 10 mm. Standard deviation and approximation of the data with Eq. (6) are also reported.

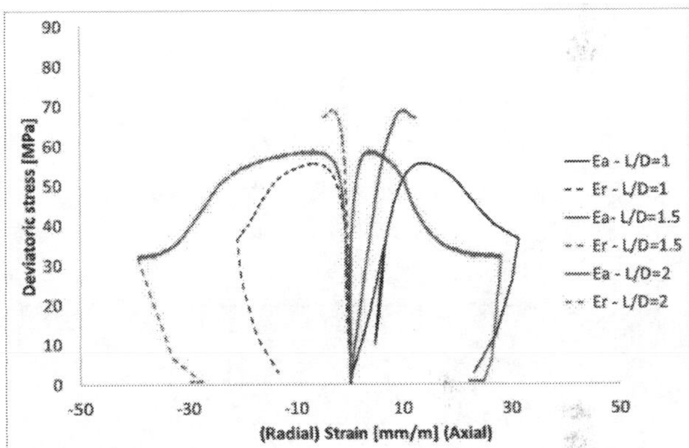


**Fig. 11.** Example of stress-strain curves for three experiments on Saltwash North sandstone at a confining pressure equal to 10 MPa.

## 5.2 Size effect

The values of normalized stiffness moduli in Figure 12 show some variability within each lithology at the various confining stresses. No clear difference is noticeable between the four rocks consistent for all confinement stresses (Figure 13). The plot does not demonstrate a clear dependency of the stiffness modulus with the size of specimen. Similar results have also been presented in [2] and [11]. However, in [15] an increase in stiffness with increasing size is shown. This behaviour can be observed also in Fig 14 where an example of stress-strain curves obtained by testing specimens of different sizes is reported for Saltwash North sandstone. According to the data, it is thus proposed to keep the stiffness modulus independent of the size of the specimen.

https://doi.org/10.1051/e3sconf/202336601014

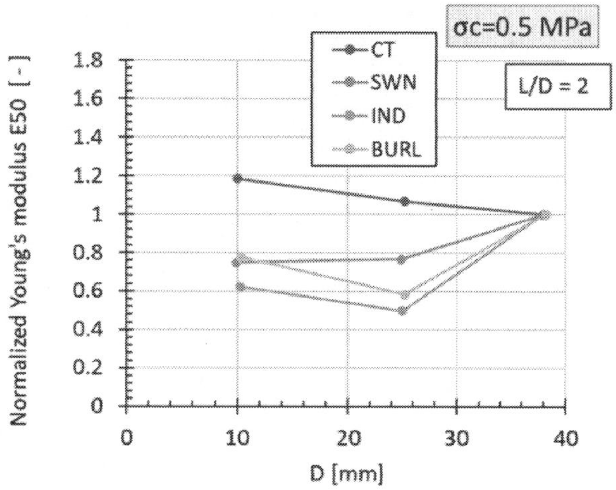


(a)

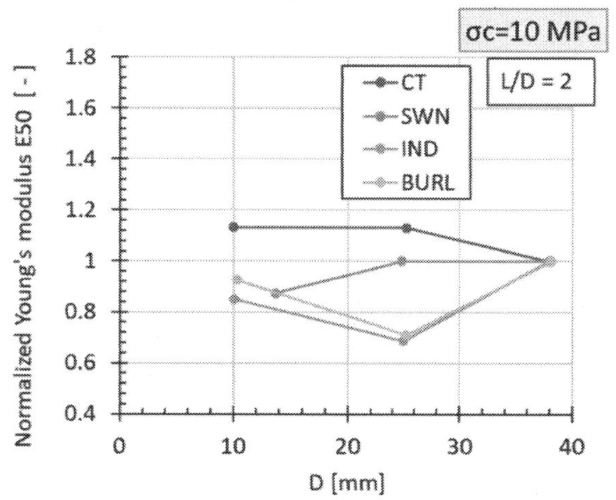


(b)

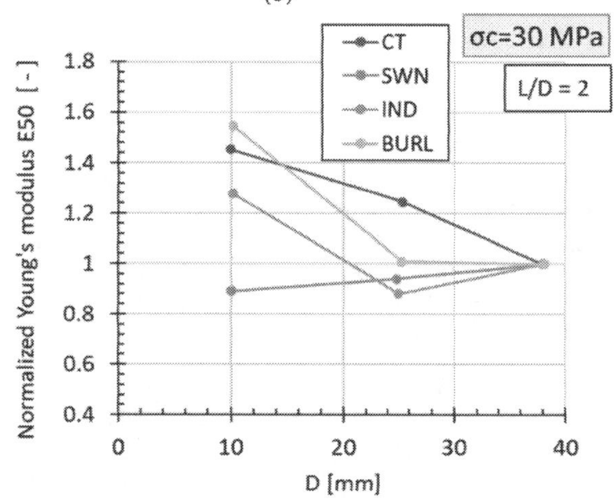


(c)

**Fig. 12.** Normalized $E_{50}$ of all rocks vs. diameter for confining stress $\sigma_c$ = (a) 0.5, (b) 10 and (c) 30 MPa. All samples have L/D ratio equal to 2.

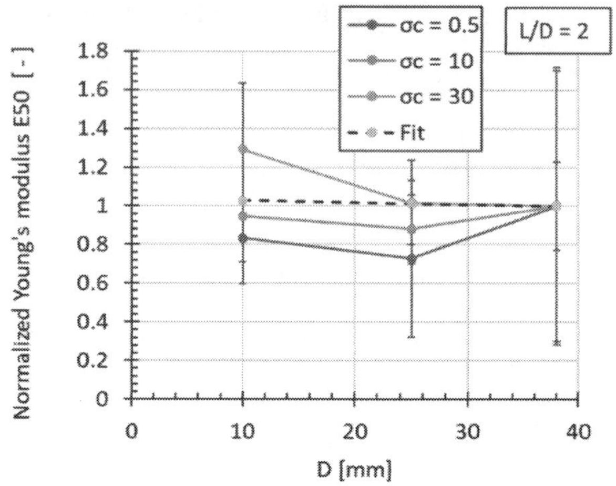


**Fig. 13.** Average normalized $E_{50}$ of all rocks vs. diameter for confining stress $\sigma_c$ = 0.5, 10, and 30 MPa. All samples have L/D ratio equal to 2. Standard deviations are also reported.

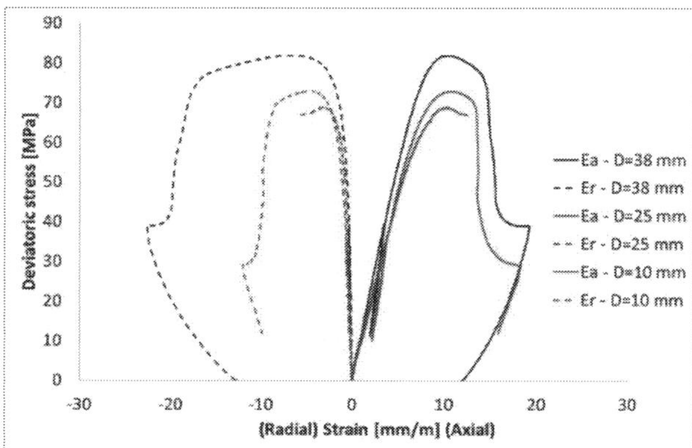


**Fig. 14.** Example of stress-strain curves for Saltwah North sandstone at 10 MPa.

### 5.3 Combined shape and size effect

In this study, the stiffness modulus appears independent of the specimen size. Therefore, only the shape effect formula Eq. (6) is necessary to obtain the $E_{50}$ value of a specimen with arbitrary shape or size.

## 6 Tangent Poisson's ratio

### 6.1 Shape effect

The values of normalized Poisson's ratio in Figure 15 show some irregular variability within each lithology at the various confining stresses. In some cases, the dilation is higher at lower L/D ratios but in other cases it is lower.

https://doi.org/10.1051/e3sconf/202336601014

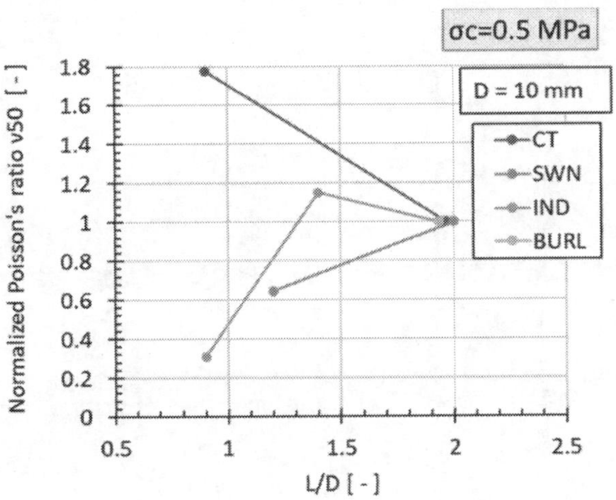


(a)

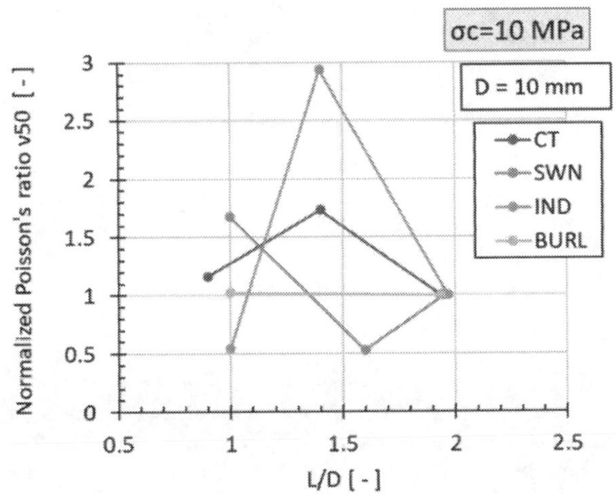


(b)

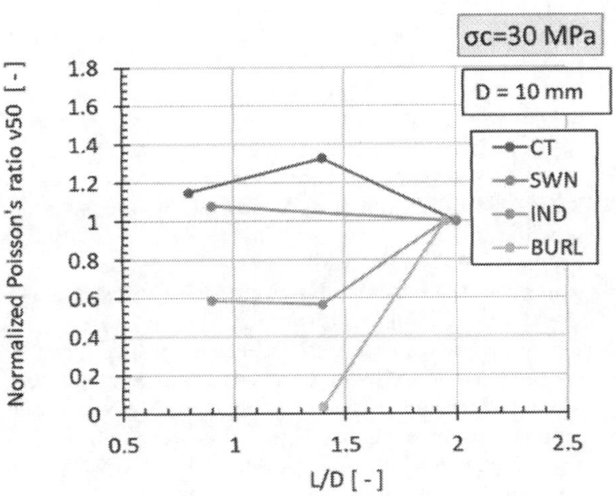


(c)

**Fig. 15.** Normalized $\nu_{50}$ of all rocks vs. L/D ratio for confining stress $\sigma_c$ = (a) 0.5, (b) 10 and (c) 30 MPa. All samples have diameter equal to 10 mm.

An increase of dilation with increasing L/D ratio may be expected due to the frictional effect in the interfaces between the platens and the specimen which may act as confinement and restrain lateral deformation. For shorter specimens, this effect may be more pronounced because the mid-height of the specimen, where the lateral deformation is measured, is closer to the loading platens. Despite that, experimental data show that Poisson's ratio increases modestly, if any at all, with increasing L/D (Figure 16). So, no correction of dilation measurement as function of L/D ratio is considered needed. It should be anyway highlighted that, as for Young's modulus, results are quite scattered for L/D = 1.5.
The shape effect has not been tested in other diameters in this study and no correlation in the literature is available.

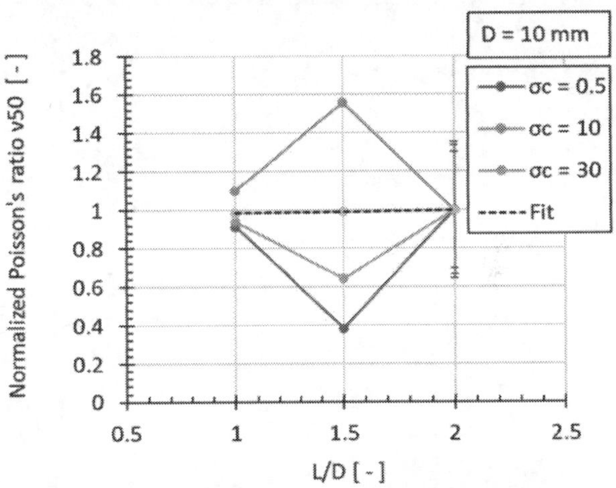


**Fig. 16.** Average normalized v50 of all rocks vs. L/D ratio for confining stress $\sigma_c$ = 0.5, 10, and 30 MPa. All samples have diameter equal to 10 mm Standard deviation is also reported.

### 6.2 Size effect

The values of normalized Poisson's ratios in Figure 17 show little variability within each lithology at the various confining stresses, more accentuated at a confining pressure of 30 MPa especially for Indiana and Burlington limestones. Indeed, in these cases, the plot shows a decrease of the Poisson's ratio with increasing diameter which however is mostly influenced by the scatter of experimental results. At other confining stresses, no size effect is observed.
The overall trend of experimental data can be regressed with a linear function (Figure 18) to model the size dependency of Poisson's ratio as:

$$\frac{\nu_{50\_L/D\_10}}{\nu_{50\_2\_38}} = 1.2 - 0.0054D \qquad (7)$$

where D is in mm.

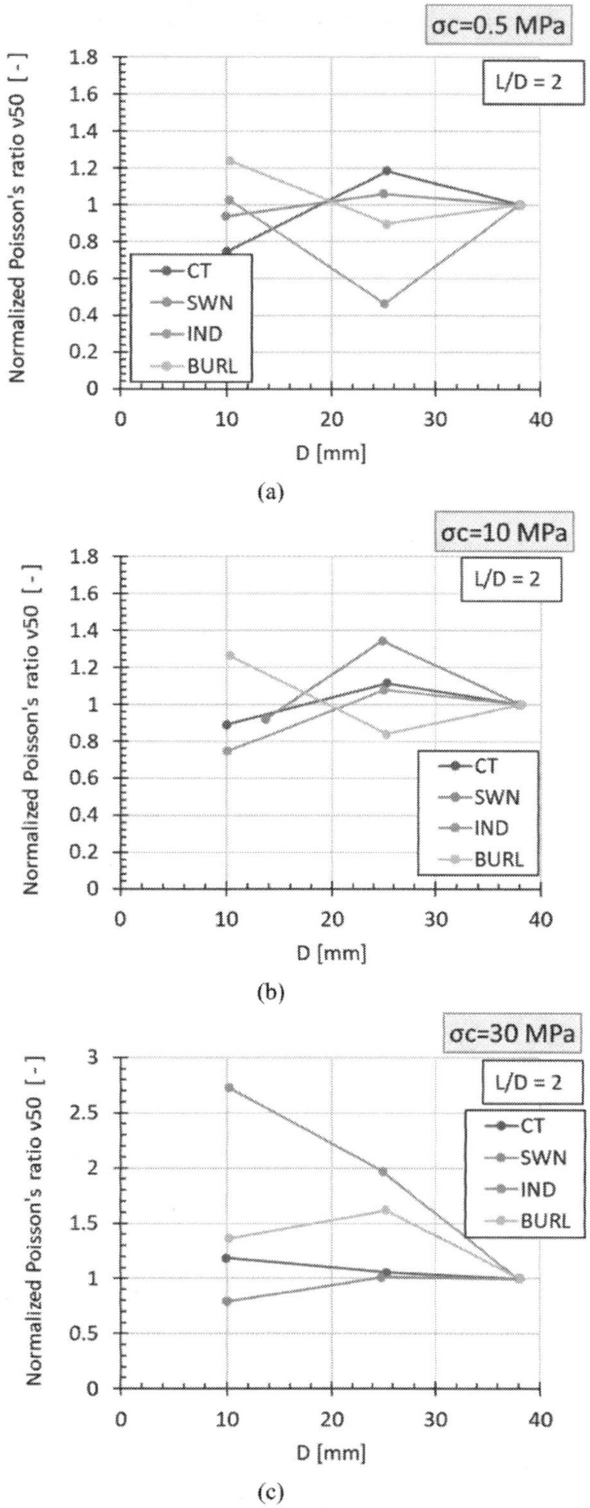


**Fig. 17.** Normalized $\nu_{50}$ of all rocks vs. Diameter (D) for confining stress $\sigma_c$ = (a) 0.5, (b) 10 and (c) 30 MPa. All samples have L/D ratio equal to 2.

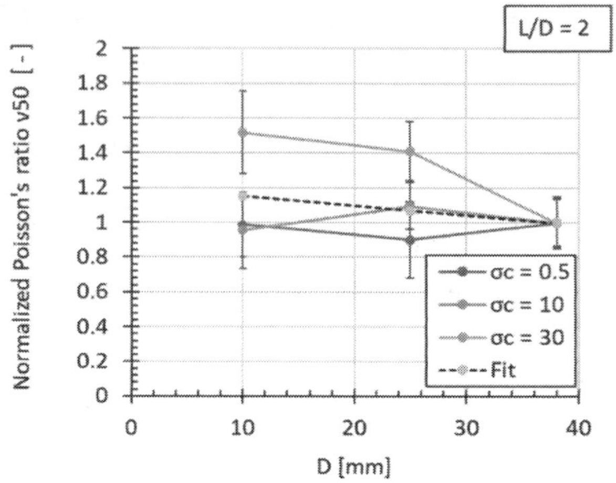


**Fig. 18.** Average normalized v50 of all rocks vs. diameter for confining stress $\sigma_c$ = 0.5, 10, and 30 MPa. All samples have L/D ratio equal to 2 Standard deviation and approximation of the data with Eq. (7) are also reported.

### 6.3 Combined shape and size effect

In this study, the tangent Poisson's ratio appears independent of the specimen shape. Therefore, only the size effect formula Eq. (7) is necessary to obtain the $\nu_{50}$ value of a specimen with arbitrary shape or size.

## 7 Conclusions

In this work, an experimental campaign has been performed to evaluate the effect of specimen size and shape on the mechanical properties measured on the small plugs that can be subsampled from field RSWCs. Indeed, few evidence are available in the literature on small specimens [9] of typical reservoir lithologies [1, 9].

Unlike the literature on larger specimens [1,2,3,4,5], rock strength for 10 mm diameter plugs decreases for decreasing L/D ratios, especially for L/D = 1 where, reasonably, damage during specimen coring and polishing can affect rock behaviour. The dependency of strength on size, instead, is coherent with evidence available in literature for diameters lower than 38-54 mm, where strength increases with size.

Stiffness, instead, is not influenced by the size of the specimen while increases with increasing L/D ratio due to the bedding effect. This problem may be eliminated if strain gauges or other local axial deformation measurements are used for plugs with low L/D ratio.

Dilation is not significantly affected by the shape of the specimen while it slightly decreases for increasing size.

The results of the tests were also quantitatively integrated to propose analytical equations that correlate strength, stiffness and dilation with both size and shape. These correlations will be useful in practical application to correlate the mechanical properties measured on small specimen retrieved from RSWCs with those that would have been measured through conventional plugs.

Previous observations suggest that small diameter specimens can be a useful tool for practical experimental rock characterizations involving RSWCs, but the use of only plugs

with L/D greater than 1.5, for which the effects of experimental artifacts is negligible, is recommended.
This study, based on experiments on dry plugs, was focused on the effects on the mechanical properties measurements of specimen geometry exclusively. Therefore, no saturation dependency was included in the proposed correlations. Anyway, future studies on saturated specimens with different shape and size could be desirable to understand the potential effect of the different allowed drainage paths during the experiments, especially for low permeability rocks.

## Appendix. Method for correcting the inherent specimen variability though density measurements.

With the final aim of removing, as much as possible, the inherent specimen variability from the strength, stiffness, and dilation measurements, a procedure has been defined to quantify this variability based on dry bulk density measurements. A series of experiments were performed on three specimens for each lithology that were prepared from blocks characterized by different density to measure the variability of mechanical properties related to density variations. Burlington limestone was not tested since all block showed similar densities. All tests were run on specimens with L/D = 2, D = 38 mm, and at confining stress $\sigma_c = 0.5$ MPa. Figure 19 plots peak axial stress, stiffness modulus and Poisson's ratio as functions of dry density together with the linear regression of the data. Due to the limited number of data for each series, the linear approximation has been considered the most suitable: the deviation with respect to measured data is anyway small within the investigated density range, only Poisson's ratio shows higher dispersion. The slopes $a_\sigma$, $a_E$ and $a_N$ of the linear approximations for the peak axial stress, the stiffness modulus and the Poisson's ratio respectively were used to define the following relations between the measured mechanical properties and the mechanical properties corresponding to a reference value of density:

$$\sigma_{ref} = \sigma_m + a_\sigma(\rho_{ref} - \rho_m)$$

$$E_{50ref} = E_{50m} + a_E(\rho_{ref} - \rho_m) \quad (8)$$

$$\nu_{50ref} = \nu_{50m} + a_N(\rho_{ref} - \rho_m)$$

where $\sigma_m$, $E_{50m}$, and $\nu_{50m}$ are, respectively, the peak axial stress, the stiffness modulus and the Poisson's ratio of a specimen with density $\rho_m$, while $\sigma_{ref}$, $E_{50ref}$, and $\nu_{50ref}$ are the corresponding values for the reference density. This reference density $\rho_{ref}$ was assumed equal to the average density of all the specimens of a given lithology. The corrected mechanical properties $\sigma_{ref}$, $E_{50ref}$, and $\nu_{50ref}$ were then used in the analysis of experimental results to directly compare values corresponding to the same density.

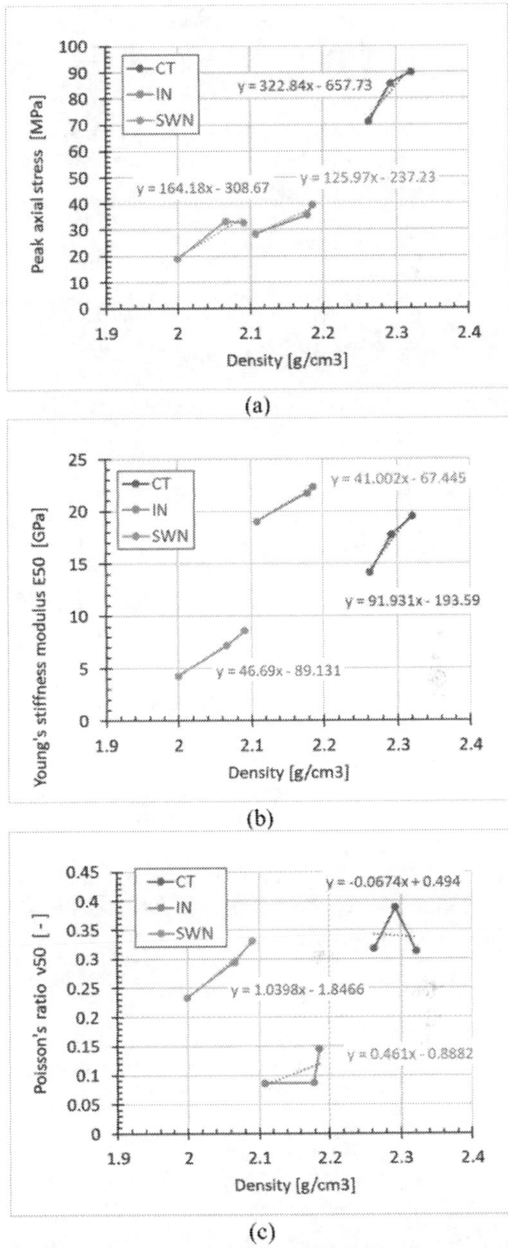


**Fig. 19.** (a) Peak axial stress, (b) Stiffness modulus $E_{50}$, and (c) Poisson's ratio $\nu_{50}$ vs. density for Carbon Tan (CT), Indiana (IN) and Saltwash North (SWN). The linear approximations to the data are also plotted.

## References

1. U. Prasad, C. Wolfe, Formation specific size correction for strength (UCS) on rotary sidewall cores. *51st US Rock Mechanics/Geomechanics Symposium*, San Francisco, CA, ARMA17-534 (2017).

2. K. Mogi, Some precise measurements of fracture strength of rocks under uniform compressive stress. *Rock Mech. Eng. Geol.* IV, 41-55, (1966).

3. L. Obert, W.I. Duvall. Rock mechanics and the design of structures in rock. New York, Wiley, (1967).

4. L.A.A. da Silva, E.C. Sansone, J. Fujii, Shape effects considerations on the uniaxial compressive strength of hard rocks. *ISRM International Symposium EUROCK 93,* 1993-092, Lisbon, Portugal, (1993).

5. D.W. Hobbs, Rock compressive strength. Colliery Engineering 41, 287-292, (1964).

6. ASTM, Standard test method of unconfined compressive strength of intact rock core specimens. D 2938: 390–391, (1986).

7. K. Mogi, The influence of the dimensions of specimens on the fracture strength of rocks – Comparison between the strength of rock specimens and that of the Earth's crust. *Bulletin of the Earthquake Research Institute,* 40, 175-185, (1962).

8. E. Hoek, E.T. Brown, Underground excavations in rock. *The Institution of Mining and Metallurgy*, London, Chapman & Hall, (1980).

9. A. B. Hawkins, Aspects of rock strength. *Bulletin of Engineering Geology and the Environment* 57, 17–30, (1998).

10. P. Dirige, J. F. Archibald, The effects of geometry on the uniaxial compressive strength properties of intact rock core specimens. *Golden Rocks 2006, 41st US Symposium on Rock Mechanics*, Golden, CO, ARMA/USRMS 06-1016 (2006).

11. K. Thuro, R.J. Plinningr, S. Zah, Scale effect in rock strength properties, *Eurock 2001: Rock Mechanics a challenge for society*. Espoo, Finland, 169-174, (2001).

12. D. A. Moronkeji, U. Prasad, J. A. Franquet, Size effects on triaxial testing from sidewall cores for petroleum geomechanics. *48th US Rock Mechanics /Geomechanics Symposium.*, Minneapolis, MN, ARMA14-7405, (2014).

13. F. Nunes, V. Cavaleiro, L. Oliveira e Sousa, Scale effect of granitic materials. *45th US Rock Mechanics/Geomechanics Symposium.*, San Francisco, CA, ARMA 11-373, (2011).

14. E. Papamichos, J. Tronvoll, I. Vardoulakis, J. F. Labuz, A. Skjærstein, T. E. Unander, J. Sulem, Constitutive testing of Red Wildmoor sandstone. *Mechanics of Cohesive-Frictional Materials* 5, 1, 1-40, (2000).

15. A.H.J. Al-Rkaby, Z. M. S. Alafandi. Size effect on the unconfined compressive strength and modulus of elasticity of limestone rock. *Electronic Journal of Geotechnical Engineering* 20, 12, 5143-5149 (2015).

16. A.S. Abou-Sayed, C. E. Brechtel, Experimental investigation of the effects of size on the uniaxial compressive strength of Cedar City quartz diorite. *17th Symposium on Rock Mechanics*, Snowbird, UT, Paper 5 D6, (1976).

17. H. Masoumi, K. J. Douglas, S. Saydam, P. Hagan, Experimental study of size effects of rock on UCS and point load tests. *46th US Rock Mechanics / Geomechanics Symposium.*, Chicago, IL, ARMA 12-215 (2012).

# Micromechanics Digital Rock: Parameterization of Consolidation Level using a Grain Contact Model

*Zhuang* Sun*, *Rafael* Salazar-Tio, *Andrew* Fager, and *Bernd* Crouse

Dassault Systèmes, 343 Sansome St San Francisco, CA, USA

**Abstract.** The mechanical behaviour of sedimentary rocks is conditioned by the interactions at the grain-grain contacts. We present a micromechanics digital rock workflow based on a cohesive contact model and introduce a general parameterization that can capture two extreme contact behaviours: free grains and fixed grains, as well as any intermediate degree of grain consolidation. With this parametric cohesive contact model, we can simulate a wide range of sedimentary rocks, from unconsolidated to well-consolidated rocks. We present a benchmark study on several samples and compare with laboratory-measured elastic moduli to calibrate its degree of consolidation. Simulations that do not include the grain contact modelling, tend to overestimate the elastic moduli, which manifests the significance of this contribution to capture well the grain contact behaviour. To demonstrate the impact of properly capturing the degree of consolidation on the rock strength and failure pattern, we present results for numerical uniaxial compression testing. This workflow provides physics-based solution to complex grain contact behaviour, which complements laboratory core analysis, and can be useful to reveal underlying grain-scale processes governing rock mechanical behaviour.

## 1 Introduction

Digital rock applications have rapidly emerged over the past decades based on X-ray micro-tomography (micro-CT) imaging technology that allows capturing pore-scale three-dimensional (3D) structures of reservoir rocks at the micro-meter scale, and virtually simulating the flow of fluids under different production conditions [1–9]. However, digital rock simulation of rock mechanical properties has been considered challenging given the intrinsic limitation of micro-CT images to capture the degree of consolidation between grains in rocks, due to the limit in imaging resolution and also because micro-CT is typically obtained at ambient conditions and does not capture the actual net confining stress (NCS) pore geometry [10–15].

Different levels of consolidation, compaction, recrystallization, and diagenesis in general, can produce a wide range of grain contact behaviours in sedimentary rocks, from loosely consolidated sand packs to completely fused grains that form a single solid structure [16]. Numerical methods in the literature typically use a single solid frame formed by the mineral grains, generally overestimating the simulated rock stiffness as compared to experimental measurements [2,10,17–19]. Some ideas have been discussed to assign different properties to contact regions and grain regions [20–22]. However, a commonality among these approaches is that grain relocation is not allowed, which limits the grain-grain contact behaviour of real rocks as well as rock failure significant deformation.

We recently introduced a micromechanics digital rock workflow based on finite element modelling of grain structures obtained from micro-CT images [23]. Some unique features of this workflow include, a robust grain-grain segmentation method, a conformal grain-grain contact meshing algorithm, and grain relocation capabilities, instead of modelling a solid framework as a whole. We are able to simulate linear elastic moduli comparable with experimental values, as well as being able to simulate significant compaction found at in-situ conditions, which are difficult to capture using micro-CT imaging at ambient conditions.

Here we present an extension of that work, based on a cohesive contact model and we introduce a general parameterization that can capture both extreme contact behaviours: free grains and fixed grains, as well as an intermediate degree of grain consolidation. The proposed workflow intends to solve problems including: (1) model properly variable grain-grain consolidation to correctly simulate mechanical rock properties, (2) being able to recover the correct NCS pore geometry to simulate all petrophysical properties under the correct in-situ conditions, (3) allow grain relocation capabilities, including failure test simulations. This paper is structured as follows. First, we describe the parametric cohesive contact model to simulate a wide range of sedimentary rocks, from unconsolidated to well-consolidated rocks. Second, we show a benchmark study on sandstone samples and compare with laboratory-measured elastic moduli to calibrate its degree of consolidation. Finally, we perform numerically uniaxial compression tests to demonstrate the impact of properly capturing the degree of consolidation on the rock strength and failure pattern.

## 2 Methodology

A micro-CT 3D voxelized image is needed as an input to identify individual grains and a connected pore geometry of rock. The limitations of micro-CT images for properly capturing a micro-mechanics rock model are related to the insufficient contrast for the identification of different mineral

* Corresponding author: zhuang.sun@3ds.com

© The Authors, published by EDP Sciences. This is an open access article distributed under the terms of the Creative Commons Attribution License 4.0 (http://creativecommons.org/licenses/by/4.0/).

grains, and for the identification of grain-grain contacts between same mineral grains. Identification of grain mineralogy and contacts can be improved by complementing 3D micro-CT imaging with 2D higher resolution and mineralogy imaging such as scanning electron microscopy (SEM) and energy dispersive x-ray spectroscopy (EDS). In order to setup a finite element (FE) micro-mechanics rock model from micro-CT data, two critical pre-processing steps are required: (1) grain-grain segmentation, and (2) conformal meshing of grain contacts. We have described these steps in detail in [23], but for the sake of completeness we include here a brief description.

For the grain-grain segmentation we start from a direct binary thresholding on the micro-CT that classifies voxels as pore or solid. After this, we want to reclassify the solid voxels with an integer index that represents at which grain it belongs. For this purpose we use a class of algorithms refer as watershed methods used for separating touching objects in binary images [24]. This algorithm computes first a distance transform 3D image, where for each solid voxel, the value of the Euclidean distance to the nearest pore/solid interface is recorded. If this is considered as a topographic depth map, the deeper parts of the image are the center of the objects. A watershed classification considers each solid voxel by the center to which it will roll down when following this inverted distance transform topographic map. In the final grain segmented 3D image each grain voxel has the value of the corresponding grain index, while pore voxels are labelled as 0.

In order to achieve an initial static solution for the segmented grains, it is of great importance to create a conformal mesh at the grain-grain contacts. A conformal mesh indicates that there is neither separation nor overclosure between any two grains of contact. The labeled voxelized 3D image is transformed into an unstructured mesh representation, where elements of the same grain are suitable for FE simulation and the elements at each side of a contact between grains conform to each other perfectly, without voids or overlaps in the contact boundary [23]. We generate the meshing for all the grains together, while maintaining the indices representing segmented grains, which results in a conformal mesh at the grain-grain contact. We can choose to have two separate surfaces or one combined interface for the conformal mesh, which corresponds to two case scenarios of free or fixed grain-grain contacts.

Finally, the meshed model is input into the finite element solver with prescribed strains/stress boundary conditions and grain-grain contact models [25,26]. A general parametrization of the cohesive contact model is used to capture contact behaviors of different levels of grain consolidation including two extreme scenarios: free grains, where only friction is modeled between grains, and fixed grains, where grains are completely fused. In terms of the FE simulations, the free grain contacts introduce duplicate nodes and elements that allows grain relocation whereas the fixed grain contacts model the solid framework as a whole.

### 2.1. Cohesive contact model

The concept of "bonded grains" has been widely adopted in discrete element method (DEM), which simulates each individual grain as a discrete rigid body [27–29]. Similarly, the FE simulations can simulate the bonded grain interface based on the cohesive contact model, with or without the possibility of damage and failure of the bond (Figure 1a). The cohesive constitutive laws include the linear elastic traction-separation model, damage initiation criterion, and damage evolution laws, which allow a precise description of the contact behavior. This work does not explicitly model the cement volume. Some processes, such as mineral segmentation, cement material property characterization, and grain meshing for cement phase, need to be developed to understand the effect of cement relative volume. We present an overview of the contact model, more details can be found in the FE solver documentation [26,30].

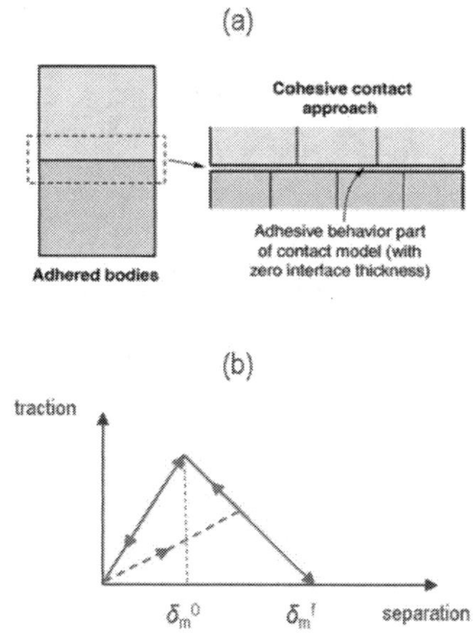


**Fig. 1.** (a) Schematic of cohesive contact behaviour [26]. (b) Typical traction-separation response.

The model assumes the traction-separation behavior initially linear elastic followed by the damage initiation and evolution. The elastic matrix $K$ relates the normal and shear stresses to the normal and shear separations:

$$\begin{bmatrix} t_R \\ t_S \end{bmatrix} = \begin{bmatrix} K_{RR} & 0 \\ 0 & K_{SS} \end{bmatrix} \begin{bmatrix} \delta_R \\ \delta_S \end{bmatrix} \qquad (1)$$

where $t_R$ and $t_S$ represent the traction stress components in normal and shear directions, $\delta_R$ and $\delta_S$ represent the corresponding separations. Normal compressive stress follows the usual contact behavior. The damage modeling consists of a damage initiation criterion and a damage evolution law. Damage initiation begins when the separations satisfy the specified initiation criteria:

$$max\left\{\frac{\delta_R}{\delta_R^0}, \frac{\delta_S}{\delta_S^0}\right\} = 1 \qquad (2)$$

where $\delta_R{}^0$ and $\delta_S{}^0$ are critical separations in normal and shear directions. Note that purely compressive stress (negative value) does not result in damage. The damage evolution law describes the degradation rate of the contact stiffness once the damage initiation starts. The contact stresses follow:

$$t_R = (1 - D)\bar{t}_R \tag{3}$$

$$t_S = (1 - D)\bar{t}_S \tag{4}$$

where $D$ is the damage variable, $\bar{t}_R$ and $\bar{t}_S$ are contact stresses predicted by the elastic behavior (Eq. (1)) without damage. The damage variable $D$ changes from 0 to 1 as the damage initiates and evolves and can be specified as a linear function of the effective separation $\delta_m$. $\delta_m^f$ and $\delta_m^0$ are the effective separation at complete failure and damage initiation, respectively. $\delta_m^{max}$ refers to the maximum value of the effective separation attained during the loading history. Figure 1b shows a typical traction-separation response for the cohesive contact model. Unloading subsequent to damage initiation is assumed to follow the history traction-separation path. Therefore, the cohesive contacts remain elastic until a bond failure occurs.

$$D = \frac{\delta_m^f(\delta_m^{max} - \delta_m^0)}{\delta_m^{max}(\delta_m^f - \delta_m^0)} \tag{5}$$

$$\delta_m = \sqrt{\delta_R^2 + 2\delta_S^2} \tag{6}$$

### 2.2 General parameterization

We describe in more detail the proposed new general parameterization of cohesive contact model that can capture both extreme contact behaviors: free grains and fixed grains, as well as any intermediate degree of grain consolidation.

The parameters of the contact model include the stiffness $K_{RR}$ and $K_{SS}$ for the linear elastic behavior, the critical separation $\delta_R^0$ and $\delta_S^0$, and the effective separation at complete failure $\delta_m^f$ for the damage behavior. In order to simplify the representation, we assume $K_{RR} = K_{SS} = K$, $\delta_R^0 = \delta_S^0 = \delta^0$, $\delta_m^f = 2.5\times\delta_m^0$ to reduce the number of parameters. The stiffness $K$ needs to be pre-determined and can be automatically computed such that the cohesive contact model of a very large $\delta^0$ can reproduce the mechanical behavior of a solid framework with no grain contacts involved (i.e. fixed grain-grain contacts). Therefore, the cohesive contact model has only one adjustable parameter, the critical separation $\delta^0$, which can be recast into a parametric relationship that represents the consolidation level as a function of a new parameter $C$.

We propose the following relationship between the critical separation $\delta^0$ and the consolidation level, new parameter $C$, as follows:

$$\delta^0 = \Delta x \frac{C}{1-C} \tag{7}$$

where $\Delta x$ is a characteristic length for the granular system (with the same unit of $\delta^0$), $C$ is dimensionless and can change from 0 to 1 with an increasing level of consolidation. Eq. (7) indicates that $\delta^0 = \infty$ when $C = 1$ and $\delta^0 = 0$ when $C = 0$, representing two extreme scenarios of fixed and free grain-grain contacts. Figure 2 shows the traction-separation responses of various consolidation levels. A characteristic length is a global variable that can be estimated for instance from the grain size distribution, or grain-grain contact area distribution, as extracted from the 3D image itself.

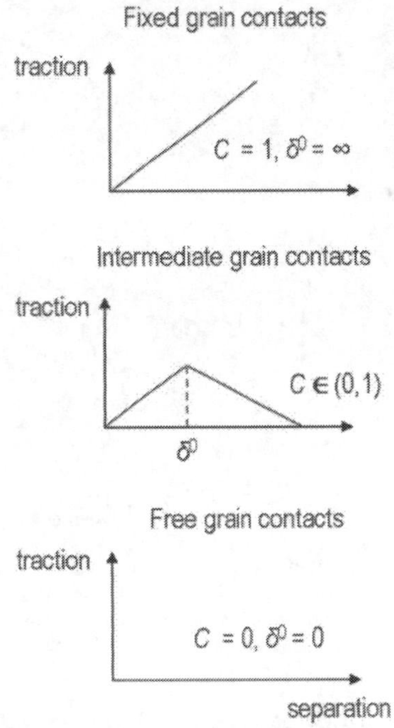


**Fig. 2.** Traction-separation responses of various consolidation levels (quantified by parameter $C$).

This single model can capture contact behaviors for different levels of grain consolidation. With this parametric cohesive contact model, we can simulate a wide range of sedimentary rocks, from unconsolidated (free grain contacts) to intermediate-consolidated rocks to well-consolidated rocks (fixed grain contacts).

### 2.3 Model validation

In order to validate this proposed simplification, first we test the parametric cohesive contact model through a simple scenario of three-grain packing. The grains are extracted from a micro-CT image of Fontainebleau sandstone model [31]. Figure 3a shows the grain mesh used in the FE simulation.

We fix the bottom of the packing and apply either a tensile or compressive stress on the top, i.e. tensile or compressive testing. The characteristic length $\Delta x$ is assumed as 0.5 µm. Figure 3b shows the tensile testing results from simulations of the cohesive contact model with different consolidation levels $C$, and fixed and free grain contacts (no cohesive contacts applied). The results indicate that the parametric cohesive contact model can reproduce the fixed and free grain contact behaviors when $C = 1$ and $C = 0$, respectively. When $C = 0$, the grains will be readily separated by the tensile stress. When $C = 1$, the three grains behave as a solid framework with no grain separation. When $C = 0.5$, the tensile force shows an increase followed by a decrease due to the damage initiates, similar to Figure 2.

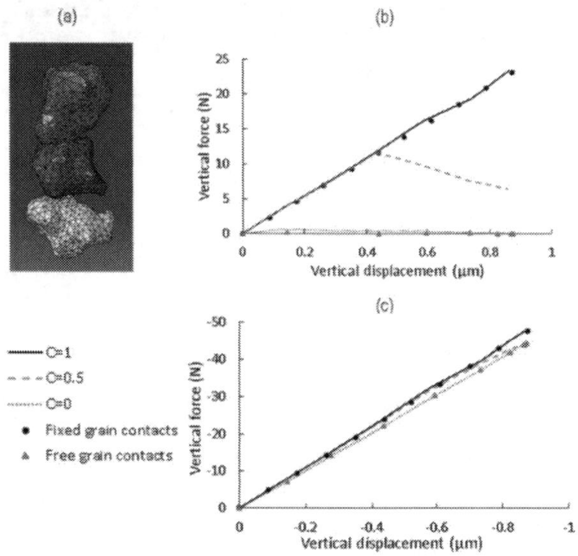


**Fig. 3.** (a) Grain mesh of a three-grain packing. Grains are differentiated by colors. Force-displacement results of (b) tensile testing, and (c) compressive testing. Parametric cohesive contact model can reproduce the fixed and free grain contact behaviors by changing the consolidation level *C*.

Figure 3c shows the simulation results of compressive testing. The compressive forces are with negative values to indicate the compaction. The parametric cohesive contact model can reproduce the fixed and free grain contact behaviors when $C = 1$ and $C = 0$, respectively. When $C = 0.5$, the force-displacement curve exhibits a transition from fixed grain contacts to free grain contacts because only two bonded contacts exist in the packing. The collective force-displacement behavior of a larger number of bonds appears to become linear as shown in the following section.

## 3 Results and discussion

We apply this new parametric cohesive contact model to three rock sample models including a sphere packing, a Grosmont Carbonate and a Fontainebleau Sandstone, each of it exhibiting different grain-grain contact behaviour. We measure the static elastic modulus by modelling the deformation of rock sample, rather than the dynamic elastic modulus (e.g. acoustic velocities). The dynamic elastic modulus is typically higher than the static modulus, and the values diverge significantly in rocks with a low elastic modulus [32,33]. We also present simulation results of a failure test on the Fontainebleau Sandstone model to demonstrate the impact of properly capturing the degree of consolidation on the rock strength and failure pattern.

### 3.1 Free grain contacts

A sphere packing is a very good example of free grain-grain contact behavior, where grain contacts are introduced duplicate nodes and elements that allows grain relocation. Sain (2010) generated the micro-CT image of a sphere packing (as shown in Figure 4a) and obtained its bulk and shear modulus as 1.17 GPa and 1.22 GPa through granular dynamic simulations [34]. The granular dynamic simulations can yield very similar elastic moduli as laboratory experiments on glass beads packs, dry Ottawa sands and dry Galveston sand [34–36]. Andrä et al. (2013) compared their numerical simulations directly based on image voxels (i.e. only allowing fixed grain-grain contacts) with the granular dynamic simulations and found a significant deviation [2]. On the contrary, our micromechanical finite element model allows grain relocation (i.e. free grain-grain contacts) therefore resulting in a very close modulus as the reference data in [34].

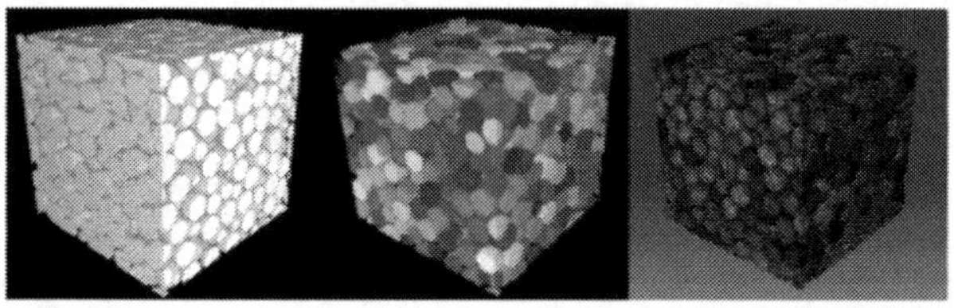

**Fig. 4.** From left to right: the original binary micro-CT image of sphere packing, segmented grains indicated by different colors, grain mesh for the FE solver.

Using this sphere pack model we demonstrate that the parametric cohesive contact model can mimic the free grain contacts. Following reference [2,23], we use a material property of quartz for the sphere packing: density is 2.65 g/cm$^3$, bulk modulus is 37.0 GPa, and shear modulus is 44.0 GPa. Table 1 summarizes the specifications for the sphere pack model.

**Table 1.** Specifications of the sphere packing.

| Rock type | Sphere packing |
|---|---|
| Sample size (voxels) | 398x319x398 |
| Voxel size (µm) | 14 |
| Porosity | 34.3% |
| Number of nodes | 693,635 |
| Number of elements | 3,306,874 |

The parametric cohesive contact model has two parameters: the consolidation level *C* and the stiffness *K*. The consolidation level *C* is equal to 0 for free grain contacts. The stiffness *K* is calibrated to reproduce the mechanical behavior of fixed grain contacts. We perform hydrostatic compression test to measure the bulk modulus and shear test to measure the shear modulus. Table 2 summarizes the results. The parametric cohesive contact model with $C = 0$ yields very similar moduli as the free grain contacts from our previous work [23]. The results are not expected to be identical considering that the numerical simulations are based on different contact mechanisms.

 https://doi.org/10.1051/e3sconf/202336601015

**Table 2.** Bulk and shear moduli of sphere packing.

| | Bulk modulus (GPa) | Shear modulus (GPa) |
|---|---|---|
| Cohesive contact model with $C = 0$ | 1.29 | 1.57 |
| Free grain contacts | 1.34 | 1.62 |
| Difference | 3.7% | 3.1% |

### 3.2 Fixed grain contacts

The Grosmont Carbonate was deposited around 380 million years ago in the late Devonian period in an open shallow limestone marine shelf environment [37]. The formation is composed of limestone and dolomite with minor amount of siltstone and shale. Grosmont Carbonate is a good example of fixed grain-grain contact behaviour, given the fact that Carbonates are typically recrystallized due to diagenetic processes. Therefore, all the grains behave as one solid framework. In this section, we perform the grain segmentation to demonstrate that cohesive contacts can mimic the behaviour of fixed grain contacts when C = 1. Please note that watershed grain segmentation is typically not applicable to complex carbonate structure. Such a segmentation is intended for validating the numerical asymptotic behaviour of the parametric cohesive contact model, but not necessarily indicates a real grain contact behaviour.

**Fig. 5.** From left to right: the original binary micro-CT image of Grosmont Carbonate, segmented grains indicated by different colors, grain mesh for the FE solver (only exterior edges are shown for better visualization).

Table 3 summarizes the specifications for the Grosmont Carbonate model. Following reference [22], we use a material property of 50% calcite and 50% dolomite for the Grosmont Carbonate: density is 2.79 g/cm$^3$, bulk modulus is 81.6 GPa, and shear modulus is 36.7 GPa. The consolidation level $C$ is equal to 1 for fixed grain contacts and the stiffness $K$ is the calibrated parameter. Table 4 shows the comparative results with our previous simulation for this model, and show that the parametric cohesive contact model with $C = 1$ gives very similar elastic moduli as the fixed grain contacts.

**Table 3.** Specifications of the Grosmont Carbonate.

| Rock type | Grosmont Carbonate |
|---|---|
| Sample size (voxels) | 400x400x400 |
| Voxel size (μm) | 2.02 |
| Porosity | 24.7% |
| Number of nodes | 11,416,848 |
| Number of elements | 50,105,905 |

**Table 4.** Bulk and shear moduli of Grosmont Carbonate.

| | Bulk modulus (GPa) | Shear modulus (GPa) |
|---|---|---|
| Cohesive contact model with $C = 1$ | 23.8 | 13.6 |
| Fixed grain contacts | 24.0 | 13.4 |
| Difference | 0.8% | 1.5% |

### 3.3 Various consolidation levels

Many sedimentary rock samples may exhibit different levels of consolidation as sediments are compacted and cemented under various geomechanical and geochemical conditions. In this section we focus on an intermediate consolidated rock, a Fontainebleau Sandstone.

For illustration purposes, we firstly apply the parametric cohesive contact model to a Fontainebleau model [31,38], with a relatively small volume of 100x100x100 voxels, and voxel size of 7.3 μm. Figure 6 shows the binary voxelized image, segmented grains, and conformal grain mesh used in the FE solver. Following reference [22], the material properties assigned are for quartz as in the sphere packing case.

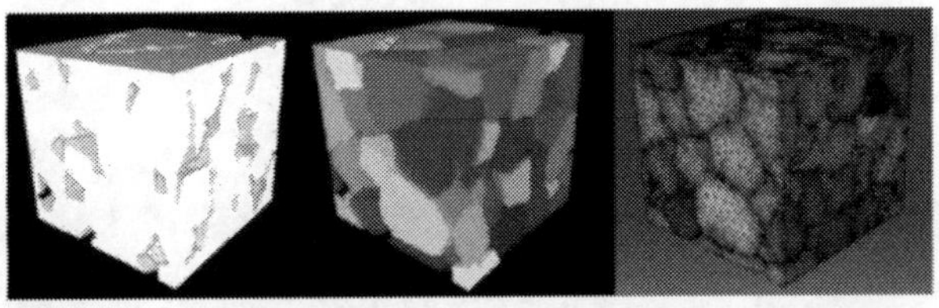

**Fig. 6.** From left to right: the original binary micro-CT image of Fontainebleau model, segmented grains indicated by different colors, and grain mesh for the FE solver.

We perform uniaxial strain tests on this Fontainebleau model, which requires an incremental strain in the vertical direction and zero strains in the horizontal directions [23]. Figure 7a shows stress-strain behaviours of various consolidation level $C$. When $C = 0$ and $C = 1$, the cohesive contact model can reproduce the moduli of free and fixed grain contacts, respectively. The modulus shows an increase as the consolidation level $C$ increases. Figure 7b shows the normalized modulus as a function of the consolidation level. The modulus tends to become asymptotic approaching to the

https://doi.org/10.1051/e3sconf/202336601015

two ends. We find that the data can be fitted well by a logistic function:

$$f(x) = \frac{1}{1+e^{-k(x-x_0)}} \quad (8)$$

where the fitted parameters $k$ and $x_0$ are 10.0 and 0.5, respectively. The purpose of using a fitting function is to minimize the number of simulations needed to create a complete contact model valid for all range of consolidation levels $0 \leq C \leq 1$ behavior. We found that we only need to run two more simulations apart from the two end-member simulations ($C = 0$ and $C = 1$) to completely capture the effect of consolidation level.

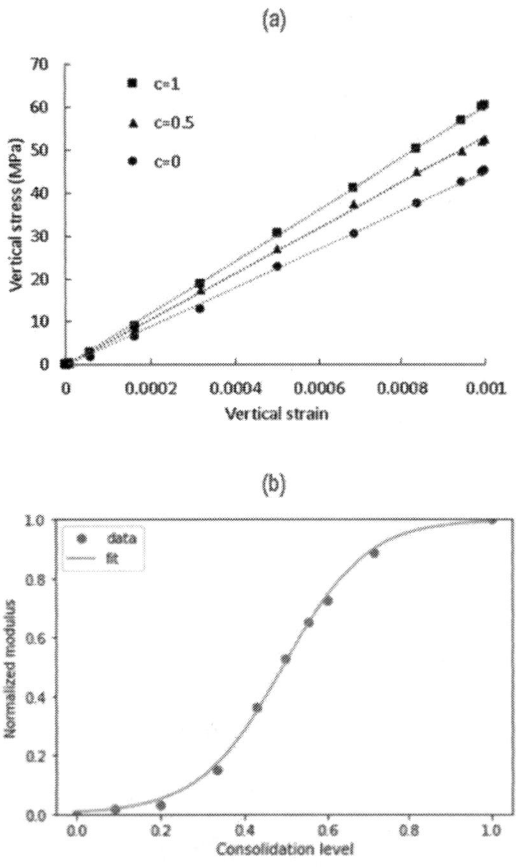


**Fig. 7.** (a) Stress-strain response for various consolidation levels $C$. (b) Normalized modulus as a function of consolidation level. The result can be fitted by a logistic function.

With the knowledge obtained from simulations on the small Fontainebleau model, we further apply the parametric cohesive contact model to a larger Fontainebleau Sandstone micro-CT model of 288x288x300 voxels [2]. The voxel size is 7.5 μm. The sample porosity is 14.7%. The material properties are assigned as quartz. Figure 8 shows the voxelized image, segmented grains, and conformal grain mesh.

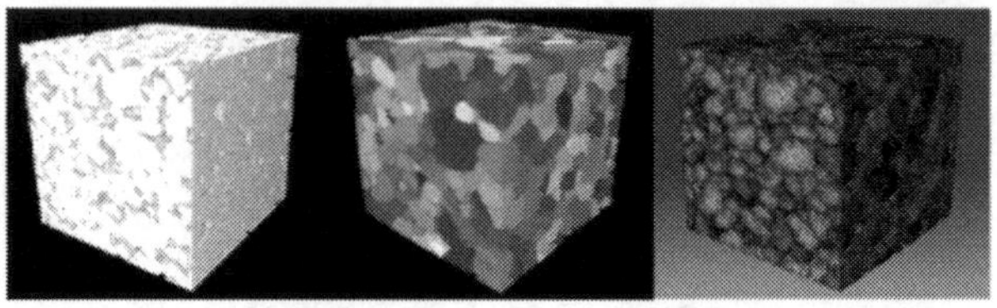

**Fig. 8.** From left to right: the original binary micro-CT image of Fontainebleau sandstone, segmented grains, and grain mesh.

We perform a hydrostatic compression test to measure the bulk modulus. Figure 9a shows the stress-strain behaviours for various consolidation levels $C$. The parametric cohesive contact model can reproduce the moduli of fixed and free grain contacts when $C = 1$ and $C = 0$, respectively. Two intermediate consolidation levels result in a modulus in between. The parameters $k$ and $x_0$ of our logistic fitting function are 15.9 and 0.5, respectively.

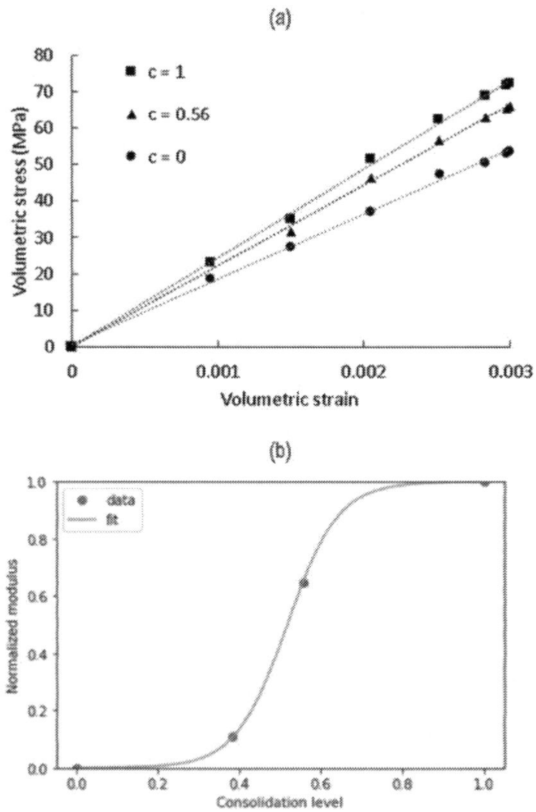


**Fig. 9.** (a) Stress-strain response for various consolidation levels $C$. (b) Normalized modulus as a function of consolidation level. The result can be fitted by a logistic function.

The experimentally measured bulk modulus is ~22.2 GPa, which corresponds to a consolidation level $C$ of 0.56 based on the simulation results in Figure 9 [2]. We further perform a shear test to measure the shear modulus when $C = 0.56$. We obtain a shear modulus of 24.9 GPa, which is also close to the laboratory-measured shear modulus. Therefore, this method can be used to estimate the contact consolidation level by finding the $C$ value so that the experiments match with simulations on a corresponding micro-CT image. The

https://doi.org/10.1051/e3sconf/202336601015

model also allows the users to select the consolidation level as an input parameter based on their knowledge of diagenesis history of sedimentary rock samples. Some ideas have been discussed about relating the consolidation level with the relative size of the grain-grain contact areas relative to the grain sizes. Small relative contact areas should correspond to lower consolidation levels, while larger relative contact areas should correspond to higher consolidation levels. A systematic study with a representative population of rock types and consolidation levels using the new parametric cohesive contact model should provide some validation of this idea.

**Table 5.** Bulk and shear moduli of Fontainebleau Sandstone.

| | **Bulk modulus (GPa)** | **Shear modulus (GPa)** |
|---|---|---|
| Experimental result | 22.2 | 23.5 |
| Numerical result ($C = 0.56$) | 22.2 | 24.9 |
| Difference | 0.0% | 5.9% |

### 3.4 Failure test

The rock failure behaviour manifests the significance to capture well the consolidation level. Uniaxial compression tests are widely used to determine the uniaxial compressive strength (UCS) and deformability of rock samples [39,40]. The testing sample is loaded in the axial direction with no confinement in the radial directions. The ratio of height/diameter of the samples is typically between 2 and 3 [41]. A ratio smaller than 2 will result in high uniaxial compressive strength [42].

In this section, we perform uniaxial compression tests on a cylindrical volume from the Fontainebleau model used in the previous section, at various consolidation levels $C$. All the other model parameters remain invariant. The sample has a height of 500 voxels and a diameter of 250 voxels. The voxel size is 7.3 μm. Figure 10 shows the voxelized image of segmented grains and the corresponding grain mesh used in the FE simulations of uniaxial compression test.

**Fig. 10.** (Left) Voxelized image of segmented grains. (Right) Grain meshing for FE simulation of failure test.

Figure 11 shows the evolution of vertical displacement field as the sample is being compressed in the vertical direction ($C = 0.2$). The Fontainebleau sample is initially intact and finally ruptures with a shear band. Figure 12 shows the vertical stress as a function of vertical strain for various consolidation levels $C$. When $C = 1$, the grains are fused together and the sample behaves elastic with an infinite strength. When $C = 0$, the grains are unconsolidated and the sample fails when the vertical stress reaches the uniaxial compressive strength, with friction as the only remaining force. An intermediate $C$ will result in stress-strain curves in between the two end-member scenarios. When $C$ is relatively small (e.g. $C = 0.2$), the stress-strain curve is very close to the result of $C = 0$ due to the asymptotic nature of Eq. (7), and shown in Figures 7b and 9b. The simulation results show that the rock failure behaviour is conditioned by the consolidation level, which is parameterized into the cohesive contact model in this work.

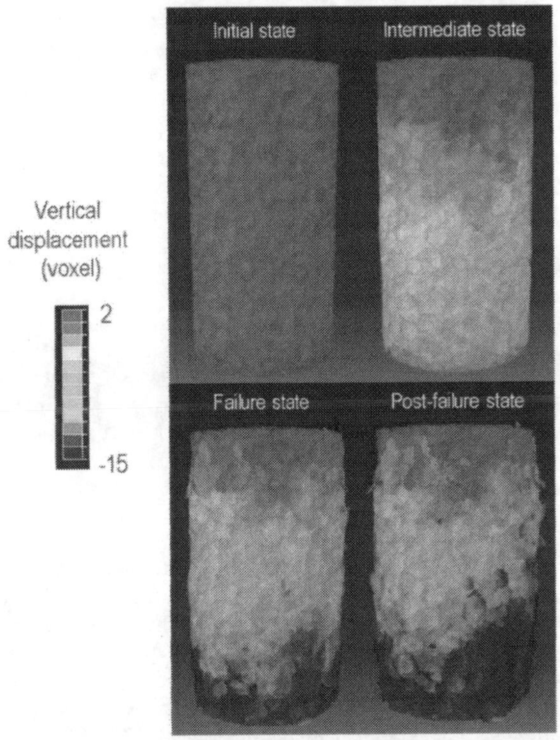


**Fig. 11.** Evolution of vertical displacement field as the Fontainebleau model is compressed in the vertical direction. The grains are shown semi-transparent for a better visualization. The sample ruptures with a shear band at the failure state.

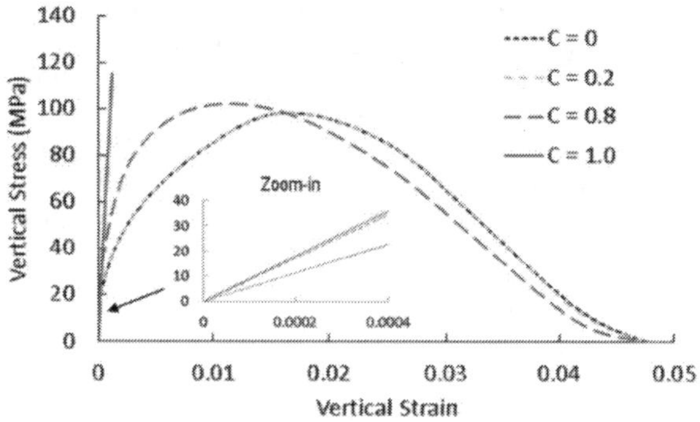


**Fig. 12.** Vertical stress as a function of vertical strain for various consolidation levels *C*. A larger C results in a larger uniaxial strength and elastic modulus.

Table 6 summarizes the Young's modulus and uniaxial compressive strength based on the results shown in Figure 12. A larger consolidation level parameter *C* will result in a larger Young's modulus and uniaxial compressive strength. Han (1987) experimentally measured the Young's modulus of several Fontainebleau sandstones and the results range from 36 to 79 GPa [43]. Baud et al. (2014) experimentally measured the uniaxial compressive strength of several Fontainebleau sandstones and the results were from 44 to 122 MPa [44]. Our numerical simulations for this particular Fontainebleau model yield reasonable values of Young's modulus and uniaxial compressive strength compared to these reference experimental data.

In addition to the consolidation level, the contact friction is another critical parameter to the rock failure process. We adopt the Coulomb friction model which relates the maximum allowable frictional stress across a grain contact to the contact pressure between the grains. The critical frictional stress at which sliding of the contacts starts is proportional to the product of friction coefficient and contact pressure. We assume a friction coefficient of 0.2 throughout this study. Besides, we find that (1) the elastic-plastic transition is smooth compared to many real loading scenarios, and (2) the post-peak behaviour is less catastrophic than real failures. This may be a result of the fact that the current meshing process is not ideal and introduces non-smooth contact boundaries that can add friction to grain contacts. The future work will be improving the grain meshing process to allow smooth contact boundaries that facilitate relative large-strain grain relocation.

**Table 6.** Bulk and shear moduli of Fontainebleau Sandstone.

| | **Young's modulus (GPa)** | **Uniaxial compressive strength (MPa)** |
|---|---|---|
| $C = 0$ | 57.0 | 98.2 |
| $C = 0.2$ | 57.4 | 98.3 |
| $C = 0.8$ | 86.1 | 102.6 |
| $C = 1.0$ | 89.8 | infinity |

## 4 Concluding remarks

In this paper, we extend our previous micromechanics workflow described in [23] to a parametric cohesive contact model. The presented workflow can capture both extreme contact behaviors: free grains and fixed grains, as well as any intermediate degree of grain consolidation. We present a benchmark study on several samples and compare with laboratory-measured elastic moduli to estimate its degree of consolidation. Simulations that do not include the grain contact modeling, tend to overestimate the elastic moduli, which manifests the significance of our contribution to capture well the grain contact behavior. To demonstrate the impact of properly capturing the degree of consolidation on the rock strength and failure pattern, we present results for numerical uniaxial compression testing consistent with measured results. This workflow provides a physics-based solution to the complex grain-grain contact behavior, which complements laboratory core analysis, and can be useful to reveal underlying grain-scale processes governing rock mechanical behavior. The importance of a more complete microscopic understanding of core-scale mechanical properties is ultimately reflected in the quality of the inputs that we use for our up-scaled geomechanics models, as well as enabling a more comprehensive framework for digital rock simulated petrophysical properties at different NCS conditions.

## 5 Nomenclature

*Notation*
$K$ = stiffness
$t$ = traction stress
$\delta$ = separation in normal direction
$D$ = damage variable
$C$ = consolidation level
$\Delta x$ = characteristic length
$k$ , $x_0$ = fitting parameters for logistic function

*Subscripts*
R = normal direction
S = shear direction
m = effective property

*Superscripts*
f = failure condition
0 = critical condition
max = maximum value

## Acknowledgments

We would like to acknowledge the support of Dassault Systèmes to allow the publication of this work.

## References

[1] H. Andrä, N. Combaret, J. Dvorkin, E. Glatt, J. Han, M. Kabel, Y. Keehm, F. Krzikalla, M. Lee, and C. Madonna, *Digital Rock Physics Benchmarks—Part I: Imaging and Segmentation*, Comput. Geosci. **50**, 25 (2013).

[2] H. Andrä, N. Combaret, J. Dvorkin, E. Glatt, J. Han, M. Kabel, Y. Keehm, F. Krzikalla, M. Lee, and C. Madonna, *Digital Rock Physics Benchmarks—Part II: Computing Effective Properties*, Comput. Geosci. **50**, 33 (2013).

[3] R. Xu, B. Crouse, D. M. Freed, A. Fager, G. R. Jerauld, N. Lane, and Q. Sheng, *Continuous vs Discontinuous Capillary Desaturation and Implications for IOR/EOR*, in *Paper SCA2018-066 Presented at at the International Symposium of the Society of Core Analysts Held in Trondheim, Norway* (2018).

[4] B. Crouse, D. M. Freed, N. Koliha, G. Balasubramanian, R. Satti, D. Bale, and S. Zuklic, *A Lattice-Boltzmann Based Method Applied to Digital Rock Characterization of Perforation Tunnel Damage*, in *Paper SCA2016-058 Presented at the International Symposium of the Society of Core Analysts Held in Snow Mass, Colorado, USA* (2016).

[5] M. J. Blunt, *Flow in Porous Media—Pore-Network Models and Multiphase Flow*, Curr. Opin. Colloid Interface Sci. **6**, 197 (2001).

[6] J. Dvorkin, M. Armbruster, C. Baldwin, Q. Fang, N. Derzhi, C. Gomez, B. Nur, and A. Nur, *The Future of Rock Physics: Computational Methods vs. Lab Testing*, First Break **26**, (2008).

[7] C. H. Arns, M. A. Knackstedt, M. V. Pinczewski, and W. B. Lindquist, *Accurate Estimation of Transport Properties from Microtomographic Images*, Geophys. Res. Lett. **28**, 3361 (2001).

[8] M. Andrew, H. Menke, M. J. Blunt, and B. Bijeljic, *The Imaging of Dynamic Multiphase Fluid Flow Using Synchrotron-Based X-Ray Microtomography at Reservoir Conditions*, Transp. Porous Media **110**, 1 (2015).

[9] A. K. Turner, F. H. Kim, D. Penumadu, and E. B. Herbold, *Meso-Scale Framework for Modeling Granular Material Using Computed Tomography*, Comput. Geotech. **76**, 140 (2016).

[10] C. H. Arns, M. A. Knackstedt, W. V. Pinczewski, and E. J. Garboczi, *Computation of Linear Elastic Properties from Microtomographic Images: Methodology and Agreement between Theory and Experiment*, Geophysics **67**, 1396 (2002).

[11] T. W. J. De Geus, J. Vondřejc, J. Zeman, R. H. J. Peerlings, and M. G. D. Geers, *Finite Strain FFT-Based Non-Linear Solvers Made Simple*, Comput. Methods Appl. Mech. Eng. **318**, 412 (2017).

[12] M. M. Hossain, J. Arns, Z. Liang, Z. Chen, and C. H. Arns, *Humidity Effects on Effective Elastic Properties of Rock: An Integrated Experimental and Numerical Study*, J. Geophys. Res. Solid Earth **124**, 7771 (2019).

[13] M. A. Knackstedt, C. H. Arns, and W. V. Pinczewski, *Velocity-Porosity Relationships, 1: Accurate Velocity Model for Clean Consolidated Sandstones*, Geophysics **68**, 1822 (2003).

[14] B. Crouse, S. Gill, J. Bautista, G. Balasubramanian, and D. Freed, *Digital Prediction of Porosity and Permeability Variation of Unconsolidated Sands Caused by Over-Burden Pressure*, in *NAFEMS World Congress* (2019).

[15] A. B. Andhumoudine, X. Nie, Q. Zhou, J. Yu, O. I. Kane, L. Jin, and R. R. Djaroun, *Investigation of Coal Elastic Properties Based on Digital Core Technology and Finite Element Method*, Adv. Geo-Energy Res. **5**, 53 (2021).

[16] S. Boggs Jr, *Principles of Sedimentology and Stratigraphy* (Pearson Education, 2014).

[17] E. J. Garboczi and J. G. Berryman, *Elastic Moduli of a Material Containing Composite Inclusions: Effective Medium Theory and Finite Element Computations*, Mech. Mater. **33**, 455 (2001).

[18] S. Meille and E. J. Garboczi, *Linear Elastic Properties of 2D and 3D Models of Porous Materials Made from Elongated Objects*, Model. Simul. Mater. Sci. Eng. **9**, 371 (2001).

[19] E. H. Saenger, *Numerical Methods to Determine Effective Elastic Properties*, Int. J. Eng. Sci. **46**, 598 (2008).

[20] C. H. Arns, M. Madadi, A. P. Sheppard, and M. A. Knackstedt, *Linear Elastic Properties of Granular Rocks Derived from X-Ray-CT Images*, in *SEG Technical Program Expanded Abstracts 2007* (Society of Exploration Geophysicists, 2007), pp. 1711–1715.

[21] N. Saxena, R. Hofmann, A. Hows, E. H. Saenger, L. Duranti, J. Stefani, A. Wiegmann, A. Kerimov, and M. Kabel, *Rock Compressibility from Microcomputed Tomography Images: Controls on Digital Rock Simulations*, Geophysics **84**, WA127 (2019).

[22] J. Fredrich, N. Lane, and J. Toms, *Image-Based Direct Numerical Simulation of Petrophysical Properties under Simulated Stress and Strain Conditions*, U.S. Patent 9,348,056 (24 May 2016).

[23] Z. Sun, R. Salazar-Tio, L. Duranti, B. Crouse, A. Fager, and G. Balasubramanian, *Prediction of Rock Elastic Moduli Based on a Micromechanical Finite Element Model*, Comput. Geotech. **135**, 104149 (2021).

[24] D. Legland, I. Arganda-Carreras, and P. Andrey, *MorphoLibJ: Integrated Library and Plugins for Mathematical Morphology with ImageJ*, Bioinformatics **32**, 3532 (2016).

[25] S. Qin, R. McLendon, V. Oancea, and A. M. Beese, *Micromechanics of Multiaxial Plasticity of DP600: Experiments and Microstructural Deformation Modeling*, Mater. Sci. Eng. A **721**, 168 (2018).

[26] Abaqus, *Abaqus 6.10 Online Documentation*, Abaqus User Subroutines Ref. Man. (2010).

[27] M. Obermayr, K. Dressler, C. Vrettos, and P. Eberhard, *A Bonded-Particle Model for Cemented Sand*, Comput. Geotech. **49**, 299 (2013).

[28] D. O. Potyondy and P. A. Cundall, *A Bonded-Particle Model for Rock*, Int. J. Rock Mech. Min. Sci. **41**, 1329 (2004).

[29] Z. Sun, D. N. Espinoza, and M. T. Balhoff, *Discrete Element Modeling of Indentation Tests to Investigate Mechanisms of CO2 -Related Chemo-Mechanical Rock Alteration*, J. Geophys. Res. Solid Earth **121**, 7867 (2016).

[30] H. Harkness, G. Ang, P. Vijalapura, and D. Cojocaru, *Contact Stress Accuracy with Robust and Broadly Applicable Implicit Contact Algorithm*, in *NWC (Nafems World Congress)* (2011).

[31] F. D. E. Latief, B. Biswal, U. Fauzi, and R. Hilfer, *Continuum Reconstruction of the Pore Scale Microstructure for Fontainebleau Sandstone*, Phys. A Stat. Mech. Its Appl. **389**, 1607 (2010).

[32] V. Brotons, R. Tomás, S. Ivorra, A. Grediaga, J. Martínez-Martínez, D. Benavente, and M. Gómez-Heras, *Improved Correlation between the Static and Dynamic Elastic Modulus of Different Types of Rocks*, Mater. Struct. **49**, 3021 (2016).

[33] J. M. Ide, *Comparison of Statically and Dynamically Determined Young's Modulus of Rocks*, Proc. Natl. Acad. Sci. U. S. A. **22**, 81 (1936).

[34] R. Sain, *Numerical Simulation of Pore-Scale Heterogeneity and Its Effects on Elastic, Electrical and Transport Properties* (Stanford University, 2010).

[35] S. N. Domenico, *Elastic Properties of Unconsolidated Porous Sand Reservoirs*, Geophysics **42**, 1339 (1977).

[36] M. A. Zimmer, *Seismic Velocities in Unconsolidated Sands: Measurements of Pressure, Sorting, and Compaction Effects* (Stanford University, 2004).

[37] B. E. Buschkuehle, F. J. Hein, and M. Grobe, *An Overview of the Geology of the Upper Devonian Grosmont Carbonate Bitumen Deposit, Northern Alberta, Canada*, Nat. Resour. Res. **16**, 3 (2007).

[38] R. Hilfer and T. Zauner, *High-Precision Synthetic Computed Tomography of Reconstructed Porous Media*, Phys. Rev. E **84**, 62301 (2011).

[39] C. E. Fairhurst and J. A. Hudson, *Draft ISRM Suggested Method for the Complete Stress-Strain Curve for Intact Rock in Uniaxial Compression*, Int. J. Rock Mech. Min. Sci. **36**, 279 (1999).

[40] Z. T. Bieniawski and M. J. Bernede, *Suggested Methods for Determining the Uniaxial Compressive Strength and Deformability of Rock Materials: Part 1. Suggested Method for Determining Deformability of Rock Materials in Uniaxial Compression*, in *International Journal of Rock Mechanics and Mining Sciences & Geomechanics Abstracts*, Vol. 16 (Elsevier, 1979), pp. 138–140.

[41] E. Tuncay and N. Hasancebi, *The Effect of Length to Diameter Ratio of Test Specimens on the Uniaxial Compressive Strength of Rock*, Bull. Eng. Geol. Environ. **68**, 491 (2009).

[42] A. B. Hawkins, *Aspects of Rock Strength*, Bull. Eng. Geol. Environ. **57**, 17 (1998).

[43] D.-H. Han, Effects of Porosity and Clay Content on Acoustic Properties of Sandstones and Unconsolidated Sediments, Stanford Univ., CA (USA), 1987.

[44] P. Baud, T. Wong, and W. Zhu, *Effects of Porosity and Crack Density on the Compressive Strength of Rocks*, Int. J. Rock Mech. Min. Sci. **67**, 202 (2014).

# Rate-Controlled Mercury Injection Experiments to Characterize Pore Space Geometry of Berea Sandstone

*Walid Mohamed* Mahmud[1]

[1]Department of Petroleum Engineering, Faculty of Engineering, University of Tripoli, Tripoli, Libya

**Abstract.** Interpretation of the relationship between heterogeneity and the flow in porous media is very important in increasing the recovery factor for an oil or gas reservoir. Capillarity for instance, controls fluids static distribution in a reservoir prior to production and remaining hydrocarbons after production commences. Therefore, capillary pressure data are used by petroleum engineers, geologists, and petrophysicists to evaluate production characteristics of petroleum accumulations. Conventional pressure-controlled mercury porosimetry produces an overall capillary pressure curve and pore throat size distribution data that provide little information about the porous medium structure and pore geometry. The present study provides information on three capillary pressure curves obtained from rate-controlled mercury injection porosimetry; one describes the larger pore spaces or pore bodies of a rock, another describes the smaller pores or pore throats that connect the larger pores, and a final curve which corresponds to the overall capillary pressure curve obtained from the conventional pressure-controlled mercury injection. An experimental constant-rate mercury injection apparatus was constructed that consists of a piston displacement pump, a computer controlled stepper motor drive and a core sample cell designed to minimize dead volume. The apparatus was placed in a glass chamber and subjected to an air bath to maintain a constant temperature of 27◦C throughout the experiments. Then constant rate mercury injection experiments were performed on three Berea Sandstone core plugs. Results show that volume-controlled or rate-controlled porosimetry provides considerably more detailed data and information on heterogeneity and the statistical nature of pore space structure than the conventional pressure-controlled porosimetry as pressure fluctuations with time reveal menisci locations in pore bodies and pore throats. Moreover, pore size distributions based on volume-accessed pores and pore radii were obtained from the pressure versus saturation relationship.

## 1 Introduction

The main principle of capillary pressure measurements by mercury porosimetry is by the incremental increase of pressure while monitoring the increase in mercury saturation being injected into a porous core plug. Pressure-controlled porosimetry method, or the conventional technique, measures the capillary pressure curves in a way that it is possible to obtain one capillary pressure curve from different distributions of pore systems. Yuan and Swanson [1] introduced a pioneering method to Wood's metal porosimetry at which mercury was injected into a core plug at a very slow and constant flow rate. It was possible, through this method, to conceptually partition pore space into pore throats; which they named rison, pore bodies or subison besides the standard capillary pressure curve identical to that obtained from the pressure-controlled mercury porosimetry. Pressure-controlled mercury intrusion method may indirectly measure significantly narrow throats but does not detect larger pores that rate-controlled method detects as it quantitatively separates and characterizes throat and pore size distributions. Therefore, more accurate information can be obtained, from the rate-controlled method, about pore space characterization and micro-pore structure characteristic parameters. Moreover, rate-controlled mercury porosimetry yields information and data on the sizes and volumes of pore bodies or nodes and pore throats or links that can be utilized to calibrate network models that, in turn, used to estimate relative permeability, capillary pressure and residual saturation. Network models may also be used to validate correlated heterogeneity in rocks by comparing measured curves obtained from rate-controlled mercury injection and those predicted from network models.
Gao et al. [2] used rate-controlled or constant-rate mercury intrusion technique on tight core plugs to determine the size of pore and throat parameters. Zhang et al. [3] utilized rate-controlled mercury injection on tight sandstone core plugs to investigate pore-throat structure characteristics. Smith and Chatzis [4] developed a method of measuring the breakthrough capillary pressure involving a constant-rate injection technique. Zhang et al. [5] analyzed pore-throat structure effect of tight rocks on the seepage and storage capacity employing, among other techniques, rate-controlled mercury injection technique. Zhang et al. [6] performed constant-rate mercury injection experimental measurements in low permeability reservoirs to investigate poor correlation while calculating permeability. Wang et al. [7] recently studied the pore throat structure characteristics of sandy conglomerate and tight sandstone using constant-rate mercury injection along with the conventional constant-pressure mercury injection and micro CT scanning data. It was shown that substantial regions of the rock remain uninvaded at low and moderate non-wetting phase saturations [8]. However, the uninvaded regions then become filled as the non-wetting phase saturation increases indicating that correlated heterogeneity might exist at the core scale. Moreover, visual studies on sandstone pore casts showed correlations between pore body sizes and adjoining pore throats [9]. Therefore, it is important to consider and model spatial and local correlations in network models [10]. Mercury porosimetry was used to study the effect of pore-throat correlations; a widening drainage capillary pressure curves were noticed on correlated networks [11].
Rate-controlled and pressure controlled porosimetries and scanning electron microscopy were utilized to study pore size distribution of bentonite [12]. The combined use of rate-controlled and pressure-controlled porosimetries was found the best for obtaining the bentonite full-scale pore size distribution while employing each method separately has its limitation.
Yao and Liu [13] compared between the conventional pressure-controlled mercury injection porosimetry, rate-controlled mercury injection porosimetry, micro focus computerized tomography and low-field nuclear magnetic resonance spectral analysis in measuring coals' pore size distribution. Pressure-controlled mercury injection method was found to have limitations such as coal sample destruction, due to high-pressure, and therefrom inaccurate estimations of coal pore size distribution, pore volume and porosity. Rate-controlled mercury injection method, on the other hand, was found effective and provided very detailed pore size distribution information of macropores but

© The Authors, published by EDP Sciences. This is an open access article distributed under the terms of the Creative Commons Attribution License 4.0 (http://creativecommons.org/licenses/by/4.0/).

insufficient information on pore size distribution of mesopores.
Adams [14] incorporated nuclear magnetic resonance and conducted mercury injection capillary pressure measurements on carbonate and sandstone core plugs to determine large, medium, and small pore sizes, and their associated pore shapes. He calculated mercury injection capillary pressure curves and distributions of pore-throat aperture and compared the results with other techniques. It was possible to discriminate between pores of different genetic origin from one another by using a combination of pore size and pore geometry. Moreover, median pore-throat apertures were found strongly correlated to permeability.
In the present study, rate-controlled mercury porosimetry equipment was built and utilized to characterize pore morphologies by quantitatively dividing pore space into pore bodies and pore throats using Berea sandstone core plugs. Constant-rate mercury injection experiments yielded important structural parameters that includes throat and pore bodies volumes, pores radii distribution, independent pore bodies, throats and total capillary pressure curves. Pore bodies and throats might be orderly or disorderly distributed. Therefore, correlations might exist between the sizes and shapes of throats and pores over extended field-scale porous media or only over a few pores [15]. Data obtained may also be utilized in network modeling studies to identify the nature of spatial and pore throat and pore body correlations and heterogeneity in rocks.

## 2 Core plugs preparation and properties

Core plugs were sampled from a full diameter Berea sandstone core that has a gas permeability of 1.31 D. Core plugs were cylindrical in shape with similar dimensions and no visual irregular external surfaces. Porosity was measured separately for each core plug using helium gas expansion porosimeter based on the principle of Boyle's Law. A sealed reference chamber in the instrument is filled with helium gas at ambient temperature to a pressure of 100 psig. The core plug is then placed in another sealed chamber, connected to the reference chamber by a two-way valve. This valve when opened allows the gas in the reference chamber to expand into the combined volume of the two chambers. From Boyle's Law, the volume of the core plug chamber can be calculated when the volume of the reference chamber, the initial pressure and the final pressure are known. The porosity and the grain density were then calculated by determining the bulk volume, and the weight of the core plug. Pore volume of the core plug equals the bulk volume minus the grain volume. Porosity equals the core plug pore volume divided by the bulk volume. Distribution of core plugs porosities ranged from 28.2 to 30.7%. The permeability and porosity values indicate good reservoir rock quality.

## 3 Experimental equipment and methodology

The experimental equipment is shown schematically in Figure 1. Unlike the conventional mercury intrusion, quasi-static capillary equilibrium injection was achieved by injecting mercury at a very low and constant flow rate of 0.00000002 cc/sec. The experimental equipment consists of a stepper motor drive controlled by computer, a cell of 11 cc capacity to contain core plug designed to minimize dead volume, a piston displacement pump that is vertically mounted to eliminate air trapping. The cell is connected at the top through a vacuum port. Each core plug was preprocessed/placed under vacuum before mercury injection commences and the core plug becomes fully surrounded and submersed in mercury. Vacuum of fifty microns of mercury was applied for eight hours before conducting each experiment. A pressure transducer measures the pressure in the cell that is powered by a twelve-volt battery to minimize signal noise. Pressure transducer's range is from 0 to 100 psi with a resolution of +/- 0.00005 psi. The equipment was put in a glass built chamber, shown in Figure 2, at constant temperature of 27 °C. In order to maintain and control constant temperature throughout the experiments and to eliminate any possible thermal effects on pressure fluctuations, electric fan heater was placed inside the chamber that is visible in Figure 2. Volume measurement resolution was 0.000031 cc equivalent to one shaft encoder count. A program was generated to read the Hewlett Packard voltmeter and to control the stepper motor that drives the mercury injection pump. The outcome of the program is readings of the cumulative mercury injected volume, pressure and time. Pore volume invaded by mercury is then obtained from the cumulative pore volume filled by mercury and the overall pore volume of the core plug.

Figure 3 shows the overall, full scale, capillary pressure versus mercury pore volume injected saturation of the rate-controlled mercury injection experiments of the present work. Figures 4 and 5 show magnified sections of the data collected during the experiment; pore volume injected represents volume of mercury injected divided by ambient helium pore volume. Constant-rate mercury injection enables the division of the overall mercury injection capillary pressure curve into pores and throats mercury injection capillary pressure curves that are utilized to characterize throat and pore size distributions [1]. It is also possible analyzing throats and pores parameters including throats and pores radii distribution and the ratio between them. Figure 5 shows the capillary pressure fluctuations and response corresponding to filling of interconnected pores. A sudden 'primary' drop in capillary pressure as mercury meniscus falls from a narrow pore throat into a wide pore body. Immediately afterwards, the capillary pressure gradually increases as mercury commences filling adjoining interconnected pore throats. This ensembles a drainage displacement at which non-wetting fluid, mercury in the present study, displaces wetting fluid, air, in the pore space of the rock. The instant pore-scale displacement events that occur are also referred to as Haines jumps [16].

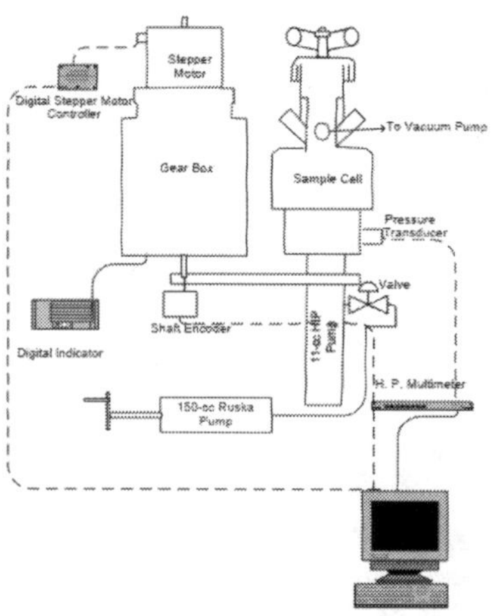


**Figure 1**: Schematic of experimental equipment.

**Figure 2**: Experimental equipment inside the glass chamber.

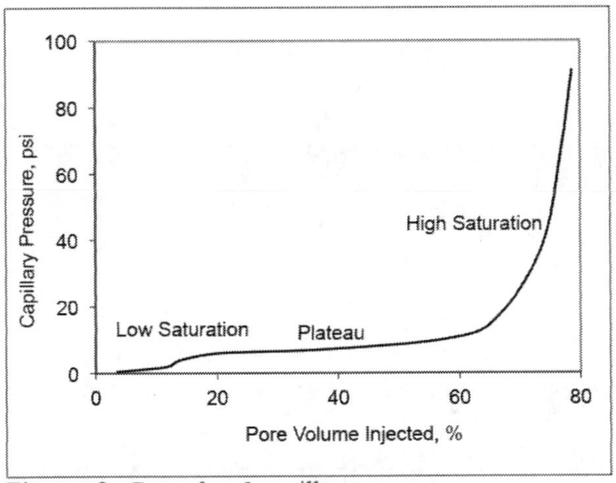


**Figure 3**: Core plug 3 capillary pressure versus mercury pore volume injected saturation.

The capillary pressure may increase to a level that has already been attained, shown as secondary in Figure 5, then it drops before increasing to a new level shown as tertiary in Figure 5. Mercury volume inserted from the primary capillary pressure drop commences until the tertiary pressure increase is reached equals the pore volume of interconnected pore bodies or system that has a threshold pressure less than or equals the pressure preceding the primary pressure drop. Compressibility tests are recommended for poor consolidated or compressible low permeability reservoir rocks where pores not yet filled by mercury might be compressed with increasing net overburden pressure as function of time. However, these tests are not relevant for the present high permeability and well-cemented Berea sandstone core plugs.

Figure 6 shows the basic principle of constant-rate mercury injection as the injected mercury pressure fluctuates with the porous media pore structure. Thus, detailed information on pore throats and pore bodies can be distinguished on the pressure fluctuations [17].

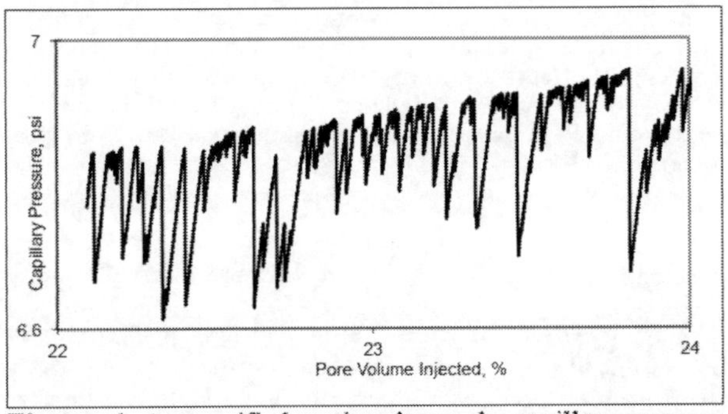


**Figure 4**: A magnified section shown the capillary pressure versus mercury volume injected divided by the ambient core plug pore volume.

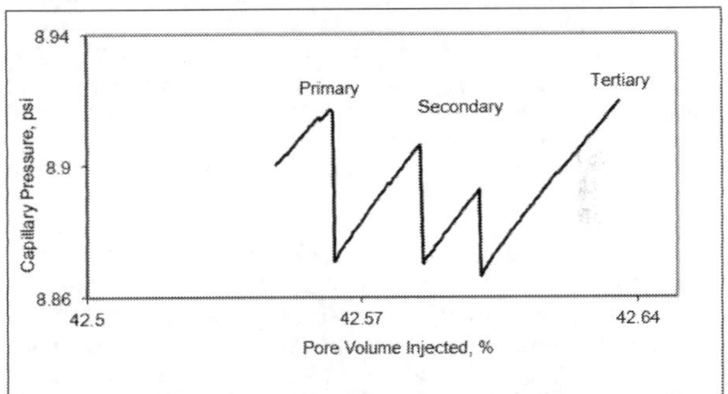


**Figure 5**: A magnified section from one of the experiments showing primary, secondary and tertiary pressure drops in an interconnected pore system.

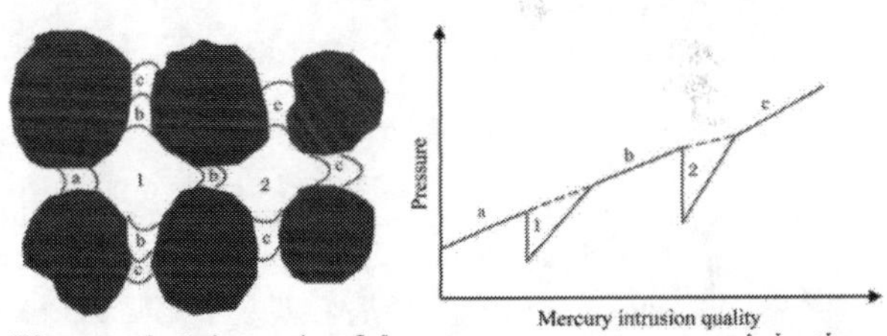


**Figure 6**: Schematic of the constant rate mercury injection process: a, b and c are throats, 1 and 2 are pores [17].

Laplace's equation is used to determine a pore throat effective entry radius:

$$r = \frac{2\,\sigma \cos \theta}{P_c} \qquad (1)$$

Where σ is the interfacial tension, θ is the contact angle, Pc is the capillary pressure obtained from the measured pressure-volume trace and r is the effective radius of the pore throat. For the calculation of the effective pore throat radius, the mercury/air contact angle equals 140· and the interfacial tension equals 480 mN/m invariant. These values were considered to be constant during the entire test because of the low rate of mercury injection applied during the experiments. As the capillary pressure increases to previously unattained levels, mercury meniscus advances into uninvaded pore throats. This analysis, therefore, resolves the rock pore space into pore throats and pore

 https://doi.org/10.1051/e3sconf/202336601016

bodies, with each throat and pore is characterized by an entry volume and pressure.

Laplace's equation assumes straight capillary tubes, while in principle the rock pore network is model as a bundle of interconnected capillary tubes. Moreover, capillary hysteresis effects are not taken into account using this model. A program was developed to determine the number and volume of pore bodies. Volume of a pore body is determined by:

$$V_{pb} = \Delta \mathrm{P} \left(\frac{\mathrm{dV}}{\mathrm{dP}}\right) \qquad (2)$$

Where $\left(\frac{\mathrm{dV}}{\mathrm{dP}}\right)$ is the rate constant determined from the pressure and time as shown in Figure 7. However, the above formula was modified to accommodate for the volumes of multiple smaller pores existing within an interconnected pore system. Volume of such system has previously been calculated by separating the pressure-volume fluctuations into compartments [18]. However, this method underestimates the invaded volume in a primary pore and overestimates the invaded volume in subsequent pores within the same system of interconnected pores [19]. Alternatively, multi-pore system was separated by extending the pressure-volume fluctuations vertically to the threshold pressure of the primary pressure drop assuming that the rate constant remains unchanged throughout the pores at a given pressure meaning $(dV/dP)_1=(dV/dP)_2$ [19]. This might be true in an individual interconnected pore system, however, it was observed in the present study that the rate constant varies from one interconnected pore system to another. Therefore, average rate constant was considered for every individual interconnected pore system. For instance, the difference between rate constants, from one of the experiments, in two neighboring interconnected pore systems was 0.000134 cc/psi. The overall pore systems volume is determined by Equation 3 and as shown in Figure 8:

$$TV_{pb} = \left(\frac{\mathrm{dV}}{\mathrm{dP}}\right)_1 (\Delta \mathrm{P}_{1r_1} + \Delta \mathrm{P}_{2r_1} + \cdots) \left(\frac{\mathrm{dV}}{\mathrm{dP}}\right)_2 (\Delta \mathrm{P}_{1r_2} + \cdots) + \left(\frac{\mathrm{dV}}{\mathrm{dP}}\right)_i (\Delta \mathrm{P}_{1r_i} + \cdots) \qquad (3)$$

Where $TV_{pb}$ is the total volume of the pore systems in the core plug, $(dV/dP)_1$ is the local rate constant in region number one, and $\Delta P_{1r_1}$ is the pressure drop in region one, and so on.

Pore space morphology, rock fluid properties, local pore geometry and topology are factors that determine the distribution and volume of residual oil ganglia in the pore space or in other words pore and core scales heterogeneities [11, 20, 21]. In a drainage displacement, as mercury, being the non-wetting phase is injected into a rock, it preferentially occupies larger pores because of their low entry threshold pressure then it occupies smaller pores and throats. The large pores with low capillarity, therefore, would most likely trap the non-wetting phase in an imbibition displacement while the smaller throats are filled by the wetting phase. Because mercury is a strongly non-wetting phase, it can be assumed that snap-off disconnects the continuous filament of mercury and traps it in all pores during reduction of capillary pressure as the experiments is terminated [22].

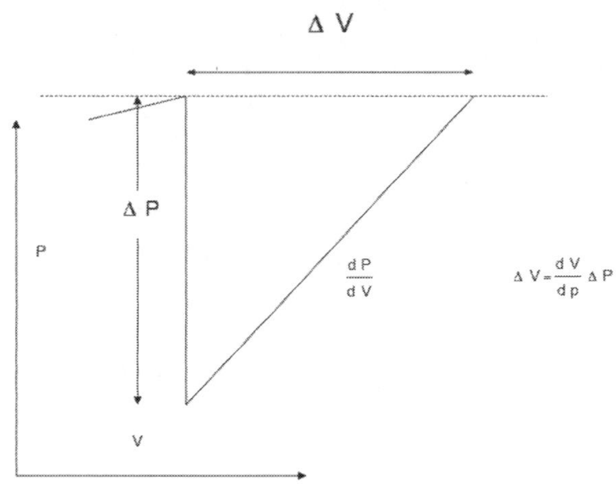


**Figure 7**: Rate constant determined from pressure and time.

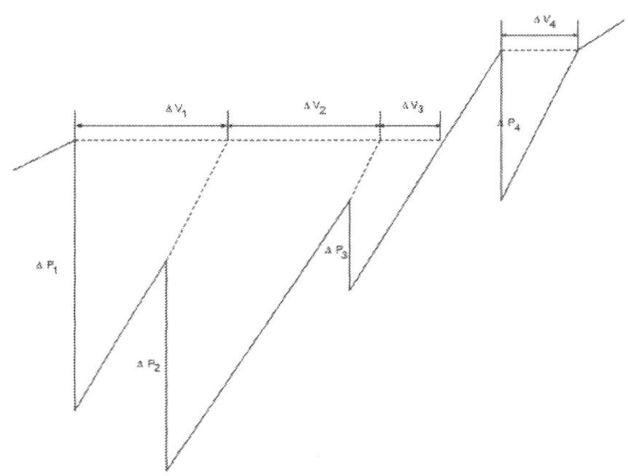


**Figure 8**: Volume determination of interconnected pore systems.

The cumulative pore bodies' saturation provides good estimates of the residual mercury/air saturation assuming that all mercury is trapped in the pore bodies upon the experiment completion and pressure reduction [18]. Pressure was gradually decreased down to atmospheric pressure of 14.7 psi and sufficient time was given for capillary pressure-mercury saturation equilibrium. Thus, an imbibition cycle displacement takes place as the wetting phase (air) displaces the non-wetting phase (mercury). Therefore, cumulative volume of pores represents the final non-wetting phase saturation $S_{nwr}$ that can be determined by:

$$\mathrm{Snwr} = \sum_{i=1}^{N} \frac{V_{rs}}{V_p} \qquad (4)$$

N is the number of pores, $V_{ri}$ is the volume of i pore and $V_p$ is the core plug pore volume. Pore geometry affects the residual non-wetting phase saturation, which is the most important factor for the estimation or reservoir ultimate recovery and in any EOR operation. The value of residual non-wetting phase saturation by mercury entrapment might not be representative of the reservoir residual fluids saturation value. Mainly due to the fact that reservoir fluids wettability behavior is not properly reproduced during the experiment, as both mercury and air are by definition non-wetting phases.

## 4 Results and discussion

Figures 9 to 11 show the total rate-controlled mercury injection capillary pressure curve along with the individual pores and throats capillary pressure curves. In Figure 9, initial steep increase in capillary pressure due to entry effect where potential surface, irregular and truncated pores are initially filled. Then at mercury saturation of approximately 15.0 to 56.0%, overlapping takes place between total and pores capillary pressure curves. Mercury penetrates and fills large number of interconnected pore bodies as the capillary pressure increases to a threshold displacement pressure of around 2 psi, which renders a plateau in the total and pore capillary pressure curves followed by sharp increase in the capillary pressure with small increase in mercury saturation. There is slight increase in the capillary pressure in the plateau region with significant increase in mercury saturation. The total number of pores, occupied by mercury, for core plug 1 is 24138 occupying volume of 0.2234 cc. The throat capillary pressure curve, on the other hand, gradually increases as mercury is injected into the core plug. By the end of the experiment, the total mercury injection saturations of the pores and throats are 60.4 and 27.5%, respectively. Therefore, higher pores saturation than the throats saturation indicates that pores mostly control the overall displacement. The pores saturation of 60.4% also represents the residual non-wetting phase saturation or the final mercury entrapment saturation as detailed in section 3. Some throats might also trap mercury as the capillary pressure is reduced to atmospheric conditions, however, volume of mercury trapped in some throats is assumed to be negligible.

The pore throats of the well-sorted and highly permeable Berea sandstone rock might have similar sizes to the pore bodies. Moreover, as shown in Figure 4, the highly sensitive pressure transducer employed in the present study detects the smallest pressure fluctuations which indicates that distinguishing between pore throats and pore bodies is considerably arbitrary. Thus, similarly the distinction between pore throat capillary pressure and pore body capillary pressure curves is also considerably arbitrary. This indicates that pore throats still and also control the displacement and have an important portion of the Berea sandstone pore space. For this reason, the distinction between pore throats and pore bodies might be considered arbitrary.

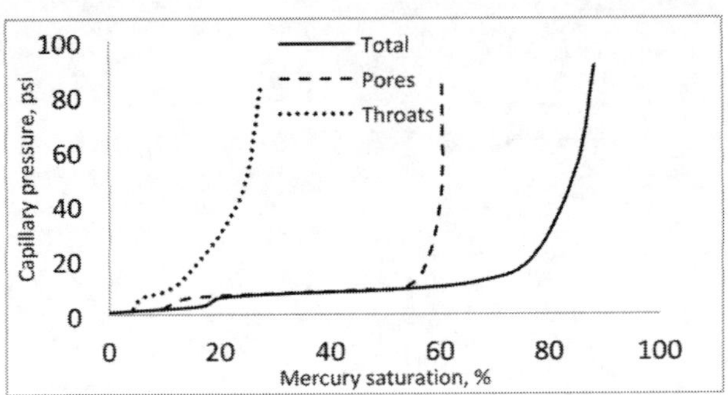


**Figure 9**: Core plug 1 mercury injection capillary pressure curves of the total, pores and throats.

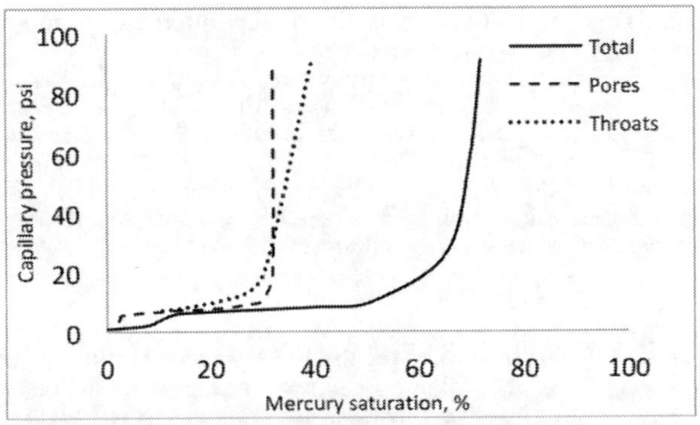


**Figure 10**: Core plug 2 mercury injection capillary pressure curves of the total, pores and throats.

Core plugs 2 and 3 shown in Figures 10 and 11, respectively have similar threshold displacement pressure of around 14 psi. The throat capillary pressure curves are higher than the pores capillary pressure curves. By the end of the experiment on core plug 2, the total mercury injection saturations of the pores and throats are 32.2 and 39.5%, respectively. The pores saturation of 32.2% also represents the residual mercury saturation.

By the end of the experiment on core plug 3, the total mercury injection saturations of the pores and throats are 32.3 and 46.3%, respectively. The pores saturation of 32.3% also represents the residual mercury saturation. The lower pores saturation than the throats saturation indicates that the throats mainly control the total mercury injection saturations of core plugs 2 and 3.

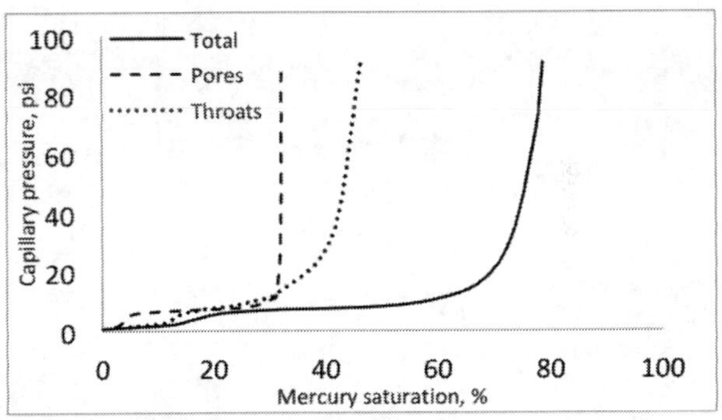


**Figure 11**: Core plug 3 mercury injection capillary pressure curves of the total, pores and throats.

The total number of pores, occupied by mercury, in core plug 2 is 21980 occupying volume of 0.2083 cc. The total number of pores, occupied by mercury, in core plug 3 is 21537 occupying volume of 0.2095 cc. The threshold displacement pressure of core plugs 2 and 3 is around 7 psi. As the threshold displacement pressure decreases, a lengthy overlap between the pore capillary pressure curve and the total capillary pressure curve exists while the overlap between the throat capillary pressure curve and the total capillary pressure curve is somewhat more evident as the threshold displacement pressure increases. Similar constant-rate mercury injection findings were reported for tight sandstone [3]. Shape of pore and throat capillary pressure curves variations as well as pore and throat occupancies between core plug 1 and core plugs 2 and 3, despite having the same lithology, textual properties and permeability, indicates that there is a cut-off limit to both spatial and local pore-throat correlations within the full diameter Berea sandstone core. Core plug 1 can be different or more heterogeneous at the pore network scale than core plugs 2 and 3. Thus, correlated heterogeneity at the pore scale may not extend to the sample size. Moreover, short-range correlations or local pore-throat correlations between pore bodies and the adjoining pore throats might not persist throughout the core plug. Therefore, large pore bodies do not necessarily adjoin large pore throats and small pore bodies do not necessarily adjoin small pore throats.

The volume and number distribution curves of the three core plugs are shown in Figure 12. Volume of pores distribution characteristics is similar for the three core plugs. It can be observed that the volume of pores ranges from 1 to more than 200 nL, however, peaking for the three core plugs at approximately 5 nL. In general, as porosity increases, the distribution of the volume of pores widens. This implies that even slightly tighter correlated sandstone may contain fewer pores and more wide and narrow throats.
Pore radius distribution curves of the three core plugs have similar characteristics as shown in Figure 13. Pore radii expand over a wide range from 0.0106 to 644 μm peaking at around 130 μm. Therefore, the pore radii of the Berea sandstone core plugs utilized contain nanopores, micropores and mesopores that are less than 1 μm, from 1 to 62.5 μm and from 62.5μm to 4 mm, respectively, according to Loucks et al. [23] pore classification. There might be no significant difference between the throat radius and pore radius distribution for large permeability core plugs [6].

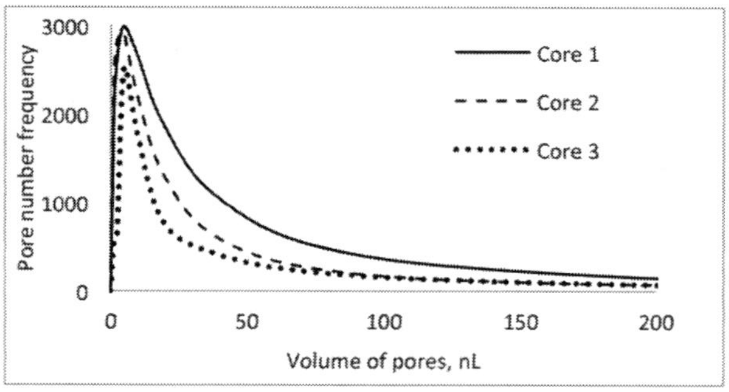


**Figure 12**: Volume and number distribution of the core plugs.

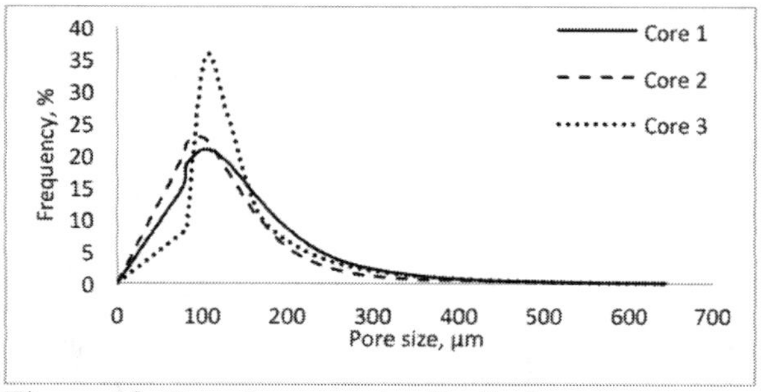


**Figure 13**: Core plugs pore size distributions.

Figure 3 shows a typical capillary pressure curve obtained from core plug 3 of the rate-controlled mercury injection experiments of the present work. The overall shape of the constant-rate mercury capillary pressure curve is similar to pressure-controlled capillary pressure curve. Equation 1 calculates the radius of a cylindrical capillary tube; however, Berea sandstone and all reservoir rocks have complex pore throats and bodies geometry and structural configurations that are very complex in geometry. Therefore, the calculated values of pore throats represent the effective radii of the pore throats that might not equal their actual dimensions. Figure 3 shows a flat and broad plateau indicating well-sorted correlated Berea sandstone dominated by throats of similar sizes. Previous studies on pore-throat correlated networks revealed that steepness of drainage capillary pressure curves decrease as correlations increase [24, 25].
The maximum pressure achieved during the experiments of the present work is around 91 psi, which was sufficient for mercury to penetrate up to around 88.0% of the pore space. Some of the previous rate-controlled mercury injection experiments underwent maximum mercury injection pressure of 899 psi especially when tight rocks were used [3]. Overall, the significant pressure increase towards the end of the experiments with small increase in mercury saturation indicates that the injected mercury only fills some unfilled small pores and throats as shown in Figure 14. Capillary pressure fluctuations in these pores and throats are very small, even when plotted on magnified scale as in Figure 14, because the relative injection rate relative to the pore volume is very high. As the relative mercury rate becomes high, the resultant capillary pressure curve is similar to that obtained by constant-pressure mercury porosimetry where pores and throats filling times are extremely short for the pressure transducer to detect or resolve.

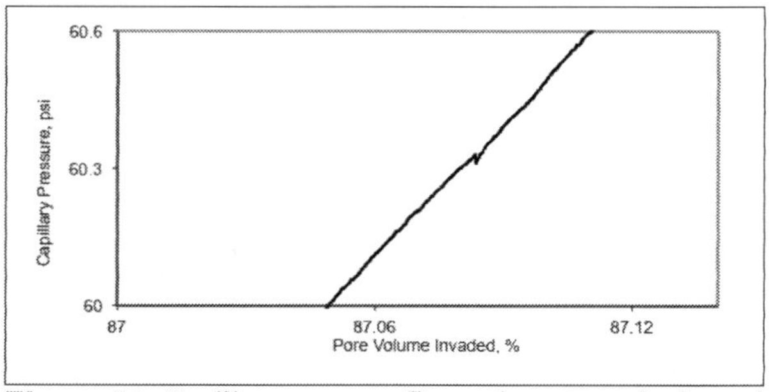


**Figure 14**: Capillary pressure fluctuations towards the end of the experiment.

Calibrations of the interpretations of the present study with petrographic observations, such as optical and electron microscopy, of the pore types, pore texture, and pore throats was not performed mainly due to health and safety reasons. However, to address the uncertainties concerning the calibration of the interpretation, some previously published data on pore geometry from petrographic observations maybe used. Figure 15 shows a back-scattered electron (BSE) image of petrographic thin section of Berea sandstone [26]; the pore space is black, quartz is medium grey, feldspar is light grey, and clays are dark grey.

**Figure 15:** Back-scattered electron image of a thin section of Berea sandstone [26].

Figure 16 shows a scanning electron microscope image of Berea sandstone; clays was found to preferentially coat the quartz grains that resulted in no correlation between the modal mineral composition of the rock and the fraction of surface area made up by the mineral [27].

**Figure 16**: Scanning electron microscope image of Berea sandstone [27].

Figure 17 shows a microtomographic image of Berea sandstone slice [28]. Berea sandstone was characterized by an increase in porosity, resulting in a higher coordination number and characterized by a dense network of numerous smaller throats and pores of radii ~1.3 and ~2.7 µm, respectively.
Figures 15 and 17 show that Berea sandstone, and all porous rocks, is composed of grains, pore bodies and throats. The pore bodies represent the voids or spaces between the rock mineral grains, and the throats are narrow tubes or channels in the rock that connect all different types of pores. The rate-controlled results of the present study reveal that the pore size ranges from 0.0106 to 644 µm. Figures 15 to 17 of the petrographic microscopic observations show that pore sizes fall within the experimentally determined range. Therefore, rate-controlled mercury experiments can effectively characterize a wide range of Berea sandstone pores. However, the accuracy of these findings can only be ensured when the petrographic microscopic thin sections are performed for the same core plugs subjected to the rate-controlled mercury injection experiments.

**Figure 17**: Microtomographic image of Berea sandstone slice [28].

## 5 Conclusions

Rate-controlled mercury porosimetry equipment was constructed and used to characterize Berea sandstone rock into pore bodies and pore throats. Three capillary pressure curves were obtained; one describes the larger pore spaces or pore bodies of the rock core plug, another describes the smaller pores or pore throats that connect the larger pores, and a final curve which corresponds to the overall capillary pressure curve, similar to that obtained from the conventional pressure-controlled mercury injection. The Berea sandstone used had good physical properties with a porosity ranged from 28.2 to 30.7% and gas permeability of 1.31 D. Information on the statistical nature of the pore structure and heterogeneity were obtained as pressure fluctuations with time revealed menisci locations of pore bodies and throats.
Pore size distributions based on volume-accessed pores and pore radii were obtained from the pressure versus saturation relationship. The radii of the pores ranged from 0.0106 to 644 µm indicating that pore radius classification included nanopores, micropores and mesopores; however, the majority of pore radii were mesopores. Results matched previously published petrographic observations data on Berea sandstone.
One of the constraints of rate-controlled mercury injection technique is that the experimental pressure is generally low; therefore, maybe limited nanopores' radii threshold entry pressure can be exceeded. Although pore bodies control the porosity, throats also constitute an appreciable portion of the pore space volume. Varying pore and throat capillary pressure curves obtained for the Berea sandstone core plugs indicate complex geometry of the throats and pores systems. Furthermore, the shapes of the total capillary curves indicate that the core plugs' local pore throats and spatial correlations persist and pore distribution can still be heterogeneous within the same core plug. It is important to mention that despite the valuable pore space morphology characteristics and hydrocarbon saturation information that

the rate-controlled mercury injection technique provides, it is also financially costly, very slow to conduct and labor intensive compared to the routine conventional pressure-controlled mercury injection techniques. The experimental procedure is also extremely time consuming and data processing is complex.

## References

1. H.H. Yuan and B.F. Swanson. SPE Form. Eval. **4**, 17-24 (1989).
2. H. Gao, T. Li and L. Yang. J Petrol Explor Prod Technol. **6**, 309–318 (2016). https://doi.org/10.1007/s13202-015-0186-6
3. W. Zhang, Z. Shi and Y.Tian. Energy Explor. Exploit. **38**(6), 2389-2412 (2020). doi: 10.1177/0144598720920729.
4. J. Smith and I. Chatzis. Canada: N. p., (2004).
5. Y. Zhang, Z. Bao, F. Yang, S. Mao, J. Song, L. Jiang. Geofluids. 2018, ID 3403026 (2018). https://doi.org/10.1155/2018/3403026
6. C. Zhang, Y. Cheng and C. Zhang. J. of Geophys. Eng. **14**, 132-142(2017).
7. Y. Wang, Y. Gao and Z. Fang. Petrol. Explor. Develop. **48**(2): 308-322(2021).
8. B.F. Swanson. J. Pet. Technol. 10-18(1979).
9. N.C. Wardlaw, N.C. Li and D. Forbes. Transp Porous Media. 2, 597-614(1987).
10. A. Makse, H. Hernan, P.S. Ivanov et al. Physica A. **233**, 587-605(1996).
11. C.D. Tsakiroglou and A.C. Payatakes. J. Colloid Interface Sci. **2**, 479-494(1991).
12. J. Liu, S. Song, X. Cao, Q. Meng, H. Pu, Y. Wang, J Liu. J. Rock Mech. Geotech. Eng. 12(2), 403-413(2020).
13. Y. Yao and D. Liu. Fuel. **95**, 152-158(2012).
14. A. J. Adams. Doctoral dissertation, Texas A&M University. Texas A&M University, (2005).
15. W.M. Mahmud. Pet. Sci. Eng. **2**(1):23-29(2017).
16. W.B. Haines. J. Agri. Sci. 20, 97-116 (1930).
17. H.-Y. Zhu, L.-Z. An and C.-Y. Jiao. Natural Gas Geoscience. **26**(7):1316-1322 (2015).
18. H.H. Yuan. SPE 19617 (1989).
19. G. Mason. SPE Form. Eval. (1991).
20. N.R. Morrow. J. Can. Pet. Technol. 35-46(1979).
21. H.K. Christenson, J.N. Israelachvili and R.M. Pashley. SPE Reserv. Eng. 155-165(1987).
22. W.M. Mahmud and V.H. Nguyen. Trans. In Por. Media. **64**, 279-300(2006).
23. R.G. Loucks, R.M. Reed, S.C. Ruppel, et al. AAPG Bulletin **96**(6): 1071–1098 (2012).
24. Y. Li, W.G. Laidlaw and N.C. Wardlaw. Adv. Colloid Interface Sci. 26,1-86(1986).
25. M.A. Ioannidis and I. Chatzis. Water Resour. Res. **29**, 1777-1785(1993).
26. P-E. Øren and S. Bakke. J PETROL SCI ENG. **39**(3-4), 177-199 (2003).
27. P. Lai, K. Moulton and S. Krevor. Chem. Geol. **411**, 260-273 (2015).
28. T. Paul-Ross, A-Z. Aizhan, H-M. Saswata. Front. Earth Sci. **6** (2018).

# Evaluation of the interFoam solver in the prediction of immiscible two-phase flow in imbibition and drainage on the pore-doublet system

*Hosein* Derijani[1], *Lesley* A. James[1] *, and *Ronald* D. Haynes[2]

[1] Hibernia EOR Research Group, Department of Mathematics[2], Memorial University of Newfoundland, St. John's, NL, Canada A1B 3X5

[2] Department of Mathematics and Statistics, Memorial University of Newfoundland, St. John's, NL, Canada A1C 5S7

**Abstract.** Drainage and imbibition at low to moderate capillary number, are very important phenomena, in which the SCAL data are derived. In this venue, there is an increasing trend to apply computational fluid dynamic models at pore scale to extract the required parameters for multiphase flow study. Before applying these methods, their abilities to capture the real performance of fluids at the pore scale with corresponding capillary numbers should be validated. Moreover, due to the dimensional difference from pore to core scale, rigorous statistical methods are required to successfully extract the corresponding SCAL parameters. Hence for any model, before constructing any SCAL data via digital methods, validation at the pore scale along with the investigation of the possible effect of the dimensional change from pore to core scale are significant priorities. In this study, the performance of interFoam, a solver for immiscible two-phase flow from OpenFOAM, at the pore scale is evaluated. An image of the pore doublet system is digitized and meshed with different resolutions to capture the contact angle movement. The average pore diameters are different, and the pore diameter pathway varies. The experimental results are extracted from an article, which extensively performed the tests for different imbibition and drainage regime on this pore doublet system. Moreover, a sensitivity analysis on mesh geometry, refinement near boundaries, and contact angle is conducted. The drainage simulation and experimental results in the pore doublet system are in good agreement with each other. As expected, the displacement happens in the larger diameter and bypasses the wetting fluid in the narrow constriction. In imbibition, the displacement first takes place in the smaller constriction and then the wetting phase starts invading the larger diameter pore. From the sensitivity analysis, it is concluded that the instability of the front is mainly related to the slip condition at the walls. Moreover, the spreading of wetting film near the wall is prevalent at very low contact angles. The results show that the solver can predict the flow behavior in imbibition and drainage on the pore doublet system and the results are comparable with the experiments.

## 1. Introduction

Understanding two-phase flow behavior especially immiscible flow is crucial in different areas of science and industry including the petroleum industry [1]. Multiphase fluid flow through porous media, at the macro scale (core and/or field), is often modeled based on the derivation of relative permeability curves which are extracted from drainage and imbibition experiments. Another approach is to extract the pore structure and apply CFD equations at the pore scale to capture the flow behavior at micro-scale. Typically, in hydrocarbon recovery the scale of fluid velocity is on order of cm/s (ft/day) and the flow regime away from the wellbore and even near the wellbore is in the laminar regime. This article aims to investigate the applicability of CFD modeling on pore scale structures and determine if they can reproduce immiscible flow behavior.

OpenFOAM, a leading, free CFD open-source software owned by OpenFOAM Foundation is being used for this purpose [2]. OpenFOAM (Open-source Field Operation and Manipulation) is an object-oriented C++ framework capable of performing a variety of computational fluid dynamics in continuum mechanics. The solver uses the finite volume discretization method [3]. This software includes solvers for incompressible, compressible flow, and covers multiphase flow modeling. Specifically, the interFoam solver is being used for this study. Deshpande et al. [4] analyzed the two-phase flow behavior at the scale of $10^{-6}$-$10^{-3}$m under capillary dominant and viscous dominant regimes. In the capillary dominant regime, they investigated the effect of perturbation at the interface of fluids in a standing column. They found interFoam can produce satisfactory results when spurious currents are controlled. Recent modifications in interFoam include an adaptive interface compression scheme to minimize parasitic current [5]. However, a study is needed to investigate the pore scale behavior of this solver for drainage and imbibition process.

We use the standard interFoam solver from OpenFOAM version 5.0. In this article, first the formulation of the underlying equations and the solution procedure are presented. Afterward, a meshed pore-doublet is used for the CFD experiment to investigate the capability of the solver to predict flow behavior in imbibition and drainage processes. The sensitivity to the contact angle, and the resolution of mesh near the surface are investigated.

## 2. Methodology

The method that is being used for the solver is presented, mostly adapted from [4]. In two phase flow, the phase fraction $\alpha_i, i = 1,2,$ is the volume of one phase over the total volume of interest:

$$\alpha_i = \int_v \frac{dv_i}{v} \quad i = 1, 2 \tag{1}$$

The mass conservation is given as:

$$\frac{\partial \rho}{\partial t} + \nabla.(\rho U) = 0 \tag{2}$$

* Corresponding author: ljames@mun.ca

© The Authors, published by EDP Sciences. This is an open access article distributed under the terms of the Creative Commons Attribution License 4.0 (http://creativecommons.org/licenses/by/4.0/).

The density of fluid (which is a mixture of immiscible phases) is calculated as:

$$\rho = \alpha_1\rho_1 + \alpha_2\rho_2 \quad (3)$$

The result of combining equation 1 and 2 yield the phase fraction equation as:

$$\frac{\partial\alpha}{\partial t} + \nabla.(\alpha U) = 0 \quad (4)$$

The above equations along with the momentum equation can be used to obtain the phase fraction and the velocity and pressure fields of interest.

The momentum equation is written as:

$$\frac{\partial(\rho U)}{\partial t} + \nabla.(\rho UU) = -\nabla p + \nabla.\left(\mu(\nabla U + \nabla U^T)\right) + \rho g + f_\sigma \quad (5)$$

The last two terms on the RHS of the above equation are forces which are related to the gravitational field and interfacial tension between the two phases. These equations can be used at the pore scale for continuous flow. The boundary conditions can be modified to account for the slip-flow regime which is prevalent near the surfaces [6].

The interfacial tension is based on the continuous surface force model adapted from [7] and is obtained as:

$$f_\sigma = \sigma\kappa\nabla.\alpha \quad (6)$$

where $\sigma$ is interfacial tension and $\kappa$ is the curvature of the interface defined as:

$$\kappa = \nabla.\left(\frac{\nabla\alpha}{|\nabla\alpha|}\right) \quad (7)$$

The above equations are coded in interFoam solver with the aid of the finite volume method. The finite volume formulation is presented below (description based on [8]).

Equation 8 is the general balance equation for any physical property being carried in the fluid.

$$\frac{\partial(\rho\phi)}{\partial t} + +\nabla.(\rho U\phi) - \nabla.(\rho\Gamma_\phi\nabla\phi) = S_\phi(\phi) \quad (8)$$

From the LHS of the equation, the first term stands for temporal evolution of property, the second one stands for the convective contribution of property by the fluid and the third one represents the contribution from diffusive distribution of the property. The RHS term represents the contribution of a property from a sink or source within the system.

Integrating over a domain of interest at specified time step yields:

$$\int_t^{t+\Delta t}\left[\frac{\partial}{\partial t}\int_{V_p}\rho\phi dV + \int_{V_p}\nabla.(\rho U\phi)dV - \int_{V_p}\nabla.(\rho\Gamma_\phi\phi)dV\right]dt = \int_t^{t+\Delta t}\int_{V_p}S_\phi(\phi)dVdt \quad (9)$$

Every volume integral can be transformed to an integral over faces using Gauss' theorem:

$$\int_{V_p}\nabla.adV = \int_S adS \quad (10)$$

A second order equation discretization in time, known as the Crank-Nicolson method is used:

$$\frac{\rho_p\phi_p^n - \rho_p\phi_p^o}{\Delta t}V_p + \frac{1}{2}\Sigma_f F\phi_f^n - \frac{1}{2}(\rho\Gamma_\phi)_f.S.(\nabla\phi_f)^n + \frac{1}{2}\Sigma_f F\phi_f^o - \frac{1}{2}(\rho\Gamma_\phi)_f.S.(\nabla\phi_f)^o = S_uV_p + \frac{1}{2}S_pV_p\phi_p^n + \frac{1}{2}S_pV_p\phi_p^o \quad (11)$$

where the F is the mass flux through the faces:

$$F = S.(\rho U)_f \quad (12)$$

This system of equations on every cell will be simplified to:

$$a_p\phi_p^N + \Sigma_N a_N\phi_N^N = R_P \quad (13)$$

These equations on every cell form a sparse matrix with $a_p$ on diagonal and *the quantities* $a_N$ on the off diagonals:

$$[A][\phi] = [R] \quad (14)$$

The momentum equation is solved by a combination of SIMPLE and PISO methods. The benefit of the combined procedure is the ability to use a larger time step with the Courant Number higher than unity [9].

The procedure for solving the equation is as follows [10]:

1. Initiate variable based on internal initial (previous time step) and boundary conditions.
2. Adjust the time step based on the Courant Number.
3. Use old time volumetric flux to solve the phase fraction equation.
4. Update physical properties; density, viscosity, face densities, unit normal vector, and curvature.
5. Solve the momentum and pressure equations using SIMPLE, PISO, or combined SIMPLE/PSIO methods.
6. Move to the time step.

## 3. Results and Discussion

In Hydrocarbon production practice, the fluid velocity is at very low orders, in a way that capillary forces gain their significant role in immiscible fluid movements. Depending on the relative importance of capillary and viscous forces, the microscopic sweep efficiency may differ [11]. Mahmoodi et al. developed a visualization method to capture multiphase fluid displacement in micro-models [12]. Saturation evolution were quantified during EOR process at pore scale [13-15]. The behavior of imbibition and drainage is visualized experimentally based on viscosity ratio and capillary number by Guo and Aryana [16]. Based on their findings for the specific microfluidic device used to perform the experiments, stable front movement is guaranteed when log M <0 and log Ca>-4. Here M and Ca are viscosity ratio and capillary number respectively. Raeini et al. [17] used a finite volume method to investigate the controlling mechanisms at pore scale in two-phase flow including the layer flow, snap-off, and the geometry effect

 https://doi.org/10.1051/e3sconf/202336601017

on trapping and mobilization of disconnected ganglia. Berg et al. [18] computed the relative permeability curves from pore scale quasi-static method and compared the results with relative permeability obtained from 3-D imaging.

The attempt here is to evaluate the performance of the interFoam solver in the prediction of two-phase immiscible flow at pore scale in drainage and imbibition scheme. A thorough investigation of the problem analytically and experimentally is presented by Chatzis and Dullien [19]. In their work, they used the conventional pore-doublet configuration to visualize the microscopic phenomena at narrow constriction on the scale of mm. As displacement may occur in imbibition or drainage schemes, numerical studies on both imbibition and drainage are performed.

### 3.1. Pore Doublet Reconstruction for Numerical Study

To reconstruct the 2-D pore doublet configuration for numerical study, the image of pore-doublet is digitized and imported to Gmsh [20] for meshing. The cells are composed of 54230 unstructured hexahedra grids. The overall domain of the pore doublet structure is 6.94 x 3.19 x 0.5 mm in x, y, z directions. Table 1 shows the overall quality of the mesh where results were obtained by checkMesh, an OpenFoam utility for evaluating mesh quality.

Table 1. Mesh Quality used for CFD Study

| Avg mesh non-orthogonality | Maximum skewness | Maximum aspect ratio | Maximum cell openness |
|---|---|---|---|
| 4.139 | 0.7876 | 2.2 | 2.4e-16 |

A view of meshed structure and representative cell configuration is presented in Figure 1.

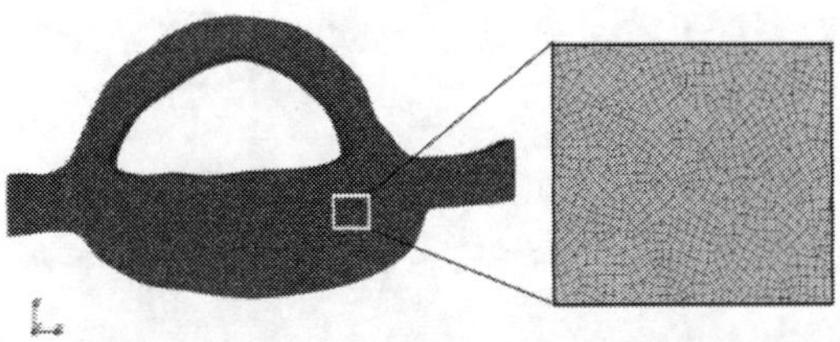

Fig. 1. Pore-Doublet, *Left*: meshed structure (54,230 cells); *Right*: zoom view of hexahedral grid mesh

### 3.2. Imbibition in a Pore-Doublet

The imbibition process is the displacement of the non-wet phase with the wetting phase due to the contribution of only capillary pressure [21]. Generally, for a pore-doublet configuration in either forced or spontaneous imbibition at very low speed when the capillary forces become significant, the wetting phase, first, fill and displace the non-wetting phase at the narrowest pore. Afterward, the process continues to push and displace the non-wet fluid in the larger constriction. The efficiency of the process is reported as 100% and all the non-wet fluid is displaced.

An imbibition experiment performed by Chatzis and Dullien [19] is selected for comparison with numerical results. The imbibition experiment is performed as displacement of n-decane (colored with an oil-soluble dye), by 2% NaCl brine.

The fluid properties for two sets of experiments are extracted from the literature and presented in Table 2. [22, 23]. The boundary and initial conditions are shown in Table 3.

Table 2. Physical Properties of the Fluids

| *Fluid* | *Density ($kg/cm^3$)* | *Dynamic Viscosity (cP)* | *Interfacial Tension (dyne/cm)* |
|---|---|---|---|
| *n-Decane* | 700 | 8.38 | 53.2 |
| *2% NaCl Brine* | 1012.5 | 1.04 | |

Table 3. Boundary & Initial Conditions in Imbibition Simulation

| | Flow rate ($cm^3/sec$) | Pressure (MPa) | Watser saturation (%) | Contact angle |
|---|---|---|---|---|
| Initial condition | 0 | 200 | 0 | - |
| inlet | 1.13e-9 | fixed flux pressure | 1 | - |
| outlet | 1.13e-9 | 200 | zero gradient | - |
| top | No-slip | fixed flux pressure | 0 | 45 |
| bottom | No-slip | fixed flux pressure | 0 | 45 |
| wall | No-slip | fixed flux pressure | 0 | 45 |

### 3.3. Numerical setup

In OpenFOAM, the numerical set up to build the discretization matrix for equation 14 is specified in a file called fvScheme. Every term in the differential equation can be discretized based on the specified scheme. The summary of discretization schemes is presented in Table 4. The 2-D solution procedure for the linear system is given in fvSolution file. The method of solving alpha, pressure, and speed are those presented in Table 5.

Table 3. Discretization Schemes used for the Set of Equations

| *Term* | *Discretization scheme* |
|---|---|
| $\partial/\partial t$ | Euler |
| $\nabla()$ | Gauss Linear |
| $\nabla.(\rho\phi U)$ | Gauss Linear Upwind |
| $\nabla.(\phi\alpha)$ | Gauss van Lear |
| $\nabla.(\alpha(1-\alpha)\phi_r)$ | Gauss Linear |
| $\nabla^2()$ | Gauss Linear |
| *interpolation* | Linear |

| *normal gradient* | corrected |
|---|---|

*$\phi$ here stands for the velocity.*

Table 4. Solution Method for Alpha, Pressure, and Velocity

| Parameter | Solver | Smoother | Pre-conditioner | Tolerance |
|---|---|---|---|---|
| $\alpha$ | smooth solver | symmetric Gauss-Seidel | | 1e-8 |
| $P$ | GAMC[a] | Gauss-Seidel | FDIC[b] | 1e-5 |
| $U$ | smooth solver | symmetric Gauss-Seidel | | 1e-5 |

[a] GAMC: Generalized geometric- algebraic multi-grid
[b] FDIC: a faster version of the diagonal based incomplete Cholesky pre-conditioner

Two iterations for the $\alpha$ loop and one iteration for its sub-cycle are selected. The compression parameter for the saturation equation is set to 1 and the PIMPLE method is employed with 3 inner corrections and one outer correction.

Figure 2 shows the visual representation of the experiment extracted from Chatzis and Dullien [19]. The water is white and the n-decane is black. The water imbibes from the right into the pore doublet and first displaces the n-decane in the smaller capillary followed by imbibition into the larger capillary. The results of OpenFOAM modeling are presented in the sequence of plots on the right where water and n-decane are depicted in blue and red, respectively.

As expected from the pore doublet experiment and also observed in the simulation, when the velocity is very low the situation becomes more similar to spontaneous imbibition and capillary force is the main contributing driving force for the system. Consequently, imbibition takes place in the smaller constriction first then displacement occurs in the larger diameter pore. The process continues until 100% displacement of the non-wetting phase by the wetting phase is accomplished, as observed from the experiment (Figure 2). It is evident that the numerical results obtained are similar to the experiment results.

However, it should be mentioned that at higher injection rates, as the viscous forces become comparable to the capillary force, the displacement in the pore with a larger diameter will start even before displacement in the smaller diameter pore is completed. The simulation results at very first stages of the imbibition process are presented in Figure 3. The front advances at both pores, and as the process continues the front in pore with larger diameter stops and it retreats back. Confirmation of the numerical results requires visual experiments to monitor the flow behavior at the early stages of its development.

### 3.4. Drainage Experiment

At the pore-scale, drainage is defined as the forced displacement of wetting fluid by a non-wetting fluid [21]. It is expected that the invading fluid occupies and pushes the fluid in the pore with a larger radius because the capillary entry pressure is lower compared to a pore with smaller radius [19]. The drainage boundary conditions are shown in table 6. All the settings for discretization and solution of equations are the same as the imbibition process which is explained in the imbibition section.

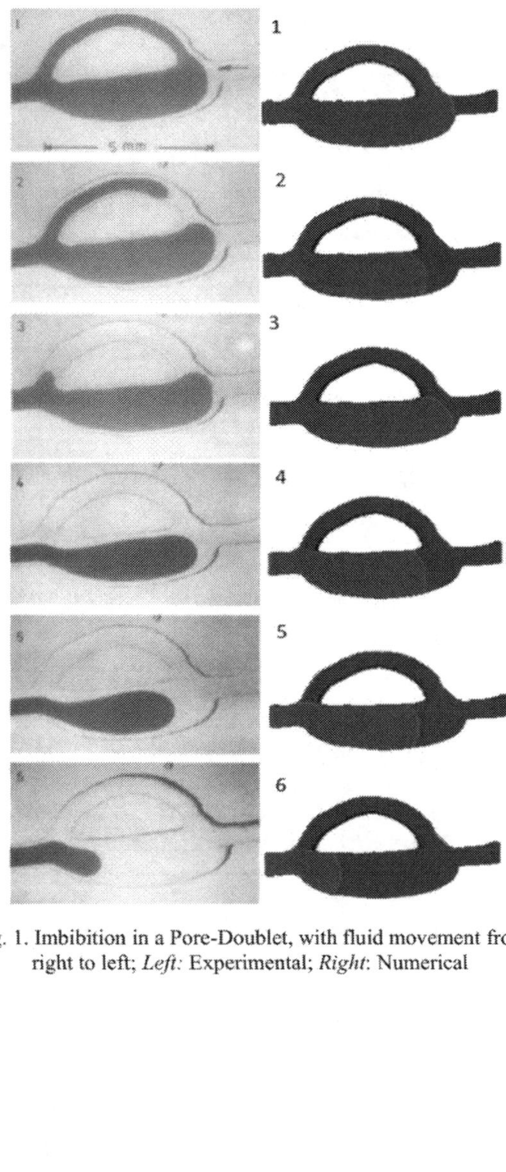


Fig. 1. Imbibition in a Pore-Doublet, with fluid movement from right to left; *Left:* Experimental; *Right*: Numerical

https://doi.org/10.1051/e3sconf/202336601017

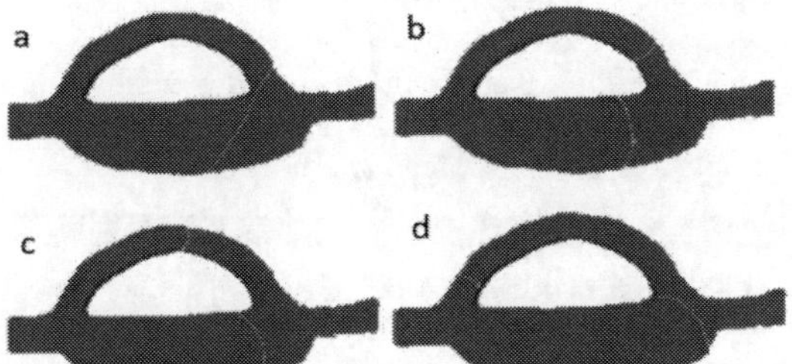


Fig. 2. Simulation results of initial imbibition stages

**Table 5**. Boundary & Initial Conditions in Drainage Simulation

| | Flow rate (cm$^3$/sec) | Pressure (MPa) | Water saturation (%) | Contact angle |
|---|---|---|---|---|
| Initial condition | 0 | 200 | 0 | - |
| inlet | 1.13e-9 | fixed flux pressure | 0 | - |
| outlet | 1.13e-9 | 200 | zero gradient | - |
| top | No-slip | fixed flux pressure | 1 | 45 |
| bottom | No-slip | fixed flux pressure | 1 | 45 |
| wall | No-slip | fixed flux pressure | 1 | 45 |

The drainage experimental results in the pore-doublet are presented in Figure 4 (left) with n-decane (black), a non-wetting fluid, injected from the left draining the water (white) to the right (downstream) preferentially through the larger diameter capillary. The numerical simulation results are presented on the right.

### 3.5. Sensitivity Analysis on Wall Boundary Conditions

To evaluate the ability of interFoam on capturing wetting film, some sensitivity analysis is performed by considering different slip conditions and contact angles. Four cases are compared as shown in the following subsections.

#### *3.5.1. Slip Condition on Walls*

In this case, all the parameters and boundary conditions are the same, just slip conditions on walls are active. The slip boundary condition erases the normal component of the variable at the patch and keeps the tangential components untouched. The results do not show any obvious difference with no-slip as boundary conditions.

The movement of the front at the juncture of two conduit, however, becomes unstable which leads to a snap-off effect. The overall behavior of front movement remains the same as the previous case. Snapshots of the simulation results are shown in Figure 5.

#### *3.5.2. Slip Condition with Dynamic Contact Angle*

Here, a dynamic contact angle was set at the boundaries. The advancing and receding dynamic contact angles are set to 45 and 60 degrees respectively. Results do not show any obvious difference compared to previous case, i.e., slip condition on walls.

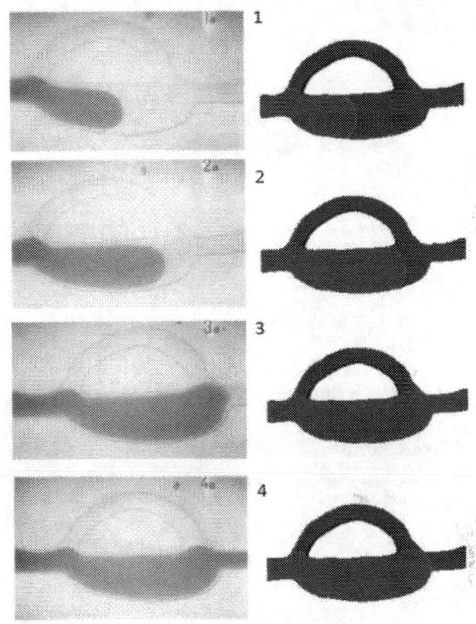


Fig. 3. Pore-Doublet Drainage; *Left:* Experimental; *Right:* Numerical

Fig. 4. Unstable drainage displacement from left to right with water (blue) displacing n-decane (red); c) shows backwards imbibition through the small capillary in the pore-doublet

#### *3.5.3. Slip Condition with Low Dynamic Contact Angle*

Here, the receding and advancing contact angles are set to 10 and 15 degrees respectively, a spread of wetting film near the walls can be seen. This is expected because, by setting the contact angle as low as 10 degrees, we assumed the walls are strongly water wet. The sequence of front advancement, in this case, is also presented in Figure 6. The advancing fluid (water) starts moving in both pore conduits, but at some point, the front retreats and stays stagnant in the pore with a larger diameter. At the same time, the front at the pore with small diameter moves and pushes the n-decane out of the pore. The simulation results are comparable to the experiment. Large diameters control imbibition and the wetting fluid imbibes the pore with a smaller diameter first.

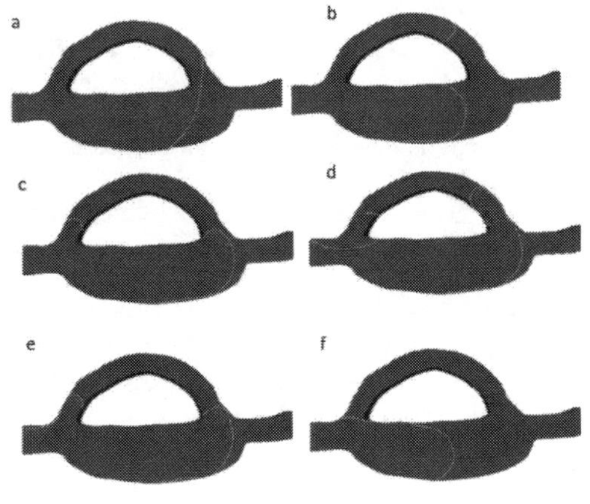


Fig. 5. Advancement of the Imbibition Front in Pore-Doublet with Low Dynamic Contact Angle

### 3.5.4. *Refinement of Mesh at Wall, Dynamic Contact Angle, 3D Flow*

Using snappyHexMesh the mesh is refined near the walls to better capture two-phase interface curvature near the wall. Consequently, the number of cells increased to 107891. The properties of cells are shown in Table 7 and a picture of the cells is shown in Figure 7.

Table 6. Mesh Properties Generated for Three Phase Flow with snappyHexMesh

| *Avg mesh non-orthogonality* | *Max skewness* | *Max aspect ratio* | *Max cell openness* |
|---|---|---|---|
| *15.75* | 1.844 | 108.1 | 2.18e-16 |

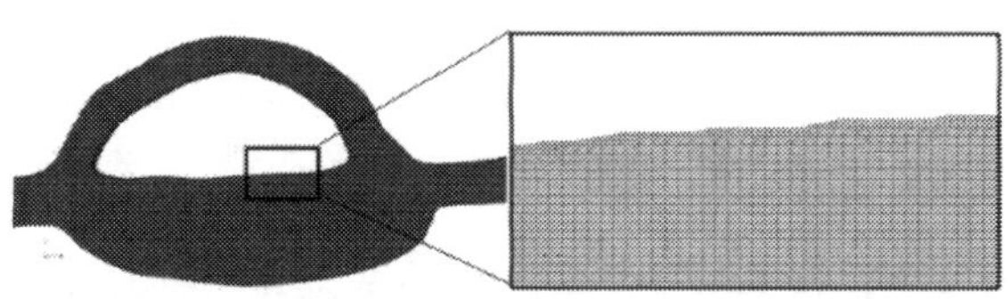

Fig. 6. Representation of Mesh with ParaView Software

The dynamic contact angle with receding and advancing angles of 10 and 15° is being used in the simulation. Three-dimensional flow simulation, despite the huge increase in run time, is being applied. Some snapshots from simulation on interface movement are being represented. The effect of thin-film flow of water that spreads over the wall is evident. As the slip condition is prevalent near the wall and water is being considered as wetting fluid, thin-film extension near the wall is expected. This phenomenon is being predicted by simulation. As demonstrated, interFoam can capture the advancement of film flow near the walls due to the high wettability of water which is being set at boundaries. Figure 8 shows a thin film of water is being constructed while water imbibes throughout the pore doublet. The results are in agreement with the simulation obtained in case 3.5.3.

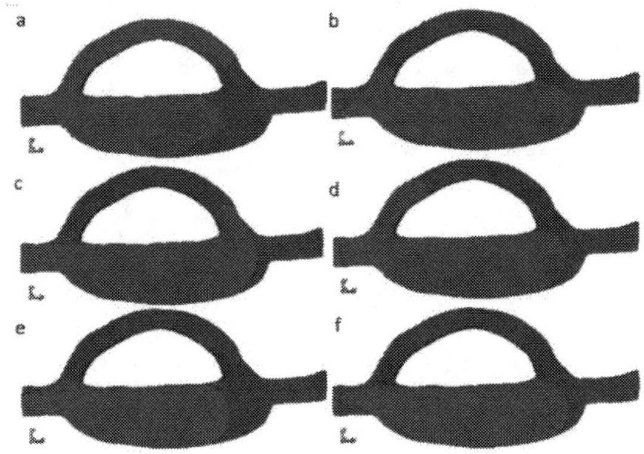


Fig. 7. Spreading of Wetting Film Near the Walls during Imbibition (right to left)

Comparing the results of slip conditions with no-slip conditions at the rock surface, the progression of film flow cannot be captured in no-slip conditions. However, at low to medium wettabilities, these conditions have no preferences in determining the movement of fluid and interface in the pore-doublet system. For strong wettability, applying slip flow is important to predict the creation and evolution of film flow near the rock surface. The effect of film flow due to the geometry of pore interconnection and their tortuosity will be more pronounced even in fluids with weak to medium wettabilities. Hence, applying slip conditions at the rock surface is important to capture the entrapment of non-wet fluids.

## 4. Conclusions

The interFoam solver for immiscible two-phase flow was evaluated by predicting imbibition and drainage in a pore doublet and qualitatively comparing the results to the experiments carried out by Chatzis and Dullien [19]. As expected, the wetting fluid first enters the smaller diameter pore, in imbibition, displacing the non-wetting fluid then continues to the larger diameter pore. The results from the experiments and simulation are, qualitatively, in good agreement with each other. The simulation shows initial displacement in the smaller pore, then front movement in the larger diameter pore due to the competition of viscous and capillary forces. The wetting fluid in the larger diameter pore then retreats to the pore inlet before displacement in the smaller diameter pore continues. These phenomena are analytically explained by Chatzis and Dullien[19].

Drainage results from the simulation and experiment are in agreement with each other. As expected, the displacement happens in the larger diameter pore first and bypasses the wetting fluid in the narrow pore.

The sensitivity analysis shows that instability of the front is mainly related to the slip condition at the walls, and spreading of wetting film near the wall is prevalent at very low contact angles or in other words, in advancing fluids with strong wettabilities. Film flow is predicted at low contact angles, as shown in sections 3.5.3 and 3.5.4. Moreover, the effect of film flow especially in fluids with strong wettabilities can be predicted by applying slip conditions at the rock surface.

In this study, the ability of interFoam in prediction of imbibition and drainage for simple pore doublet system is evaluated. Numerical analysis must be conducted on a pore scale system, representative of real porous media for validation in the next step.

Overall, interFoam solver has the acceptable capability to follow the behavior of two-phase immiscible flow in either drainage or imbibition for pore doublet system. Other CFD tools can be evaluated and be compared with each other.

The authors would like to thank the Hibernia Management and Development Company (HMDC), Chevron Canada Ltd, Petroleum Research Newfoundland and Labrador (PRNL), the Natural Sciences and Engineering Research Council of Canada (NSERC), the Province of Newfoundland and Labrador, and Mitacs for financial support. We thank our colleagues in the Hibernia EOR Research Group for their technical support.

## References

1. S. Pasquier, M. Quintard and Y. Davit, "Modeling Two-Phase Flow of Immiscible Fluids in Porous: Bukeley Leverett Theory with Explicit Coupling Terms," *Physical Review Fluids,* vol. **2**, no. 10, (2017).
2. "OpenFOAM," 2019. [Online]. Available: https://openfoam.org/.
3. F. Moukalled, L. Mangani, and M. Darwish *The Finite Volume Method in Computational Fluid Dynamics*, Springer, (2016), p. 129.
4. S. S. Deshpande, L. Anumolu and M. F. Trujillo, "Evaluating the Performance of the Two-Phase Flow Solver in InterFoam," *Computationl Science and Discovery,* (2012).
5. M. Aboukhader, A. Georgoulas, M. Marengo, M. Gavaises and K. Vogiatzaki, "Simulation of Micro-Flow Dynamics at Low Capillary Numbers Using Adaptive Interface Compression," *Computers and Fluids,* vol. **165**, pp. 13-32, (2018).
6. M. Gad-El-Hak, "The Fluid Mechanics of Microdevices—the Freeman Scholar Lecture," *J. Fluids Eng. Trans. ASME*, vol. **121**, no. 1, pp. 5–33, (1999).
7. J. U. Brackbill, D. B. Kothe and C. Zemach, "A Continuum Method for Modeling Surface Tesion," *Computational Physics,* vol. **100**, pp. 335-354, (1992).
8. S. M. Damian, PhD Thesis: An Extended Mixture Model for the Simultaneous Treatment of Short and Long Term Scale, (2013).
9. T. Holzman, Mathematics, Numerics, Derivations and OpenFOAM, (2016).
10. H. Rusche, Computational Fluid Dynamics of Dispersed Two Phase Flow at High Phase Fractions, University of LONDON, (2002).
11. R. Lenormand and E. Touboul, "Numerical Models and Experiments on Immiscible Displacements in Porous Media," *Journal of Fluid Mechanics,* vol. **189**, pp. 165-187, (1988).
12. M. Mahmoodi, L. A. James and T. Johansen, "Automated Advanced Image Processing for Micromodel Flow Experiments: An application Using LabView," *Journal of Petroleum Science and Engineering,* no. **167**, pp. 829-843, (2018).
13. S. Mahdavi and L.A. James, "Micro and Macro Analysis of Carbonated Water Injection (CWI) in Homogeneous and Heterogeneous Porous Media," *Fuel*, vol. **257**, no. June, p. 115916, (2019).
14. P. Nguyen, H. Fadaei, and D. Sinton, "Pore-scale Assessment of Nanoparticle-Stabilized $CO_2$ Foam for Enhanced Oil Recovery," *Energy and Fuels*, vol. **28**, no. 10, pp. 6221–6227, (2014).
15. H. Khorshidian, L. A. James, and S. D. Butt, "Pore-Level Investigation of the Influence of Wettability and Production Rate on the Recovery of Waterflood Residual Oil with a Gas Assisted Gravity Drainage Process," *Energy and Fuels*, vol. **32**, no. 6, pp. 6438–6451, (2018).
16 F. Guo and S. A. Aryana, "An Experimental Investigation of Flow Regimes in Imbibition and Drainage Using a Microfluidic Platform," *Energies,* vol. **12**, pp. 1-13, April (2019).
17 A. Q. Raeini, B. Bijeljic, and M. J. Blunt, "Numerical Modelling of Sub-pore Scale Events in Two-Phase Flow Through Porous Media," *Transport in Porous Media*, vol. **101**, no. 2, pp. 191–213, (2014).
18 S. Berg *et al.*, "Modelling of Imbibition Relative Permeability by Direct Quasi-Static Approach," *Internatioal Symposium of the Society of Core Analysts*, pp. 1–12, (2016).
19. I. Chatzis and F. A. Dullien, "Dynamic Immiscible Displacement Mechanisms in Pore Doublets: Theory Versus Experiments," *Journal of Colloid and Interface Science,* pp. 199-222, January (1983).

https://doi.org/10.1051/e3sconf/202336601017

20. C. Geuzaine and J. F. Remacle, "Gmsh: a Three Dimensional Finite Element Mesh Generator with Built-in Pre- and Post-Processing Facilities," *International Journal for Numerical Methods in Engineering,* vol. **79**, no. 11, pp. 1309-1331, (2009).
21. J. Bear, Dynamics of Fluids in Porous Media, New Yourk: Elsevier, (1972).
22. A. Goebel and K. Lunkenheimer, "Interfacial Tension of Water/n-Alkane Interface," *Langmuir,* vol. **13**, pp. 396-372, (1997).
23. National Center for Biotechnology Information. PubChem Database. Decane CID=15600, "https://pubchem.ncbi.nlm.nih.gov/compound/Decane," [Online]. [Accessed July 2019].

# Dynamic pore-scale modeling of residual fluid configurations in disordered porous media

*Yanbin* Gong*, *Mohammad* Sedghi, and *Mohammad* Piri

Center of Innovation for Flow through Porous Media, Department of Petroleum Engineering, University of Wyoming, Laramie, Wyoming 82071, United States

**Abstract.** Fluid trapping in porous media is important in many subsurface flow processes such as enhanced oil recovery and geological sequestration of carbon dioxide. To achieve optimal performance in such applications, a fundamental understanding of residual trapping mechanisms at the pore scale is necessary. In this work, we present a computational study of fluid trapping behaviors in natural porous media under different flow regimes by employing a dynamic pore-network modeling approach. The model incorporates many advanced features that have not been collectively used in previous dynamic platforms. For instance, it rigorously solves for fluid pressure fields from two-phase mass balance equations in each pore element, incorporates a detailed description of pore-scale fluid displacement dynamics of piston-like advance, snap-off, and pore-body filling, and explicitly accounts for flow through wetting layers forming in corners and rough surfaces of pore spaces. Moreover, we extend the ability of our model by including contact angle hysteresis, which is often neglected in existing dynamic models. A heavily-parallelized implementation of this platform is further advanced to achieve an efficient computational performance. We first conduct primary drainage and imbibition simulations in pore networks representing Bentheimer and Berea sandstones. We show that the predicted two-phase relative permeability curves agree well with their experimental counterparts reported in the literature. Afterwards, the validated model is used to systematically probe fluid trapping behaviors in a core-sized pore network that is constructed from high-resolution micro-computed tomography images of a Berea sandstone core sample. The effects of dynamic flow conditions and fluid properties on core-scale two-phase displacement pattern, residual-fluid configuration, and residual oil saturations are examined in detail. Fluid trapping properties such as maximum and average residual-fluid cluster size and capillary-controlled invasion selectivity at the pore scale are analyzed under both capillary- and viscous-dominated flow regimes.

## 1 Introduction

Residual trapping of the non-wetting fluid in porous media is important in many science and engineering applications such as enhanced oil recovery (EOR), geological sequestration of carbon dioxide, and contaminant removal from groundwater [1–3]. To achieve the optimal performance in these applications, it is crucial to accurately model the flow behaviors and predict the residual fluid configurations. However, multiphase flow systems are exceedingly complex due to, for instance, the complicated pore space geometries and topologies, fluid-fluid capillary interaction, and wettability effects [4–11]. Among many other efforts to better understand the residual trapping behaviors, extensive numerical simulation studies have been conducted at multiple scales. In the conventional continuum models, nevertheless, most of the complex pore-scale processes are significantly simplified and taken into consideration using a series of input functions, e.g., relative permeability, with little physical basis. Yet these properties are expensive and time-consuming to obtain experimentally. Therefore, physics-based models that can predict fluid trapping properties and bridge the gap between the pore and continuum scales flow processes are desperately needed.

Over the years, with the advances in imaging techniques, e.g., micro-computed tomography (micro-CT), different pore-scale simulation techniques have been proposed and investigated to fulfill different purposes. In general, pore-scale models can be categorized into direct pore-level models and pore-scale network models. In the first group, computational fluid dynamics (CFD) methods, such as level-set [12], lattice Boltzmann [13, 14], and volume-of-fluid [15], are utilized to solve the governing flow equations in the pore space obtained through the imaging techniques. The direct models preserve the detailed pore space geometries. However, the complex pore space geometries of naturally-occurring rocks make their computations expensive. The complicated pore-scale fluid dynamics of two-phase system carry the modeling works into a further level of complexity.

In the second group of pore-scale models, fluid flow and solute transport are solved in a network of inter-connected pores and throats with idealized geometries. Through this approximation, the fluid interfacial movements can be described by a set of displacement rules with closed-form mathematical expressions to substantially reduce the computational costs. In 1956, Fatt [16] made the pioneering work of modeling drainage process in porous media by using a two-dimensional (2D) network of cylindrical tubes. Since then, tremendous progress has been made in developing different network models, which mainly include quasi-static and dynamic models. In quasi-static models, different fluids are assumed to be at capillary equilibrium and solely the

* Corresponding author: ygong@uwyo.edu

© The Authors, published by EDP Sciences. This is an open access article distributed under the terms of the Creative Commons Attribution License 4.0 (http://creativecommons.org/licenses/by/4.0/).

capillary pressure determines the fluid displacement sequence [17–19]. Therefore, the quasi-static models are only applicable to slow flow processes where the capillary number is approximately in the order of $10^{-7}$ or lower.

To account for the viscous effect, the dynamic models solve the fluid pressure field by writing mass conservation equations for all volume-occupying phases in every pore spaces. In this regard, the computational demand of the dynamic models is orders of magnitude higher than that of the quasi-static simulation. Koplik et al. [20], Dias and Payatakes [21], and Lenormand et al. [22] are among the first to develop the dynamic algorithm for two-phase flow in porous media. They wrote the Washburn equation at the fluid–fluid invasion front to account for the capillary and viscous forces. In these early studies, many pore-scale fluid dynamics such as fluid flow through the wetting layers are neglected, even though they are crucial in maintaining the connectivity of the wetting phase and affecting the residual trapping behaviors of the non-wetting fluid during two-phase flow [23, 24]. Later, Singh and Mohanty [24] proposed a rule-based scheme to implement layer flow for primary drainage displacement. They calculated the wetting phase pressure in layers using a spatial extrapolation approximation and updated the wetting layer volume following a heuristic rule that the amount of wetting phase removed from a layer was proportional to its local capillary pressure drop. Nguyen et al. [25] studied the effect of flow rate on imbibition displacement properties assuming that no viscous pressure drop was associated with the non-wetting phase and the pressure of the wetting phase was constant behind the displacement front. With a similar approach, Wang et al. [26] simulated two-phase imbibition in a regular 2D network consisting of 800 pores and 1540 throats. Thompson [27] introduced a fully dynamic model, in which the local capillary pressure is approximated as a function of the local water saturation in regular cubic pore spaces. To extend the Thompson's approach, Sheng and Thompson [28] conducted steady-state flow simulations to predict two-phase relative permeability; Joekar-Niasar and Hassanizadeh [29] investigated the non-equilibrium capillary pressure during two-phase flow. More recently, Li et al. [30] studied the effect of viscosity ratio on two-phase imbibition inside a regular 2D pore network consisting of $100 \times 50$ pores and a regular three-dimensional (3D) pore network with $20 \times 20 \times 20$ pores. Boujelben et al. [31] presented a new dynamic model to study low-salinity water injection, where the wetting layer was treated as volumeless. Qin et al. [32] extended the model of Joekar-Niasar and Hassanizadeh [29] and studied air-water drainage flow with phase change in two pore networks with 1,163 and 10,619 pores, respectively. Yelkhovsky and Pinczewski [33] modeled two-phase capillary rise in both round and polygonal capillary tubes by solving an averaged Navier-Stokes equation.

While the geometries of the pore spaces of naturally occurring porous media can be complex [34, 35], to the best of our knowledge, existing fully dynamic models always assume the pore elements have square cross sections. Moreover, factors such as neglecting wettability hysteresis, ignoring throat volumes, and neglecting snap-off displacements in pore bodies further limit the predictability of the previous developed models.

This study aims at developing a new, generalized dynamic pore-network modeling platform to study fluid trapping behaviors in natural porous media under different flow regimes. We start with an introduction of the representative pore networks used in this work. This is followed by a detailed description of the new dynamic pore-scale modeling algorithm and its parallel implementation. Then, quantitative and qualitative validations of the modeling platform are presented. Afterwards, the model is used to examine the effects of dynamic flow conditions on two-phase displacement patterns, residual oil cluster configurations, and final residual oil saturation during different imbibition displacements. In addition, the impacts of fluid properties including viscosity ratio and wettability are discussed in detail. Finally, the main findings of this work are summarized in the conclusion section.

## 2 Pore network constructions

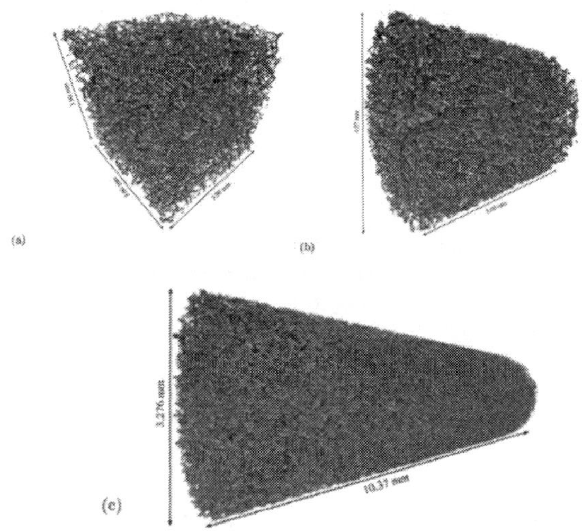


**Figure 1**. Visualization of (a) the small Berea pore network, (b) the Bentheimer pore network, and (c) the large Berea pore network. Orange and gray colors represent the pores and throats, respectively.

In this work, we utilize four networks for different purposes. For the purpose of validation, we first generate a 2D lattice pore network that is statistically equivalent to the micro-model reported in the experimental study of Lenormand et al. [22]. This 2D pore network consists of $100 \times 100$ pores with cross-sectional shape of square. Then, we utilize two pore networks being representative of the Berea and Bentheimer sandstones to predict their two-phase relative permeabilities. The Berea network (see Fig. 1a), having a dimension of $3 \times 3 \times 3$ mm$^3$, is constructed from a process-based technique [36] and has been successfully tested in previous modeling studies [37–39]. The Bentheimer network (see Fig. 1b), having a length of 5.00 mm and a cross-sectional diameter of 4.97 mm, is extracted from the high-resolution (scanning resolution 2.6 μm) micro-CT images of a Bentheimer sandstone rock sample [40]. The Bentheimer rock sample has a measured porosity of 23.42% and an absolute permeability of 2568 mD. To investigate two-phase flow behaviors and residual fluid configurations in disordered porous media, we further extract a longer, cylindrical pore network directly from high-resolution (scanning resolution 2.49 μm) micro-CT images of a Berea sandstone sample with a measured porosity of 21.69% and an absolute permeability of 623 mD [40]. The network (see Fig. 1c) has a length of 10.374 mm and a cross-sectional diameter of 3.276 mm. Hereinafter, we will refer to this network as the large Berea network to distinguish it from the smaller network constructed using the process-based technique. Pores and throats of the three pore networks can have different cross-sectional shapes of square, equilateral and scalene triangles.

## 3 Model description

### 3.1 Model assumptions

Fluids are assumed to be Newtonian, incompressible, and immiscible. The flow process is isothermal and there is no interaction between the fluids and the rock material. Also, clay content is assumed to be fully saturated with immobile water. It is worth noting that in our model, both pore and throat elements have their own volume and conductivity. This is distinct from the existing dynamic models, where zero volume or zero conductance are often imposed on either throat elements or pore bodies to simplify the computations. Hereinafter, we will use the terms "water" and "oil" to represent the wetting and non-wetting phases, respectively, and "pore elements" to denote both pore bodies and pore throats.

### 3.2 Pore-level physics of two-phase flow

Based on the experimental observations of two-phase flow in porous media [41–43], three fundamental pore-level displacing mechanisms are included in this dynamic model: piston-like, cooperative pore-body filling, and snap-off. Their detailed descriptions are presented in our previous work [44]. However, it is worth mentioning that snap-off corresponds to the displacement of the non-wetting phase in the center of a pore element by arc menisci (*AMs*) formed in the corners and on the rough surfaces of the pore space. When local capillary pressure decreases, the relevant wetting corners gradually swell. Eventually, at a critical threshold capillary pressure, oil at the center is snapped off and the element is spontaneous filled by water. This displacement mechanism significantly contributes to oil entrapment in porous media [45, 46].

### 3.3 Phase hydraulic conductance

The hydraulic conductance of a pore element filled with a single phase is calculated based on its geometries [18, 47]:

$$g = \begin{cases} \dfrac{3R^2A}{20\mu}, & for\ triangular\ elements \\ \dfrac{0.5623GA^2}{\mu}, & for\ square\ elements \end{cases} \quad (1)$$

where $R$, $A$, and $G$ are the element's inscribed radius, cross-sectional area, and shape factor, respectively; and μ is the fluid viscosity.

After a local drainage, it is impossible for oil to completely drain water out of an angular pore element as a certain amount of water remains in the corners. In such fluid configurations, the flow area open to the corner water and the center oil can be computed as:

$$A_w = \begin{cases} r^2\left[\cos\theta\left(\cot\alpha\cos\theta - \sin\theta\right) + \theta + \alpha - \dfrac{\pi}{2}\right], & if\ \alpha+\theta \neq \dfrac{\pi}{2} \\ \dfrac{b^2}{\tan\alpha}, & if\ \alpha+\theta = \dfrac{\pi}{2} \end{cases} \quad (2)$$

$$A_o = A - A_w \quad (3)$$

$$r = \frac{\sigma_{ow}}{p_c} \quad (4)$$

where $\alpha$ is the corner half angle of the pore element; $\theta_{ow}$ is the contact angle of the oil-water-solid system; $r$ is the curvature radius of the *AMs*; $\sigma_{ow}$ is the oil-water interfacial tension; $p_c$ is the local capillary pressure; and $o$ and $w$ are the phase indexes denoting oil and water phases. The center oil conductance is then evaluated using $A_o$ and Eq. 1, while the calculation of the corner water conductance can be found in [44, 48, 49].

One should note that at pore scale, the oil conductivity of an element could instantly change from a finite value to zero immediately after the oil phase is snapped off or displaced by water.

### 3.4 Governing equations

The volume conservation equation of each fluid phase is applied at every pore and throat:

$$V_i\frac{dS_{\alpha,i}}{dt} = \sum_{j=1}^{N_i}\frac{g_{\alpha,ij}}{L_{ij}}\left(p_{\alpha,j} - p_{\alpha,i}\right), \quad \alpha = o, w \quad (5)$$

where $V_i$ and $N_i$ are the respective total volume and coordination number of element $i$; $S_\alpha$ and $p_\alpha$ are the saturation and pressure of phase α. The equivalent phase conductance $\frac{g_{\alpha,ij}}{L_{ij}}$ through two neighboring elements, $i$ and $j$, is considered as the harmonic mean of the phase conductance of $i$ and $j$.

In addition, we know that for any element containing two phases, the pressures of oil and water are related by the local capillary pressure, as:

$$p_{c,i} = p_{o,i} - p_{w,i} \quad (6)$$

and the summation of phase saturations in $i$ follows:

$$S_{o,i} + S_{w,i} = 1 \quad (7)$$

Combining Eqs. 5–7, we obtain the following pressure equation, as:

$$\sum_{j=1}^{N_i}\left(\frac{g_{w,ij}}{L_{ij}}+\frac{g_{o,ij}}{L_{ij}}\right)(p_{w,j}-p_{w,i})=-\sum_{j=1}^{N_i}\left(\frac{g_{o,ij}}{L_{ij}}\right)(p_{c,j}-p_{c,i}) \quad (8)$$

We choose $p_w$ as the principal unknown because water is hydraulically connected across the entire pore network.

### 3.5 Local rules

#### *3.5.1 Local displacement*

At pore scale, the occurrence of the local displacement is controlled by the threshold capillary pressure, $p_c^{th}$. In this work, the threshold capillary pressure for piston-like displacement is calculated using the Mayer-Stowe-Princen (MSP) theory [50, 51]. Interested readers are referred to [37, 52] to find the detailed computations of the $p_c^{th}$ for piston-like, cooperative pore-body filling, and snap-off displacements. However, it is worth mentioning that for a given pore element, the threshold capillary pressure for piston-like displacement is always larger than that of the snap-off event. Therefore, piston-like displacement is more favored when the fluid configuration is allowed. However, snap-off does not require the presence of a main terminal meniscus (*MTM*) to occur, meaning that it can take place anywhere in the porous media as long as the fluids are hydraulically connected.

In this work, for a partially invaded element, we assume its local capillary pressure equals to the threshold capillary pressure of the local displacement (i.e., local drainage or imbibition) until the element is fully invaded. By the end of a local imbibition, oil is fully displaced and the local capillary pressure disappears. In contrast, after a local drainage, water layers form in the corners of the angular pore space. In this case, we define a critical saturation corresponding to the point when the entire central pore space is occupied by oil under the threshold drainage capillary pressure of $p_{c,piston}^{th,dra}$. Denoted as $S_{w,piston}^{th,dra}$, this critical saturation can be obtained using Eqs. 2–4 and $p_{c,piston}^{th,dra}$.

*Snap-off handling*. Immediately after a water element is invaded by oil, we compute its threshold capillary pressure for snap-off, $p_{c,snap}^{th,imb}$, and its critical saturation $S_{w,snap}^{th,imb}$ corresponding to $p_{c,snap}^{th,imb}$ using Eqs. 2–4. In our simulations, as fluid saturations are updated after every time step, a snap-off checking process is performed for all elements containing two phases; if the updated water saturation in an element $i$ exceeds $S_{w,snap}^{th,imb}$, spontaneous snap-off will occur and the oil phase will break into $n$ parts, where $n$ is the coordination number of $i$. Note that during this process, new *MTM*s will form, resulting in an instant increase in the local capillary pressure. This, in turn, accelerates the retreat of oil from $i$ to its neighbors [53]. Accordingly, we update the local capillary pressure in element $i$ based on the curvature of the newly formed *MTMs*:

$$p_{c,i}=\frac{2\sigma\cos\theta_{ow,i}^{adv}}{R_i} \quad (9)$$

where $R_i$ and $\theta_{ow,i}^{adv}$ are the inscribed radius and advancing contact angle of $i$.

#### *3.5.2 Contact angle hysteresis and corner interface handling*

We assign each pore element a receding ($\theta_{ow}^{rec}$) and advancing ($\theta_{ow}^{adv}$) contact angle for the local drainage and imbibition displacement, respectively. Moreover, within each element, different parts of the solid surface can exhibit different wettabilities. In specific, each element is initially full of water and is considered as strongly water-wet. After it is invaded by oil, the water-filled corners will remain strongly water wet, while the rest of the solid surfaces contacted by oil will become less water-wet (i.e., have higher contact angles).

When a new *AM* forms, its initial location can be uniquely determined by the threshold capillary pressure, contact angle, and the corner half angle of the pore element through a geometrical analysis [37]. As fluid flow continues, the configuration of the *AM* is then subject to the dynamic change of the local capillary pressure. It can stay pinned with a fixed apex-meniscus distance $b$ as the contact angle, $\theta_{ow}^{h}$, hinges between $\theta_{ow}^{rec}$ and $\theta_{ow}^{adv}$ or it can move at a constant $\theta_{ow}^{rec}$ or $\theta_{ow}^{adv}$. The detailed calculation of the corner interface configuration can be found in [44]. It is evident that such wettability hysteresis behavior results in a non-smooth local capillary pressure–water saturation function.

#### *3.5.3 Local capillary pressure*

We note that for a pore element containing two phases, the fluid interface configuration and the phase saturation can be uniquely determined under a given local capillary pressure, as illustrated in Appendix C in [37]. In addition, at a new time step $n$+1, the local capillary pressure, $p_{c,i}^{n+1}$, will be either in the range of $[p_{c,i}^{n},\ p_{c,i}^{max}]$ (or $[p_{c,snap,i}^{th,imb}, p_{c,i}^{n}]$) for a local drainage (or imbibition) displacement, where $p_{c,i}^{n}$ is known from the last time step $n$ and $p_{c,i}^{max}$ is a local maximum. In view of these, we propose a bi-sectional iterative method to find the $p_{c,i}^{n+1}$ in any pore element using the newly updated local saturation, $S_{w,i}^{n+1}$. The detailed calculation procedure is illustrated in Fig. 2. It is important to note that this iterative numerical scheme allows our dynamic pore-network modeling platform to work with pore elements with irregular and arbitrary cross-sectional shapes.

### 3.6 Boundary and initial conditions

At the inlet of the pore network, a Neumann boundary condition is employed to accommodate a constant fluid injection rate. Specifically, the following two equations are added to the linear system of the pressure equations:

$$\sum_{j=1}^{N_{in}}\frac{g_{w,ij}}{\mu_w}(p_{w,j}-p_{w,in})=-Q_{w,in} \quad (10)$$

$$\sum_{j=1}^{N_{in}}\frac{g_{o,ij}}{\mu_o}(p_{o,j}-p_{o,in})=-Q_{o,in} \quad (11)$$

https://doi.org/10.1051/e3sconf/202336601018

where $N_{in}$ is the total number of inlet-connected throats; and $Q_{\alpha,in}$ and $p_{\alpha,in}$ are the fluid injection rate and injection pressure, respectively. At the outlet of the pore network, a constant production pressure $p_{w,out}$ is enforced for the water phase. When oil is also present at the outlet, we adopt the assumption proposed by Joekar-Niasar and Hassanizadeh [29] and set the local capillary pressure gradient in the outlet-connected throats as zero, namely, $\left.\frac{\partial p_c}{\partial x}\right|_{outlet} = 0$. A no-flow condition is applied at the side boundaries of the network.
In our simulations, the pore networks are initially full of water and are gradually drained by oil during the primary drainage displacement. After a steady-state condition (or a targeted initial oil saturation) is reached, the simulation proceeds to the imbibition displacement, during which water is injected at the inlet of the pore network under a constant volumetric flow rate.

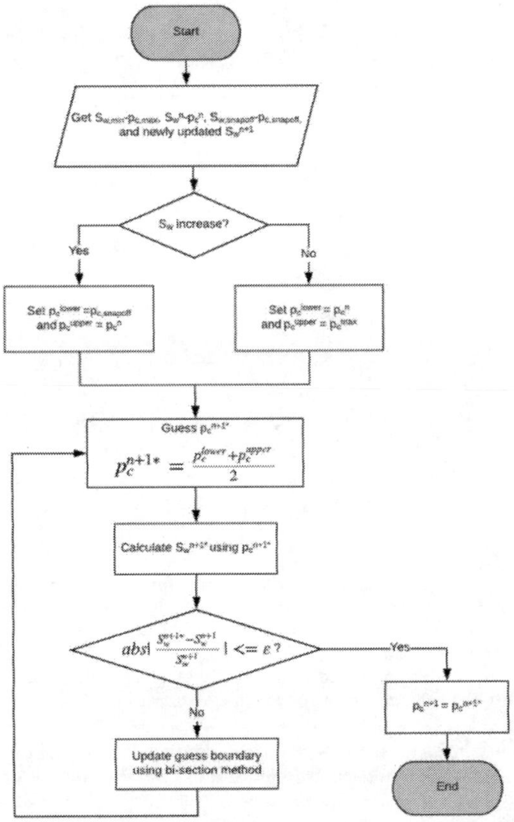


**Figure 2.** Flow chart of the iterative local capillary pressure finding algorithm.

### 3.7 Numerical implementation

At a new time step $n + 1$, we first update the fluid pressure field implicitly by listing the pressure equation (Eq. 8) for all connected pores and throats, as:

$$\sum_{j=1}^{N_i}\left(\frac{g^n_{w,ij}}{L_{ij}}+\frac{g^n_{o,ij}}{L_{ij}}\right)\left(p^{n+1}_{w,j}-p^{n+1}_{w,i}\right) = -\sum_{j=1}^{N_i}\left(\frac{g^n_{o,ij}}{L_{ij}}\right)\left(p^n_{c,j}-p^n_{c,i}\right) \quad (12)$$

A linear system of pressure equations is then obtained and solved with an algebraic multigrid preconditioned conjugated gradient method using PETSc parallel solver package [54]. Then, we explicitly update the fluid saturations in pores and throats by discretizing the volume balance equation (Eq. 5), as:

$$V_i\frac{S^{n+1}_{\alpha,i}-S^n_{\alpha,i}}{\Delta t^{n+1}} = \sum_{j=1}^{N_i}\frac{g^n_{\alpha,ij}}{L_{ij}}\left(p^{n+1}_{\alpha,j}-p^{n+1}_{\alpha,i}\right), \quad \alpha = o, w. \quad (13)$$

To keep the numerical stability, the time step size ($\Delta t$) is chosen in a way that at most one pore element can be fully invaded at every time step. Moreover, after every certain number of invasions, we adjust the pore-scale capillary pressures in accordance with the macroscopic capillary pressure distribution to avoid possible numerical oscillations in local $p_c$. The inherent assumption here is that in a porous medium, the variation of the capillary pressure along the principal flow direction is smooth and continuous. To this end, the macroscopic $p_c$ distribution along the length of the pore network is first calculated as the averages of the pore-scale capillary pressures over the cross-sections of the network. Then, the capillary pressures in different pore elements are updated according to the element's distance from the inlet of the network and the macroscopic $p_c$ distribution.

### 3.8 Domain decomposition and parallelization

For a fully dynamic pore network model, the computation of phase pressure updating, local capillary pressure iteration, and fluid interface tracking can become significant, particularly for multiphase displacements in a large pore network. In this work, the modeling platform is designed to run on massively parallel computing architectures. To this end, the entire large Berea network is divided into a number of sub-domains. The MVAPICH2 library, a high performance message passing interface (MPI) implementation, is used to communicate information between different computing processors. Please refer to our previous works [44] for more details of parallel computing.

## 4 Results and discussions

In section 4.1, we first validate our dynamic pore network model. Afterwards, we conduct a series of two-phase displacement simulations to examine the effects of flow rate, viscosity ratio ($M$), and wettability on dynamic displacement behaviors and residual fluid configurations. Specifically, we conduct five imbibition simulations (simulations 1–5) under elevated water injection rates. Then, another two imbibition simulations (simulations 6 and 7) are performed with enhanced water viscosity to study the viscosity ratio effect. We define the $M$ as the ratio of the viscosity of the invading phase to that of the defending phase. Lastly, four more imbibition simulations (simulations 8–11) are designed with various oil-water contact angles to study the wettability effect. In all simulations, the imbibition displacements are initiated at an initial oil saturation of $S_{oi}$ = 0.9. Properties of the fluids used in these simulations are listed in Table 1. The detailed simulation conditions and residual oil saturations ($S_{or}$) are summarized in Table 2.

**Table 1.** Properties of the fluids used in simulations 1–11.

| Fluid | No. | Viscosity (cp) | $\theta_{ow}^{rec}$ | $\theta_{ow}^{adv}$ |
|---|---|---|---|---|
| Oil | 1-11 | 0.83 | | |
| Water | 1-5 | 1.01 | 0-15 | 15-30 |
| | 6 | 10.1 | 0-15 | 15-30 |
| | 7 | 101 | 0-15 | 15-30 |
| | 8 | 1.01 | 0-15 | 0-15 |
| | 9 | 1.01 | 15-30 | 30-45 |
| | 10 | 1.01 | 30-45 | 45-60 |
| | 11 | 1.01 | 45-60 | 60-75 |

**Table 2.** Fluid injection rates, capillary numbers, and residual oil saturations of simulations 1–11.

| No. | $Q_{w,inlet}$ (cc/min) | $N_c$ | $S_{or}$ (%) |
|---|---|---|---|
| 1 | 0.001 | $2\times10^{-7}$ | 43.9 |
| 2 | 0.01 | $2\times10^{-6}$ | 43.3 |
| 3 | 0.1 | $2\times10^{-5}$ | 42.2 |
| 4 | 1.0 | $2\times10^{-4}$ | 34.8 |
| 5 | 10.0 | $2\times10^{-3}$ | 27.0 |
| 6 | 0.1 | $2\times10^{-4}$ | 29.7 |
| 7 | 0.1 | $2\times10^{-3}$ | 9.3 |
| 8 | 0.001 | $2\times10^{-7}$ | 51.2 |
| 9 | 0.001 | $2\times10^{-7}$ | 39.4 |
| 10 | 0.001 | $2\times10^{-7}$ | 36.6 |
| 11 | 0.001 | $2\times10^{-7}$ | 36.3 |

### 4.1 Model validation

To validate our model, we first carry out a series of primary drainage simulations in the 2D network over a wide range of capillary numbers and viscosity ratios, following the experimental conditions reported in Lenormand et al. [22]. Figs. 3a–3c show the development of the simulated oil flow paths in the 2D network at flow conditions of (a) log$M$ = 2.0 and log$N_c$ = −0.9, (b) log$M$ = −4.7 and log$N_c$ = −3.0, and (c) log$M$ = −4.7 and log$N_c$ = −9.3. Three important two-phase displacement configurations of stable displacement (Fig. 3a), viscous fingering (Fig. 3b), and capillary figuring (Fig. 3c) are clearly identified and they agree well with the experimental observations presented in [22]. In the stable displacement scenario with a viscosity ratio of $M >> 1$ and a large capillary number, the invasion front between the fluids is flat and oil invades nearly all of the pore spaces behind of it. In this case, the pore-scale displacements exclusively take place at the invasion front because the viscous pressure drop associated with the invading phase dominates the displacing process (Fig. 3a). As the viscosity ratio becomes much smaller than one, there is almost no viscous resistance to the flow of the invading phase, while the viscous defending phase consumes nearly all the pressure drop. In consequence, as the invading oil creates a small number of fingers with relatively high conductivities at the inlet, a few of them grow rapidly toward the outlet as the invasions continues, resulting in a quick breakthrough and leaving most of the network uninvaded (Fig. 3b). With the decrease in $N_c$, the viscous force becomes negligible in both fluids and the pore-scale displacement sequence is largely controlled by the capillary force. In this case, the invading phase selectively takes paths with more favorable capillary pressures and develops capillary fingering, as shown in Fig. 3c. Note that the configurations of the viscous fingering and capillary fingering have distinct differences. The former invasion configuration primarily orient towards the outlet, while the latter one grows anomalously inside the network. This is also in line with the experimental observations reported in [22].

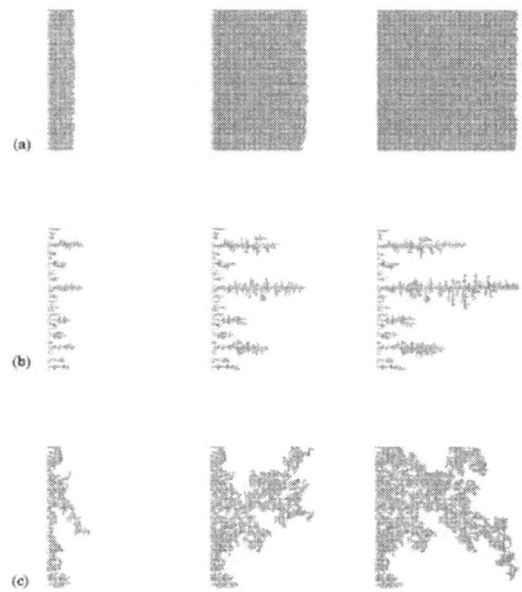


**Figure 3.** Visualization of the oil configurations in the 100×100 two-dimensional (2D) pore network during primary drainage simulations under different flow regimes: (a) log$M$ = 2.0 and log$N_c$ = −0.9, (b) log$M$ = −4.7 and log$N_c$ = −3.0, and (c) log$M$ = −4.7 and log$N_c$ = −9.3. Pictures are placed from left to right as injection continues. Only elements invaded by oil are shown in red.

We further conduct primary drainage and imbibition simulations in the small Berea and the Bentheimer networks to predict their two-phase relative permeability curves at a low flow rate. During the simulations, the phase relative permeability is computed as:

$$k_{r\alpha} = \frac{\mu_\alpha Q_{\alpha,in} \Delta p_{s\alpha}}{\mu_s Q_{s\alpha,in} \Delta p_{t\alpha}} \quad (14)$$

where the subscripts $s$ and $t$ represent the single- and two-phase flow condition, respectively, and $\Delta p_\alpha$ is the [33]pressure drop of phase α across the network. A receding contact angle of 0° is used for the primary drainage displacements, whereas advancing contact angles in the

 https://doi.org/10.1051/e3sconf/202336601018

ranges of 60–75° and 30–50° are assigned for the Berea and the Bentheimer networks, respectively, following the works of [37] and [18].

In Figs. 4 and 5, the computed oil and water relative permeabilities versus water saturation for the small Berea and the Bentheimer networks are compared quantitatively against their corresponding measured data collected from the work of Oak [55] and Oren et al. [18]. As shown in these two figures, our predicted relative permeability curves agree closely with the experimental data for both primary drainage and imbibition displacements. Moreover, the dynamic model accurately predicts the residual oil saturations at the end of the imbibition displacements in both sandstones. We also note that at the early stage of the primary drainage, the predicted water relative permeability underestimates the experimental data. This discrepancy is caused by the fact that phase saturations along the network are not uniform before oil breakthrough. Overall, the simulation results illustrate that our dynamic pore network model is capable of capturing two-phase flow properties under a wide range of flow conditions.

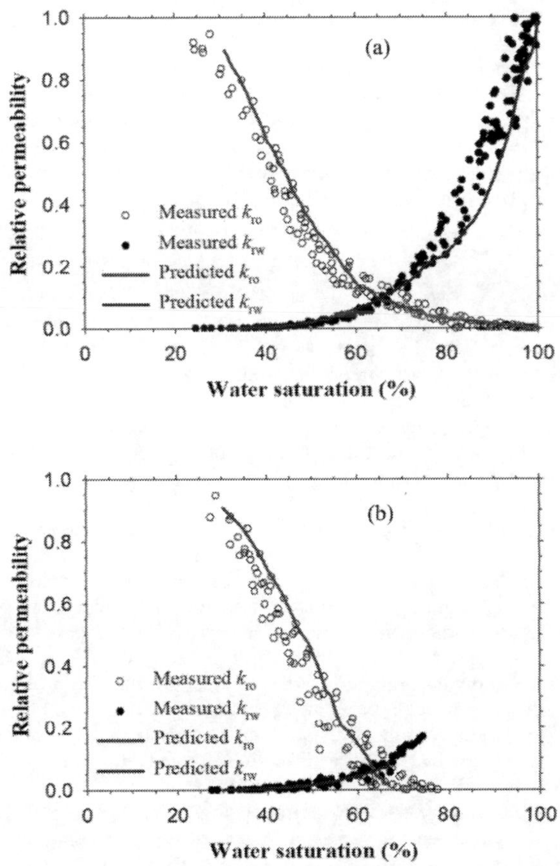


**Figure 4.** Predicted relative permeabilities ($k_r$) versus water saturation ($S_w$) compared against experimental data collected from the literature [55] for (a) primary drainage and (b) imbibition displacements. Both simulations are conducted in the small Berea network.

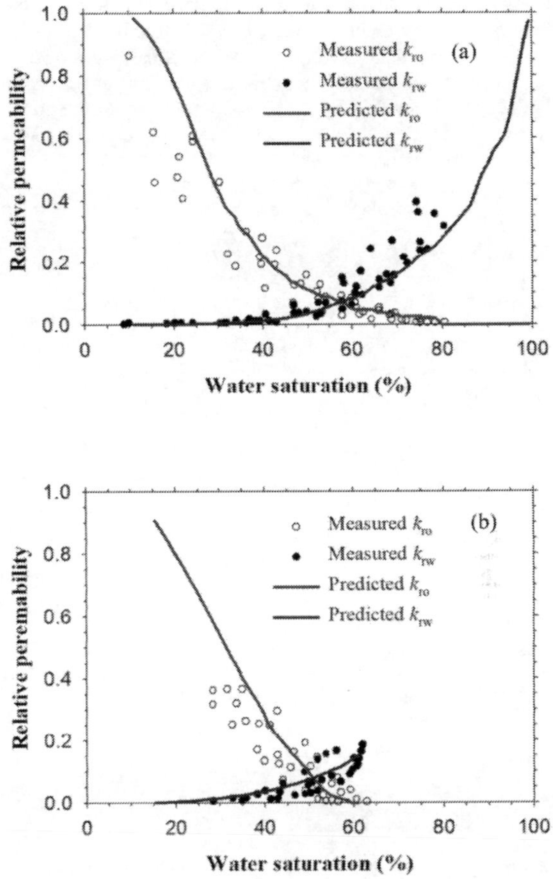


**Figure 5.** Predicted relative permeabilities ($k_r$) versus water saturation ($S_w$) compared against experimental data from the literature [18] for the (a) primary drainage and (b) imbibition displacements. Both simulations are conducted with the Bentheimer network.

### 4.2 Flow rate effect

#### *4.2.1 Fluid displacement patterns*

In this section, simulations 1–5 are conducted with elevated water injection flow rates to achieve different capillary numbers, $N_c$, ranging from $2.0 \times 10^{-7}$ to $2.0 \times 10^{-3}$. Fig. 6 shows the fluid pressure distribution along the Berea network at early stage of the imbibition process in simulations 1–5.

As shown in Fig. 6, the oil pressure is higher than the water pressure in all five simulations due to the positive capillary pressure under the water-wet condition. However, at low capillary numbers ($N_c \leq 2.0 \times 10^{-5}$), the capillary pressure difference ($\Delta p_{cap}$) between oil and water is much higher than the viscous pressure drop ($\Delta p_{vis}$). In this case, pore-scale displacement is driven by the capillary force, meaning that water prefers to fill the elements with more favorable threshold capillary pressures. Moreover, the low flow rate allows sufficient time for water to penetrate through the wetting layers and do displacement in elements that are ahead of the invasion front. The low capillary number and the hydraulically connected water channels jointly lead to a capillary-dominated imbibition pattern: local invasion can take place throughout the network and they proceeds largely in the order controlled by the size and geometry of the pores, regardless of their locations along the flow direction. As a result, water saturation profile shows a uniform increase across the network in Fig. 7b.

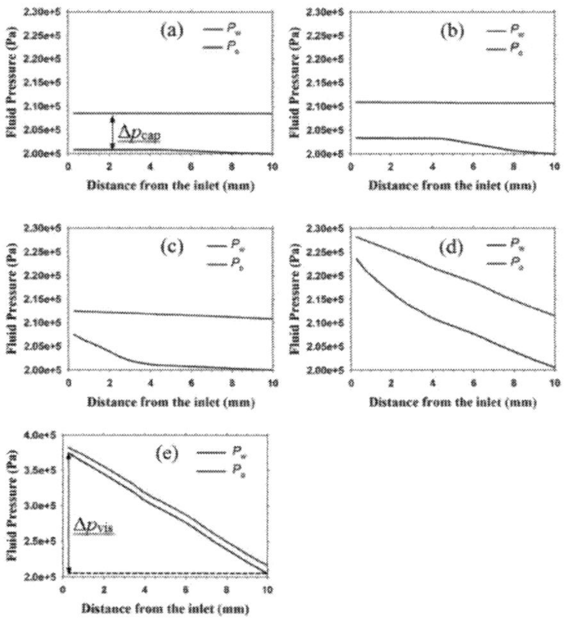


**Figure 6.** Distributions of water and oil along the large Berea network during the imbibition displacement in simulation (a) #1 ($N_c = 2.0\times10^{-7}$), (b) #2 ($N_c = 2.0\times10^{-6}$), (c) #3 ($N_c = 2.0\times10^{-5}$), (d) #4 ($N_c = 2.0\times10^{-4}$), and (e) #5 ($N_c = 2.0\times10^{-3}$).

In contrast, at high capillary numbers of $N_c \geq 2.0 \times 10^{-4}$, the viscous pressure drop surpasses the capillary pressure difference. In particular, simulation 5 has a pronounced viscous pressure drop of $\Delta p_{vis}$ = 172 kPa compared to a capillary pressure difference about $\Delta p_{cap}$ = 10 kPa, as depicted in Fig. 6e. This significant viscous pressure gradient favors frontal displacement (piston-like and pore-body filling) close to the water invasion front. Meanwhile, the rapid water injection rate leaves limited time for water flowing through the wetting layers with restricted flow conductance. This further suppresses water invasions ahead of the invasion front. Therefore, a piston-like invasion front develops and gradually moves towards the outlet of the network as the injection continues, as shown in Fig. 7c.

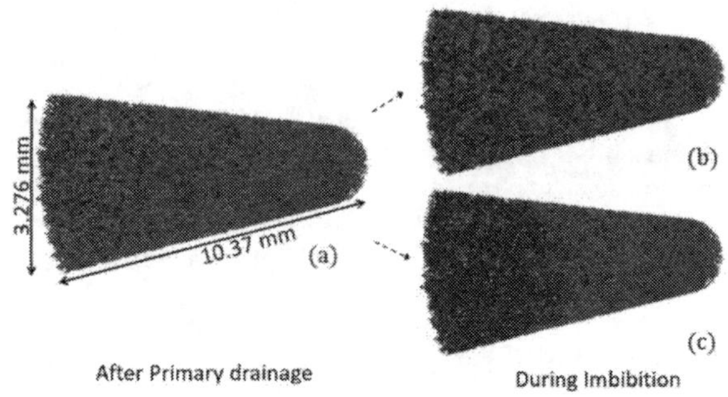


**Figure 7.** Snapshots of fluid distributions in the large Berea network (a) after the primary drainage process and (b, c) during the imbibition processes with the capillary number of $N_c = 2.0 \times 10^{-7}$ (simulation 1) and $N_c = 2.0 \times 10^{-3}$ (simulation 5), respectively. Oil elements are depicted in red and water elements are depicted in blue. Partially invaded elements are shown in orange.

### *4.2.2 Trapped oil clusters and residual oil saturations*

During imbibition in a water-wet rock, bypassing and snap-off displacement can lead to entrapment of oil and formation of oil clusters with complex structures and a wide range of sizes [23, 56]. This section presents characterizations of the trapped oil clusters, including the volumetric cluster size distribution and the final residual oil saturation, under different capillary numbers. At the end of each imbibition displacement, we calculate a total amount of the trapped oil clusters ($V_{o,total}^{trap}$) and a cumulative volumetric distribution to the $V_{o,total}^{trap}$ from clusters that are smaller than a volume size of $V_{o,i}^{trap}$. Fig. 8 plots the computed cumulative volumetric distribution of trapped oil clusters after the imbibition processes in simulations 1–5. Fig. 9 summarizes the final residual oil saturation of simulations 1–5.

Under the capillary-dominated flow conditions, the injected water has sufficient time to flow through the wetting layers and to swell in the corners of the angular porous space across the network, causing snap-offs in small pore elements. This results in a large amount of hydraulically disconnected oil regions. From Fig. 8, we find that the contributions of oil clusters with sizes up to 0.1 mm$^3$ are similar in simulations 1–3. Accordingly, Fig. 9 shows that up to a critical capillary number about $2 \times 10^{-5}$, there is a minor drop in the $S_{or}$ with the increase of the capillary number. As $N_c$ increases, the viscous pressure gradient is sufficient to restrain snap-off throughout the imbibition process and favors more piston-like advances, which produce many small-to-medium sized blobs in the vicinity of the invasion front. As a result, in simulations 4 and 5 conducted under the viscous-dominated flow regime, the overall oil cluster sizes significantly decrease (see Fig. 8) and the final residual oil saturation drops sharply (see Fig. 9). The same tendency was also described in experimental studies [57, 58]. Note that all five simulations create many blobs with volumes in the range of $1.0 \times 10^{-9} \sim 1.0 \times 10^{-8}$ mm$^3$, which is even smaller than the order of element size. The formation of the sub-pore-sized blobs during imbibition was also observed in previous experimental work [58, 59]. Such isolated sub-pore-sized or pore-sized blobs creates local obstacles for water flow through the narrow pore or throat spaces across the network.

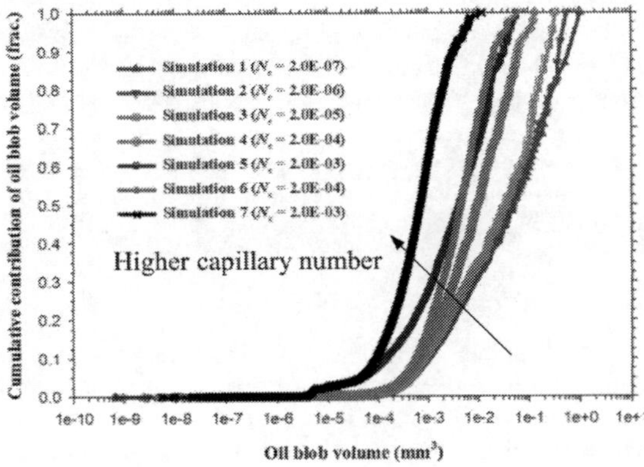


**Figure 8.** Cumulative volumetric distribution of trapped oil blobs to the overall trapped oil volume after the imbibition displacements in the large Berea pore network in simulations 1–5.

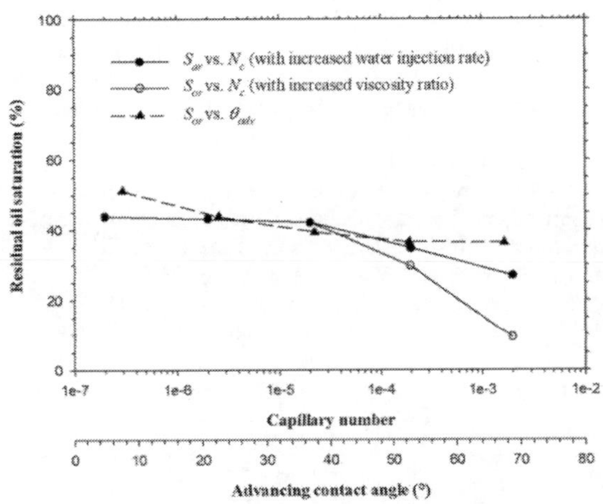


**Figure 9.** Computed residual oil saturation ($S_{or}$) versus capillary number ($N_c$) in simulations 1–7, and computed $S_{or}$ versus contact angle ($\theta_{ow}^{adv}$) in simulations 1 and 8–10. Simulations 1–7 are conducted with a same advancing contact angle distribution ranging from 15° to 30°, while simulations 1 and 8–10 are performed under a same capillary number of $N_c = 2.0 \times 10^{-7}$.

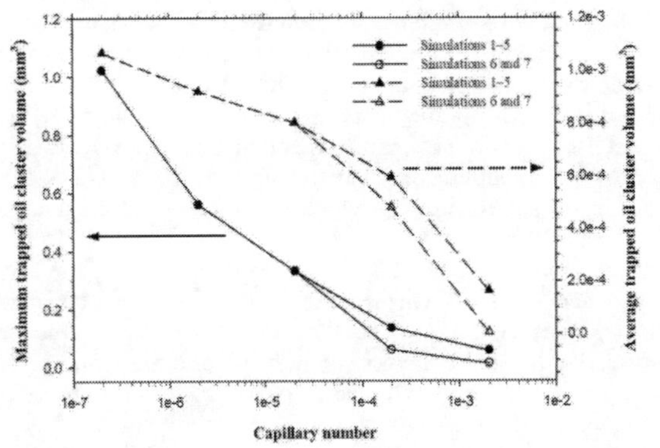


**Figure 10**. Calculated maximum and average trapped oil cluster sizes after imbibition in the large Berea network in simulations 1–7.

We further plot the maximum trapped oil cluster size, $V_{o,max}^{trap}$, and average trapped oil cluster size, $V_{o,ave}^{trap}$, of simulations 1–5 in Fig. 10. It is found that simulation 1 has a maximum trapped oil cluster volume of $V_{o,max}^{trap}$ = 1.022 mm³, which is the largest among the five simulations. As the capillary number increases, both the maximum and the average trapped oil cluster size decreases as the high viscous pressure drop breaks the larger oil clusters. However, it is interesting to note that the variations of the maximum and the average trapped oil cluster sizes with the capillary number are different from the $S_{or}$–$N_c$ curve shown in Fig. 9; as capillary number increases, both $V_{o,max}^{trap}$ and $V_{o,ave}^{trap}$ show a continuous discernible decrease, although the drop in $S_{or}$ is minor until the capillary number reaches about $2 \times 10^{-5}$. This suggests that even under the capillary-dominated flow regime, the configurations of the residual oil clusters experience evident changes as the viscous pressure gradually builds up, although the magnitude of the viscous force is insufficient to push the oil clusters out of the porous media. Previous studies [58, 60] have suggested that the driving force needed to mobilize a trapped oil cluster is inversely proportional to the cluster length. In this respect, the knowledge of the cluster size distribution of the residual oil is an important aid for optimizing the design of EOR processes.

### 4.3 Fluid property effect

In this section, we first investigate the effects of viscosity ratio on the residual oil saturation by performing simulations 6 and 7. Both simulations are conducted with the same fluid injection rate as that of simulation 3. However, the viscosity of water is increased to10.10 and 101.0 cP in simulations 6 and 7, respectively. In simulations 8–11, different oil-water contact angles are conducted to examine the wettability effect. In these four simulations, a low capillary number of $2.0 \times 10^{-7}$ is used to keep them under the capillary-dominated flow regime. The detailed fluid properties, simulation conditions, and the final $S_{or}$ are tabulated in Tables 1 and 2.

#### *4.3.1 Viscosity ratio effect*

The cumulative volumetric contribution of trapped oil clusters, and the maximum and the average trapped oil cluster sizes for simulations 6 and 7 are included in Figs. 8 and 10. As the viscosity ratio increases, more contributions to the final $S_{or}$ are from the small-to-medium-sized oil clusters and both the maximum and the average oil cluster sizes are decreased significantly. A closer examination of the spatial distribution map of the trapped oil clusters along the length of the network in simulations 3, 6, and 7 is presented in Fig. 11, where each dot represents the volume-weighted centroid location of an oil cluster. We find that the oil globules distribute uniformly across the network and the majority of the globules' size is in the range of $\delta = 10^{-6} \sim 10^{-3}$ mm$^3$. As the viscosity ratio becomes higher, there is an evident narrowing in the range of $\delta$ because of the magnified viscous pressure gradient of water. Notably, the population of the sub-pore- and pore-sized oil globules reduces as $\delta$ becomes narrower. This can be attributed to the fact that partial invasions are more likely to be completed as the invading water becomes more viscous.

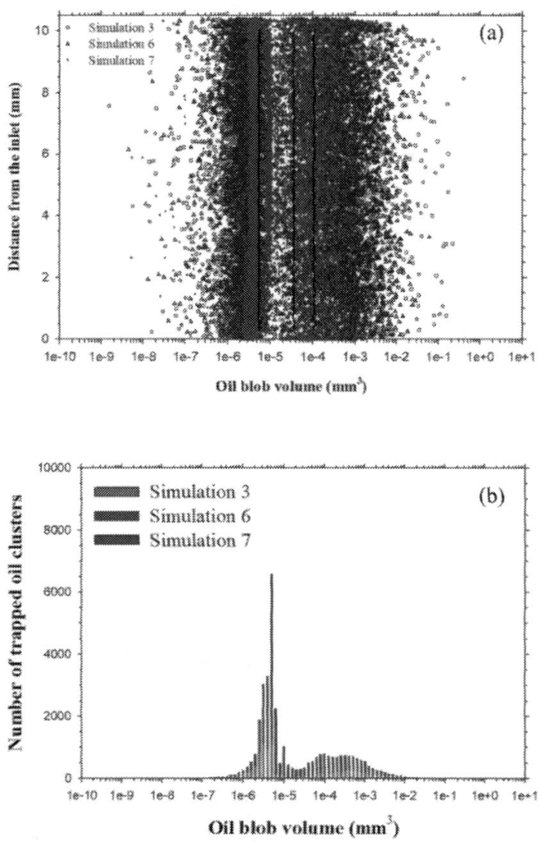


**Figure 11.** (a) Distribution map of the trapped oil clusters along the length of the large Berea network in simulations 3 (viscosity ratio of $M = 1.22$), 6 ($M = 12.2$), and 7 ($M = 122$). Each dot represents the location of the volume-weighted centroid of an oil cluster. The three vertical lines from left to right show the average volumes of the pore throats, pore elements, and pore bodies, respectively. Stacked histograms of the size distribution of the trapped oil clusters are shown in (b).

The above trend also agrees with that observed from simulations 3–5 as we increase the water injection flow rate. Nevertheless, even under a same capillary number, simulations 6 and 7 exhibit lower final $S_{or}$ and smaller $V_{o,max}^{trap}$ and $V_{o,ave}^{trap}$ than simulations 4 and 5 (see Figs. 9 and 10). This is because the higher water–oil viscosity ratio further confines the occurrence of water invasion within the vicinity of the invasion front and thereby, creates a more stable displacement that leaves less residual oil by the end of the imbibition.

*4.3.2 Wettability effect*

The residual oil saturations of simulations 8–11 are summarized in Fig. 9. As indicated in Fig. 9, when porous medium is strongly water-wet ($\theta_{ow}^{adv} \leq 45°$), there is a clear reduction in the residual oil saturation as the contact angle increases. When the contact angles are lower than 15° (simulation 8), a large number of snap-off events take place across the network under the capillary-dominated flow condition. Spontaneous snap-offs and water bypassing [23, 61] jointly leave a significant amount of residual oil. As the contact angle increases, it becomes more difficult for snap-off invasion to occur. This is particularly the case for weakly water-wet porous media (simulations 10 and 11), where the majority of oil entrapment solely results from water bypassing. Consequently, much less residual oil is trapped in these two simulations. Similar results have also been reported in the experimental studies of [62, 63].

To further explore the rules of wettability on the residual trapping behavior, we categorize the overall pore elements into three groups based on their inscribed radius ($R$) and analyze the water saturation of each group. More specifically, water saturations of small ($R < 10$ $\mu$m), medium ($10 \leq R \leq 45$ $\mu$m), and large ($R > 45$ $\mu$m) pore elements are calculated at the end of drainage and imbibition displacements. The results are plotted and compared in Fig. 12.

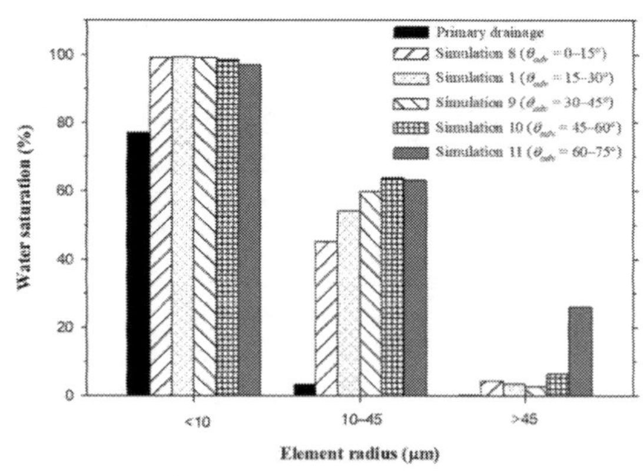


**Figure 12.** Comparison of the calculated water saturations in various size ranges of elements at the end of the primary drainage and imbibition displacements in simulations 1 and 8–10, all of which are conducted under a same capillary number of $N_c = 2.0 \times 10^{-7}$.

In general, the magnitude of the threshold capillary pressure is lower for larger pore elements assuming uniform contact angles and interfacial tension. Therefore, it is easier for the non-wetting phase to invade into large pores and throats. Accordingly, in Fig. 12, we find that nearly all the medium-to-large elements are invaded by oil while the water saturation is still as high as 77% in small elements after primary drainage. On the other hand, in the following imbibition process, the wetting water phase preferably fill the small elements due to their higher threshold capillary pressures. As shown in Fig. 12, water occupies almost all of the small elements, a fair amount of the medium-sized elements, and a limited amount of the large elements at the end of the imbibition in these five simulations. This selective pore-scale displacement leads to capillary fingering (see section 4.1), resulting in water bypassing and oil entrapment. From Fig. 12, we find that after imbibition displacement in water-wet porous media, the water saturations are always high in small elements. However, as the wettability condition is altered towards neutral wetness and the absolute difference in threshold invasion capillary pressures of pore elements across the pore network shrinks, we see a clear increase in the water saturation (i.e., reduced residual oil trapping) in medium-sized pore elements. In other words, the suppression of the capillary invasion selectivity is more pronounced in the medium-sized pores than in the small elements. For the large pores, however, improved oil recovery is only observed when the porous media have large contact angles of $\theta_{ow}^{adv}$ = 60−75°. Note that only a limited number of piston-like invasions can occur in large pores under the strongly water-wet conditions. There is even a slight drop in the water saturation as the contact angle increases from 0−15° to 30−45°, which can be caused by the changes in the water-corner configurations. In summary, the residual oil saturation becomes lower as the fluids-rock system become more neutral wet. The additional oil recovery is mainly attributed to restrained snap-off invasions and suppressed capillary invasion selectivity, particularly in medium-to-large-sized elements.

## 5 Conclusions

In this work, we developed a new dynamic pore network model that is designed for simulating two-phase flow in disordered porous media. The model has several advantages in comparison with the existing dynamic models: it iteratively solves for local capillary pressure in pore elements with irregular cross-sectional shapes, explicitly accounts for contact angle hysteresis, and rigorously considers pore-scale two-phase flow dynamics including frontal displacement, snap-off, and flow through wetting layers in both pore bodies and throat elements. A heavily-parallelized computing scheme of this model was implemented to achieve computational efficiency.

We performed a series of two-phase primary drainage and imbibition simulations in different pore networks representing two-dimensional (2D) micro-model, three-dimensional (3D) Bentheimer and Berea sandstones. We illustrate that our model is capable of capturing fluid invasion behaviors of stable displacement, viscous fingering, and capillary fingering over a wide range of fluid–fluid viscosity ratios and capillary numbers. The predicted primary drainage and imbibition relative permeability curves of Bentheimer and Berea sandstones show satisfactory matches with their experimental counterparts.

The validated model is then applied to systematically study the effects of dynamic flow conditions and fluid properties on two-phase displacement patterns, residual oil configurations, and final residual oil saturations. A series of dynamic imbibition simulations are conducted in a core-sized Berea sandstone pore network that is constructed from high-resolution micro-computed tomography (micro-CT) images of a Berea rock sample. Results show that after capillary-dominated imbibition displacement, a large amount of oil is trapped in the form of large-sized oil clusters because of the abundant snap-offs and bypassing. As the water injection flow rate increases up to a critical capillary number of about $N_c = 2 \times 10^{-5}$, there is a minor change in the final residual oil saturation. However, both the maximum and the average residual-oil cluster sizes reduce considerably. With a further increase in the water injection flow rate, a clear piston-like invasion front is formed at the core scale. Under such viscous-dominated flow regime, the significant viscous pressure drop favors frontal displacement in the vicinity of the invasion front, resulting in less amount of large-sized residual oil clusters and a much lower residual oil saturation. Finally, the effects of fluid–fluid viscosity ratio and wettability on residual trapping are investigated. It is found that the overall residual oil cluster sizes are further decreased by enhancing the water-to-oil viscosity ratio, owing to a strengthened frontal displacement. The amount of residual trapping can also be reduced by altering the wettability condition toward neutral wetness, where the pore-scale capillary invasion selectivity is suppressed, particularly in medium-to-large-sized pore elements. In summary, our parallelized dynamic pore network model provides novel insights into residual trapping behaviors in disordered porous media under a wide range of flow conditions and fluid properties. This is vital for optimizing subsurface applications such as enhanced oil recovery, carbon sequestration, and soil remediation.

We gratefully acknowledge the financial support of Thermo Fisher Scientific and the School of Energy Resources at the University of Wyoming.

References

[1] A. L. Herring, L. Andersson and D. Wildenschild, Geophysical Research Letters, vol. 43, no. 18, pp. 9677-9685.

[2] S. Iglauer, A. Paluszny and M. J. Blunt, Fuel, vol. 103, pp. 905-914, 2013.

[3] H. Ni, B. M. C. Garing and S. M. Benson, International Journal of Greenhouse Gas Control, vol. 86, pp. 158-176, 2019.

[4] Y. Zaretskiy, S. Geiger, K. Sorbie and M. Förster, Advances in Water Resources, vol. 33, no. 12, pp. 1508-1516, 2010.

[5] A. Mehmani, R. Verma and M. Prodanović, Marine and Petroleum Geology, vol. 114, p. 104141, 2020.

[6] J. Cai, T. Jin, J. Kou, S. Zou, J. Xiao and Q. Meng, Langmuir, vol. 37, no. 5, pp. 1623-1636, 2021.

[7] N. Alyafei and M. J. Blunt, Advances in Water Resources, vol. 90, pp. 36-50, 2016.

[8] S. Iglauer, T. Rahman, M. Sarmadivaleh, A. Al-Hinai, M. A. Fernø and M. Lebedev, Society of Petroleum Engineers Journal, vol. 21, no. 06, pp. 1916-1929, 2016.

[9] Y. Gong, M. Sedghi and M. Piri, Transport in Porous Media, vol. 140, no. 1, pp. 143-179, 2021.

[10] Y. Gong, M. Sedghi and M. Piri, Water Resources Research, vol. 57, no. 12, e2021WR030104, 2021.

[11] A. Zankoor, M. Khishvand, A. Mohamed, R. Wang and M. Piri, Journal of Colloid and Interface Science, vol. 603, p. 356–369, 2021.

[12] M. Prodanović and S. L. Bryant, Journal of Colloid and Interface Science, vol. 304, p. 442–458, 2006.

[13] Q. Kang, P. C. Lichtner and D. Zhang, Journal of Geophysical Research: Solid Earth, vol. 111, 2006.

[14] S. Bakhshian, S. A. Hosseini and N. Shokri, Scientific Reports, vol. 9, p. 3377, 2019.

[15] A. Q. Raeini, M. J. Blunt and B. Bijeljic, Journal of Computational Physics, vol. 231, p. 5653–5668, 2012.

[16] I. Fatt, Trans. AIME, vol. 207, no. 01, pp. 144-181, 1956.

[17] S. Bryant and M. Blunt, Physical Review A, vol. 46, p. 2004, 1992.

[18] P.-E. Øren, S. Bakke and O. J. Arntzen, Society of Petroleum Engineers Journal, vol. 3, p. 324–336, 1998.

[19] T. W. Patzek, Society of Petroleum Engineers Journal, vol. 6, no. 02, pp. 144-156, 2001.

[20] J. Koplik and T. J. Lasseter, Society of Petroleum Engineers Journal, vol. 25, p. 89–100, 1985.

[21] M. M. Dias and A. C. Payatakes, Journal of Fluid Mechanics, vol. 164, p. 305–336, 1986.

[22] R. Lenormand, E. Touboul and C. Zarcone, Journal of Fluid Mechanics, vol. 189, p. 165–187, 1988.

[23] M. J. Blunt, Cambridge University Press, 2017.

[24] M. Singh and K. K. Mohanty, Chemical Engineering Science, vol. 58, p. 1–18, 2003.

[25] V. H. Nguyen, A. P. Sheppard, M. A. Knackstedt and W. V. Pinczewski, Journal of Petroleum Science and Engineering, vol. 52, p. 54–70, 2006.

[26] S. Wang, Q. Feng, Y. Dong, X. Han and S. Wang, Journal of Natural Gas Science and Engineering, vol. 26, p. 118–129, 2015.

[27] K. E. Thompson, AIChE Journal, vol. 48, p. 1369–1389, 2002.

[28] Q. Sheng and K. Thompson, Advances in Water Resources, vol. 95, p. 92–108, 2016.

[29] V. Joekar-Niasar and S. M. Hassanizadeh, International Journal of Multiphase Flow, vol. 37, p. 198–214, 2011.

[30] J. Li, S. R. McDougall and K. S. Sorbie, Advances in Water Resources, vol. 107, p. 191–211, 2017.

[31] A. Boujelben, S. McDougall, M. Watson, I. Bondino and N. Agenet, Journal of Petroleum Science and Engineering, vol. 165, p. 462–476, 2018.

[32] C.-Z. Qin, B. Guo, M. Celia and R. Wu, Chemical Engineering Science, vol. 202, pp. 194-207, 2019.

[33] A. Yelkhovsky and W. Pinczewski, Physical Review Fluids, vol. 3, no. 4, p. 044003, 2018.

[34] C. H. Arns, M. A. Knackstedt and K. R. Mecke, in Morphology of Condensed Matter, Berlin, Heidelberg, Springer, 2002, pp. 37-74.

[35] Y. Gong and Y. Gu, Energy & Fuels, vol. 29, no. 10, p. 6213–6223, 2015.

[36] P.-E. Øren and S. Bakke, Transport in Porous Media, vol. 46, p. 311–343, 2002.

[37] M. Piri and M. J. Blunt, Physical Review E, vol. 71, p. 026301, 2005.

[38] N. A. Idowu and M. J. Blunt, Transport in Porous Media, vol. 83, p. 151–169, 2010.

[39] A. Zolfaghari and M. Piri, Transport in Porous Media, vol. 116, p. 1139–1165, 2017.

[40] A. Aghaei and M. Piri, Journal of Hydrology, vol. 522, p. 488–509, 2015.

[41] R. Lenormand, C. Zarcone and A. Sarr, Journal of Fluid Mechanics, vol. 135, p. 337–353, 1983.

[42] G. Mason and N. R. Morrow, Journal of Colloid and Interface Science, vol. 141, p. 262–274, 1991.

[43] S. Berg, H. Ott, S. A. Klapp, A. Schwing, R. Neiteler, N. Brussee, A. Makurat, L. Leu, F. Enzmann, J.-O. Schwarz and others, Proceedings of the National Academy of Sciences, vol. 110, p. 3755–3759, 2013.

[44] Y. Gong and M. Piri, Transport in Porous Media, vol. 135, no. 1, pp. 181-218, 2020.

[45] J. G. Roof, Society of Petroleum Engineers Journal, vol. 10, p. 85–90, 1970.

[46] Y. Liu, S. Iglauer, J. Cai, M. A. Amooie and C. Qin, Capillarity, vol. 2, no. 1, pp. 1-7, 2019.

[47] T. W. Patzek and D. B. Silin, Journal of Colloid and Interface Science, vol. 236, p. 295–304, 2001.

[48] D. Zhou, M. Blunt and F. M. Orr Jr, Journal of Colloid and Interface Science, vol. 187, p. 11–21, 1997.

[49] M.-H. Hui and M. J. Blunt, The Journal of Physical Chemistry B, vol. 104, p. 3833–3845, 2000.

[50] R. P. Mayer and R. A. Stowe, Journal of Colloid Science, vol. 20, pp. 893-911, 1965.

[51] H. M. Princen, Journal of Colloid and Interface Science, vol. 30, pp. 69-75, 1969.

[52] M. Piri and M. J. Blunt, Physical Review E, vol. 70, p. 061603, 2004.

[53] M. S. Al-Gharbi and M. J. Blunt, Physical Review E, vol. 71, p. 016308, 2005.

[54] S. Balay, S. Abhyankar, M. F. Adams, J. Brown, P. Brune, K. Buschelman, L. Dalcin, A. Dener, V.

Eijkhout, W. D. Gropp, D. Karpeyev, D. Kaushik, M. G. Knepley, D. A. May, L. C. McInnes, R. T. Mills, T. Munson, K. Rupp, P. Sanan, B. F. Smith, S. Zampini, H. Zhang and H. Zhang, Technical Report ANL-95/11-Revision 3.11, 2019.

[55] M. J. Oak, in SPE/DOE Enhanced Oil Recovery Symposium, 1990.

[56] S. Iglauer, A. Paluszny, C. H. Pentland and M. J. Blunt, Geophysical Research Letters, vol. 38, 2011.

[57] M. Khishvand, M. Akbarabadi and M. Piri, Advances in Water Resources, vol. 94, p. 379–399, 2016.

[58] I. Chatzis, M. S. Kuntamukkula and N. R. Morrow, SPE Reservoir Engineering, vol. 3, p. 902–912, 1988.

[59] Z. T. Karpyn, M. Piri and G. Singh, Water Resources Research, vol. 46, 2010.

[60] N. R. Morrow, Journal of Canadian Petroleum Technology, vol. 18, no. 3, 1979.

[61] M. J. Blunt, Journal of Petroleum Science and Engineering, vol. 20, p. 117–125, 1998.

[62] H. T. Kennedy, E. O. Burja and R. S. Boykin, The Journal of Physical Chemistry, vol. 59, p. 867–869, 1955.

[63] N. R. Morrow, H. T. Lim and J. S. Ward, SPE Formation Evaluation, vol. 1, p. 89–103, 1986.

# A Mechanistic Study of Wettability Alterations in Sandstone by Low Salinity Water Injection (LSWI) and $CO_2$ Low Salinity Water-Alternating-Gas (WAG) Injection

*Shijia* Ma[1] and *L.A.* James[1,*]

[1]Hibernia EOR Research Group, Department of Process Engineering, Memorial University of Newfoundland, St. John's, NL, Canada, A1B 3X5

**Abstract.** Low salinity water injection (LSWI), an emerging Enhanced Oil Recovery (EOR) method, has proven to be effective in increasing oil recovery by wettability alteration. As low salinity water is injected into the reservoir, the pre-established equilibrium is disturbed. The chemical reactions among the oil/brine/rock system alters the existing wettability, resulting in enhanced oil recovery. Water-alternating-gas (WAG) injection is also a leading EOR flooding process in light to medium oil sandstone and carbonate reservoirs. A recently proposed hybrid EOR method, $CO_2$ low salinity (LS) WAG injection, shows promise based on experimental and simulation studies, compared to LSWI or $CO_2$ injection alone. Wettability alteration is considered as the dominant mechanism for $CO_2$ LSWAG injection. In this study, a new displacement contact angle measurement which better mimics the actual displacement process taking place in a reservoir is used, aiming to investigate the effect of monovalent and divalent cations, $CO_2$, and injection schemes. It is found that the injection of NaCl low salinity water alters the wettability towards slightly water-wet, and the injection of $CaCl_2$ low salinity water alters the wettability towards slightly oil-wet. The injection of $CO_2$ promotes water-wetness and geochemical reactions between oil and brine. Injection scheme of $CO_2$ and NaCl low salinity water is more efficient than WAG cycle of $CO_2$/NaCl in wettability alteration towards more water-wet. However, the opposite trend is observed with $CaCl_2$ low salinity water, of which WAG cycle of $CO_2/CaCl_2$ is more efficient in altering wettability towards water-wet. The oil drop deformation process during LSWI resembles the process of oil removal using surfactant. As $CO_2$ is introduced, due to the acidic effect of $CO_2$ and ion exchange, it acts to wet the rock surface, leading to a more water-wet state. With introduction of $CO_2$, the oil drop deformation resembles the "roll-up" oil removal process.

## 1 Introduction

Low salinity water injection (LSWI) has been widely investigated and recognized as an effective enhanced oil recovery (EOR) method in both secondary and tertiary mode [1-3]. Compared to other chemical EOR methods, such as polymer or surfactant flooding, LSWI is advantageous due to its lower cost and reduced impact on the environment. Another advantage for LSWI is that it can be combined with other EOR methods to further improve oil recovery [4-6]. According to economic evaluations on chemical EOR methods by Al-Murayri et al. [7] and Muriel et al. [8], LSWI and $CO_2$ injection generate the highest net present value (NPV) and both methods are effective in increasing oil recovery. Therefore, a hybrid technique termed $CO_2$ salinity water-alternating-gas (LSWAG) injection, which combines the EOR effect of both methods, has been developed over the last 15 years. $CO_2$ LSWAG injection has been studied through core flooding experiments, contact angle and interfacial tension (IFT) measurements, primarily with sandstone, at ambient or reservoir conditions. Most results confirm improved oil recovery using this hybrid technique, in both secondary and tertiary modes, with some exceptions [9-11]. Studies with negative or neutral outcomes are mainly due to the fact that the cores are strongly water-wet or contain very small amount of clay minerals. Clean water-wet sandstones may not be the most favourable reservoir conditions for $CO_2$ LSWAG injection [12, 13].

The proposed mechanisms of $CO_2$ LSWAG injection are a combination of LSWI and $CO_2$ WAG injection. Al-Abri et al. [14] proposed that the improved oil recovery by immiscible $CO_2$ LSWAG injection is due to mobility control and wettability alteration. The IFT between high salinity brine and oil reduces as $CO_2$ is introduced. However, changes in the IFT of low salinity brine and oil are not noticeable, indicating that IFT reduction is not a dominant mechanism in this process. They also suggest multi-component ionic exchange (MIE) in which $Na^+$ substitutes the divalent cations ($Mg^{2+}$) accounts for the higher oil recovery when injecting monovalent NaCl brine compared to injection of $MgCl_2$ brine. Teklu et al. [15] claimed that $CO_2$ LSWAG injection improved oil recovery of conventional $CO_2$ WAG injection by forming in-situ carbonated water of higher $CO_2$ saturation in the brine phase due to the higher $CO_2$ solubility in low salinity water. This in-situ carbonated water promotes wettability alteration towards more water-wet and $CO_2$-brine IFT reduction, hence improved oil recovery. They also compared the $CO_2$-brine solubility model developed by Enick and Klara [16] and Li and Nghiem [17] with fresh water and

* Corresponding author: ljames@mun.ca

© The Authors, published by EDP Sciences. This is an open access article distributed under the terms of the Creative Commons Attribution License 4.0 (http://creativecommons.org/licenses/by/4.0/).

100,000 ppm NaCl at 71°C from 0 to 41 MPa. Both models show that $CO_2$ solubility in brine increases with pressure and $CO_2$ solubility is higher in fresh water. Chaturvedi et al. [18] and AlQuraishi et al. [11] suggest that fines migration and wettability alteration, mechanisms of LSWI, might be the dominant mechanisms for increased oil recovery by $CO_2$ LSWAG injection. The presence of clay minerals, especially kaolinite, is considered essential. However, this proposed mechanism is questioned by Zolfaghari et al. [19] as they achieved oil recovery in sandstone without kaolinite. Wettability alteration towards more water-wet was suggested by Al-Saedi et al. [20-22]. Based on the proposed mechanisms, wettability alteration and mobility control may be considered as the dominant mechanisms in $CO_2$ LSWAG injection.

Wettability alteration taking place during $CO_2$ LSWAG injection could be ascribed to LSWI or the acidic effect of $CO_2$ [23, 24] or a combination of both. Drummond and Israelachvili [25] demonstrate wettability alteration, indicated by contact angle measurements at ambient conditions for low salinity water, varying from oil-wet to water-wet as pH is lower than 9 and from water-wet to intermediate-wet as pH is greater than 9. The pH during a LSWI is mostly below 9 [26], indicating the wettability alteration is more likely to be from water-wet to intermediate-wet [27]. The main functions of injecting $CO_2$ are oil swelling, viscosity reduction due to $CO_2$ solubility in oil, miscibility with oil if pressure is above minimum miscible pressure and wettability modification [21]. Since low salinity water and $CO_2$ both impact wettability, the question remains as to whether the introduction of $CO_2$ in low salinity water will assist in promoting the geochemical reactions and low salinity effect. Generally, monovalent cations ($Na^+$) and divalent cations ($Ca^{2+}$ or $Mg^{2+}$) have different impacts on the rock surface. The MIE mechanism by LSWI proposed by Lager et al. [28] demonstrates that multivalent cations, such as $Ca^{2+}$, act as bridges between the oil polar components and the negatively charged rock surface, promoting oil-wetness. The mechanism of electrical double layer expansion by LSWI suggested by Ligthelm et al. [29] indicates that lowering the electrolyte content, especially reducing the content of multivalent cations, yields expansion of the electrical double layer surrounding the clay and oil, and an increase in zeta potential. This leads to wettability modification of the rock surface towards more water-wet. Wettability characterization is uncertain for $CO_2$ introduced with low salinity water in the sandstone/oil/brine/rock system since previous experimental data is limited.

To investigate wettability of the rock and fluid systems, direct or indirect and qualitative or quantitative methods can be used. Indirect and qualitative methods for characterizing rock wettability are capillarimetric method [30], spontaneous imbibition [31, 32], capillary pressure curves, and relative permeability method [33, 34]. Indirect and quantitative methods include the Amott and Amott-Harvey index method, USBM (U.S. Bureau of Mines) method, and the combined Amott/USBM method, which measure the average wettability of the rock samples, i.e., a macroscopic mean value of the rock wettability to a given fluid. Direct and quantitative method for characterizing the wettability of a specific surface is the contact angle measurement [35, 36], as well as the new SEM-MLA method introduced [37, 38]. Contact angle measurement works the best with pure fluids and well-prepared surfaces [39]. It can also be used to determine the effect of crude oil, brine chemistry, temperature and pressure on wettability. According to Arif et al. [40], direct contact angle measurement is a widely recognized technique for wettability characterization of rock/$CO_2$/brine or rock/oil/$CO_2$-enriched-brine systems. In this research we used contact angle measurement as a method to evaluate wettability alteration. Our particular interest is understanding the ion exchange and chemical reactions of the rock/oil/brine/ and rock/oil/brine/$CO_2$ systems by comparing the effect of monovalent and divalent ions and the effect of $CO_2$ on water-wet and oil-wet sandstone under different injection sequences. Contact angle measurements may indeed be the best choice to differentiate any wettability changes as we are able to 1) conduct measurements with high temperature and pressure, under which the commonly used Amott or USBM method is not applicable [40, 41] and 2) still appreciate the core scale aspects of rock mineralogy and some differences in water-wet and oil-wet "real" surfaces [42].

The data for contact angle measurements on sandstone/crude oil/$CO_2$-enriched brine system is very limited. Jaeger et al. [43] performed captive bubble contact angle measurements with sandstone samples which were previously aged at room temperature with 1.5 wt% cyclohexanepentanoic acid in decane for two months. They reported a strongly water-wet condition, contact angle of 46°, of such system under 50°C and 20.7 MPa with 32,000 ppm synthetic seawater. Ameri et al. [44] conducted contact angle measurements on Bentheimer sandstones that are initially water-wet and oil-wet at elevated pressure from 0.2 to 14 MPa and with NaCl brine salinity ranging from 5000 to 35,000 ppm. They found that in initially water-wet sandstones, contact angle increases with pressure, and decreases with increasing NaCl brine concentration. The overall contact angle for using NaCl brine salinity ranging from 20,000 to 35,000 ppm is very low, less than 20°. The average contact angle for using 5000 ppm NaCl brine is 40° and the value when distilled water is used is 57°. This indicates that decreasing salinity of NaCl brine leads to a less water-wet state, which differs from the results obtained by Espinoza et al [45]. For initially oil-wet sandstone, samples were aged for 22 months with crude oil at 60°C. In the sub-critical and near-critial state region of $CO_2$ (0-9 MPa), contact angle increase is slightly higher with 35,000 ppm brine than with distilled water. However, the opposite is observed for super-critical state region of $CO_2$ (9-13 MPa), Seyyedi et al. [46] reported that contact angle values of the quartz with $CO_2$-enriched brine are slightly higher than that with brine phase alone, indicating that injection of $CO_2$ alters wettability towards slightly less water-wet. However, Al-Abri et al. [14] reported contradictory results, showing that the contact angle was reduced with the addition of $CO_2$, indicating a wettability alteration towards more water-wet. They conducted their contact angle measurements with Berea sandstone discs aged with crude oil at 60° for three weeks. They also found that changes in wettability are greater with divalent ions ($Mg^{2+}$) than monovalent ions ($Na^+$ and $K^+$).

It is worth mentioning that there are limitations on these contact angle measurements that would result in misleading interpretations on the effect of $CO_2$ and low salinity water, and the working mechanisms of $CO_2$ LSWAG injection. In these measurements, the oil drop is introduced after the rock surface has been in contact with injection brine and $CO_2$, which is not representative of the actual displacement in a reservoir, where oil exists before the injection fluids. Therefore, in order to better capture and mimic the real displacement procedure taking place in a reservoir to investigate what triggers the wettability alteration during LSWI and $CO_2$ LSWAG injection, a displacement method developed by Sofla et al. [42] for measuring contact angle was employed in this study. We investigated the dynamic contact angle changes during the displacement process of LSWI, continuous $CO_2$ and LSWI, and $CO_2$ LSWAG injection, respectively. With this method, the interactions among crude oil, brine and $CO_2$ were investigated without the additional effect of capillary imbibition and drainage [47]. The objective is to compare the resulting wettability alteration (through contact angle measurements) due to multi-component ion exchange (MIE), chemical reactions, and injection sequence of low salinity water. This paper addresses the question as to whether or not the ionic charge and injection scheme play a role in differentiating LSWI and $CO_2$ LSWAG injection at the fluid-rock interaction level.

## 2 Materials

***Fluids.*** The oil phase used in this experiment is an offshore Newfoundland and Labrador (NL) light crude oil. Synthetic brines were prepared to mimic the Hibernia formation water and Grand Banks seawater. The two low salinity brines are 2000 mg/L NaCl and 2000 mg/L $CaCl_2$. Their chemical compositions and basic properties are presented in Table 1.

**Table 1.** Compositions and properties of synthetic brines and oil

| Component | FB | SW | NaCl | $CaCl_2$ |
|---|---|---|---|---|
| $Na^+$, mg/L | 35,671 | 10,974 | 786 | / |
| $Mg^{2+}$, mg/L | 330 | 1,310 | / | / |
| $Ca^{2+}$, mg/L | 3,599 | 420 | / | 721 |
| $K^+$, mg/L | 255 | 407 | / | / |
| $Cl^-$, mg/L | 62,371 | 19,740 | 1,214 | 1,279 |
| $SO_4^{2-}$, mg/L | 233 | 2,766 | / | / |
| $HCO_3^-$, mg/L | / | 129 | / | / |
| **Total** | **102,430** | **35,746** | **2,000** | **2,000** |
| pH@22° C | 5.9 | 7.9 | 6.2 | 5.8 |
| Density, g/cm³ | 1.074 | 1.023 | 1.0 | 1.0 |
| Oil Viscosity, cP | 5.0 | | | |
| Oil Density, g/cm³ | 0.878 | | | |

**Legends:** FB – formation brine; SW – seawater; / - not included

***Rock sample.*** Berea sandstone with 80% quartz content and <2% clay content [37] was used in this study. The core samples with an approximate diameter of 2 cm were cut into 5-mm thin slices using MK-370EXP Tile Saw. The dust was blown off with pressurized nitrogen and the core slices were dried in oven overnight. The water-wet samples were immersed in formation brine for one day prior to conducting contact angle measurements. To obtain an oil-wet initial wettability, the core slices initially immersed in formation brine were removed to a beaker containing NL crude oil and aged at 98°C in the oven for six weeks as suggested by Sripal et al. [37] to obtain oil-wet conditions. Subsequently, the surface of the oil-wet samples was cleaned and immersed in formation water for one day before contact angle measurements.

## 3 Methods

Figure 1 illustrates the experimental setup used to perform contact angle measurements in this study. All experiments are conducted at ambient conditions. In order to mimic the actual displacement taking place in a reservoir, the cell is initially filled with formation brine. Subsequently, an oil drop with radius ranging from 1.8 to 2.0 mm is introduced through the needle at the bottom of the cell and adheres onto the rock surface. The initial contact angles are measured after the system reached equilibrium, which is 1 hour after the oil drop was introduced into the system. This indicates the initial wettability of the rock surface and initial condition of an oil droplet in a reservoir.

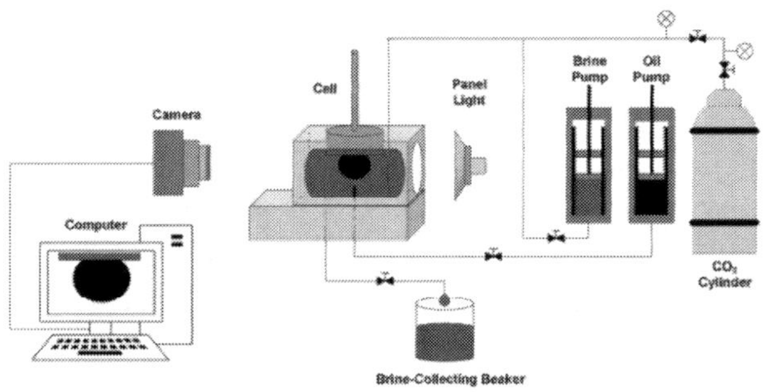


**Fig. 1** Schematic diagram of contact angle measurement.

A total of 13 experiments were completed, using the injection schemes outlined in Table 2. Each scenario was completed twice, once using NaCl as the LSW and again using $CaCl_2$ as the LSW.

**Table 2.** Injection scheme of experiments

| Scenario # | Injection Scheme | | | |
|---|---|---|---|---|
| | Cycle 1 (60 mL) | Cycle 2 (20 mL) | Cycle 3 (20 mL) | Cycle 4 (20 mL) |
| 1 | SW | SW | SW | SW |
| 2 | SW | LSW | LSW | LSW |
| 3 | LSW | LSW | LSW | LSW |
| 4 | SW | $CO_2$ (10ml) | LSW (25mL) | LSW (25mL) |
| 5 | SW | $CO_2$/LSW | $CO_2$/LSW | $CO_2$/LSW |
| 6 | LSW | $CO_2$ (10mL) | LSW (25mL) | LSW (25mL) |
| 7 | LSW | $CO_2$/LSW | $CO_2$/LSW | $CO_2$/LSW |

Scenario #1 represents seawater injection. Scenario #2 and #3 represent LSWI. Scenarios #4 and #6 represents seawater or low salinity water injection, followed by continuous $CO_2$ injection and LSWI. Scenario #5 and #7 represent seawater or low salinity water injection, followed by LSWAG injection. Brines and $CO_2$ are injected through the injection inlet into the cell to displace the existing fluid. The injection speed is controlled so that the oil drop remains attached on the rock surface throughout the experiment.

The total volume of the cell is 20 mL. In cycle 1, 60 mL of seawater or low salinity water is injected to ensure that the initial formation brine is fully displaced. The system is allowed to set for equilibrium for half an hour after every 20 mL of injection fluid and the reading at equilibrium state is taken. Figure 2, as an example, shows the contact angle changes during the half-an-hour equilibrium time of scenario #1, indicating that an equilibrium was gradually established.

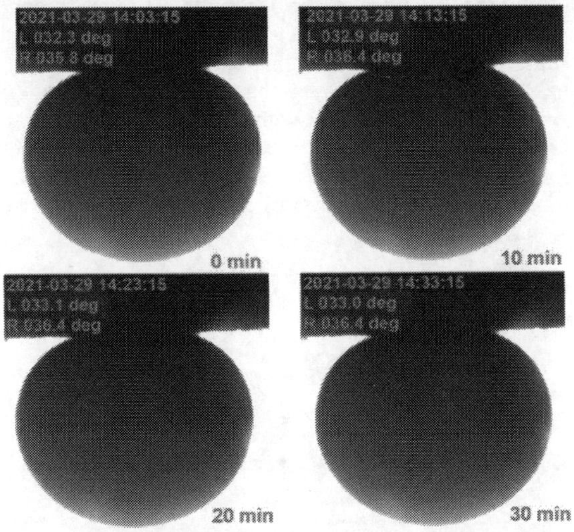


**Fig. 2** Contact angle changes after seawater injection during half-an-hour equilibrium time (scenario #1).

After cycle 1, another 60 mL of $CO_2$ (g), low salinity water and a combination of both are further injected in cycles 2-4 representing the injection schemes of continuous $CO_2$ and low salinity water injection, and the $CO_2$ LSWAG process. Contact angles are measured dynamically for each injection cycle and measurements are taken half an hour after each injection cycle. Three distinct measurements are carried out to monitor repeatability. Contact angles are reported as averages of the three measurements. The change in contact angle is calculated using the equation below.

$$\Delta\theta\,[\%] = (\theta - \theta_{initial})/\theta_{initial} \times 100 \quad (1)$$

where $\Delta\theta$ refers to the change in contact angle, $\theta$ is the value of contact angle measured after each injection cycle, and $\theta_{initial}$ is the initial contact angle measured with the presence of formation water. The reason for comparing changes instead of absolute contact angles is to avoid the influence of the samples and each scenario starts from the same point. Initial contact angle is also reported.

To calculate the uncertainty, or error propagation of $\Delta\theta$, the root-sum square method proposed by Kline and McClintock is used [48]. The effect of uncertainty $\sigma_{\Delta\theta}$ on the calculated $\Delta\theta$ can be expressed as follow:

$$\sigma_{\Delta\theta} = \sqrt{\sigma_\theta^2 * \left(\frac{\partial(\Delta\theta)}{\partial\theta}\right)^2 + \sigma_{\theta_{initial}}^2 * \left(\frac{\partial(\Delta\theta)}{\partial\theta_{initial}}\right)^2} \quad (2)$$

Subsequently, changes in contact angle with calculated uncertainty are plotted against injected volume to investigate the effect of low salinity water, injection of $CO_2$ and WAG injection schemes.

Moreover, in order to validate that the measured contact angle changes are mainly due to the chemical reactions (intermolecular forces) in the oil/brine/rock system, rather than gravitational force, we have estimated the Bond number ($B_o$) of the oil/seawater/brine system using equation from Li et al. [49].

$$B_o = \frac{\Delta\rho g L^2}{\gamma} \quad (3)$$

where $\Delta\rho$ is the density difference of oil and brine (kg/m$^3$), $g$ is gravitational acceleration (m/s$^2$), $L$ refers to the radius of curvature of oil drop (m), $\gamma$ is surface tension (N/m). With the measured surface tension (31.5 mN/m), and oil drop radius in seawater (1.86 mm), Bond number is calculated to be 0.154, which is lower than 1, indicating that surface tension dominates.

## 4 Results and Discussion

As shown in Table 3, section 4.1 investigates the effect of seawater and low salinity water (scenario #1, #2 and #3) on wettability alteration of water-wet and oil-wet Berea sandstone samples. Section 4.2 discusses the effect of $CO_2$ by comparing scenario #2 and #4, and #3 and #6. Subsequently, the deformation process of the oil drops during the injection of low salinity water and $CO_2$ is investigated in section 4.3. In the end, section 4.4 studies the effect of different injection schemes by comparing $CO_2$ + LSW injection scheme to $CO_2$/LS WAG injection scheme (#4 and #5, and #6 and #7). The effect of monovalent and divalent cations is discussed and compared in all sections.

**Table 3.** Comparison of different scenarios.

| Section | Comparison of different scenarios | |
|---|---|---|
| 4.1 Effect of Low Salinity Water | #1 | SW + SW |
| | #2 | SW + LSW (NaCl and $CaCl_2$) |
| | #3 | LSW + LSW (NaCl and $CaCl_2$) |
| 4.2 Effect of $CO_2$ | #2 | SW + LSW |
| | #4 | SW + $CO_2$ + LSW |
| | #3 | LSW + LSW |
| | #6 | LSW + $CO_2$ + LSW |
| 4.4 Effect of Injection Scheme | #4 | SW + $CO_2$ + LSW |
| | #5 | SW + $CO_2$/LS WAG |
| | #6 | LSW + $CO_2$ + LSW |
| | #7 | LSW + $CO_2$/LS WAG |

### 4.1 Effect of Low Salinity Water

Contact angle changes due to the injection of seawater alone, low salinity waters alone, and combinations of seawater and low salinity water are shown in Figures 3 and 4 for water-wet and oil-wet sandstones, respectively. These injection schemes mimic the displacement process of (1) seawater injection, (2) secondary seawater and tertiary LSWI, and (3) LSWI. Overall, changes in contact angle in the oil-wet samples are not as significant as in water-wet samples. However, it is

https://doi.org/10.1051/e3sconf/202336601019

worth comparing and understand the changing trend after each injection cycle, which could be an estimation for the potential changes in a core scale experiment.

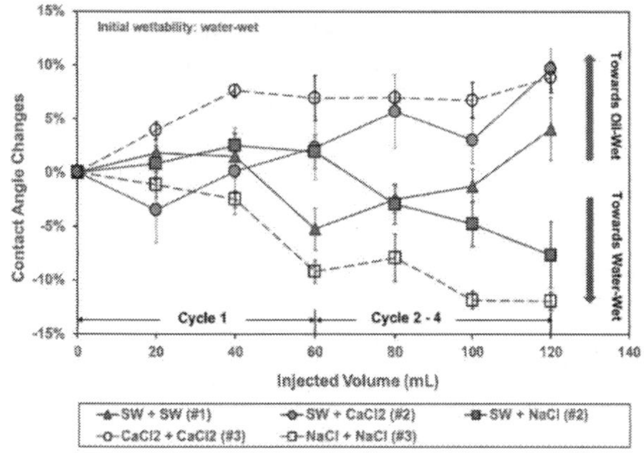


**Fig. 3** Contact angle changes during seawater and low salinity water injection in water-wet sandstone (scenario #1, #2, #3).

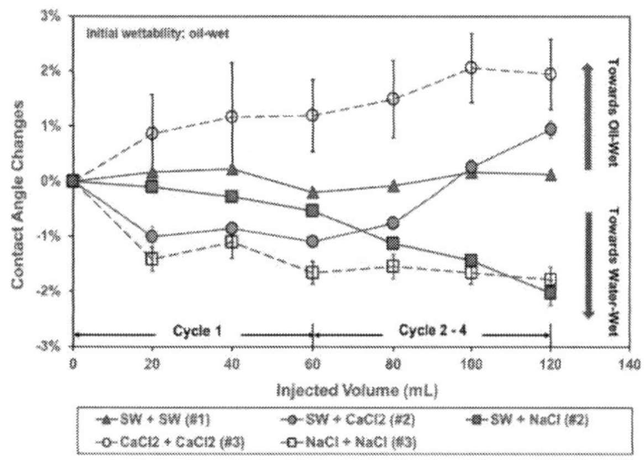


**Fig. 4** Contact angle changes during seawater and low salinity water injection in oil-wet sandstone (scenario #1, #2, #3).

In Figure 3, the initial wettability of the rock sample is water-wet, with measured contact angles varying from 34° to 50° (average: 40.6°±5.0°). Contact angle changes due to the injection of seawater are within 5%, which is not very significant. This indicates that the injection of seawater has negligible effect on the rock wettability. The trend of using NaCl and $CaCl_2$ in LSWI shows different impacts on wettability. The red arrow in Figure 2 indicates changing towards more oil-wet and the blue arrow suggests changing towards more water-wet. It is seen that NaCl alters the wettability towards more water-wet, around 10% less compared to initial contact angle, whereas $CaCl_2$ results in wettability alteration moving to less water-wet. A similar trend is also observed in the combined seawater and LSWI process.

In Figure 4, the initial wettability of the rock sample is oil-wet, with measured contact angle varying from 117° to 155° (average: 133.0°±13.5°). For seawater injection (SW + SW), the contact angle remains almost constant throughout the process. The injection of NaCl LSW alters the rock wettability towards slightly less oil-wet (SW + NaCl, NaCl + NaCl) and use of $CaCl_2$ (SW + $CaCl_2$, $CaCl_2$ + $CaCl_2$) alters the wettability towards more oil-wet. This observation agrees with that in the water-wet samples where NaCl promotes water-wetness and $CaCl_2$ promotes oil-wetness.

Generally, the configuration of water on rock mineral surfaces exist in two ways: (1) pendular-ring on contact points of grains; and (2) thin film on the mineral surfaces [50]. In this study, the oil drop is introduced after formation water and is kept attached to the surface throughout the experiment. Therefore, the model proposed is as shown in Figure 5, where a thin water film is formed between the rock and oil drop. A similar model was also proposed by Lee et al. [51]. They manufactured sand/clay like silica particles using simple anionic surface similar to sand grain and measured the thickness of this water film to be roughly 9-15 nm. According to their measurements on the simple wet system (fabricated simple anionic surface, similar to a sand grain) [51], the thickness of the water film on the silica/clay (sandstone-like) surface is thicker in brines with lower salinities (except for pure water). Therefore, in a system where the substrate is initially oil-wet, in order to alter the wettability from oil-wet to intermediate-wet or water-wet, a thicker water film along the pore wall is needed.

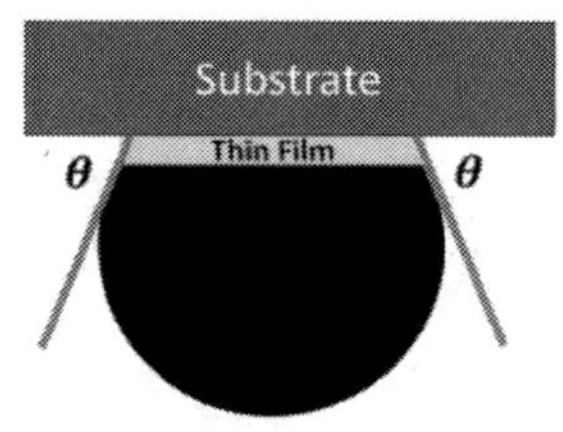


**Fig. 5** A proposed model with water thin film forming between the rock/brine and oil/brine interface (Adapted from Lee et al. [51]).

Based on the results from Figure 3 and Figure 4, it is observed that the use of monovalent cations as injection brine alters the wettability towards more water-wet, which agrees with the finding from Xie et al. [52] that monovalent cations ($Na^+$) give rise to positive disjoining pressure; however, divalent cations ($Ca^{2+}$) lead to negative disjoining pressure at the same concentration. Negative disjoining pressure between rock surface and oil droplet indicates the attractive force is dominant; thus, more oil-wet is expected for the rock surface. On the other hand, positive disjoining pressure suggests the repulsive force between the rock surface and oil droplet, leading to more water-wet.

### 4.2 Effect of $CO_2$

The wettability changes caused by $CO_2$ after seawater injection and LSWI (NaCl or $CaCl_2$) are investigated by comparing the contact angle changes in scenarios with $CO_2$ (#4 and #6) and without $CO_2$ (#2 and #3). Contact angle changes during cycle 2 – 4 are studied. For LSWI (#2 and #3), cycle 2-4 are injection of LSW. For $CO_2$ + LSW (#4 and #6), cycle 2 is injection of $CO_2$, cycle 3-4 are injection of LSW. To calculate the changes, $\theta_{initial}$ in Eq. (1) is not the initial value in cycle 1, but the equilibrium contact angle measured after cycle 1 ($\theta_{cycle\ 1,eq}$). Hence, Eq. (4) is used to calculate contact angle changes ($\Delta\theta$) and uncertainty is calculated according to Eq. (2).

$$\Delta\theta\ [\%] = \left(\theta - \theta_{cycle\ 1,eq}\right)/\theta_{cycle\ 1,eq} \times 100 \quad (4)$$

Based on this, all the scenarios investigated in this section will start from the same point in cycle 2 with respect to contact angle change.

Figures 6 and 7 show the results of LSWI and $CO_2$ + LSWI after seawater injection in water-wet and oil-wet samples respectively. When comparing scenario #2 (SW + LSW) and #4 (SW + $CO_2$ + LSW), the addition of $CO_2$ after seawater promotes water-wetness for both water-wet and oil-wet samples. After $CO_2$ injection, further injection of $CaCl_2$ changes the wettability towards more oil-wet, and the injection of NaCl changes further more towards water-wet.

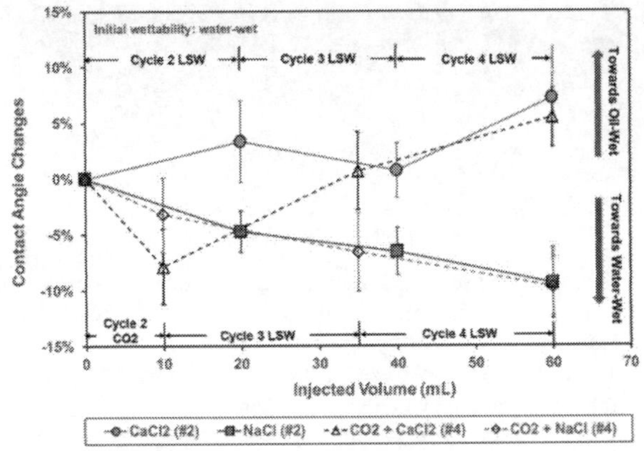


**Fig. 6** Comparison of contact angle changes during LSWI and $CO_2$ + LSWI in water-wet sandstone after 1st cycle of seawater injection (scenario #2 and #4).

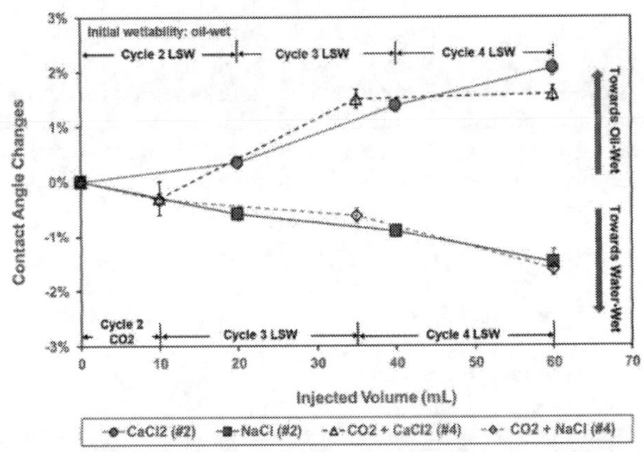


**Fig. 7** Comparison of contact angle changes during LSWI and $CO_2$ + LSWI in oil-wet sandstone after 1st cycle of seawater injection (scenario #2 and #4).

The contact angle changes of LSWI (scenario #3) and $CO_2$ + LSWI (scenario #6) after LSWI in cycle 1 are shown in Figure 8 and Figure 9. Injection of LSW in cycle 2-4 has no significant impact on contact angle after the 1st cycle of LSWI. However, with the injection of $CO_2$ in cycle 2 and NaCl in cycle 3-4, $CO_2$ + NaCl alters wettability towards more water-wet in both water-wet and oil-wet samples (Figure 8 and Figure 9). For scenario #6 ($CO_2$ + $CaCl_2$), $CO_2$ alters wettability towards more water-wet, whereas further injection of $CaCl_2$ changes the wettability to more oil-wet.

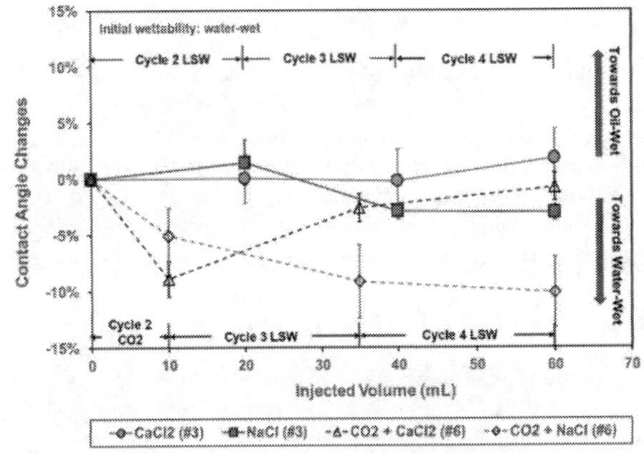


**Fig. 8** Comparison of contact angle changes during LSWI and $CO_2$ + LSWI in water-wet sandstone after 1st cycle of LSWI (scenario #3 and #6).

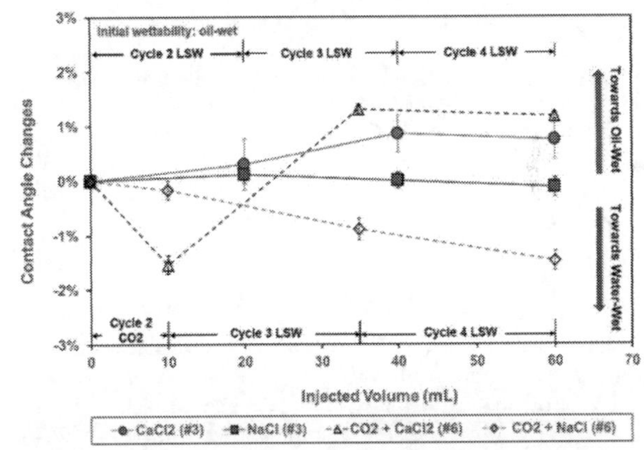


**Fig. 9** Comparison of contact angle changes during LSWI and $CO_2$ + LSWI in oil-wet sandstone after 1st cycle of LSWI (scenario #3 and #6).

It is also observed from Figure 8 and 9 that $CO_2$ injected after $CaCl_2$ low salinity water alters wettability towards more water-wet compared to that injected after NaCl low salinity water. As suggested by Lager et al. [53], it is possible that divalent cations are exchanged for monovalent cations during LSWI. Therefore, in our case, more $Ca^{2+}$ on the rock surface is expected in scenario #6 with $CaCl_2$.

When $CO_2$ is in contact with water, it first dissolves according to reaction (5):

$$CO_2(g) \rightleftharpoons CO_2(aq) \tag{5}$$

At room temperature, solubility of $CO_2(g)$ is 0.034 mol/L. Subsequently, reaction (6) takes place to form $H_2CO_3$. This reaction is kinetically slow and only a small fraction (0.2 – 1.0%) of dissolved $CO_2$, $CO_2(aq)$, is converted to $H_2CO_3$.

$$CO_2(aq) + H_2O \rightleftharpoons H_2CO_3(aq) \tag{6}$$

However, this carbonic acid dissociates very rapidly at ambient conditions to bicarbonate [54], as shown in reaction (7). The bicarbonate electrolyte in the solution can also form $CO_3^{2-}$ as shown by reaction (8).

$$H_2CO_3(aq) \rightleftharpoons HCO_3^- + H^+ \tag{7}$$

 https://doi.org/10.1051/e3sconf/202336601019

$$HCO_3^- \rightleftharpoons CO_3^{2-} + H^+ \quad (8)$$

With the injection of $CO_2$, some of the produced $CO_3^{2-}$ would potentially react with the existing $Ca^{2+}$ ions, forming $CaCO_3$, which results in equation (8) to move to the right direction, leading to slight increase in $H^+$. Based on the selectivity of cation affinity to negatively charged surfaces from Velde [55], as shown below, the proton $H^+$ has the strongest affinity to be adsorbed onto a negatively charged surface.

$$Li^+ < Na^+ < K^+ < Mg^{2+} < Ca^{2+} < H^+$$

Therefore, the generated $H^+$ is likely to replace the pre-attached divalent cations, resulting in more water-wetness. In this way, the injection of $CO_2$ after $CaCl_2$ low salinity water alters wettability towards more water-wet compared to injection of $CO_2$ after NaCl low salinity water.

### 4.3 Surfactant-Like Behavior of Oil Drops

During the injection of LSW and a combination of $CO_2$ and LSW in the water-wet and oil-wet samples, a surfactant-like deformation process of the oil drop is constantly observed when the initial equilibrium of the system is disturbed. The oil drop deformation with and without $CO_2$ is discussed respectively in the subsections.

#### *4.3.1 Deformation in the absence of $CO_2$*

Figure 10 and Figure 11 illustrate the deformation process during the injection of NaCl LSW in water-wet and $CaCl_2$ LSW in oil-wet sandstones, respectively. For NaCl LSWI in water-wet sample, the contact angle varies from water-wet to intermediate-wet and then back to more water-wet while reaching equilibrium. For $CaCl_2$ LSWI in oil-wet sample, contact angle changes from oil-wet to intermediate-wet and then back to more oil-wet in the end.

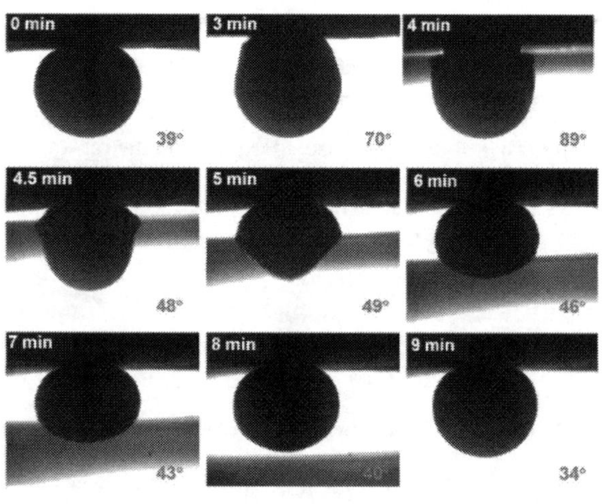


**Fig. 10** Oil drop deformation process during NaCl LSWI in water-wet sandstone (scenario #3).

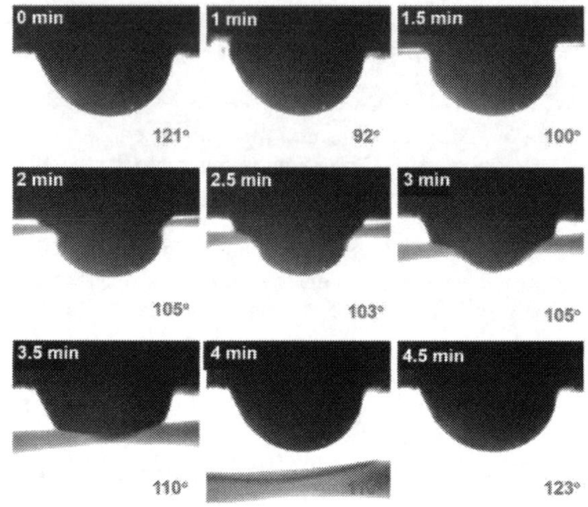


**Fig. 11** Oil drop deformation process in time during $CaCl_2$ LSWI in an oil-wet sandstone system (scenario #3).

Based on Figures 10 and 11, the deformation process during LSWI resembles a surfactant-like behaviour. The potential removal of the droplet exhibits a "necking" or emulsification mechanism. One mechanism for LSWI proposed by McGuire et al. [56] suggests that the changes in wettability during low salinity water injection appear to be similar to the observations from alkaline and surfactant flooding. In this study, the interactions between the oil drop and injection fluids are more dominant due to the presence of just one oil drop. As listed in Table 1, the pH of the injection fluids is higher compared to that of the initial formation water. During the injection of low salinity water, in-situ "surfactants" are generated, as shown in Eq. (9), when the oil drop is in contact with the elevated pH fluid near the rock and oil surfaces. This improves oil recovery [56]. In this way, low salinity water injection is similar to micellar or surfactant flooding.

$$(R_1COO)_3R_2 + 3NaOH \leftrightarrow 3(R_1COONa) + R_2(OH)_3 \quad (9)$$

where $R_1$ and $R_2$ represent the R group, which consists of a group of carbon and hydrogen atoms.

According to the study of oil removal from soil surfaces by Miller and Raney [57], two approaches are proposed as mechanisms for oil removal from hydrophobic surfaces using surfactants: (1) roll-up resulting from wetting; and (2) emulsification resulting from reduction in interfacial tension (Figure 12).

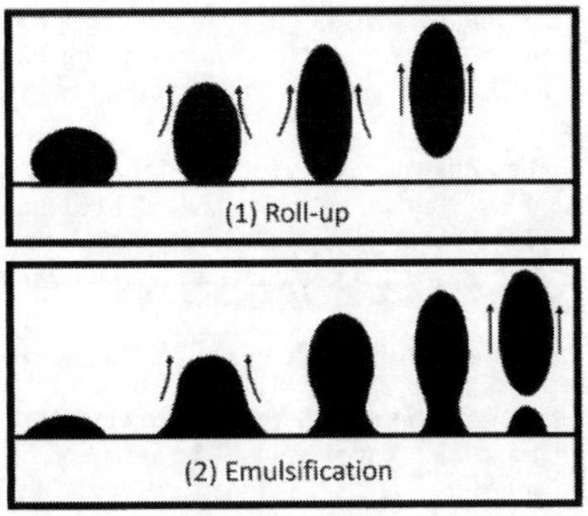


**Fig. 12** Mechanisms of oil removal from surface by (1) roll-up and (2) emulsification (adapted from Miller and Raney [57]).

### 4.3.2 *Deformation with $CO_2$ present*

The top two pictures shown in Figure 13 are the oil drop deformation during injection of $CO_2$ + $CaCl_2$, (scenario #4) and the bottom two pictures are during $CaCl_2$ LSWI (scenario #2). These two deformation processes resemble the two approaches in Figure 12. Without addition of $CO_2$, the detachment of the oil drop is a saponification or emulsification process. However, the roll-up process is expected with $CO_2$ due to the geochemical reactions that change the wetting state of the contact point on the rock surface.

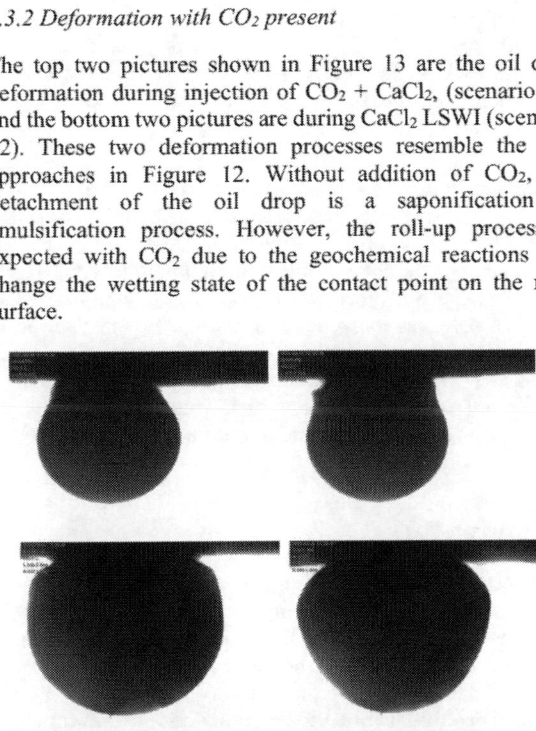

**Fig. 13** Comparison of oil drop deformation process: top (scenario #4): during $CO_2$ and $CaCl_2$ injection (roll-up); bottom (scenario #2): during $CaCl_2$ injection (emulsification) after seawater injection in water-wet sandstone.

## 4.4 Effect of Injection Scheme

The impact of different injection schemes with respect to $CO_2$ is explored by comparing the scenarios of SW + $CO_2$ + LSW (#4) and SW + LSWAG (#5), and scenarios of LSW + $CO_2$ + LSW (#6) and LSW + LSWAG (#7).

Contact angle changes of scenario #4 and #5 are shown in Figure 14 and Figure 15. It is observed that after 1st cycle of seawater injection, further injection of $CO_2$ + $CaCl_2$ alters wettability in the direction of more oil-wet, however, $CO_2$/$CaCl_2$ WAG injection alters rock wettability towards more water-wet. Injection schemes of $CO_2$ + NaCl and $CO_2$/NaCl WAG both change the rock wettability to more water-wet, whereas changes are more significant for $CO_2$ + NaCl. The wettability changes by $CO_2$/LSWAG processes (for both NaCl and $CaCl_2$) are not as significant as $CO_2$ + LSW injection when implemented after seawater injection (Figure 14 and Figure 15).

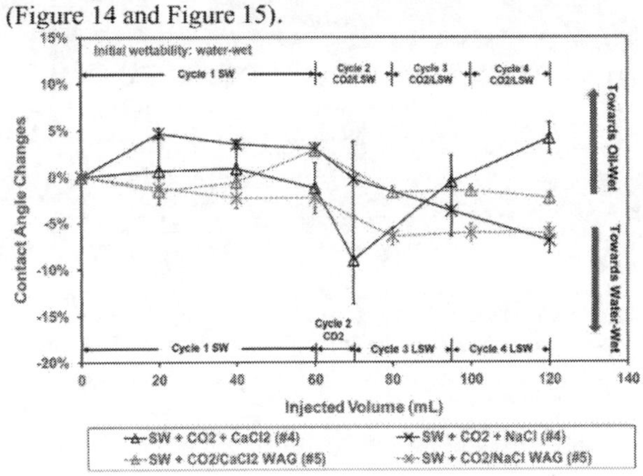


**Fig. 14** Comparison of contact angle changes during SW + $CO_2$ + LSWI and SW + $CO_2$/LS WAG injection in water-wet sandstone (scenario #4 and #5).

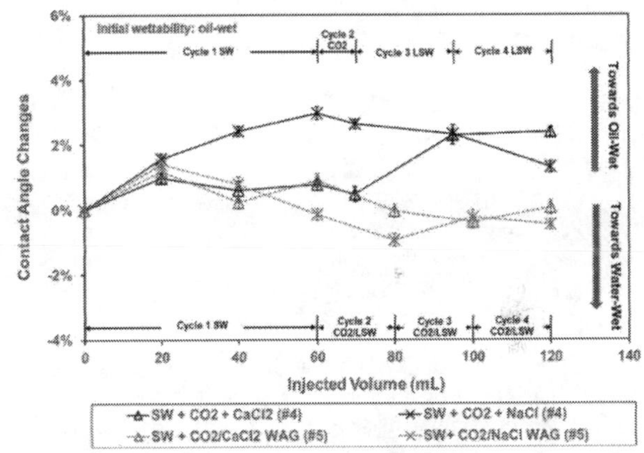


**Fig. 15** Comparison of contact angle changes during SW + $CO_2$ + LSWI and SW + $CO_2$/LSWAG injection in oil-wet sandstone (scenario #4 and #5), error bars are too small to be seen.

Figure 16 and Figure 17 show the comparison between #6 (LSW + $CO_2$ + LSW) and #7 (LSW + $CO_2$/LSWAG) in water-wet and oil-wet samples. The results in the figures show that the addition of $CO_2$ has a minor effect on the wettability for the $CO_2$/NaCl WAG process. When comparing $CO_2$ + $CaCl_2$ with $CO_2$/$CaCl_2$ WAG injection in both water-wet and oil-wet samples, the trend for wettability change is different. For the scheme of $CO_2$ + $CaCl_2$ injection, the addition of $CO_2$ promotes water-wetness of the rock, and the chasing $CaCl_2$ low salinity water changes the wettability back to more oil-wet. However, in the $CO_2$/$CaCl_2$ WAG process, the wettability is altered towards more water-wet.

 https://doi.org/10.1051/e3sconf/202336601019

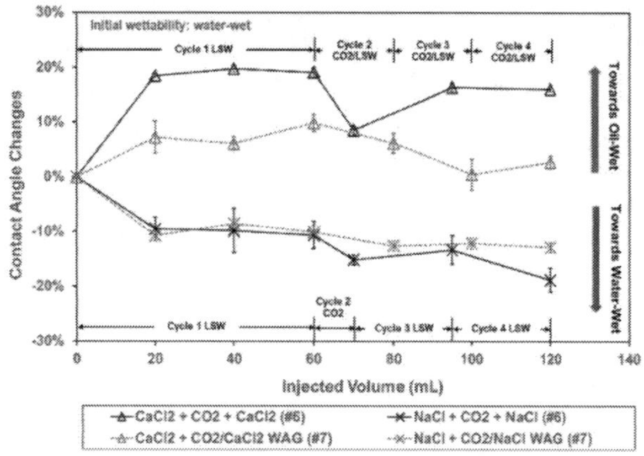


**Fig. 16** Comparison of contact angle changes during LSWI, $CO_2$ and LSWI, and $CO_2$ LSWAG injection in water-wet sandstone (scenario #6 and #7), some error bars are too small to be seen.

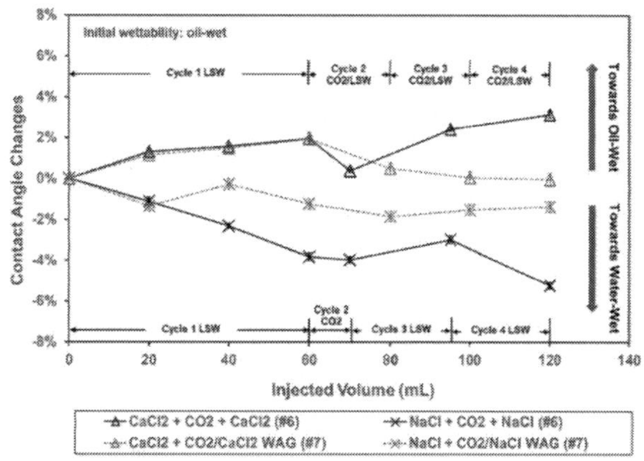


**Fig. 17** Comparison of contact angle changes during LSWI, $CO_2$ and LSWI, and $CO_2$ LSWAG injection in oil-wet sandstone (scenario #6 and #7), error bars are too small to be seen.

The WAG process of $CO_2$ and $CaCl_2$ low salinity water leads to wettability alteration to slightly water-wet. With respect to the scenario of $CO_2$ + $CaCl_2$ (#6), even though $CO_2$ changes wettability to be more water-wet, the generated $H^+$ is not sufficient. Thus, subsequent injection of $CaCl_2$ replaces the monovalent cations and alters the rock wettability towards more oil-wet.

Summarising, if NaCl LSW is used, the continuous $CO_2$ + NaCl injection scheme is more efficient than WAG cycle of $CO_2$/NaCl in achieving a more water-wet condition of sandstone. However, if $CaCl_2$ LSW is used, WAG cycle of $CO_2/CaCl_2$ can alter the rock wettability to be more water-wet compared to continuous $CO_2$ + $CaCl_2$ injection.

## 5 Conclusion

In this study, a displacement method for measuring contact angle changes during the process of seawater injection, LSWI, $CO_2$ and LSWI, and $CO_2$ LSWAG injection has been conducted. Seawater, low salinity water with only monovalent and divalent cations are selected as the injection aqueous phases. The effect of these ions, oil drop deformation process, and the effect of $CO_2$ and injection scheme have been investigated.

1. It is found that for our Berea sandstone with an initial wettability of either water-wet and oil-wet[1], the injection of 2000 ppm NaCl water alters the wettability towards slightly water-wet, and the injection of 2000 ppm $CaCl_2$ alters the wettability towards slightly oil-wet. Low salinity water with divalent cation could increase the attraction forces between the oil/rock and oil/brine interfaces, promoting oil-wetness. However, low salinity with monovalent cation reduces the attraction forces, i.e., repulsive force increases, therefore, resulting in more water-wet.
2. The deformation process during LSWI resembles the process of oil removal using surfactant. This "surfactant-like" behaviour lowers the interfacial tension and contributes to increased oil recovery. As $CO_2$ is introduced, due to the acidic effect of $CO_2$, it acts to wet the rock surface, leading to a more water-wet state. Therefore, the oil removal or oil drop deformation resembles the "roll-up" oil removal process.
3. The injection of $CO_2$ promotes water-wetness and geochemical reactions between oil and brine. In the WAG process, more interactions between injection brine, $CO_2$ and pre-existing brine are expected, and this leads to different wettability alteration trend compared to $CO_2$ + LSWI. When NaCl LSW is used, continuous $CO_2$ + NaCl injection scheme is more efficient than WAG cycle of $CO_2$/NaCl in wettability alteration towards more water-wet. However, with $CaCl_2$ LSW, WAG cycle of $CO_2/CaCl_2$ can alter the rock wettability to be more water-wet compared to continuous $CO_2$ + $CaCl_2$ injection.

## 6 Future Work

In this study, all the measurements are conducted at ambient condition. The effect of temperature and pressure is not considered. As the temperature and pressure exceeds the critical point for $CO_2$, the state of $CO_2$ will become supercritical, with properties midway between a gas and a liquid. Therefore, in order to better understand the wettability alterations with supercritical $CO_2$, more research with respect to elevated temperature and pressure should be carried out in the future. If wettability alteration is considered as the main mechanism for LSWI or $CO_2$ LSWAG injection, this displacement contact angle measurement which mimics the real reservoir displacement process could be used as a preliminary screening for brine concentration and composition, as well as injection schemes. However, to achieve a systematic evaluation process, more experimental data with respect to temperature and pressure are required.

[1] Note: wettability was inferred from the contact angle and was not independently verified by measurements like USBM or Amott.

## 7 Acknowledgements

*The authors would like to thank the Hibernia Management and Development Company (HMDC), Chevron Canada Ltd, Petroleum Research Newfoundland and Labrador (PRNL), the Natural Sciences and Engineering Research Council of Canada (NSERC), the Province of Newfoundland and Labrador, and Mitacs for financial support. We thank our colleagues in the Hibernia EOR Research Group for their technical support.*

## References

1. N. Morrow, and J. Buckley, Improved oil recovery by low-salinity waterflooding. Journal of petroleum Technology, **63**(05), pp.106-112, SPE-129421-JPT. (2011).
2. K.J. Webb, C.A. Black, H. Al-Ajeel, Low salinity oil recovery-log-inject-log. In SPE/DOE Symposium on Improved Oil Recovery, SPE-89379-MS. (2004).
3. K.J. Webb, C.J.J Black, I.J. Edmonds, Low salinity oil recovery - The role of reservoir condition corefloods. In IOR 2005-13th European Symposium on Improved Oil Recovery (pp. cp-12). European Association of Geoscientists & Engineers. (2005).
4. L. Moghadasi, P. Pisicchio, M. Bartosek, E. Braccalenti, P. Albonico, I. Moroni, R. Veschi, F. Masserano, S. Scagliotti, L. Del Gaudio, M. De Simoni, Laboratory investigation on synergy effect of low salinity-polymer water injection on sandstone porous media. In Offshore Mediterranean Conference and Exhibition, OMC-2019-0868. (2019).
5. E. Alagic, A. Skauge, Combined low salinity brine injection and surfactant flooding in mixedwet sandstone cores. Energy & fuels, **24**(6), pp.3551-3559. (2010).
6. T.W. Teklu, W. Alameri, H. Kazemi, R.M. Graves, A.M. AlSumaiti, Low salinity water - Surfactant - CO2 EOR. Petroleum, **3**(3), pp.309-320. (2017).
7. M.T. Al-Murayri, H.E. Al-Mayyan, K. Moudi, F. Al-Ajmi, D. Pitts, M.J. Wyatt, K.French, J. Surtek, E. Dean, Chemical EOR Economic Evaluation in a Low Oil Price Environment: Sabriyah Lower Burgan Reservoir Case Study. In SPE EOR Conference at Oil and Gas West Asia, SPE-190337-MS. (2018).
8. H. Muriel, S. Ma, S.J.D. Sofla, L.A. James, Technical and Economical Screening of Chemical EOR Methods for the Offshore. In Offshore Technology Conference, OTC-30740-MS. (2020).
9. H. Jiang, L. Nuryaningsih, H. Adidharma, The effect of salinity of injection brine on water alternating gas performance in tertiary miscible carbon dioxide flooding: experimental study. In SPE Western Regional Meeting, SPE-132369-MS. (2010).
10. R. Ramanathan, A.M. Shehata, H.A. Nasr-El-Din, Water Alternating CO2 Injection Process-Does Modifying the Salinity of Injected Brine Improve Oil Recovery?. In OTC Brasil, OTC-26253-MS. (2015).
11. A.A. AlQuraishi, A.M.Amao, N.I. Al-Zahrani, M.T. AlQarni, S.A. AlShamrani, Low salinity water and CO2 miscible flooding in Berea and Bentheimer sandstones. Journal of King Saud University-Engineering Sciences, **31**(3), pp.286-295. (2019).
12. N.R. Morrow, G.Q. Tang, M. Valat, X. Xie, Prospects of improved oil recovery related to wettability and brine composition. Journal of Petroleum science and Engineering, **20**(3-4), pp.267-276. (1998).
13. N.R. Morrow, Wettability and its effect on oil recovery. Journal of petroleum technology, **42**(12), pp.1476-1484. (1990).
14. H. Al-Abri, P. Pourafshary, N. Mosavat, H. Al Hadhrami, A study of the performance of the LSWA CO2 EOR technique on improvement of oil recovery in sandstones. Petroleum, **5**(1), pp.58-66. (2019).
15. T.W. Teklu, W. Alameri, R.M. Graves, H Kazemi, A.M. AlSumaiti, Low-salinity water-alternating-CO2 flooding enhanced oil recovery: theory and experiments. In Abu Dhabi International Petroleum Exhibition and Conference, SPE-171767-MS. (2014).
16. R.M. Enick, S.M. Klara, CO2 solubility in water and brine under reservoir conditions. Chemical Engineering Communications, **90**(1), pp.23-33. (1990).
17. Y.K. Li, L.X.Nghiem, Phase equilibria of oil, gas and water/brine mixtures from a cubic equation of state and Henry's law. The Canadian Journal of Chemical Engineering, **64**(3), pp.486-496. (1986).
18. K.R Chaturvedi, D. Ravilla, W. Kaleem, P. Jadhawar, T. Sharma, Impact of low salinity water injection on CO2 storage and oil Recovery for improved CO2 utilization. Chemical Engineering Science, 229, p.116127. (2021).
19. H. Zolfaghari, A. Zebarjadi, O. Shahrokhi, M.H. Ghazanfari, An experimental study of CO2-low salinity water alternating gas injection in sandstone heavy oil reservoirs. Iranian Journal of Oil & Gas Science and Technology, **2**(3), pp.37-47. (2013).
20. H.N. Al-Saedi, S.K. Al-Jaberi, R.E.Flori, W. Al-Bazzaz, Y. Long, A new design of low salinity-CO2-different chemical matters. In Abu Dhabi International Petroleum Exhibition & Conference, SPE-197118-MS. (2019).
21. H.N. Al-Saedi, R.E. Flori, Novel coupling smart water-CO2 flooding for sandstone reservoirs. Petrophysics-The SPWLA Journal of Formation Evaluation and Reservoir Description, **60**(04), pp.525-535. (2019).
22. H.N. Al-Saedi, Y. Long, R.E. Flori, B. Bai, Coupling smart seawater flooding and CO2 flooding for sandstone reservoirs: smart seawater alternating CO2 flooding (SMSW-AGF). Energy & Fuels, **33**(10), pp.9644-9653. (2019).
23. L.W. Holm, V.A .Josendal, Mechanisms of oil displacement by carbon dioxide. Journal of petroleum Technology, **26**(12), pp.1427-1438. (1974).
24. M.S.A. Perera, R.P. Gamage, T.D. Rathnaweera, A.S. Ranathunga, A. Koay, X. Choi, A review of CO2-enhanced oil recovery with a simulated sensitivity analysis. Energies, **9**(7), p.481. (2016).

25. C. Drummond, J. Israelachvili, Surface forces and wettability. Journal of Petroleum Science and Engineering, **33**(1-3), pp.123-133. (2002).
26. J.J. Sheng, Critical review of low-salinity waterflooding. Journal of Petroleum Science and Engineering, **120**, pp.216-224. (2014).
27. P.P. Jadhunandan, N.R. Morrow, Effect of wettability on waterflood recovery for crude-oil/brine/rock systems. SPE reservoir engineering, **10**(01), pp.40-46, SPE-22597-PA. (1995).
28. A. Lager, K.J. Webb, C.J.J. Black, Impact of brine chemistry on oil recovery. In IOR 2007-14th European symposium on improved oil recovery (pp. cp-24). European Associ. of Geoscientists & Engineers. (2007).
29. D.J. Ligthelm, J. Gronsveld, J. Hofman, N. Brussee, F. Marcelis, H. van der Linde, Novel waterflooding strategy by manipulation of injection brine composition. In EUROPEC/EAGE conference and exhibition, SPE-119835-MS. (2009).
30. R.T. Johansen, H.N. Dunning, Relative wetting tendencies of crude oils by capillarimetric method, **5752**. US Department of the Interior, Bureau of Mines. (1961).
31. J.W. Graham, J.G. Richardson, Theory and application of imbibition phenomena in recovery of oil. Journal of Petroleum Technology, **11**(02), pp.65-69. (1959).
32. N.R. Morrow, G. Mason, Recovery of oil by spontaneous imbibition. Current Opinion in Colloid & Interface Science, **6**(4), pp.321-337. (2001).
33. L.E. Treiber, W.W. Owens, A laboratory evaluation of the wettability of fifty oil-producing reservoirs. Society of petroleum engineers journal, **12**(06), pp.531-540, SPE-3526-PA. (1972).
34. J.P. Batycky, F.G. McCaffery, P.K. Hodgins, D.B. Fisher, Interpreting relative permeability and wettability from unsteady-state displacement measurements. Society of Petroleum Engineers Journal, **21**(03), pp.296-308, SPE-9403-PA. (1981).
35. A.B.D. Cassie, S. Baxter, Wettability of porous surfaces. Transactions of the Faraday society, **40**, pp.546-551. (1944).
36. N. Siemons, H. Bruining, H. Castelijns, K.H. Wolf, Pressure dependence of the contact angle in a CO2 - H2O - coal system. Journal of colloid and interface science, **297**(2), pp.755-761. (2006).
37. E. Sripal, L.A. James, Application of an optimization method for the restoration of core samples for SCAL experiments. Petrophysics-The SPWLA Journal of Formation Evaluation and Reservoir Description, **59**(01), pp.72-81. (2018).
38. E. Sripal, D. Grant, L. James, Application of SEM Imaging and MLA Mapping Method as a Tool for Wettability Restoration in Reservoir Core Samples for SCAL Experiments. Minerals, **11**(3), p.285. (2021).
39. W. Anderson, Wettability literature survey-part 2: Wettability measurement. Journal of petroleum technology, **38**(11), pp.1246-1262. (1986).
40. M. Arif, S.A. Abu-Khamsin, S. Iglauer, Wettability of rock/$CO_2$/brine and rock/oil/$CO_2$-enriched-brine systems: Critical parametric analysis and future outlook. Advances in colloid and interface science, **268**, pp.91-113. (2019).
41. N.S. Kaveh, E.S.J. Rudolph, P. Van Hemert, W.R.Rossen, K.H. Wolf, Wettability evaluation of a CO2/water/bentheimer sandstone system: contact angle, dissolution, and bubble size. Energy & Fuels, **28**(6), pp.4002-4020. (2014).
42. S.J.D. Sofla, L.A. James, Y. Zhang, Toward a mechanistic understanding of wettability alteration in reservoir rocks using silica nanoparticles. In E3S Web of Conferences, **89**, p. 03004. EDP Sciences. (2019).
43. P.T. Jaeger, M.B. Alotaibi, H.A. Nasr-El-Din, Influence of compressed carbon dioxide on the capillarity of the gas– crude oil– reservoir water system. Journal of Chemical & Engineering Data, **55**(11), pp.5246-5251. (2010).
44. A. Ameri, N.S. Kaveh, E.S.J. Rudolph, K.H. Wolf, R. Farajzadeh, J. Bruining, Investigation on interfacial interactions among crude oil - brine - sandstone rock - CO2 by contact angle measurements. Energy & fuels, **27**(2), pp.1015-1025. (2013).
45. D.N. Espinoza, J.C. Santamarina, Water - CO2 - mineral systems: Interfacial tension, contact angle, and diffusion — Implications to CO2 geological storage. Water resources research, **46**(7). (2010).
46. M. Seyyedi, M. Sohrabi, A. Farzaneh, Investigation of rock wettability alteration by carbonated water through contact angle measurements. Energy & Fuels, **29**(9), pp.5544-5553. (2015).
47. A.B. Dixit, S.R. McDougall, K.S. Sorbie, J.S. Buckley, Pore scale modelling of wettability effects and their influence on oil recovery. In SPE/DOE Improved Oil Recovery Symposium, SPE-35451-MS. (1996).
48. S.J. Kline, F.A. McClintock, Describing uncertainties in single sample experiments. Mech. Engineering, 75, pp.3-8. (1953).
49. S. Li, M. Liu, D. Hanaor, Y. Gan, Dynamics of viscous entrapped saturated zones in partially wetted porous media. Transport in Porous Media, **125**(2), pp.193-210. (2018).
50. H. Ding, S. Rahman, Experimental and theoretical study of wettability alteration during low salinity water flooding-an state of the art review. Colloids and Surfaces A: Physicochemical and Engineering Aspects, **520**, pp.622-639. (2017).
51. S.Y. Lee, K.J. Webb, I.R. Collins, A. Lager, S.M. Clarke, M. O'Sullivan, A.F. Routh, X. Wang, Low salinity oil recovery - Increasing understanding of the underlying mechanisms. In SPE Improved Oil Recovery Symposium. SPE-129722-MS. (2010).
52. Q. Xie, A. Saeedi, E. Pooryousefy, Y. Liu, Extended DLVO-based estimates of surface force in low salinity water flooding. Journal of Molecular Liquids, **221**, pp.658-665. (2016).

53. A. Lager, K.J. Webb, C.J.J. Black, M. Singleton, K.S. Sorbie, Low salinity oil recovery-an experimental investigation1. Petrophysics-The SPWLA Journal of Formation Evaluation and Res. Descrip. **49**(01). (2008).
54. H. Wang, J. Zeuschner, M. Eremets, I. Troyan, J. Willams, Stable solid and aqueous H2CO3 from CO2 and H2O at high pressure and high temperature. Scientific Reports, **6**(1), pp.1-8. (2016).
55. B. Velde, *Origin and mineralogy of clays: clays and the environment*. Springer Sci. & Business Media. (2013)
56. P.L McGuire, J.R. Chatham, F.K. Paskvan, D.M. Sommer, F.H. Carini, Low salinity oil recovery: An exciting new EOR opportunity for Alaska's North Slope. SPE-93903-MS. (2005).
57. C.A. Miller, K.H. Raney, Solubilization - emulsification mechanisms of detergency. Colloids and Surfaces A: Physicochemical and Engineering Aspects, **74**(2-3), pp.169-215. (1993).

# Oscillating Back Pressure Regulator (OBPR) for High-Pressure Core Flooding

*Sanjay* Dubey[1], *Hossein* Khorshidian[2], *Maziyar* Mahmoodi, [1]*Edison* Sripal[1], *Lesley* James[1*]

[1] Hibernia EOR Laboratory, Memorial University of Newfoundland, St. John's, Canada
[2]Opla Energy, Calgary, Canada

**Abstract.** The challenge in the core analysis industry is the availability of a versatile back pressure regulator (BPR) that can operate efficiently at ultra-low or no-flow conditions at very high pressure. The existing models of back pressure regulators are limited to a specific range of operating conditions, automation, and robust core flooding environment. In this work, we developed an initial stage prototype of an Oscillating BPR (OBPR), emphasizing detailed instrumentation and implemented automation. The critical component in an OBPR is the valve that needs to operate within defined close and open position based on the feedback received from an encoder rather than the oscillation created by the flow system or by using a square wave oscillator. The paper investigates the static and dynamic characteristics of the newly designed OBPR instrument. The paper discusses the graphical programming LabVIEW software used to create the feedback control loop using the PID controller. Initial results of the test run were done with the different core flooding systems like Water, Brine, and $CO_2$. The study also included comparing OBPR pressure performance with the standard back pressure regulator in the market.

## 1 Introduction

A novel oscillating back pressure regulator (OBPR) was developed to control the backpressure downstream of a two-phase or three-phase core flooding experiment. Core flood experiments are used to investigate the multiphase behavior of fluid flow in rock cores at reservoir operating conditions. Core flooding technology and instrumentation play a significant role in accurately measuring physical characteristics (relative permeability, connate water content, fluid saturation, pore-space, etc.) and better analyzing the core data. BPRs applied in core flood experiments are subjected to very high pressure (~68 MPa) and temperature conditions (~200°C). Commonly, mechanical regulators accompany all core flooding experiments to provide back pressure and control flow/pressure. The primary purpose of the BPR is to maintain reservoir operating conditions in the core holder and keep a steady low-flow and stable backpressure for in-situ reservoir analysis, and reliable displacement of fluid. As fluid at the core outlet are produced, the backpressure decreases. The goal is to keep a constant backpressure while collecting sampling fluid at core outlet. BPRs can also be used to enhance the downhole pump performance and well productivity in large oil reservoirs [1]. Role of BPRs in the different core flood experiements is disucsssed in the following paragraphs. BPRs are also especially important in gas injection core flood experiments, core flooding involving live fluids (oil or Brine), or EOR applications such as carbonated water injection (CWI) as in-situ pressures maintained to ensure appropriate phase saturations and relative fractional flow during oil production. BPR are used in the Ion Chromatography analysis of advanced ion management carbonate core-flood experiments [2]. Foam stability is an essential factor in understanding the displacement of the oil efficiency in porous media, BPR adopted in this chemical flooding experiment allows setting the desired pressure and temperature [3]. BPR is applied to build the pressure of 98066.5 Pa in the studies of microbial enhanced oil recovery for the core flood utilizing the model ex-situ bioaugmenting a thermo-, and halo-tolerant rhamnolipid generate using Pseudomonas aeruginosa [4]. Gas drainage strategy, such as enhanced coalbed methane, is used to recover the coalbed methane to provide safe conditions for underground coal mining. The lab-scale adsorption investigation was used to observe the displacement of methane by injecting gas ($CO_2$ or $N_2$) into intact coal core samples. BPR was used to hold the constant pore pressure during the core flooding experiment [5]. The above applications illustrate that BPR is an important component in core flooding research and justify the need for better BPR design.

Zero flow rate can be challenging to achieve using the conventional BPRs. The high flow coefficient makes the system inefficacious for the core flood application [6]. The backpressure regulator used in the core flood experiments must be able handle multiphase flow. Perfectly designed BPR must minimize the fluctuation in core outlet pressure. Designing the high-precision feedback control system for OBPR can provide highly reliable pressure regulation for core flood applications. Low-precision BPRs cannot maintain the core pressure holder at a specific pressure. The precision of BPR is essential in measuring relative permeability, MMP in WAG, apparent viscosity of foam, and many other applications. Imperfect BPRs can cause drastic measurement error, such as distorted breakthrough concentration profiles during the measurement of dispersion coefficient of $CO_2$ in $CH_4$ as part of the $CO_2$ sequestration for the enhanced gas recovery

* Corresponding author: ljames@mun.ca

© The Authors, published by EDP Sciences. This is an open access article distributed under the terms of the Creative Commons Attribution License 4.0 (http://creativecommons.org/licenses/by/4.0/).

[7]. High dead volume for the backpressure regulator will allow more process fluid to be stagnant, which can be a problem in specific applications. Most of the samples are collected and analyzed by using chromatography; hence this can be a severe problem for chemical-sensitive core flood experiments. During the experiment to study the effect of oil saturation on foam propagation, the diaphragm BPR requires to clean the regulator since the experiment is susceptible to undesired chemical reactions and could cause measurement error [8]. Commercial BPRs available in the market are complicated in design and to operate during the core flood experiments. The newly designed OBPR is easy to use, maintain, clean, and troubleshoot. The newly designed OBPR is an improved version of the predecessor OBPR model[6]. Old design uses the square wave oscillator to create oscillation, whereas this OBPR uses nitrogen in the flow system, which causes the feedback loop into the oscillation. This makes the nitrogen to act as a damper for avoiding large pressure fluctuations in the system. According to the literature review, the OBPR design is indistinguishable from another back pressure regulator in the market. The newly designed OBPR is only the adjustable active system in the market. It means that once the set point is given and it is actively controlling the pressure. OBPR automation design can regulate the pressure up to 9,500psi. The BPRs in the market is primarily are passive types. The active system of the OBPR automation is based on the PID LabVIEW design. Fig.1 illustrates a research flow diagram for the new OBPR design completed at the Hibernia EOR lab.

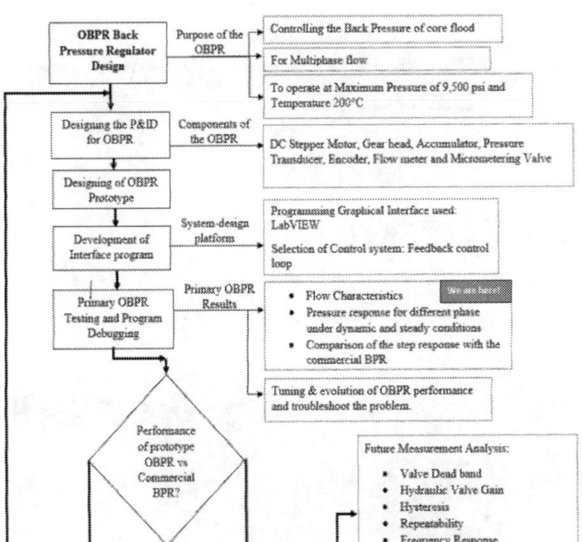


Figure 1a. OBPR Design Algorithm

Initial steps include understanding the desired operating conditions of the OBPR and construction and assembling the components into the OBPR prototype. The preliminary results obtained at the debugging level were used to compare the OBPR performance of the regulator to the commercial BPRs available in the market and to improve the original design. Preliminary results obtained during the initial run include the core outlet pressure response at different flow rates around 2000 psi, core outlet pressure at different pressure, and core outlet pressure response for the gaseous phase. Understanding the strengths and weaknesses of the OBPR feedback control system will improve its application to core flooding. Future work shown in the design algorithm includes a sensitivity analysis of different components affecting the control performance of the OBPR.

### 1.1 Back Pressure Regulators

Regulators are differ from control valves; generally, a control valve uses an external power system to control the process variable [3]. BPRs are self-actuating valves used for large and small-scale fluid flow experiments. BPRs are classified based on the mechanism (spring load, dome load, air load and diaphragm, etc.), construction material (steel and elastomer), type of fluid, and application. Selection criteria for a BPRs are based on the pressure and temperature operating conditions, accuracy and precision, piping sizing, and desired flow rate. Based on the design, the elementary regulator comprises three essential elements: restricting, sensing, and loading. Depending on the type of regulator, the downstream or upstream port communicates with the sensing element like a diaphragm, piston, and spring. All the regulators are self-powered and considered throttling valves as they modulate according to fluid or system pressure.

There are several types of regulators: pressure reducing, pressure relief, pressure switching, vacuum, etc. It is important to understand the BPRs and why it is different from other types of valves. Fluids are accumulated at the sampling point using manual, fractional or rotatory collectors used for the various displacement study, chemical analyses, which gives more details about the fluid-rock interaction within the core[10, 11]. BPRs can be used as a Low-Pressure Safety Relief Valve (SRV), but the primary goal of an SRV is to self-actuate and safeguard the equipment and life. An SRV only opens when the system's pressure exceeds the design pressure of the material [9]. It is also difficult to control fluid discharge flow through several SRV during blowdowns, making it an inappropriate choice for core flood experiments. The OBPR shown in Figure 2 was designed based on the feedback control system, which may provide a better result for the physical measurement for core flooding experiments. It allows a better pressure dynamic response and handling of for open and closed-loop controllers [10]. This paper focuses our discussion on the design, automation, and preliminary pressure response results of this new back pressure regulator.

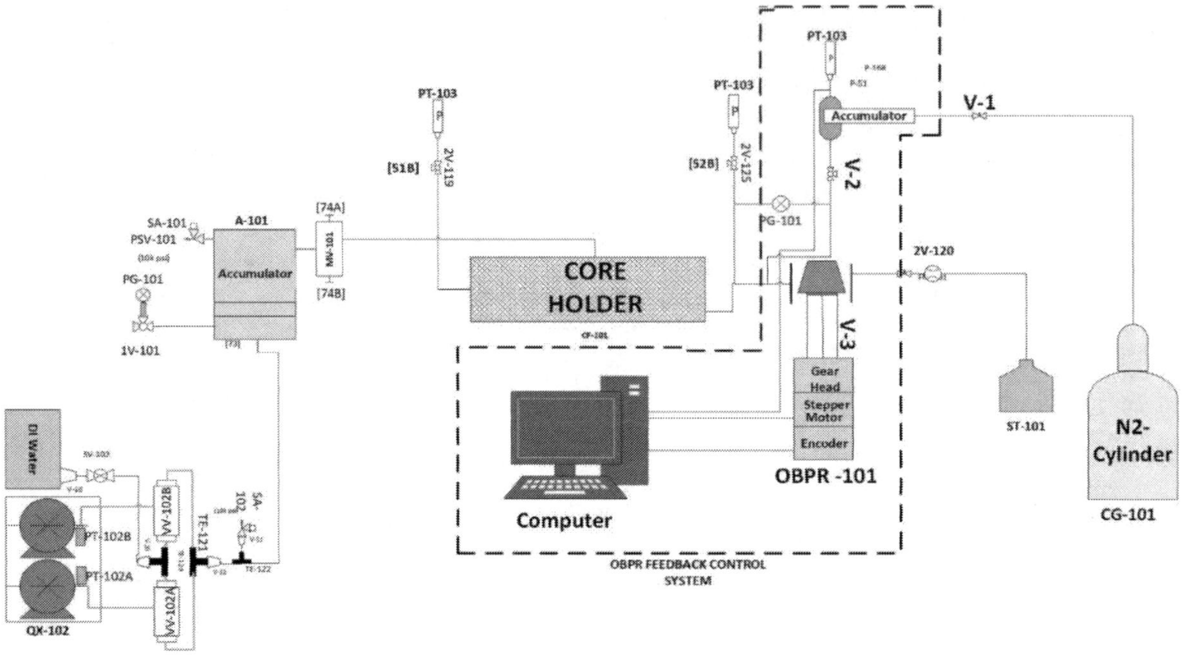


Fig 1b. Hibernia core flood experimental setup with newly OBPR feedback control system [13]

As shown in Figure. 1b, BPRs are generally employed at the downstream position of the core flooding experimental setup[11]. The core flooding assembly for this experiment was designed to conduct experiments with single-core plugs for water, Brine, and $CO_2$. The essential components of the setup are a hassler core holder, accumulators, Quizix pumps, quartz pressure sensor, flow meter, oscillating back pressure regulator (OBPR), and sample collector. The flooding system was designed to operate at 150 to 200°C, pore pressure of 9500 psi and overburdened pressure of 10000 psi. Water and brine were injected using the Quizix pump through the floating piston accumulator to the core holder inlet port. $CO_2$ was injected into the core holder using the booster pump. Core outlet pressure was maintained constant using the OBPR feedback control system, as shown in the dotted box of Fig 1b. Flowmeter was used to measure gas flow, and liquid manual measurement was performed using the standard graduated cylinder. A nitrogen gas tank is placed at the corner of the accumulator to use nitrogen gas as the pressuring gas for the OBPR operation. The biaxial core holder was used with no overburden pressure. Experiments were conducted from ambient conditions to 2000 psi pressure. The initial pressure of the sample core (sandstone core, length: 5.29 cm, diameter: 3.75 cm, area: 11.04 $cm^2$, total porosity: 16% sample obtained at a depth of 3862m of Hibernia formation, no overburden pressure) enclosed within in rubber sleeve. Core outlet pressure response was measured at 500, 1000 and 1500 psi. The core flood experiment was designed to understand the core outlet pressure response under static and dynamic conditions. Core outlet pressure response obtained from the OBPR was compare with the core outlet pressure response behavior of the commercial BPR.

### 1.2 Oscillating Back Pressure Regulator (OBPR) Design

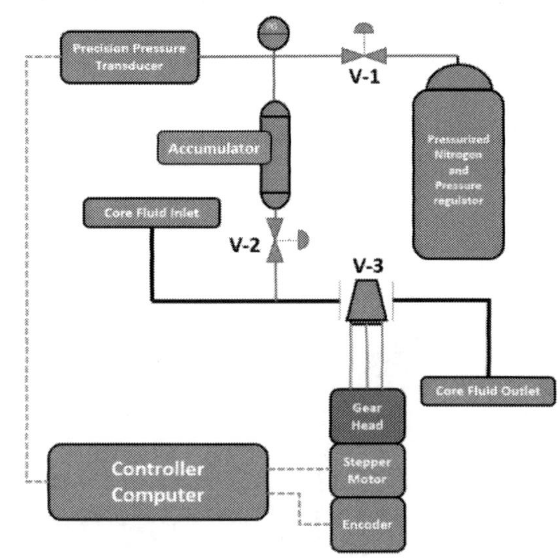


Fig 2. OBPR P&ID Instrumentation Diagram

The OBPR apparatus has been developed to control the backpressure of the core holders during the core flooding experiments and routine core analysis. Figure 2 represents a typical embodiment of the OBPR piping and instrumentation diagram. The OBPR consists of a precision needle valve (micro-metering valve), which is actuated by a high-resolution stepper motor (20,000 steps per revolution), to build up intended backpressure by restricting the flow area.

https://doi.org/10.1051/e3sconf/202336601020

During the start of the experiment V-3 micro metering valve is closed, and the nitrogen has been charged while keeping the valve V-2 open. As the core fluid pump through the core holder, the pressure at the outlet increases through the line between the V-2 and V-3. The primary fluid enters the accumulator tube, which will increase the nitrogen pressure to the back pressure set point. As the desired back pressure is achieved, the V-3 starts to open and let the excess pressure down by letting some fluid flow out and closing the valve V-3 again to maintain the pressure. This open and close phenomena case at a particular valve location causes the oscillation hence the name of the valve, motor, and feedback control system is given as Oscillating Back Pressure Regulator. Once the desired set pressure is achieved, the flow rate is being controlled. The generated back pressure is measured by a precision pressure transducer, and the data is sent to a controller that sends commands to drive the stepper motor based on the processed data. The stepper motor actuates the stem of the needle valve by a gear head that increases the stepper motor torque and enhances the resolution. The LabVIEW program executes the controlling code using the National Instruments data acquisition system (model no NI PXle 8430/16).

### 1.3 OBPR Components and Assembly

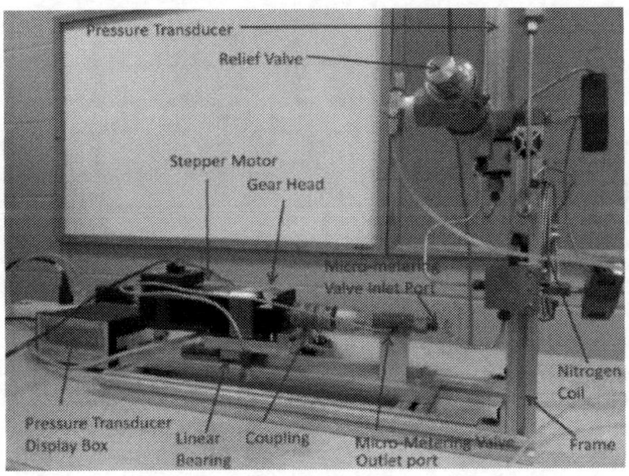


Fig. 3. Assembled OBPR Components

Figure 3 shows the assembled OBPR component. The motor power system provides electricity for the stepper motor. The gearhead is installed on the stepper motor shaft to increase torque and reduce motor speed (70:1). The gearhead shaft is connected to the valve stem using a precision coupling. The coupling should be disengaged when the valve positioning is implemented. The linear bearing is mounted beneath the motor since the motor moves back and forth by opening and closing the micro-metering valve. The micro-metering valve inlet (1/8" Autoclave Speed-bite) is connected to a flow line (e.g., the core holder outlet) to control its backpressure.

A tee and valve V-2 in Figure 2 connect the micro-metering valve inlet to the accumulator coil. The coil can be charged with nitrogen up to 10,000 PSIG with a pressurized gas cylinder or a gas booster. The set pressure must be 100 PSIG (50-150 psi is acceptable) lower than the intended backpressure. This causes the controller to keep the micro-metering valve closed before running and receiving any fluid from the core holder. Valve 2 can be opened when nitrogen charging is complete and valve 1 is closed. Valve 1 disconnects the nitrogen source from the coil accumulator. The pressure gauge and pressure transducer display box can read the coil pressure. The OBPR software also displays real-time pressure. The pressure relief valve provides additional safety for the relief of pressure in the case of OBPR failure. The relief valve cracking pressure is at 9500 PSIG. The OBPR valve can operate up to 5000 psi in a close position, but the maximum allowable pressure is 12,500 psi. The valve stem is not closed too tightly to guard the valve seat against getting damage. It is recommended to add an O-ring sealed check valve at the equipment outlet (e.g., core flood unit) to avoid backflow of pressurized nitrogen toward the equipment (e.g., core holder) due to any OBPR malfunction. In addition, using a diaphragm-type accumulator may avoid contact of fluid in the flow line and nitrogen in the coil. The composition of the oil will affect the calculation of the $N_2$ miscibility with the crude oil. Miscibility of $N_2$ is relatively high, especially 5000 to 6000 psi for an Alaskan oil[12]. As mentioned earlier, the core flooding experiments were performed with water, brine, and gas ($CO_2$).

### 1.4 OPBPR Operation Procedure

The needle valve is positioned in a near-closed condition to build up the intended back pressure inside an accumulator to create backpressure. The accumulator is a long 20ft coiled tube (0.125" OD, 0.04" ID) filled with nitrogen and positioned behind the micro-metering valve. Since pressurized nitrogen is compressible, it allows the core fluids to be charged or discharged in the coil. A higher motor speed reduces the time of the actuation of the system lag. However, it is recommended to set the rate at an intermediate value (e.g., 5 revs/sec) to avoid premature failure of the stepper motor. The pressure transducer at its standard mode provides high-quality pressure data (± 0.25 PSIG) within an acceptable period. The initial valve positioning will dramatically reduce the lag time of the valve operation. This will entirely minimize the time required to shut in a valve from its maximum opened position. The valve closed position should build up the pressure of 5000 PSIG when the core holder is flooded by water at the rate of 5 cc/hr. Although further tightening of the valve stem builds a higher pressure at low flow rates, it will damage the needle valve stem as the valve touches the seat every time it is fully closed. The valve has to be operated within a defined closed and opened position based on the feedback received from an encoder. One rotation of the valve stem is sufficient to increase the flow area for a low-pressure fluid discharge. The controlling parameters (proportional and derivative terms) must be set regarding the working conditions. An increase in the volume of the accumulator produces a smoother pressure profile. However, it increases the response time of the pressure transducer and the dead volume. The coil length is currently 20 ft (ID: 0.02"), leading to a magnitude of 1.2 ml. To run the OBPR, the following programs have to run by LabVIEW software. The required adjustments and commands must be executed in each program sequentially. In addition, the OBPR hardware has to be prepared to keep the backpressure at the desired value. The valve positioning is adjusted after the software LabVIEW adjustment before charging the accumulator with pressurized Nitrogen. In addition, the test script recommends a range of control parameters that the user can execute to obtain the desired back pressure profile. The OBPR performance is affected by different parameters and factors such as:

1. Speed and acceleration of the stepper motor
2. Accuracy of the pressure transducer data
3. Valve opened and closed positions
4. Controlling parameters
5. The volume of the accumulator
6. Pressure response

## 2 Automation Implementation

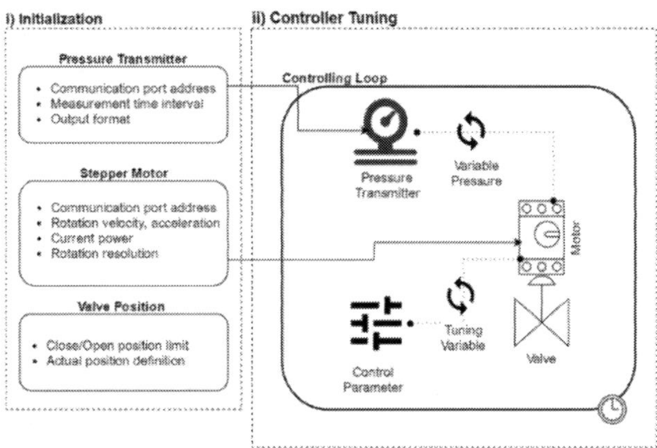


Fig. 4. OBPR Basic Layout Using LabVIEW

The design was automated using the LabVIEW program to make the system work smoothly, as shown in Figure 4. The feedback control system was developed to operate the OBPR. Application of the PID. (Proportional Integral Derivative) a controller can render optimal and robust performance for stable, unstable, and non-linear processes [15]. The programming platform used in this project is LabVIEW 2015, commercially offered by National Instruments, which has extensive application in the automation and instrumentation industries. The primary controlling code is compatible with the proposed mechanical setup (pressure transmitter and stepper motor) and provides open source for further upgrades. The logical algorithm is explained as shown in Figure 4. The overall process flow in this algorithm comprises two main steps; I) Initialization and II) Controller Tuning.

The earliest stage in initialization is to establish serial communication with active serial ports of the stepper motor and pressure sensor. The initialization task also includes the primary setting for the pressure sensor and stepper motor parts. As presented in Figure 4, the next phase is to adjust the controlling parameters determined in advance based on flowing phases, flow rate, and operating pressure. The initialization step can be broken into three subsections. First, the pressure sensor properties should be set into desired values. Second, the operation parameters of the stepper motor need to be adjusted before running the main controlling loop. Finally, the valve position, a mechanical part of the equipment, needs to be initially closed and appropriately connected to the motor shaft. After implementing these sequences, the primary controller LabVIEW program is ready to run and execute rotation commands to the motor port based on the pressure difference between set value and recorded value from the pressure sensor. The immediate critical information that constantly is needed through controlling time is the actual encoder and positions. Consequently, the calling program is used for requesting position data and sending it to PID.

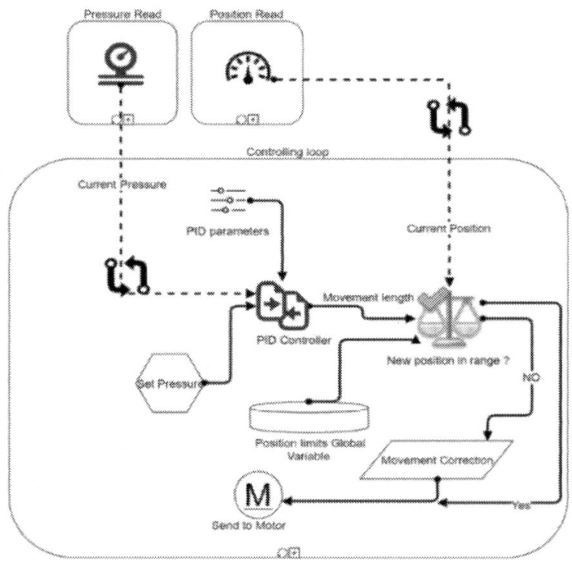


Fig. 5. OBPR Feedback Control System

Controller function Figure 5 shows the feedback control system design and its working mechanism. The initial stage of automation for this technique is to obtain the pressure information from the transducer and calculate the difference between the current. **Table 1.** Comparison of different commercial BPR model pressures and reference pressure. PID is used to calculate the proper response based on the pressure difference. Position and pressure variables are read from motor and pressure sensors and inserted into the LabVIEW control panel. The proportional, integral, and derivative terms are also adjustable in this front panel. Derivative of the pressure between two consecutive time steps is measured; it assists in creating controlling command over the stepper motor based on the differential pressure and derivative of the pressure. To correct maximum allowable displacement, the limit test is performed more or less like a trial and error to get the correct limit. The final valve position for the specific command is calculated to check if the valve position is still within the allowable range. Computed movement is then sent to the motor for execution. In the final step, the motor waits for the new pressure value

## . 3 Commercial BPR.

Several studies use pressure response to analyze a regulator's performance [12, 13, 14, 15, 16, 17, 18]. To understand the initial pressure response of the OBPR, the model was compared with the pressure response of two commercial BPRs that are used for similar applications. Water, Brine, and $CO_2$ were chosen as the primary fluids for these experiments as they are commonly used for experimental studies in geology, oil recovery, and hydrogeology [19].

**Table 1. Comparison of different commercial BPR models**

| BPR Model | EBPR | VBPR | OBPR |
|---|---|---|---|
| **BPR Type** | Metal Diaphragm | Piston | Metering valve |
| **Max Temperature (°C)** | 200 | 150 | 200 |
| **Max Pressure (psi)** | 10,000 | 10,000 | 9,500 |
| **BPR Material** | SS316L | SS316L | SS316L |
| **Flow coefficient** | $1*10^{-09}$ | $2.64*10^{-07}$ | $1*10^{-04}$ |

Details of the two commercial BPRs, a diaphragm BPR and a piston dome-loaded BPR, are provided below and summarized in Table 1. The first commercial BPR used in the comparison is a dome-loaded diaphragm-based regulator (model: ULF EHP SB8) manufactured by Equilibar, referred to as EBPR. The bottom and top dome have O-rings with high-yield strain material that renders the liquid-tight seal. In the EBPR case, the elastomer material for the O-ring is stainless steel SS316L. Its sustainability at high pressure and temperature may also affect the regulator selection and internal flow pattern. The diaphragm is placed coplanar to the bottom of the dome surface. The choice of the diaphragm is essential in chosen based on the core flood experiments. The top dome (reference cap) has a reference fluid that can create uniform hydrostatic pressure over the diaphragm. When the process pressure exceeds the determined pressure, the diaphragm will lift, allowing the fluid to pass across the system.

The second commercial BPR used for the comparison is a needle valve-based BPR (model number C06-003-1), manufactured by Vinci Technologies and referred to as VBPR. It is comprised of two chambers, i.e., top and bottom chambers or dome. The bottom section consists of the needle enclosing the process fluid passage. When the bottom fluid pressure is higher than the top dome's reference pressure, the fluid's passage is open, or else it is closed. Uninterrupted flow is secured once the process and reference pressure equilibrium is attained. The newly developed OBPR feedback control system was used to study the regulator performance under dynamic and steady-state conditions. This is a preliminary study of the OBPR performance based on its design and pressure response. BPR models used for all the core flood systems have been subjected to the limited operating condition (~ up to 1500 psi) to analyze the initial functionality of the OBPR and to ensure safe working conditions. The initial pressure response of the OBPR is a BPR regulator for high-pressure core flood experiments is investigated. Before the experiments, it is necessary to ensure no leakage of the high-pressure fluid from the steam of the valves or the system. This can cause pressure loss resulting in measurement errors and malfunctioning of the valve. It is required to update the position of the stepper motor during the closing of the valve, as over-rotation of the valve handle can cause damage to the valve seating. Motor heating, power outage, or surge can increase or decrease the motor's torque ability, affecting the pressure response. LabVIEW program complications can also cause trouble in the experimental procedure. Measurement begins with the inspection of the reference pressure. Pressure measurement is continuously streamed on display rendering the real-time value of the pressure in the system. Measurement resolution is essential for this study since our pressure difference across the core holder is relatively low; it is best to use the IIR (infinite impulse response) nano-resolution filter device to report accurate pressure measurement up to 12 digits and acts as an effective in anti-aliasing. To make the stepper motor work properly, it was necessary to configure it or tune the motor. The stepper motor was initially disengaged from the gearhead coupling from the micro metering valve; the motor was also connected to the computer using an RS232 cable. Immediate and buffer types of commands were utilized through the LabVIEW program to communicate with the stepper motor. Critical parameters such as the maximum current, data format, velocity encoder ratio, encoder function, and motor ratio were initially set. The motor response was noted before it was connected for the actual experiment run. The metering valve was adjusted to the closed position to build up the pressure to 5000 PSIG. The position limit controller, which defines the incremental steps that the valve is turned for the maximum opening position (400,000 steps were optimal, sets the entire valve stem rotation at 0.56 turn), was switched on. Once the set position controller defines the current valve position, executes the run, it will communicate the motor and save the initial position of the micro metering valve when the system is at rest with no charge nitrogen flow, the PID. Constant value was input. Once the nitrogen begins to charge into the accumulator coil, the OBPR is set at the desired pressure value. The pressure source is removed once the system is ultimately charged with the Nitrogen. As the feedback control system detects the reference pressure, the gearhead rotates to set the new opening position of the micro metering valve. During the initial operation, the valve stem moves in an opening place, creating a sudden pressure drop in the system.

## 4. Results and Discussion

A feedback control approach was applied to the OBPR system consisting of a micro-metering valve, stepper motor, pressure transducer, and control architecture like PID. During initial experiments, the efficiency of the feedback control system must satisfied the primary goal of holding desired set backpressure under safe conditions. Future work might focus on the detailed studies of the feedback control system, cost, applicability to the multiphase system, and significant OBPR performance parameters. An ideal feedback control system must have zero steady-state error and constant stability throughout the performance. OBPR results are confined to a controlled variable (pressure response) offset and PID in this discussion. The set point value which was chosen to provide the suitable control performance for given operating conditions. The results show the application of OBPR at ultra-low flow rates initially and preliminary pressure response.

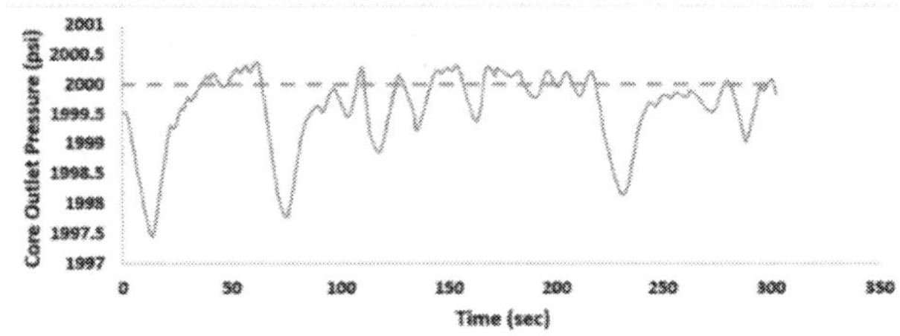


Fig. 6A. OBPR Pressure Response under 2000 psi with Maximum Fluctuation ±0.125% at 50 ml/hr (Primary Fluid: Water)

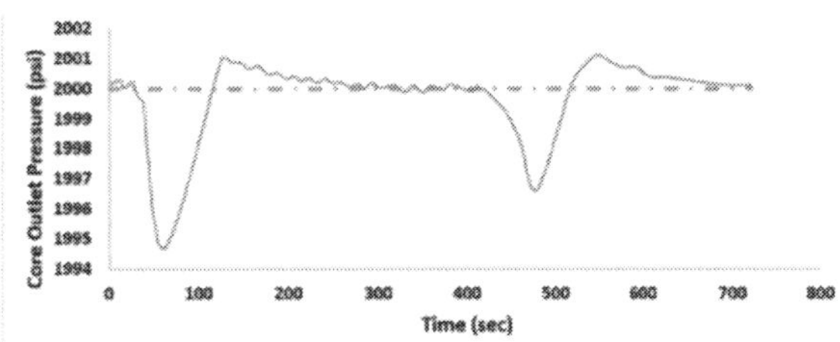


Fig. 6B. OBPR Pressure Response under 2000 psi with Maximum Fluctuation ±0.25% at 10 ml/hr (Primary Fluid: Water)

The OBPR outlet pressure response at different flow rates experiments was completed with water as the primary fluid and using two flow rates, 50 ml/hr (Figure 6A) and 10 ml/hr (Figure 6B). Figure 6A and 6B show the deviation of the core outlet pressure from the set point of the OBPR system that is 2000 psi. This can happen due to various types of elements present in the OBPR feedback control system, such as stem of the valve, pressure sensor, manual tuning, charging gas (nitrogen), stepper motor, gas pressure, etc. Disturbance or deviation seems to be non-periodic and looks like an approximated sine wave function. It is interesting to see that this disturbance frequency is too fast because of quick response from the stepper motor. Figure. 6 A and 6B shows a similar pressure response to the OBPR pressure response shown in Fig.7, but in these experiments, the primary fluid used in core flooding was water. Both the water system in Fig 6 A and B fluctuates around ±0.25% of the reference value. Figure 6B indicates a small number of fluctuations compared to the low volumetric flow rate. High flow rates might cause the system to respond more quickly to maintain the reference point pressure. Hence higher pressure fluctuation can be observed compared to a low flow rate. It was observed that OBPR feedback control system that response too quickly also results in poor performance. The amplitude ratio for the consecutive trough of the pressure response curve in Fig.6A and 6B declines as time progresses; it indicates that the OBPR feedback system reduces the steady-state error between the set point and control process variable. The decay ratio between the neighboring peaks was getting smaller for low flow rates compared to the high flow rate at the same pressure. This indicates that water testing with low flow rates at high pressure has a less settling time than high flow rates. This can cause the outlet core pressure response to reach a steady-state condition with reduced offset error. Fig. 6A & 6B both show the controlled variable overshoot which can cause the long-term degradation of valve.

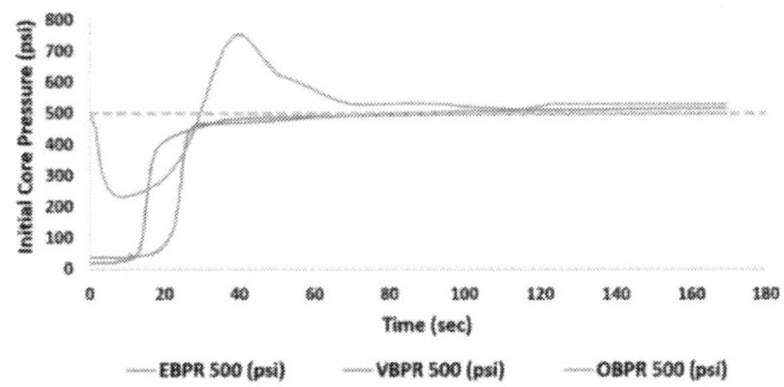


Fig. 7A. BPR Pressure Response for Single Phase Brine at 500 psi

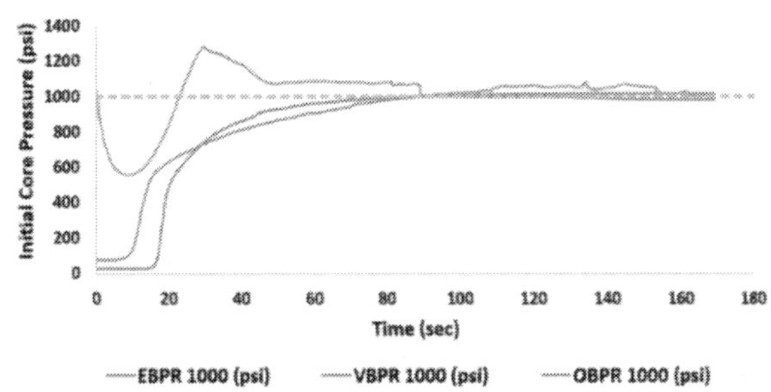


Fig. 7B. BPR Pressure Response for Single Phase Brine at 1000 psi

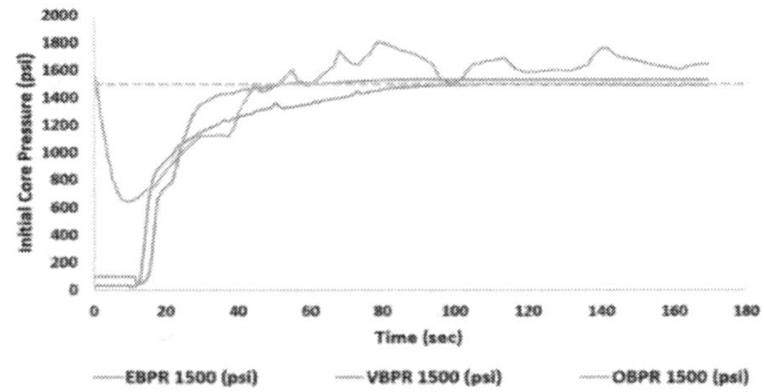


Fig. 7C. BPR Pressure Response for Single Phase Brine at 1500 psi

https://doi.org/10.1051/e3sconf/202336601020

Dynamic pressure response from the two commercial BPRs (EBPR and VBPR) and the OBPR with initial core pressures of 500, 1000, and 1500 psi, using Brine, are shown in Figures 7A, 7B and 7C. A better understanding of the effects of pressure response dynamics can lead to a better design for the OBPR model with optimizing control and design improvement. The OBPR feedback control system is subjected to a step change disturbance, and the initially controlled pressure variable deviates for few seconds. This category is termed a step response. The pressure response vs time was graphed in Fig 7A, 7B & 7C. It can be seen from all the Fig.7 that the rise time for the OBPR pressure response varies for different pressure, Fig.7C have longest rise time. To judge the control performance of the OBPR with least disturbance, the volumetric flow was kept constant for all the pressure response. The sudden fall in the pressure for all the OBPR flow system at the early seconds of the operation is shown in Figures 7-A, B and C. Pressure drop can happen to overcome the lock-up conditions which require a fixed amount of torque to rotate the stem of the valve. Lock up is the no-flow state where the regulator has difficulty maintain the initial set pressure [20]. As the feedback control system regains, it re-adjusts the regulator opening position and runs to achieve desired reference value. From Fig. 7A, it is seen that settling time was quite less than another similar system because of the system less overshoot than other two. The core outlet pressure will start to build up and converge around the desired value making the motor run in an oscillatory mode compared to the square wave generator in the predecessor model. It can be observed that OBPR feedback adjustment had an immediate impact on the pressure variable for the initial period as compared to the VBPR and EBPR, which have dead time causing the controller to take action slowly or to get first-order delay response. OBPR feedback control system is quick and holds the least dead time. This can perform better than other BPRs designs for the step disturbance. The algorithm was used for the OBPR feedback control system to get better control performance. However, it can still result in excessive fluctuation or instability for the pressure variable. The final element that can also affect the stability and control performance is the valve. The OBPR pressure instability dampens over the period as the feedback control system approaches the steady-state pressure value. The settling time for the OBPR feed control system starts to get prolonged as the pressure increases due to high rise time—the proportional and integral mode of the PID. Controller tends to get the oscillatory and slow dynamic response. This slight pressure fluctuation can depends on various parameters, such as equilibrium condition between the inlet fluid and nitrogen pressure in the accumulator, regulator opening, PID. Value, phase of the primary fluid, stepper motor, and type BPR regulator. From Figure 7 A, B and C it can be seen that EBPR and VBPR show slow and stable dynamic responses compared to the OBPR. One of the reasons it might be that VBPR and EBPR are better self-regulating the disturbance and approach quickly to attain the desired set point. The sensing parameters (diaphragm and piston) are not significantly affected by the sudden decrease in the dome pressure. The core outlet pressure starts to build up due to restricting elements of the valve. The dome side was already pressurized EBPR and VBPR, causing slow fluid displacement and less pressure drop to cause the instability. In Fig.7 all the set points are represented in the dotted form to understand the system's dynamic and steady state response compared to a reference point. Every system has unique characteristics; the value of PID is obtained on the type of tuning methods. Manual tuning was done on the OBPR system, keeping the oscillation fluctuation at constant amplitude. The stepper motor also adds, such as oscillation or unstable phenomena to overall OBPR feedback control system [21]. It can be observed from Figure 7 that the OBPR is unstable (overshoot) during the initial phase of the operation of the system. This overshoot problem could be resolved with a better PID tuning method. The experimental results illustrated in Figure 7 help us understand the pressure's dynamic response compared to the market's commercial BPRs.

To understand the OBPR performance on a gaseous core flood system, experiments were completed using $CO_2$ as the primary fluid. Figure 8 shows the steady-state pressure response to the gaseous system. A higher oscillation rate is observed compared to the water and brine systems.

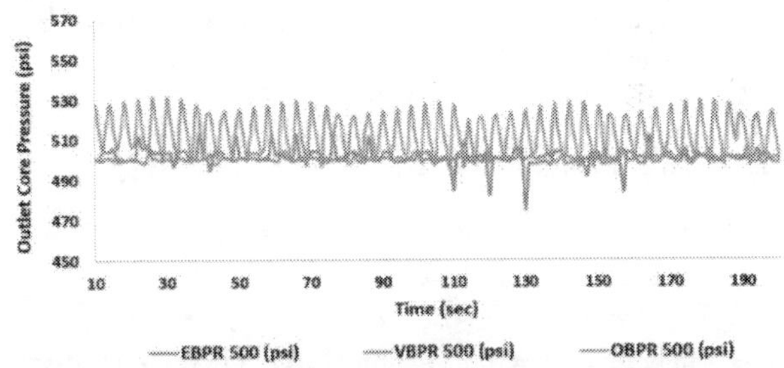


Fig. 8. BPR Pressure Response with Single Phase Gaseous $CO_2$

The OBPR pressure response was observed to be quick for the gaseous system, and a more periodic disturbance is kept for the OBPR than the commercial BPRs. The PID was designed to reduce the error signal fixing the PID gain and will not reduce the noise signal. This noise signal may arise from the booster pump that is used to pressurize the core holder. Along with this noise fluctuation, the OBPR was able to maintain the steady pressure with $\pm 50$ psi. The OBPR will render limitations for close measurement applications with such high delta P across. A more experimental study is required to check the performance of the OBPR for the gaseous system.

The conventional core flood experiment was designed to understand the OBPR feedback control response behavior under static and dynamic conditions. The core outlet pressure response obtained from the OBPR was compared with the core outlet pressure response behavior of the commercial BPR. The pressure fluctuation in the outlet pressure impact the comparison test result for the three BPR, if the OBPR pressure response is compared to the commercial BPR response, it can clearly be seen that the OBPR pressure fluctuation are unstable and respond aggressively to step input disturbance compared to the other two BPR. These fluctuations in OBPR feedback control system are caused by various types of elements present in the OBPR model, such as the stem of the valve, pressure sensor, manual tuning, stepper motor, charging gas, primary fluid pressure, etc. All the dynamic responses for three different BPRs were recorded under uniform test conditions except at different pressure. For the high permeable core sample, this small fluctuation can be impactful, but the all the initial experiments were plan to test the preliminary design of the OBPR model and its ability to achieve the desired set backpressure value. This fluctuation also affects the accuracy characteristics of the instrument to measure the core outlet pressure response. Even though the systematic error is reduced, there still will be some inherent errors in the instrument design. Core outlet pressure was measured with $\pm 0.25\%$ of the setpoint value. These are good initial results. The number of replicates for each experiment was around three or more. There might be a chance of precision error due to analyzing the sets of pressure measurement data for several experiments. It was found that core outlet measurements were reproducible and nearly consistent for desired set point until the tuning parameters were affected. Especially in gas phase, the OBPR system can measure the outlet core pressure response with the tolerance of around $\pm 6\%$ of the original pressure. All the core outlet pressure measurements were only valid under the controlled condition of pressure and temperature. Any variation in the ambient temperature can change the sensitivity for pressure measurement. It's better to ignore such systematic and statistical random error at the initial stage of the designing and testing process.

## 5. Conclusions

A newly designed oscillating back pressure regulator model for core flood analysis was developed. An experimental study was undertaken to understand the pressure response of the OBPR using different core flood experiments. The OBPR was tested in three fluid flow scenarios using brine, water and $CO_2$. Results included are the dynamic and steady-state response of the pressure. The OBPR was found to have a faster response than the EBPR and VBPR for the gas and brine systems. The OBPR took some time to achieve the steady-state compared to the EBPR and VBPR for the liquid system. The OBPR was found to suffer from instability in pressure response compared to the commercial regulators. It can also be concluded that the OBPR feedback control performance was assessed based on the magnitude of the step input. The results from the OBPR and comparison with different commercial BPR show that this device can respond well to liquid. More experimental investigation is needed for gas phase system. OBPR system can hold the backpressure similar to another commercial device for core flood application.

## 6. Future works:

The OBPR requires more rigorous experimental study before applying it successfully to the core flood application. Some of the future works include to:

a. Evaluate the OBPR feedback control performance at maximum pressure up to 9,500 psi to and high temperature.

b. Improve the feedback control system to reduce the pressure offset and instability.

c. Test the OBPR on different gases and multiphase flow systems.

d. Use the diaphragm accumulator or floating piston type accumulator to avoid mixing of nitrogen with the primary core fluid such as brine, water, gas and oil.

e. Perform the sensitivity analysis of different parameter on the OBPR feedback control performance.

f. Tune the OBPR using the first principle and to use the Mat lab Simulink."

g. Develop other embodiments for the accurately controlling of the flow and pressure eliminating the comingle of Nitrogen with the process fluids.

h. Implement the well-developed OBPR design to measure the physical parameter for core flood applications such as Relative Permeability.

The authors would like to thank Chevron Canada, Hibernia Management and Development (HMDC), Department of Tourism, Culture Industry and Innovation (TCII), Natural Sciences and Engineering Research Council of Canada (NSERC) for funding my research project. Thanks are extended to Equilibar and Vinci Technologies for providing their BPR for our research study.

## References

1 F. Torabi, B. Yadali Jamaloei, S. Y. Hong, and N. K. Bakhsh, "Improving downhole pump efficiency and well productivity in heavy oil reservoirs utilizing back pressure regulator," *J. Pet. Explor. Prod. Technol.*, vol. 2, no. 4, pp. 181–195, (2012).

2 L. T. Vo, R. Gupta, and O. J. Hehmeyer, "Ion Chromatography Analysis of Advanced Ion Management Carbonate Coreflood Experiments," no. Ic, (2012).

3 A. Shokrollahi, M. H. Ghazanfari, and A. Badakhshan, "Application of foam floods for enhancing heavy oil recovery through stability analysis and core flood experiments," *Can. J. Chem. Eng.*, vol. 92, no. 11, pp. 1975–1987, (2014).

4 S. J. Varjani and V. N. Upasani, "Core Flood study for enhanced oil recovery through ex-situ bioaugmentation with thermo- and halo-tolerant rhamnolipid produced by Pseudomonas aeruginosa NCIM 5514," *Bioresour. Technol.*, vol. 220, pp. 175–182, (2016).

5 L. D. Connell, R. Sander, Z. Pan, M. Camilleri, and D. Heryanto, "History matching of enhanced coal bed methane laboratory core flood tests," *Int. J. Coal Geol.*, vol. 87, no. 2, pp. 128–138, (2011).

6 W. O. Lease and C. F. Van Egmond, "Oscillating Backpressure Regulator for Low Flow Rate, High-Pressure Coreflooding Experiments.," *Soc. Pet. Eng. J.*, vol. 22, no. 6, pp. 899–901, (1982).

7 T. J. Hughes, A. Honari, B. F. Graham, A. S. Chauhan, M. L. Johns, and E. F. May, "$CO_2$ sequestration for enhanced gas recovery: New measurements of supercritical CO 2-CH 4 dispersion in porous media and a review of recent research," *Int. J. Greenh. Gas Control*, vol. 9, pp. 457–468, (2012).

8 K. Mannhardt and I. Svorstøl, "Effect of oil saturation on foam propagation in Snorre reservoir core," *J. Pet. Sci. Eng.*, vol. 23, no. 3–4, pp. 189–200, (1999).

9 A. P. Spence and R. W. Watkins, "Effect of Microscopic Core Heterogeneity on Miscible Flood Residual Oil Saturation.," *Soc. Pet. Eng. AIME, SPE*, (1980).

10 R. J. Hwang and J. Ortiz, "Effect of $CO_2$ flood on geochemistry of McElroy oil," *Org. Geochem.*, vol. 29, no. 1-3–3 pt 1, pp. 485–503, (1998).

11 M. Hellemans, "Overpressure protection," *Saf. Reli. valve Handb. Des. use Process Saf. valves to ASME Int. codes Stand.*, no. 1st edition, pp. 327–434, (2009).

12 A. Morar, "Stepper Motor Model for Dynamic Simulation," *Acta Electroteh.*, vol. 44, no. 2, pp. 117–122, (2003).

13 R. A. Nasralla and H. A. Nasr-El-Din, "Coreflood study of low salinity water injection in sandstone reservoirs," *Soc. Pet. Eng. - SPE/DGS Saudi Arab. Sect. Tech. Symp. Exhib. 2011*, no. May, pp. 187–199, (2018).

14 H. M. Sebastian and D. D. Lawrence, "Nitrogen minimum miscibility pressures," *Eighth Symp. Enhanc. Oil Recover.*, pp. 299–310, (1992).

15 V. Rajinikanth and K. Latha, "Tuning and Retuning of PID Controller for Unstable Systems Using Evolutionary Algorithm," *ISRN Chem. Eng.*, vol. 2012, pp. 1–11, (2012).

16 M. Mutlu, T. Wassar, M. A. Franchek, and J. A. Gutierrez, "Condition and performance analysis of a subsea BOP control system pressure regulator," *Proc. Annu. Offshore Technol. Conf.*, vol. 2, pp. 1193–1203, (2018).

17 R. El Golli *et al.*, "Modelling of a pressure regulator To cite this version : HAL Id : hal-01845776," vol. 84, no. 4, pp. 234–243, (2018).

18 B. Delenne, L. Mode, and M. Blaudez, "Modelling and simulation of a gas pressure regulator," in *European simulation symposium. Erlangen, Germany*, (1999).

19 H. Ott *et al.*, "Core-flood experiment for transport of reactive fluids in rocks," *Rev. Sci. Instrum.*, vol. 83, no. 8, (2012).

20 Bill Menz, "Pressure Regulator Selection Strategy," *Swagelok*, no. August, 2011

21 Pickup, I. E. D., and A. P. Russell, "Dynamic instability in permanent-magnet synchronous/stepping motors.," *IEE Proc. B (Electric Power Appl.*, vol. 134, no. 2, pp. 91–100, (1987)

# AUTHOR INDEX

**EDP Sciences – Web of Conferences**
17, Avenue du Hoggar
Parc d'Activité de Courtabouf
BP 112
F-91944 Les Ulis Cedex A
France

ISBN 978-1-7138-6642-8

CURRAN ASSOCIATES INC.
proceedings.com